ANALYZING THE THERMAL HISTORY OF SEDIMENTARY BASINS: METHODS AND CASE STUDIES

EDITED BY:

NICHOLAS B. HARRIS
UNIVERSITY OF ALBERTA

KENNETH E. PETERS
SCHLUMBERGER
and
STANFORD UNIVERSITY

SEPM SPECIAL PUBLICATION NO. 103

SEPM and the authors are grateful to the following
for their generous contribution to the cost of publishing

IPRA, Université de Pau et des Pays de l'Adour
University of Kansas
US Geological Survey

This book includes a CD/DVD which contains a complete digital copy of this volume, with many color figures substituted for printed black and white versions.

SEPM (Society for Sedimentary Geology) is an international not-for-profit Society based in Tulsa, Oklahoma. Through its network of international members, the Society is dedicated to the dissemination of scientific information on sedimentology, stratigraphy, paleontology, environmental sciences, marine geology, hydrogeology, and many additional related specialties.

The Society supports members in their professional objectives by publication of two major scientific journals, the Journal of Sedimentary Research (JSR) and PALAIOS, in addition to producing technical conferences, short courses, and Special Publications. Through SEPM's Continuing Education, Publications, Meetings, and other programs, members can both gain and exchange information pertinent to their geologic specialties.

For more information about SEPM, please visit **www.sepm.org**

Gary J. Nichols and Brian Ricketts, SEPM Special Publication Editors
Michele Tomlinson, SEPM Publications Coordinator
SEPM Special Publication 103

ISBN 978-1-56576-315-9
CD/DVD version 978-1-56576-317-3

4111 S Darlington, Suite 100
Tulsa, Oklahoma 74135-6373, USA

Analyzing the Thermal History of Sedimentary Basins: Methods and Case Studies

Editors: Nicholas B. Harris and Kenneth E. Peters

CONTENTS

ANALYZING THERMAL HISTORIES OF SEDIMENTARY BASINS: METHODS AND CASE STUDIES—INTRODUCTION

NICHOLAS B. HARRIS

Department of Earth and Atmospheric Sciences, 1-26 ESB, University of Alberta, Edmonton, Alberta T6G 2E3, Canada

AND

KENNETH E. PETERS

Schlumberger Information Solutions, 18 Manzanita Place, Mill Valley, California, 94941, USA and Department of Geological and Environmental Sciences, Stanford University, Stanford, California, 94305, USA

INTRODUCTION

Thermal histories of sedimentary basins are important sources of scientific and practical information. They provide us with windows into past and present tectonic processes and the configuration of the crust and mantle. With records of present and past temperature distributions, we can identify and constrain interpretations of tectonic events, distinguish different basin types, and interpret pathways of fluid flow. These insights can, in turn, be used to interpret and predict the distribution of minerals and hydrocarbons.

Thermal histories can be obtained either from petroleum system or tectonic models or from paleothermometers. Modeling approaches require complex sets of assumptions about the processes of heat transfer and boundary conditions, and although a sound theoretical basis can be found for all of these, the accuracy of modeling is limited without paleotemperature records. Paleothermometers are components of rock for which the chemical composition or physical properties were determined by past temperatures. By measuring those properties, parts of the temperature history can be reconstructed. The most powerful approach to determining thermal histories has proven to be the integration of models and data from paleothermometers, where model results are calibrated by paleothermometers.

This volume sets out to summarize the current state of the art for many of the modern approaches used to estimate paleotemperatures. In the past, only a few tools were available to calibrate thermal models, most notably corrected borehole temperatures and vitrinite reflectance; there are now many techniques available based on both organic and inorganic components of the rock. Even those techniques that are now many years old, such as apatite fission-track analysis, have undergone significant advances in the past decade.

SIGNIFICANCE OF THERMAL HISTORY ANALYSIS

The relationship between basin type and thermal history is particularly important because thermal histories provide insight into the timing and nature of basin evolution, which in turn provide us with key information about the distribution of hydrocarbons and ores. In terms of thermal structure, two end-member basin types are recognized: rift systems, with high heat flows and geothermal gradients; and foreland basins, with low heat flows and geothermal gradients. Rift basins by definition involve thinned crust and shallow mantle; consequently, they exhibit heat flows that are two times the global average. Not surprisingly, heat flows are higher in rift systems with greater amounts of stretching. Heat flow in rift basins also varies with time; it is highest during the initial stretching and decreases with time. Foreland basins, on the other hand, form as the result of crustal processes, where the crust is loaded and subsides due to tectonic thickening. Here, the crust and mantle are driven deeper instead of rising up, and, as a result, heat flows are lower.

The thermal state of the crust significantly influences geomechanical properties and topography of sedimentary basins. Geomechanical properties refer to characteristics that determine the ways in which rocks respond to stress; at high temperatures, rock may deform in plastic or ductile fashion, whereas at lower temperatures, the response may be more brittle (Ord and Hobbs 1989). Topography is a response to the interrelated crustal variables of temperature, thickness, mechanical strength, and density. Topography in turn affects how a basin fills.

The thermal state of sedimentary basins is also critical to understanding fluid flow and the development of accurate reaction transport models. Fluids respond to pressure fields by flowing; temperature and fluid composition control density, which is a partial control on pressure. Accurate temperature models are necessary to understand the ways in which fluids might have flowed in the past and the possible ways in which problems of mass balance and mass transport might be solved. Chemical components move in sedimentary basins on a large scale by advection, transported by moving fluid. Chemical equilibria and reaction rates also depend on temperature.

APPLICATIONS TO PETROLEUM GEOLOGY

Thermal histories impact many aspects of petroleum geology, perhaps most importantly the generation of hydrocarbons. Both primary generation of oil and gas from kerogen and the secondary generation of gas from oil are described by the Arrhenius equation:

$$\varepsilon = [A_o]_* exp^{-Ea/RK}{}_* t$$

where ε is the extent of reaction, A_o is the pre-exponential or frequency factor, E_a is the activation energy, R is the universal gas constant, K is temperature in Kelvin, and t is time.

Both temperature and time are variables in this equation; the reaction rate varies exponentially as a function of temperature, and extent of reaction varies linearly with time. Thus, developing a time–temperature relationship is critical for accurate modeling of petroleum generation (Hantschel and Kauerauf 2009).

Time–temperature histories are also important to understanding reservoir quality. Recent studies (Walderhaug 2000) have shown that quartz cementation in sandstone can be effectively modeled as a function of temperature. Various researchers (Harris 1992, Giles et al. 2000) have noted that the transition to quartz cementation with depth is commonly abrupt; sandstones at one depth may have little cement and

Analyzing the Thermal History of Sedimentary Basins: Methods and Case Studies
SEPM Special Publication No. 103, Copyright © 2012
SEPM (Society for Sedimentary Geology), Print ISBN 978-1-56576-315-9, CD/DVD ISBN 978-1-56576-317-3, p. 1–4.

high porosity, while the same formation only slightly deeper and hotter may have considerable cement and reduced porosity.

Another significant reaction in clastic diagenesis is the mixed-layer clay reaction. Smectite, the sodium-rich mixed-layer clay, reacts at elevated temperature to form potassium-rich illite. This reaction typically takes place over a range of depths and temperatures and has been successfully modeled by application of the Arrhenius equation (Velde and Vasseur 1992, Elliott and Matisoff 1996). As the reaction proceeds, the expandability of the clay decreases, with implications for borehole stability and sensitivity of the rock matrix to drilling fluids.

APPLICATION IN BASIN MODELS

Basin models are numerical simulations that quantify the evolution of rocks and fluids during basin history. These quantitative models evolved from the petroleum system concept, as described by Magoon and Dow (1994). While basin modeling quantifies the evolution of the rocks, petroleum system modeling addresses the evolution of the hydrocarbon fluids (Peters 2009). Readers should be aware that "basin modeling" is used by many as a general term that includes both basin and petroleum system modeling.

Thermal histories can be reconstructed in quantitative basin models. The more sophisticated of these models combine conductive heat transport (the transfer of thermal energy between substances in contact) and advective heat transport (the transfer of heat by movement of fluid). Conductive heat transport uses as input basal heat flow, surface temperatures, burial histories, and thermal conductivities of formations in the stratigraphic column. Advective heat transfer is accomplished by movement of formation water or other fluids through pore spaces as sediments compact and requires as input the heat capacity of fluids and variables describing fluid flow (porosity, permeability, pressure). The decay of radioactive isotopes of uranium, thorium, and potassium provides an additional source of heat and is commonly incorporated into models. This effect is most significant in basins underpinned by young granitic crust or those containing sediments with large amounts of potassium-rich clays (illite) or feldspar (e.g., orthoclase).

Two- and three-dimensional models are required to describe lateral transfer of heat. This can occur as conductive heat transfer when rocks of different thermal conductivity are laterally juxtaposed. Such scenarios occur due to stratigraphic relationships, for example, where a carbonate pinnacle reef is juxtaposed against basinal shales. Lateral heat transfer can also occur when structural relationships juxtapose lithologies with different properties, for example, due to fault juxtaposition associated with salt and shale diapirs (Corrigan and Sweat 1995, Nagihara 2003). The former is particularly important in salt provinces such as the Gulf of Mexico or the South Atlantic salt basins. Lateral heat transport can occur through convective transfer. Fluids move laterally within formations and fluid can flow laterally through fault and fracture systems. In either case, when hot water moves up dip to shallower, cooler levels in a basin, heat is also transported, for example, in the Western Canada Sedimentary Basin (Bachu and Burwash 1994). Anomalously low geothermal gradients may develop when cold groundwater penetrates deep into a basin, for example, when recharge areas for aquifers outcrop at high elevation (Majorowicz et al. 1986).

WHY CALIBRATE BASIN MODELS?

Basin models require calibration in order to make reasonably accurate predictions. Calibration can take two forms: (1) calibration to present-day geothermal parameters, such as heat flow, thermal conductivities, and surface and downhole temperatures, or to fluid-flow parameters such as pressure, porosity, and permeability; and (2) calibration to measurements of past temperatures. Calibration to present-day parameters is critical; in particular, good models have well-defined thermal conductivity structures that reflect both rock composition and porosity of stratigraphic units within the sedimentary basin.

Even calibration to present-day temperatures and heat flow is a nontrivial exercise that requires thought and diligence. The paper in this volume by Peters and Nelson describes the application of downhole temperature data to determine accurate present-day geothermal gradients.

Many critical parameters in thermal modeling vary with time:

- Thermal conductivities typically increase as a formation subsides, the rock compacts, and fluids are expelled.
- Surface or sediment–water interface temperatures vary as a location changes geographic position due to plate-tectonic movement or as water depth changes in a marine sedimentary basin.
- Heat flow changes due to the tectonic evolution of a basin, for example, decreasing with time in rift basins.

Rocks do not preserve time records of heat flow, thermal conductivity, and surface temperature, but these controls on paleo-temperature can be predicted from various types of models. Models for crustal evolution can be used to estimate crustal thickness. Tectonic models for sedimentary basins can be used to estimate heat flow: For example, rift models relate the timing and amount of extension to heat flow (McKenzie 1978, Jarvis and McKenzie 1980). Other models predict porosity evolution with depth (e.g., Sclater and Christie 1980), which can be incorporated into calculations of thermal conductivity. Finally, detailed paleogeographic models can be applied to the estimations of paleo-surface temperature. However, the uncertainties associated these parameters are significant, particularly with regard to predicting paleo-heat flow from tectonic models and thermal conductivities from compaction models.

Petroleum system models are especially sensitive to estimates of paleotemperature with regard to hydrocarbon generation and expulsion, where, for example, a 20° C range in a paleotemperature estimate can mean the difference between predicting little or considerable amounts of oil generation. That level of accuracy is generally impossible in petroleum system models based on tectonic models for heat flow or porosity models for thermal conductivity. A high level of accuracy in petroleum system modeling requires calibration to paleotemperature records.

TYPES OF CALIBRATION

Calibration of basin models is accomplished by adjusting model variables until a match is achieved between independently estimated paleotemperature records and model predictions. There has been a proliferation of thermal calibration parameters in the past 40 years. Of these approaches, only fluid inclusion analysis provides a direct measure of paleotemperature, because inclusions sample fluids at a geologically instantaneous time in the past. Measurements of fluid composition can, under the right circumstances, be used to calculate trapping temperatures.

All other calibration methods are measures of the integrated time–temperature history and are not direct samples of temperature at some past time. These methods measure the combined effect of temperature and the period of time a rock was exposed to that temperature, a parameter generally called thermal maturity. For individual samples, multiple time–temperature paths can achieve the same thermal maturity; for example, a short high-temperature pulse or a long, lower-temperature pulse can both result in the same thermal maturity. On the other hand, a series of linked samples—for example, from a single well bore—can yield temperature models that are fairly well constrained. Modeling is almost always required to arrive at a solution and to test the limits of possible satisfactory models.

A SUMMARY OF PALEOTHERMOMETRIC TECHNIQUES

Paleothermometric Techniques that Apply to Organic Substances

Techniques for extracting thermal histories fall into two broad groups: measurements on organic matter (including certain fossils) and measurements on minerals. The history of organic thermal parameters is long, and several have been developed. Among these, the most notable for the availability of data are vitrinite reflectance and the Rock-Eval pyrolysis parameter Tmax. Vitrinite reflectance relies on chemical changes in vitrinite, which is a product of the woody tissue of plants. These changes result from increased thermal maturity and produce increased reflectivity of the material on polished surfaces. Rock-Eval pyrolysis involves the rapid heating of rock samples in an inert atmosphere, which expels existing hydrocarbons and pyrolyzes kerogen in the rock. The temperature at which most of the kerogen pyrolyzes (Tmax) has been shown to be a fairly sensitive and predictable measure of maturity (Peters 1986).

Other approaches to determining thermal maturity using organic substances include the spore coloration index (Staplin 1969, Marshall 1991), also called thermal alteration index (TAI) and conodont color index (CAI; Epstein et al. 1977). In both cases, chemical reactions that occur in response to increased temperature result in darkening of the material. A standard set of colors has been developed to categorize samples, and quantitative basin models are capable of incorporating TAI results. Other, less commonly used optical maturity scales include the transmittance color index (TCI; Robison et al. 2000), and acritarch fluorescence (Obermajer et al. 1997).

Biomarkers, so-called molecular fossils, include compounds that change in abundance with increasing temperature and thermal maturity (Peters et al. 2005). Three different classes of response can occur: (1) isomerization—changes in the ratios of compounds that are identical except for their geometric configuration; (2) aromatization—the formation of unsaturated carbon ring structures; and (3) cracking—the formation of smaller molecules from larger precursors. The paper by Walters, Lillis, and Peters in this volume provides a detailed review of the application of biomarker data to maturation estimates.

Biomarker applications have been further extended by the incorporation of hydrogen isotope data. Natural biological compounds tend to be strongly enriched in the light isotopes of carbon and hydrogen, and consequently many biomarkers are characterized by low values of δD. With exposure to higher temperature, certain biomarkers become enriched in deuterium, and the δD of these organic substances increases as a result (Schoell 1984), either through the preferential loss of H during cracking reactions or by exchange of hydrogen with deuterium-enriched formation water. The paper by Maslen, Grice, Dawson, Wang, and Horsfield explores this phenomenon and shows that δD of pristane and phytane correlates with more traditional measures of thermal maturity such as vitrinite reflectance.

Stable isotope geochemistry has also been applied to natural gases. Here, $\delta^{13}C$ and δD of hydrocarbon gases are sensitive to the temperatures at which cracking reactions in kerogen and oil take place, as well as the initial composition of the kerogen or oil. This approach first gained widespread attention through the work of Schoell (1983). Since that time, applications of gas composition to paleothermometry have gained a sound theoretical basis, summarized in a paper in this volume by Xia and Tang.

One additional technique, applied much less frequently but nonetheless worth noting, utilizes the color and intensity of fluorescence of certain types of organic matter resulting from ultraviolet excitation (Thompson-Rizer and Woods 1987).

Paleothermometric Techniques that Apply to Minerals

Mineral recorders of thermal maturity rely on a variety of properties. Fluid inclusions are effectively a sample from an instant in time, and their properties reflect the temperature and pressure at the time of trapping. The paper in this volume by Robert Goldstein provides a detailed summary of this method. Other approaches rely on reaction rates that vary as a function of temperature, for example, the conversion of smectite to illite, where the extent of reaction is a measure of thermal maturity. Finally, some techniques rely on a phenomenon known as blocking temperature, in which certain components of a mineral can escape from the crystal lattice only when temperatures exceed a thermal threshold. When that component is radiogenic, the length of time the mineral was exposed to temperatures lower than the blocking temperature can be calculated, for example, in the (U-Th)/He system.

Among the approaches that rely on variable reaction rates, there is the mixed-layer clay illite–smectite reaction. The phenomenon of smectite reacting to illite over a range of temperatures was first documented in a classic paper by Hower et al. (1976). The paper in this volume by Clauer and Lehrman summarizes and extends that work by incorporating isotope geochemistry.

The mechanism of blocking temperature has been particularly well demonstrated by the diffusion of helium gas from silicate crystal lattices (Reiners 2005). At relatively high temperatures, helium diffuses from the crystal lattice, but at lower temperatures, the diffusion rate effectively decreases to zero. This process has been extensively studied in apatite and, to a lesser extent, in other common accessory minerals in sandstones, such as zircon, titanite, monazite, and xenotime. The paper in this volume by Ketcham provides a modern review of the topic.

Fission-track analysis is an approach to paleothermometry that is very similar to the application of blocking temperatures. This method relies on two phenomena: the production of damage trails in minerals from the radioactive decay of uranium, and the annealing of these tracks upon exposure to elevated temperatures. Early descriptions are found in Fleischer et al. (1975) and Naeser (1979). The number of tracks produced in a crystal is a function of the uranium content and the crystal age and can thus provide an age of formation if the uranium content is known. Annealing can be complete, in which case a new set of tracks forms and is preserved upon cooling, or partial, in which both the original tracks, now shortened, and a new set of tracks are preserved. The paper in this volume by Green and Duddy provides a detailed explanation of the method and new developments.

CASE STUDIES

The practical derivation of thermal histories from paleothermometers and models is a complex process, typically requiring multiple iterations and calibrations of models to converge on a best solution. Four papers that illustrate the challenges and rewards of such efforts are presented in this volume.

Houseknecht, Burns, and Bird present a study of the thermal maturation history of the North Slope, Alaska. This study relies largely on modeling, calibrated by vitrinite reflectance and downhole temperature data, to investigate a tectonically complex region that includes both rift and foreland basin elements in the geologic history.

Scotti and Fantoni present a case study from the southern Alps, Europe, another tectonically complex region that involved rifting and continental drift, overprinted in much of the region by the Alpine orogeny. Vitrinite reflectance and Tmax data are integrated with basin models to show that extremely high heat flow must have accompanied the rift phase, decreasing during the passive-margin phase. This is an important instance in which early high heat flow was not completely overprinted during later passive-margin subsidence.

Gonzales, Ferket, Callot, Guilhaumou, Ortuno, and Roure present a thermal maturity analysis from yet another complex terrane, the southernmost extension of the Laramide thrust belt and foreland basin in eastern Mexico. The authors employ vitrinite reflectance, Tmax, bottom-hole temperature, and fluid inclusion data to calibrate petroleum system models and then use the model to describe the generation and migration of hydrocarbons.

Finally, Lampe and Schwark present a paper on thermal maturity in part of the Rhine Graben, Germany. The authors investigate a rare instance where thermal maturities, as determined from vitrinite reflectance, Tmax, and biomarker data, were high at shallow depths and exhibit no clear vertical trend. They conclude that hydrothermal activity was responsible for the anomalous maturities in the shallow part of the section.

ACKNOWLEDGMENTS

This volume benefited from the considerable work of many reviewers. We thank Richard Alley, Joe Curiale, Harry Dembicki, Daniel Dawson, Kevin Furlong, Robert Harris, Barry Katz, Michael Lewan, Larry Meckel, and Kate Weissenburger. We thank the staff of the Society for Sedimentary Geology (SEPM), in particular Michele Tomlinson and Howard Harper, for their efforts and diligence. We note with appreciation the review of this introduction by Jacek Majorowicz. Finally, we thank the authors for both their hard work and their patience as this volume progressed to publication.

REFERENCES

Bachu S, Burwash RA. 1994. Geothermal regime in the Western Canada Sedimentary Basin. *In* Mossop GD, Shetsen I (Editors). *Geological Atlas of the Western Canada Sedimentary Basin*: Geological Survey of Canada, Calgary, Canada. p. 447–454.

Corrigan J, Sweat M. 1995. Heat flow and gravity responses over salt bodies: a comparative model analysis. *Geophysics* 60:1029–1037.

Elliott WC, Matisoff G. 1996. Evaluation of kinetic models for the smectite to illite transformation. *Clays and Clay Minerals* 44:77–87.

Epstein AG, Epstein JB, Harris LD. 1977. Conodont Color Alteration—An Index to Organic Metamorphism: US Geological Survey, Professional Paper 995, 27 p.

Fleischer RL, Price PB, Walker RM. 1975. *Nuclear Tracks in Solids*: University of California Press, Berkeley, California. 605 p.

Giles MR, Indrelid SL, Beynon G, Amthor J. 2000. The origin of large-scale quartz cementation: evidence from large data sets and coupled heat–fluid mass transport modelling. *In* Worden RH, Morad S (Editors). *Quartz Cementation in Sandstones*: International Association of Sedimentologists, Blackwell, Oxford, UK. Special Publication 29. p. 21–38.

Hantschel T, Kauerauf AI. 2009. *Fundamentals of Basin Modeling*: Springer, New York. 476 p.

Harris NB. 1992. Burial diagenesis of Brent sandstones: a study of Statfjord, Hutton and Lyell fields. *In* Morton AC, Haszeldine RS, Giles MR, Brown S (Editors). *Geology of the Brent Group*: Geological Society of London, London. Special Publication 61. p. 351–375.

Hower J, Eslinger E, Hower ME, Perry EA. 1976. The mechanism of burial diagenetic reactions in argillaceous sediments, 1. Mineralogical and chemical evidence. *Geological Society of America Bulletin* 87:725–737.

Jarvis GT, McKenzie D. 1980. Sedimentary basin formation with finite extension rates. *Earth and Planetary Science Letters* 48:42–52.

Magoon LB, Dow WG. 1994. *The Petroleum System—From Source to Trap*: American Association of Petroleum Geologists, Tulsa, Oklahoma. Memoir 60. 655 p.

Majorowicz JA, Jones FW, Jessop AM. 1986. Geothermics of the Williston Basin in Canada in relation to hydrodynamics and hydrocarbon occurrences. *Geophysics* 51:767–779.

Marshall JEA. 1991. Quantitative spore colour. *Journal of the Geological Society of London* 148:223–233.

McKenzie D. 1978. Some remarks on the development of sedimentary basins. *Earth and Planetary Science Letters* 40:25–32.

Naeser CW. 1979. Fission-track dating and geologic annealing of fission tracks. *In* Jäger E, Hunziker JC (Editors). *Lectures in Isotope Geology*: Springer-Verlag, New York. p. 154–169.

Nagihara S. 2003. Three-dimensional inverse modeling of the refractive heat-flow anomaly associated with salt diapirism. *American Association of Petroleum Geologists Bulletin* 87:1207–1222.

Obermajer M, Stasiuk LD, Fowler MG, Osadetz KG. 1997. Acritarch fluorescence as a new thermal maturity indicator. *American Association of Petroleum Geologists Bulletin* 81:1561.

Ord A, Hobbs BE. 1989. The strength of the continental crust, detachment zones and the development of plastic instabilities. *Tectonophysics* 158:269–289.

Peters KE. 1986. Guidelines for evaluating petroleum source rock using programmed pyrolysis. *American Association of Petroleum Geologists Bulletin* 70:318–329.

Peters KE. 2009. *Basin and Petroleum System Modeling*: AAPG/Datapages, Tulsa, Oklahoma. AAPG Getting Started Series No. 16.

Peters KE, Walters CC, Moldowan JM. 2005. *The Biomarker Guide,* 2nd ed.: Cambridge University Press, Cambridge, UK. 1155 p.

Reiners PW. 2005. Zircon (U-Th)/He thermochronometry. *Reviews in Mineralogy and Geochemistry* 58:151–179.

Robison CR, Van Gijzel P, Darnell LM. 2000. The transmittance color index of amorphous organic matter: a thermal maturity indicator for petroleum source rocks. *International Journal of Coal Geology* 43:83–103.

Schoell M. 1983. Genetic characterization of natural gases. *American Association of Petroleum Geologists Bulletin* 67:2225–2238.

Schoell M. 1984. Recent advances in petroleum isotope geochemistry. *Organic Geochemistry* 6:645–663.

Sclater JG, Christie PAF. 1980. Continental stretching: an explanation of the post-mid-Cretaceous subsidence of the central North Sea basin. *Journal of Geophysical Research* 85:3711–3739.

Staplin FL. 1969. Sedimentary organic matter, organic metamorphism, and oil and gas occurrence. *Canadian Petroleum Geologists Bulletin* 17:47–66.

Thompson-Rizer CL, Woods RA. 1987. Microspectrofluorescence measurements of coals and petroleum source rocks. *International Journal of Coal Geology* 7:85–104.

Velde B, Vasseur G. 1992. Estimation of the diagenetic smectite-to-illite transformation in time–temperature space. *American Mineralogist* 77:967–976.

Walderhaug O. 2000. Modeling quartz cementation and porosity loss in Middle Jurassic Brent Group sandstones of the Kvitebjørn field, northern North Sea. *American Association of Petroleum Geologists Bulletin* 84:1325–1339.

CRITERIA TO DETERMINE BOREHOLE FORMATION TEMPERATURES FOR CALIBRATION OF BASIN AND PETROLEUM SYSTEM MODELS

KENNETH E. PETERS
Schlumberger, 18 Manzanita Place, Mill Valley, California 94941, USA
e-mail: kpeters2@slb.com

AND

PHILIP H. NELSON
US Geological Survey, Denver Federal Center, Box 25046 MS 939, Denver, Colorado 80225-0046, USA

Abstract: Bottom-hole temperatures (BHT) from well log headers are common, but they require correction because they are biased lower than true formation temperature. Care must be taken to avoid short static times, recirculation between measurements, and spurious records of times or temperatures from logs. Criteria for reliable Horner-corrected BHT data include a minimum of three logging runs that record time and temperature for each run, temperature extrapolation less than the range of temperature data, and deviations from the least-squares regression line that are less than measurement uncertainty (±1–3° C or ±2–5° F).

Based on published comparisons of drill-stem test (DST) and Horner-corrected BHT data from the same depths, the standard deviation of corrected bottom-hole temperatures is about ±8° C (±14° F). Some studies show that corrected data may still be systematically biased lower than true formation temperature. For a one-dimensional PetroMod® basin and petroleum system model of the upper Cook Inlet in Alaska, error of ±8° C resulted in calculated depth to top of the oil window in the Jurassic Tuxedni Group source rock of as much as 305 m (1001 ft) above and 6.2 million years earlier or 231 m (758 ft) below and 4.5 million years later than that calculated using a corrected BHT formation temperature of 92.4° C. Error in maturity assessment associated with shallow BHT data can propagate to the greater depth of the source rock where temperature and thermal maturity are to be estimated. In summary, BHT data are an important source of uncertainty that needs to be considered when calibrating basin and petroleum system models.

Key words: Basin and petroleum system modeling, Horner correction, bottom-hole temperature, BHT, Cook Inlet

INTRODUCTION

The purpose of this paper is to describe the magnitude of potential error that can result from using formation temperatures estimated from bottom-hole temperature (BHT) measurements to calibrate "basin and petroleum system models." We highlight this term to emphasize that basin modeling and petroleum system modeling differ. Petroleum systems (Magoon and Dow 1994) occur in sedimentary basins, but not all sedimentary basins contain petroleum systems. Basin modeling reconstructs the geohistory of the rocks and is a prerequisite for petroleum system modeling, which calculates the evolution of the fluids (i.e., oil and gas). The target audience for this paper consists mainly of basin and petroleum system modelers who may be unfamiliar with drilling practices and unaware of the impact of uncertainty associated with interpreted formation temperatures. An additional goal of this paper is to provide practical guidelines and examples of the application of Horner corrections of BHT data.

Accurate or "true" formation temperatures are important because temperature affects many key parameters for petroleum geology, including (1) water resistivity, needed to calculate oil saturation from resistivity logs, (2) reservoir-fluid formation volume factors, needed for petroleum reserve estimates, and (3) geothermal gradients, needed to estimate temperatures and pressures in undrilled deep zones. In addition, accurate temperature measurements from boreholes can be used in exploration to calibrate one-, two-, or three-dimensional (1D, 2D, or 3D) basin and petroleum system models (e.g., Welte et al. 1997, Peters et al. 2008). As discussed herein, temperature calibration can have a major impact on the calculated timing, depth, and yield of petroleum from source rocks in these models.

However, most temperatures recorded during drilling, logging, or well maintenance operations differ from undisturbed, static, or true formation temperature for several reasons. Drilling mud removes cuttings but also cools and lubricates the drill bit and the rock near the borehole (Hearst et al. 2000). Thus, borehole temperature measured during or soon after circulation of drilling mud is typically lower than true formation temperature prior to drilling. Thermal diffusivity, which is the ratio of thermal conductivity to heat capacity per unit volume, controls how fast temperature can change with time and distance from the wellbore in response to thermal perturbation. Material with low thermal diffusivity, such as water, requires more time to equilibrate after temperature change than that with high diffusivity, such as a high-silicate rock. Because of varying diffusivities and geometries of rock strata and wellbore components, such as mud, casing, and cement, one cannot expect to achieve an exact solution to thermal recovery (i.e., reequilibration of a well bore to true formation temperature), and consequently all temperature correction methods should be regarded as approximations.

A summary by Deming (1989, p. 783) of difficulties related to the use of BHT data is worth repeating here: "(1) BHT data are generally cooler than true formation temperatures by an amount that, on the average, may be about 10 to 15° C, but probably varies from area to area, (2) the amount of correction needed increases with depth, but this increase may not be linear, and the correction may even decrease at great depths, (3) the amount of correction needed decreases with increasing shut-in time, and (4) any single BHT may be spurious."

TYPES OF TEMPERATURE DATA

For the reasons outlined here, it is necessary to identify the measured temperatures that are suitable for correction to best reflect true subsurface temperatures prior to the drilling disturbance. Production test or DST data are rare, but they provide generally reliable

Analyzing the Thermal History of Sedimentary Basins: Methods and Case Studies
SEPM Special Publication No. 103, Copyright © 2012
SEPM (Society for Sedimentary Geology), Print ISBN 978-1-56576-315-9, CD/DVD ISBN 978-1-56576-317-3, p. 5–15.

temperatures based on downhole gauge measurements of flowing formation fluid. BHT data are abundant, but they commonly underestimate true formation temperature and must be corrected to be useful for calibration of basin and petroleum system models. These data types are discussed here.

Production, Drill-stem, and Static Test Temperatures

Flowing tests, such as production tests or DST measurements, determine the temperature of formation fluids flowing into the wellbore, whereas static tests measure the temperature in the well when not flowing. Flowing and static tests can yield temperatures that are useful to calibrate basin and petroleum system models when measured with a downhole gauge within or near the producing interval. These temperatures are used in the calculation of bottom-hole pressure and for other engineering purposes, but they are also generally the best data for calibration of basin and petroleum system models. Temperature recorded during a DST generally approaches true formation temperature as long as the recovered fluid (>100 barrels, bbl) comes from beyond the zone of thermal disturbance around the borehole.

Long static tests where a well is shut-in for weeks or months can be very accurate. Although long static tests provide excellent formation temperature data for calibration, they are rare because shut-in wells produce no petroleum or income.

Temperature Logs

Temperature logs continuously record data during flowing or static conditions by moving the tool up and down the borehole. They can be run in exploration or production wells and in wells with or without casing. Casing consists of steel pipe that is cemented in the well as drilling progresses or when the well is completed. Temperature logs may be inaccurate because they are usually run in wells having problems that can obscure formation temperature, such as leakage of formation fluids behind casing. Lachenbruch et al. (1988) provided examples of successive temperature logs that show the gradual dissipation of thermal drilling disturbance (e.g., temperature profiles observed 125, 876, and 1200 days after completion of drilling in the Husky Oil Awuna-1 well, Alaska).

Temperature logs can be difficult to interpret in gas-producing wells. The Joule-Thompson effect is produced by movement of gas at higher pressure through porous rock to an environment of lower pressure, such as a borehole. During this process, hydrocarbon gases expand and lose heat, causing a drop in borehole temperature. When shut-in for long periods, gas-producing zones exhibit a slow recovery to true formation temperature. Adjacent static zones away from perforations and the producing zone commonly yield more reliable formation temperatures under both flowing and static conditions, although these kinds of data are rare.

Accurate temperature measurements are of great interest to investigators studying heat flow in Earth's crust, and temperature logs acquired for this purpose may be available. However, such logs are generally rare and cover limited depth ranges. Temperature logs acquired in inactive cased wells before a workover or before being placed into production can provide data representative of true formation temperature. Such data may be available in the oil and gas literature (e.g., Carter et al. 1998) or in the heat-flow literature (e.g., Jessop 1990).

Other Temperature Measurements

Heat Flow: Except for bottom-hole temperatures and temperatures from DST or static tests, most other temperature measurements in the literature are less useful for calibration of basin and petroleum system models. For example, heat-flow data can be obtained from shallow wells or tunnels. Peters et al. (2008) created a present-day surface heat-flow map using temperature and thermal conductivity measurements from many shallow wells and aqueduct tunnels in and around the San Joaquin Basin in California as part of a 3D basin and petroleum system model. However, heat-flow data are generally not used to infer formation temperatures, except in data-poor areas where other temperature data are unavailable. Heat flow is the product of rock thermal conductivity and thermal gradient (i.e., the increase of temperature with depth; Hantschel and Kauerauf 2009). Thermal conductivity is related to the mineralogy and texture of the rock, especially porosity and fluid content. If we assume constant heat flow, then heterogeneous basin fill results in a succession of rock units having different compositions and thermal conductivities; i.e., the geothermal gradient is piecewise linear in various segments at the scale of meters to tens of meters, as shown by the following expression for the temperature (T) in layered strata with constant heat flow (Q):

$$T(z) = T(z_o) + Q\sum(\Delta z_i/k_i) \tag{1}$$

In Equation 1, z is depth, $T(z_o)$ is the average annual temperature at Earth's surface, Δz_i is the thickness of the ith stratum, and k_i is the thermal conductivity of the ith stratum. Carter et al. (1998) and Conaway and Beck (1977) showed examples of the effect of lithology on local temperature gradients. Because of the piecewise linear character of the temperature–depth profile in sedimentary basins, the temperature of a specific formation cannot be accurately estimated from an average thermal gradient; Equation 1 must be used.

Water Chemistry: Various measures of formation water chemistry, such as the concentration of silica or ratios of certain cations, can roughly indicate subsurface temperature (e.g., Bowen 1986). However, these measurements are rare, they are most successful in high-temperature geothermal systems, and their use to predict formation temperature is not recommended.

Wire Line Temperature: Many formation test or log tools, such as RFT and MDT (repeat formation tester and modular dynamic formation tester), are lowered into the borehole on a wire line. Their primary function is to measure pressure and recover small fluid samples, but they also measure temperature. However, these wire line tools commonly do not sample enough fluid to allow the temperature to equilibrate with that of the formation, unlike high-volume flowing well tests. Like uncorrected BHT data, formation test or log tool data are generally biased lower than formation temperature, but they are not as readily corrected. Such temperatures are only useful to indicate a lower limit for formation temperature. If necessary, one can locally calibrate some temperatures determined by RFT/MDT wire line tools using the more accurate production or drill-stem test temperatures.

Mud Circulation Temperature: At regular intervals, mud loggers on the floor of the drilling rig examine and sample mud returns, which are mixtures of drilling fluid and cuttings that circulate from the drill bit up the hole to the surface. They log various properties, such as drilling rate, mud weight, natural gas and oil content, pump pressure, lithology of the cuttings, and circulation temperature. Mud log circulation temperature depends on many poorly controlled variables and cannot be used to determine true formation temperature.

Bottom-Hole Temperature

Before casing the hole, one or more open-hole well logging surveys are normally conducted to determine properties of the borehole and exposed formations. Casing normally prevents the formation walls from caving into the hole during drilling, protects the formation from damage (fluid invasion and fracturing) when heavy mud weights are required, and restricts mixing of fluids between permeable formations and between formation fluids and drilling mud. During most open-hole logging runs, maximum-recording thermometers attached to the tool measure the temperature of the drilling mud, thereby recording the bottom-hole temperature or BHT. These thermometers provide uncorrected data that are generally biased lower than true formation temperature determined using DST by as much as ~20 to 30° C (36–54° F; e.g., Kehle 1972, Jessop 1990, Goutorbe et al. 2007), but they are also the most widely available temperature data. For this reason, several methods have been developed to correct BHT data, as discussed later in this paper.

The typical sequence is to drill to depth, condition the hole by circulating the mud without drilling, stop circulation of the drilling fluid after the cuttings have been brought to the surface, withdraw the drill string out of the borehole, insert the logging tool into the hole and log with successive runs using different wire line tools, and then case the hole if another stage of drilling is to follow. Maximum temperature is recorded in each logging run near the bottom of the wellbore. Borehole conditions may require re-entry of the drill string and circulation of mud between logging runs.

The time component is the difference between the time circulation stopped and the time when the logging tool records maximum temperature, usually near the bottom of the hole. The last rock layer drilled is exposed to the mud only during the short period of circulation after drilling ceases and before logging begins at the bottom of the borehole. Therefore, mud temperature at the bottom of the well after circulation stops typically approaches true formation temperature more rapidly than at other depths. Many wells are drilled in two or more stages, resulting in multiple strings of casing and sets of successive measured bottom-hole depth–temperature pairs. Multiple logging runs give rise to different time intervals between the last mud circulation and the time when temperature was recorded. This provides the basis for establishing a complex temperature profile. For example, since wells are commonly cased just above overpressured zones, it is often possible to establish different geothermal gradients in normally pressured and overpressured regimes.

CORRECTING BOTTOM-HOLE TEMPERATURES

The following discussion focuses on the Horner correction of BHT measurements because it is the most widely used method in the petroleum industry. Many other methods have been developed to correct BHT data; their assumptions and required parameters are described by Hermanrud et al. (1990) and Goutorbe et al. (2007). They found that estimates of true formation temperature did not necessarily improve as the logging correction method became more sophisticated. For example, the Bullard (1947) correction is commonly thought to be the most reliable approach to correct BHT data. However, this method requires borehole diameter and effective thermal diffusivity of the rock–borehole system as input. The latter is complex and not generally available, so this and many of the other approaches described by Hermanrud et al. (1990) and Goutorbe et al. (2007) are rarely used. Where data are sufficient for both methods, Horner and Bullard corrections give approximately the same answer (e.g., Hermanrud et al. 1990). Before discussing the Horner correction, we provide a brief review of area-specific empirical methods that can be used to correct BHT data.

Empirical Bottom-Hole Temperature Corrections

A multiwell statistical approach to correct BHT data, determined as part of a geothermal survey of North America sponsored by the American Association of Petroleum Geologists, is commonly referred to as the AAPG or Kehle correction (Kehle 1972). Equilibrium

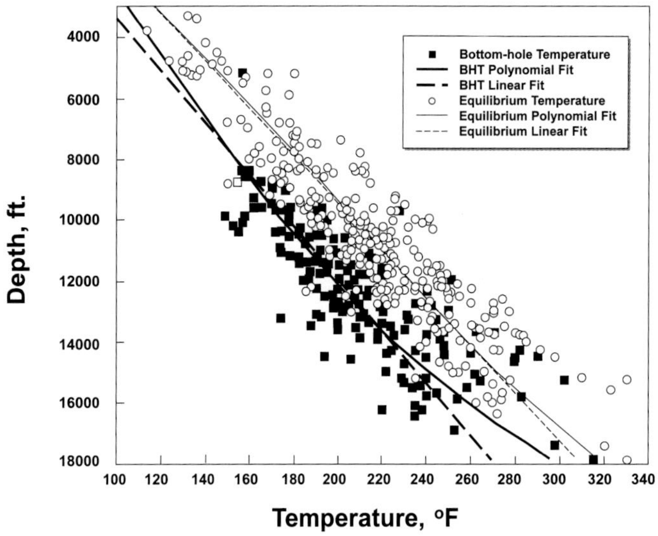

FIG. 1.—Bottom-hole temperature from well log headers, formation equilibrium temperature from reservoir data, and linear and polynomial fits to both data sets (raw data from Kehle 1972). All temperature data are from southern Louisiana.

temperatures in 336 wells in southern Louisiana were obtained from engineers of the operating companies, and BHT data were obtained from well logs in the same wells (Fig. 1). The figure shows least-square fits of the equilibrium temperatures and BHT data to linear and polynomial curves, with depth as the only independent variable. Temperature correction algorithms, also in polynomial form, were then derived from the fits to the data (e.g., Deming 1989). The resulting algorithm (not shown) imposes an additive correction that increases from zero at the surface to a maximum of 18.3° C (33° F) at 3657 m (11,998 ft). The correction diminishes below this depth. It appears that below a depth of 4267 m (14,000 ft) in Figure 1, the curve fits are not well constrained by the data, and the polynomial correction should not be used. Most southern Louisiana reservoirs from which the equilibrium temperatures were obtained occur close to the top of overpressure, which affects the temperature gradient, and many reservoirs are in close proximity to salt domes, which are thermally conductive and perturb the local temperature field. Thus, some of the scatter in equilibrium and BHT temperatures in Figure 1 probably results from geological aspects of the study area. A test of the algorithm shows that the correction is invalid in the Mackenzie Delta area of Canada (Majorowicz et al. 1990), and its shortcomings have also been demonstrated in the deep Tuscaloosa trend of central Louisiana (Nelson 2003). Simple depth dependence is attractive, but the relationship is generally unique to each area where it is derived and cannot be universally applied.

Temperature estimation is improved if the time between logging and cessation of circulation is incorporated into the empirical correction algorithm. The general form of the algorithm is as follows:

$$T_{true} - T_{log} = (f - 1) \times (T_{log} - T_{surf}) \quad (2)$$

T_{true} is true formation temperature, T_{log} is the measured BHT, T_{surf} is surface temperature, and f is a factor that incorporates the time since circulation stopped and the depth. Alternately, depth can be incorporated in a second term rather than included in f. The functional form and coefficients of factor f are determined so as to minimize the difference between the corrected and equilibrium temperatures, which is determined from drill-stem tests. Various implementations of

Equation 2 have been reported for the Malay Basin (Waples and Ramly 2001), the Gulf of Mexico (Waples et al. 2004), and the Danish Central Graben (Waples and Pederson 2004). In all three cases, the time since circulation stopped is a more important parameter than depth in determining the magnitude of the correction. The uncertainty in the temperature estimate also strongly depends on the time since circulation stopped. According to the authors, temperatures in one area can be corrected fairly well using the algorithm from one of the two other areas, indicating that incorporation of time since circulation results in a robust correction algorithm.

Horner Correction of Bottom-Hole Temperature

The following discussion identifies criteria that can be used to correct temperatures for short-duration shut-in tests, in particular BHT. Uncorrected BHT data are abundant, but of low quality. Because time spent logging exploration wells is expensive, operators continue to find ways to log faster and run fewer tool strings after drilling. Logging-while-drilling (LWD) tools are also available, where petrophysical measurements similar to those acquired by open-hole logging can be obtained without shut-in of the well. A consequence of faster log runs is that mud in the borehole has less time to equilibrate with formation temperature when circulation stops and the log run begins. The maximum temperature from these logs is lower than formation temperature and must be corrected to determine true formation temperature.

Four criteria are recommended to select the BHT data most suitable for correction:

1. circulation time, t (cooling of the formation by circulating mud), prior to BHT measurement must be of short duration (<24 hours; data from drilling report);
2. time-since-circulation stopped or shut-in time, Δt (warming of the static mud column by the formation), prior to temperature measurement must be of long duration (>12 hours; data from well log header);
3. no recirculation can occur between logging runs; and
4. BHT data having three or more time–temperature measurements at the same depth are best, as discussed below.

Reliable Horner corrections of BHT require a suite of at least three time–temperature pairs at each depth in the well, as shown in an example from the Husky Oil Inigok well, North Slope, Alaska (Fig. 2). Each Horner suite records the temperature history of a static, passively reequilibrating point at specific depth within the mud column and wellbore. The correction can be made using only two time–temperature pairs, but three or more pairs allows a check on linearity of the data. For reliable correction, the well must not be circulated during logging or between the several logging runs of a Horner suite. The Horner plot is a semilog graph that shows temperature on the vertical axis and a logarithmic function of time on the horizontal axis. Many Horner plots differ in how they calculate the time axis. Short times after the end of circulation are to the right and longer times are to the left. The extrapolated temperature at infinite time after circulation, i.e., the time of full temperature equilibrium, plots on the far left of the horizontal axis.

The Horner correction assumes that BHT rises in the same way as pressure (Horner 1951, Timko and Fertl 1972). The method is based on the Horner plot as modified by Lachenbruch and Brewer (1959) and Dowdle and Cobb (1975). Although not mathematically rigorous, this modified plot was shown by Dowdle and Cobb (1975) to give generally reliable estimates of true formation temperature:

$$T = T_i - K_1[\log(t + \Delta t)/\Delta t] \quad (3)$$

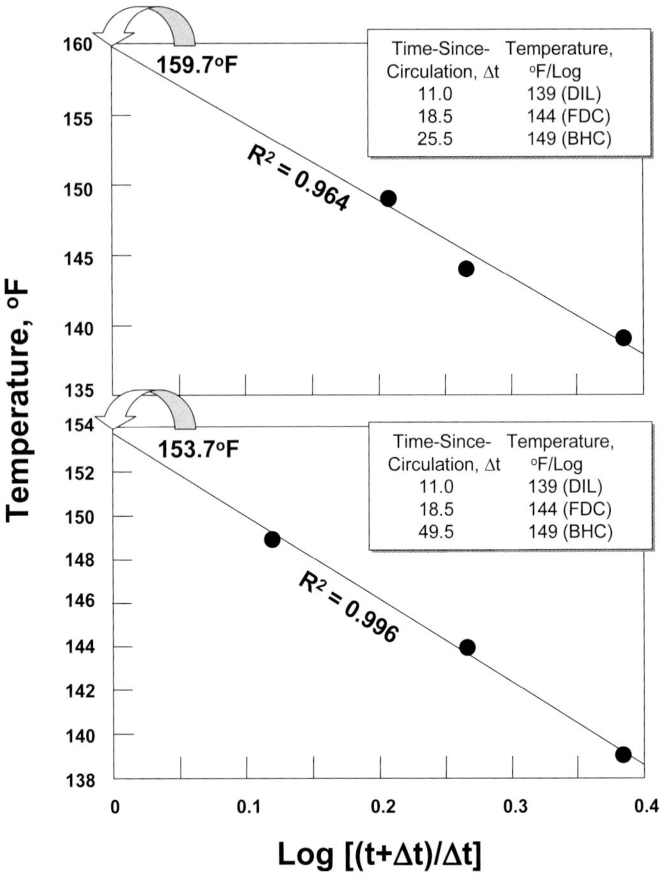

FIG. 2.—Horner plots of three time–temperature pairs (insets; Table 1) at ~8260 ft (~2518 m) in the Husky Oil Inigok-1 well, onshore North Slope, Alaska. DIL = dual induction-laterolog, FDC = compensated formation density log, BHC = borehole compensated sonic log, t = circulation time (16 hours), and Δt = time-since-circulation. Curved arrows indicate least-square extrapolated true formation temperature. The top and bottom parts of the figure compare Horner predictions based on two interpretations of the recorded time-since-circulation, as discussed in the text.

where T = true formation temperature, T_i = observed log temperature at time $t + \Delta t$, t = circulation time (cooling time), and Δt = time-since-circulation (warming time).

Therefore, a semilog plot of T_i vs. $\log(t + \Delta t)/\Delta t$ can be extrapolated by least-squares fit along a line of slope K_1 to a time ratio of unity (or zero on a logarithmic plot) to yield T, as illustrated in Figure 2. Blanchard and Tailleur (1982) used a weighted least-squares fit that is biased toward small values of $\log(t + \Delta t)/\Delta t$ because shorter circulation times (t) and longer times since circulation (Δt) more closely approach equilibrium and give the most reliable estimates of T.

Three wire line log headers for the Husky Oil Inigok-1 well, onshore North Slope, Alaska (Figs. 3–5), exemplify the difficulties that interpreters may encounter when attempting to derive useful BHT data for Horner correction. Although not available on the log headers, well history information can be used to derive circulation time (cooling). For this well, the mud log drilling rate curve and driller records indicate that the circulation time was about 16 hours prior to the wire line logging run #2 (Table 1). After mud circulation was stopped, each of the three logging tools was separately lowered into the hole and raised

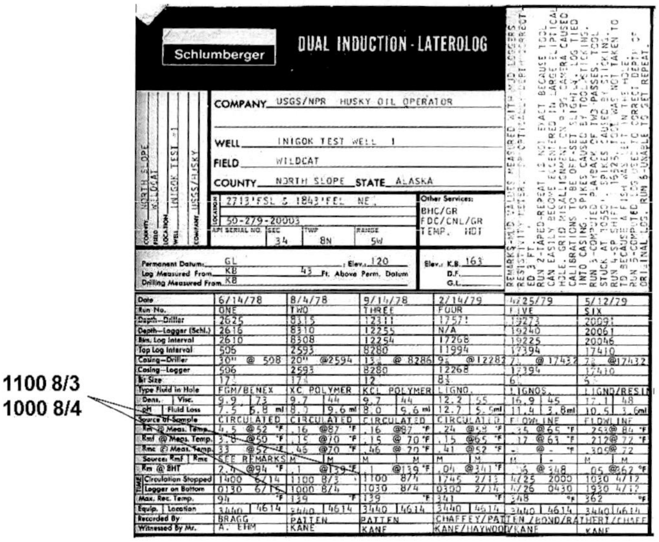

FIG. 3.—Dual induction-laterolog (DIL) well log header for the Husky Oil Inigok-1 well, North Slope, Alaska. Highlighted times are for circulation stopped (top) and logger-on-bottom (bottom) data used to calculate time-since-circulation (Table 1).

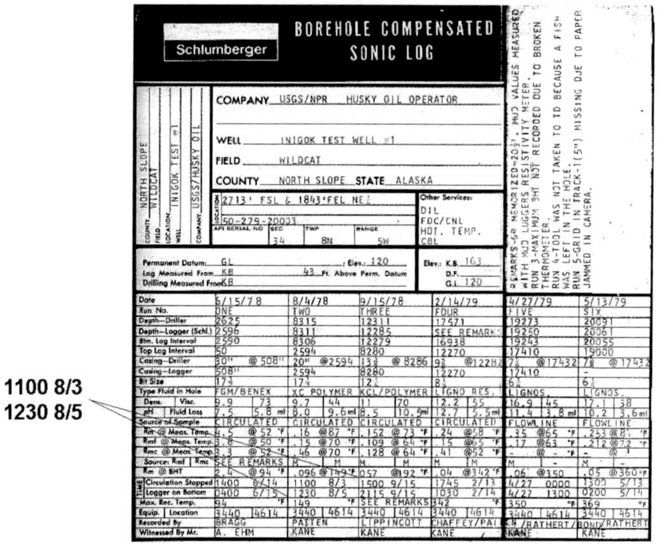

FIG. 5.—Borehole compensated sonic (BHC) well log header for the Husky Oil Inigok-1 well, North Slope, Alaska. Highlighted times are for circulation stopped (top) and logger-on-bottom (bottom) data used to calculate time-since-circulation (Table 1).

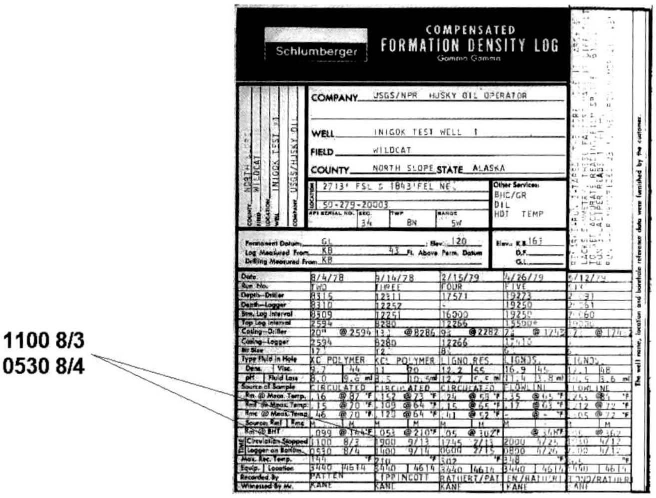

FIG. 4.—Compensated formation density (FDC) well log header for the Husky Oil Inigok-1 well, North Slope, Alaska. Highlighted times are for circulation stopped (top) and logger-on-bottom (bottom) data used to calculate time-since-circulation (Table 1).

to the surface. Time-since-circulation was calculated as the difference between the time that circulation stopped and the recorded logger-on-bottom data from the final typed log headers. As discussed later herein, some of the recorded logger-on-bottom data are questionable and were reinterpreted (interpreted logger-on-bottom column, Table 1).

The dual induction-laterolog header (DIL; Fig. 3) for logging run #2 on August 4, 1978, indicates a "depth-logger" of 8310 ft (2533 m). If we assume ~50 ft (15 m) between the bottom of the hole and the location of the temperature gauge on the tool (e.g., Blanchard and Tailleur 1982), then the corrected BHT depth is ~8260 ft (~2518 m). Circulation stopped at 1100 h (11 am) on 8/3/78 (August 3, 1978), and according to the record, the logger reached the bottom of the hole at 1000 h (10 am) on August 4, 1978, resulting in a calculated time-since-circulation (warming) of 23 hours. The corresponding maximum recorded temperature was 139° F (59° C). The same circulation stopped time (1100 on August 3, 1978) is recorded for logging run #2 on the compensated formation density log header (FDC) and the borehole compensated sonic log header (BHC), which tends to validate the accuracy of the recorded data. However, the logger-on-bottom time is suspect because very similar data for circulation stopped and logger-on-bottom occur in the column for DIL run #3 on September 14, 1978, with times of 1100 h and 1030 h and a recorded date for each of August 4, 1978, which conflicts with the September 14, 1978, date at the top of the column. This inconsistency in the log header information suggests that the logger-on-bottom data for runs #2 and #3 are spurious or fabricated. A further indication of error is that the maximum recorded temperature of 139° F (59° C) for DIL logging run #2 is lower than that for the FDC log run #2 run (144° F [62° C]) discussed later, even though the FDC run had a shorter time-since-circulation (18.5 hours vs. 23 hours; Table 1). Note that all of the time calculations assume that both the logger and the interpreter used a 24-hour clock, e.g., "1000" means 10 am and not 10 pm. Based on our interpretation of the data, it appears that the logger erroneously entered 1000 on August 4, 1978, when in fact, the correct information was 2200 h (10 pm) on August 3, 1978. Experience indicates that these kinds of errors that occur near the end of a long work shift are fairly common.

The compensated formation density log header (FDC; Fig. 4) for August 4, 1978, indicates a depth-logger of 8310 ft (2533 m), corresponding to a corrected BHT depth of ~8260 ft (2518 m). Circulation stopped at 1100 h (11 am) on August 3, 1978, and the logger reached bottom at 0530 (5:30 am) on August 4, 1978, resulting in a calculated time-since-circulation of 18.5 hours. The maximum recorded temperature was 144° F (62° C; Table 1).

Run #2 from the borehole compensated sonic log header (BHC; Fig. 5) at a corrected BHT depth of ~8261 ft (2518 m) indicates that circulation stopped at 1100 h (11 am) on August 3, 1978, and that the logger reached bottom at 1230 h on August 5, 1978, resulting in a calculated time-since-circulation of 49.5 hours. Since the date at the

TABLE 1.—Summary data for logging run #2 from compensated formation density (FDC), dual induction-laterolog (DIL), and borehole compensated sonic logs (BHC) at ~8260 ft (~2518 m) in the Husky Oil Inigok-1 well, North Slope, Alaska (Figs. 3–5). Values in parentheses are based on reinterpreted logger-on-bottom times as discussed in the text.

	Circulation					
Log tool	Time (hours)	Stopped (time, date)	Recorded logger-on-bottom (time, date)	Interpreted logger-on-bottom (time, date)	Time-since-circulation (hours)	Temperature (°F/°C)
DIL	16	1100, 8/3	1000 8/4	2200 8/3	23 (11)	139/59
FDC			0530 8/4	0530 8/4	18.5 (18.5)	144/62
BHC			1230 8/5	1230 8/4	49.5 (25.5)	149/65

top of this column is 8/4/78, one might interpret that the August 5, 1978, date recorded for logger-on-bottom is incorrect and was actually August 4, 1978, resulting in a time-since-circulation of 25.5 hours rather than 49.5 hours (Table 1). These two possible interpretations of the recorded data are compared in Figure 2, as discussed later herein. Blanchard and Tailleur (1982) calculated the correct time-since-circulation for this logging run at 25.5 hours, although their table for the well erroneously indicates that logger-on-bottom occurred at 0030 h on August 5, 1978.

Figure 2 compares Horner plots using data selected from Table 1. Time-since-circulation from the DIL was interpreted to be 11 rather than 23 hours because of the apparently spurious data on the log and because the BHT of 139° F (59° C) suggests that the associated time must be less than 18.5 hours (the FDC time corresponding to 144° F or 62° C). The top and bottom parts of Figure 2 predict true formation temperatures of 159.7 and 153.7° F (70.9 and 67.6° C) using 25.5 and 49.5 hours, respectively, for the time-since-circulation from the BHC. As discussed above, either of these times is possible based on interpretation of the recorded data.

Much of the nonlinear character of the best-fit line for three or more points on Horner plots such as Figure 2 is due to inaccuracy in the calculated time-since-circulation and temperature. However, part of this nonlinearity may also result from differences in the exact depths where temperatures were recorded. Bottom-hole tools record the maximum temperature on each wire line logging run down the column of static drilling fluid to a somewhat variable depth above drilled bottom. Imprecise bottom-hole depths can result from (1) location of the thermometer some distance above the bottom of the logging tool, (2) variable lengths of single or coupled logging tools, and (3) cavings or other obstructions that keep the base of the tool from reaching drilled bottom. Blanchard and Tailleur (1982) assumed that the temperature gauge for BHT measurements averaged 50 ft (15.24 m) above the maximum recorded depth reached by the logging tool.

The quality of Horner corrections requires three criteria to be met:

1. A minimum of three logging runs that record time and temperature for each run is required.
2. The temperature extrapolation should be less than the range of temperature data.
3. Deviations from the least-squares regression line should be less than measurement uncertainty (<5° F or 3° C).

For example, both Horner plots in Figure 2 satisfy the requirements for three logging runs (criterion 1), and deviations from the best fit line are less than uncertainty (criterion 3). However, the temperature extrapolations for the top and bottom plots in the figure (criterion 2) are ~11 and ~5° F (~6 and ~3° C) respectively, while the range of temperature data is ~10° F (~6° C). Thus, one might consider the top Horner plot in Figure 2 to be less reliable than that at the bottom, although neither plot need necessarily be rejected. However, logging run #5 at corrected depth of ~19,200 ft (5852 m; Figs. 3–5) is a clear example of a time–temperature pair data that does not satisfy the criteria. The DIL, FDC, and BHC time–temperature data from the figures are 8.5 hours–348° F (176° C), 12 hours–348° F (176° C), and 13 hours–350° F (177° C), respectively, resulting in a predicted true formation temperature from the Horner plot of 353.5° F (196° C). The temperature extrapolation (3.5° F, 2° C) exceeds the range of the data (2° F, 1° C). No other data from these logging runs allow three time–temperature pairs to be calculated.

Error Estimates: Two types of error affect the usefulness of corrected BHT data. Random errors produce scatter in various corrected BHT values around the true formation temperature. Systematic errors manifest themselves as an offset or bias of the average corrected BHT value from true formation temperature. Systematic error is more difficult to evaluate than random error, and correction of systematic error requires special situations where one has good independent estimates of true formation temperature.

Although error assessment of Horner-corrected temperatures can be complex due to unreliable information on log headers or other factors, two studies suggest a standard (random) error of about ±8° C (14° F) regardless of depth or true formation temperature. Based on a study of 983 corrected BHT and associated equilibrium temperature estimate (Teq) pairs conducted by J. Corrigan, the Teq uncertainty (1 sigma) using the Horner correction is ±14° F (±8° C; ZetaWare, 2007). The error estimate was obtained using DST temperatures and adjacent (±500 ft, 152 m) Horner-corrected BHT data. It provides an estimate of the uncertainty associated with Horner-corrected temperatures. A similar study was based on 18 wells from the Oseberg field in Norway (Hermanrud et al. 1990). The measurements were adjusted to compensate for different depths of burial, and a true formation temperature at 2700 m (8858 ft) of 98.9 ± 8° C (209.8 ±14° F) was estimated.

Some studies indicate that Horner-corrected temperatures are lower than true formation temperatures inferred from DST data; i.e., systematic errors exist. For example, Hermanrud et al. (1990) found that the Horner-plot estimate of 98.9 ±8° C (209.8 ±14° F) did not agree with a mean depth-adjusted estimate of 107.2 ±2.5° C derived from DST measurements in the Oseberg field. Waples and Pedersen (2004) obtained excellent agreement between Horner-plot temperatures (T_{HP}) and true formation temperatures (T_{true}) based on DST data for wells in the Danish Central Graben, when Horner temperatures were corrected upward by an amount proportional to the slope (A) of the Horner plot using the following equation.

$$T_{true} = T_{HP} + 26 \times \exp(-0.0052A) \qquad (4)$$

Estimates of systematic error in one basin probably cannot be applied to another basin because drilling conditions vary. For example,

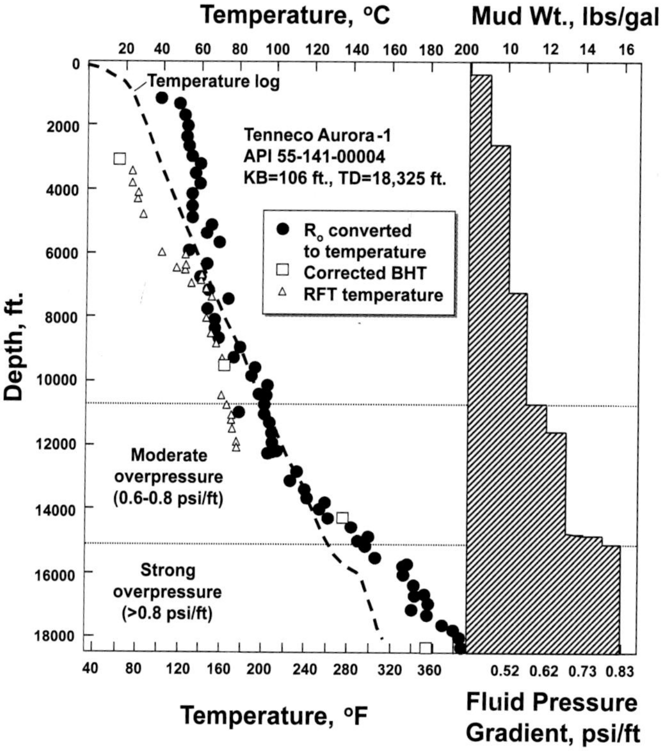

FIG. 6.—Comparison of corrected bottom-hole temperature (BHT, open squares), temperature log (dashed line), repeat formation test temperature (RFT, open triangles), temperature calculated from vitrinite reflectance (R_o, solid dots; method of Barker and Pawlewicz [1994], where $° C = [1.68 + \ln R_o]/0.0124$), and mud weight for the Tenneco Aurora-1 well, offshore North Slope, Alaska (modified from Bird et al. 1999).

temperature perturbations produced by circulating mud in offshore vs. onshore locations are expected to differ due to differences in heat flux and thermal conductivity profiles and consequent differences in temperature gradient, and also due to the near-constant temperature encountered by drilling mud as it passes through the riser in offshore wells.

TEMPERATURE ESTIMATES IN THE AURORA-1 WELL, ALASKA

The Tenneco Aurora-1 well from offshore North Slope, Alaska (Fig. 6), has abundant temperature data (Paul et al. 1994, Bird et al. 1999) that serve as an example of the complexity that may be encountered when multiple data sets are available. Temperature log, BHT, and RFT temperatures were recorded, as well as vitrinite reflectance, which can be converted to maximum paleotemperature, as discussed later herein. Temperature log data obtained under static conditions were reported at 1000-ft (305-m) intervals from 1000 to 18,000 ft (305 to 5486 m) and were measured in two partly overlapping log runs: run #4 = 10 to 16,424 ft (3 to 5006 m) and run #5 = 15,847 to 18,172 ft (4830 to 5539 m). The apparent sharp decrease in temperature at less than 1000-ft (305-m) depth on the temperature log curve in Figure 6 is the result of connecting points for the temperature log measurement at 1000 ft (305 m; 66° F) with a temperature measured at the mud line (33° F) in 68 ft (21 m) of water below the drill floor.

Figure 6 compares the present-day log temperature curve with maximum paleotemperature inferred from vitrinite reflectance in the Tenneco Aurora-1 well using the method of Barker and Pawlewicz (1994). The traditional method used to predict vitrinite reflectance sums the extent of reflectance evolution based on vitrinite kinetic parameters for all intervals throughout the burial history of a rock unit. This method assumes that both temperature and time affect reflectance evolution, even after temperature decrease due to uplift or long time intervals at constant temperature. However, Barker and Pawlewicz (1994) considered time to be a significant factor only as the peak burial temperature (T_{peak}) is approached. Predicted temperatures for various published cases of burial heating based on their method closely agree with those using the traditional method based on burial history reconstruction and kinetic modeling.

Figure 6 contains four sets of BHT time–temperature pairs that allowed calculation of corrected BHT. The corrected BHT at 3000 ft (914 m) in the well represents a more accurate estimate of true formation temperature than the temperature log at that depth. Conditioning of the well by circulation of the mud prior to log run #4 (10–14,864 ft, 3 to 4530 m) brought warm mud from depth to the surface. Circulation was stopped for an unrecorded time prior to the temperature log (10–16,424 ft, 3 to 5006 m) during log run #4 (Paul et al. 1994). The warm drilling mud had insufficient time during the temperature log run to equilibrate with the cooler formation. Therefore, the drilling mud was warmer than the formation at shallow depth (<6000 ft, <1829 m), resulting in higher temperature from the temperature log (measured in the mud column) compared to RFTs (uncorrected) and the corrected BHT. The RFT data are similar to the corrected BHT, apparently because they sampled the formation fluid at or near true formation temperature.

The maximum paleotemperature inferred from vitrinite reflectance at shallow depth (<6000 ft, <1829 m; solid dots in Fig. 6) in the Aurora-1 well is higher than that from the temperature log (dashed line) and higher than the true formation temperature at these depths, probably because of recycled vitrinite in these shallow Tertiary rocks. Recycled vitrinite was exposed to higher thermal maturity during initial burial, followed by uplift, erosion, and subsequent redeposition in less mature sediments. This is consistent with the generalization that the North Slope is a large Mesozoic–Cenozoic basin surrounded by exposed Paleozoic–Mesozoic highlands. The dogleg in the calculated temperature profile at 6000 ft (1829 m) in the well (Fig. 6) probably indicates the time when the Sadlerochit Mountains began to contribute recycled vitrinite to the coastal plain near the present-day Aurora-1 well. Apatite fission-track evidence indicates that the most severe uplift and erosion events occurred in the Sadlerochit Mountains about 45 Ma and 23 Ma (Eocene and Oligocene–Miocene, respectively; O'Sullivan et al. 1993).

In the depth range 6000 to 14,000 ft (1829–4267 m), present-day temperatures from a corrected BHT (9400 ft, 2865 m) and multiple RFTs remain lower than those from the temperature log, but the discrepancy is generally not as great as that in the overlying range (<6000 ft or <1829 m; Fig. 6). Although the magnitude of these temperature discrepancies between corrected BHT and temperature log differs between the two depth ranges, the same explanation applies. The temperature of the drilling mud measured by the temperature log remained higher than the true formation temperature approximated by the corrected BHT and RFT temperatures in the depth range 6000 to 14,000 ft (1829–4267 m) because it was exposed to warmer, deeper parts of the well during conditioning and prior to the temperature log. The close correlation between temperature log and vitrinite-derived temperatures in the range 6000 to 14,000 ft (1829–4267 m) compared to <6000 ft (<1829 m) is apparently fortuitous because the temperature log still had not achieved equilibrium with formation temperature.

At depths greater than 14,000 ft (4267 m; Fig. 6), the temperature log records lower temperatures than the two corrected BHT

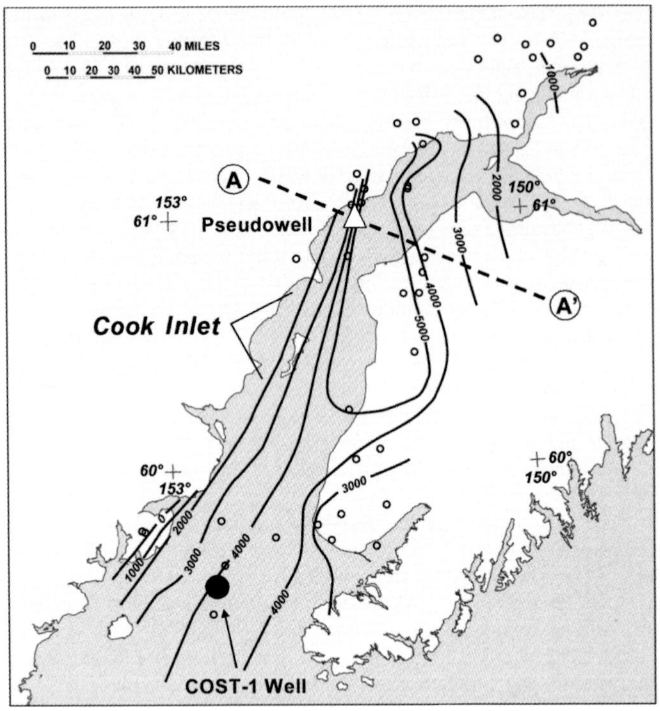

FIG. 7.—Map of Cook Inlet, Alaska, showing location of the COST-1 well (solid dot) and pseudowell (open triangle) along transect A-A′ defined by Magoon (1994). Solid lines are depth contours to the 0.6% vitrinite reflectance isograd in meters below sea level based on data from other wells (open dots; contours from Johnsson et al. 1993).

measurements and the reflectance-converted temperatures. At these depths, the drilling mud is cooler than the formation, resulting in lower temperature from the temperature log than that from corrected BHT data at 14,400 and 18,350 ft (4389 and 5593 m). The higher temperature gradient is evident in the paleotemperature profile. Bird et al. (1999) postulated that overpressures and an associated "thermal blanketing effect" related to enhanced porosity and lower thermal conductivity, typical of overpressured zones (Hunt 1996), were responsible for the abrupt increase in vitrinite reflectance and inferred paleotemperature below 14,000 ft (4267 m). The cause of overpressure is unclear, but it could result from petroleum generation, disequilibrium compaction, and/or tectonic stress.

INFLUENCE OF BHT ERROR ON A 1D MODEL: UPPER COOK INLET

One-dimensional basin and petroleum system models are commonly used in sedimentary basins having minimal data where rapid assessment is required to predict the timing of source-rock thermal maturity and petroleum generation at one location as a function of depth. A 1D model can be constructed using only well logs and calibration data, such as vitrinite reflectance and corrected bottom-hole temperatures. In poorly explored basins, pseudowells can be constructed using only seismic data and inferences about the imaged strata to provide tentative results, commonly without calibration. Obviously, such poorly constrained models need to be interpreted with care, but in many cases they are nonetheless useful. The following example of a 1D model aims to illustrate potential error in model results associated with calibration using a single corrected BHT value. For this purpose, we will assume that the BHT value represents the true formation temperature, but that it is subject to a standard error of ±8° C, as discussed previously.

Vitrinite reflectance data from wells in the area of the upper Cook Inlet, Alaska (open dots, Fig. 7), allowed construction of a depth contour map to the top of the oil window (0.6% R_o). However, these wells lacked suitable data for estimates of true formation temperature in the deepest part of the basin (open triangle, Fig. 7). A pseudowell at that location (Peters et al. 2006) provides an example of calibration for a PetroMod® (version 9.0; Schlumberger, Aachen, Germany) 1D basin and petroleum system model. PetroMod geohistory reconstruction accounts for changes in porosity and thermal conductivity during burial for each lithology using decompaction and backstripping algorithms (Hantschel and Kauerauf 2009). The pseudowell is an imaginary well located in the depocenter near the Middle Ground Shoal field along a cross-section defined by Magoon (1994), where stratigraphy was determined from seismic data. The pseudowell was assigned corrected BHT (92.4° C at 3382 m or 198.3° F at 11,096 ft; Magoon 1986) and vitrinite reflectance values ($R_o = 0.32$ and 0.59% at 31 and 3746 m or 102 and 12,290 ft; Johnsson and Howell 1996) based on data from the COST-1 well in the lower Cook Inlet, ~175 km (~109 mi) to the southwest and on-strike with the pseudowell. In this example, we assume that the corrected BHT in the COST-1 well represents true formation temperature and that error in the model is due only to that associated with the BHT measurement.

At least two independent calibration tools, such as Horner-corrected BHT data (left in Fig. 8) and vitrinite reflectance (R_o in percent; right in Fig. 8), are recommended for reliable temperature history reconstruction (Welte et al. 1997). The PetroMod® simulator uses stratigraphy, burial history, boundary conditions (paleo–water depth, sediment–water interface temperature, heat flow), and decompaction algorithms to calculate vitrinite reflectance values based on the EASY%R_o method (Sweeney and Burnham 1990). In this simple model, heat flow was assumed to be constant through time. Calibration was accomplished by varying the value of heat flow until the calculated temperature and R_o curves intersected the measured BHT and R_o values, respectively.

The inferred BHT in the pseudowell at 3382 m (11,096 ft) is shallow compared to the top of the Middle Jurassic Tuxedni Group source rock (10,144 m or 33,281 ft); in this case, about one-third as deep. This is typical of most wells in the area, which target shallow reservoir rocks and do not penetrate to depths where the source rock is thermally mature. The Tuxedni source rock is the thin unit between the Upper Jurassic Naknek Formation and the Middle Jurassic Tuxedni Group in Figure 8. Error associated with determination of true formation temperature at the depth of the BHT propagates to the greater depth of the source rock where temperature and thermal maturity are to be estimated. The upper Cook Inlet example has a surface temperature of 4° C and only one Horner-corrected BHT rather than two or three, which might better define a line for extrapolation. Therefore, the example gives a sense of the maximum error that might be expected in predictions of thermal maturity of the deep source rock.

Time or depth to the top of the oil and gas windows (0.6% or 1.2% R_o, respectively) for Tuxedni Group source rock in the pseudowell varies systematically with the adjusted heat flow required to fit BHT data that span plausible error in the measurement (Table 2). For example, a BHT of 92.4° C (best-fit scenario) at present-day depth of 3382 m results in top oil window for the source rock at 82.50 Ma and 2409 m paleodepth (Fig. 9). BHT values 8° C above or below 92.4° C (one standard deviation based on previous discussion) result in top oil window 6.2 million years earlier and 305 m shallower or 4.5 million years later and 231 m deeper than the measured BHT, respectively (10.7 million years and 536 m range). Likewise, a BHT of 92.4° C results in top gas window for the Tuxedni Group source rock at 22.73 Ma and 4187 m depth (Table 2). BHT values 8° C above or below 92.4° C result in top gas window 1.8 million years earlier and 312 m shallower or

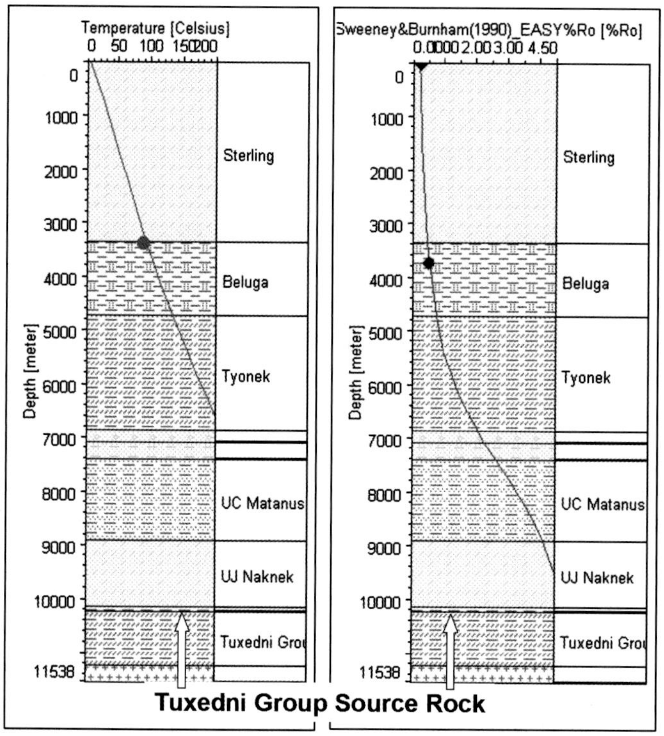

FIG. 8.—Example shows calibration of the 1D PetroMod® basin and petroleum system model for a pseudowell in the upper Cook Inlet, Alaska, based on one Horner-corrected BHT (red dot, left column) and two measured vitrinite reflectance (R_o) values (black dots, right column) in the COST-1 well (Peters et al. 2006). In this case, optimal fit of the calculated temperature and vitrinite reflectance curves was achieved using a constant 67.65 mW/m² heat flow (Table 2).

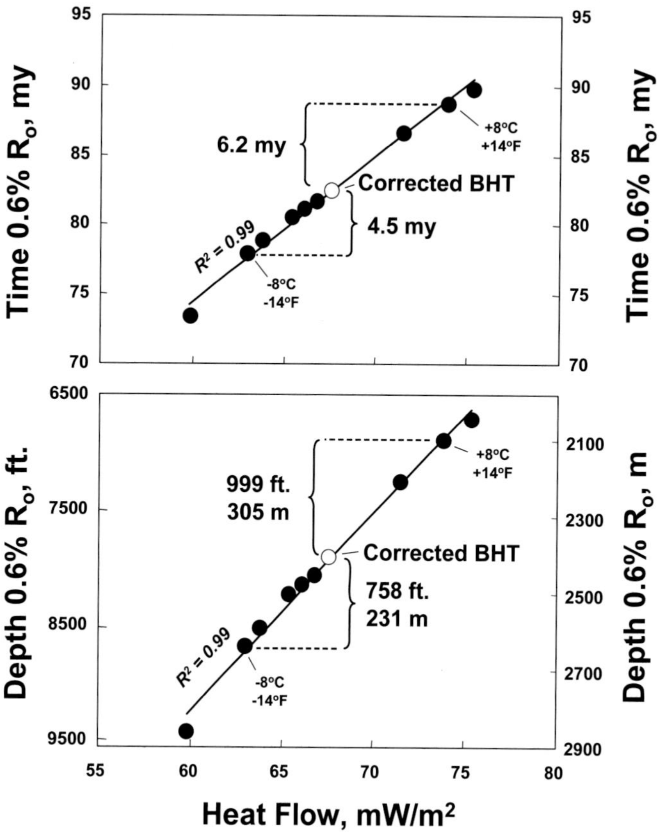

FIG. 9.—Influence of BHT error (±8° C or 14° F) on calculated age (million years, top) and burial depth (ft) of the Jurassic Tuxedni Group source rock at 0.6% vitrinite reflectance (top of oil window) in a pseudowell from the upper Cook Inlet, Alaska (see Fig. 7). The ten data points are from Table 2. Open dot is the best-fit scenario where PetroMod® calibration to a BHT of 92.4° C results in calculated top oil window for the source rock at 82.50 Ma and 2409 m paleodepth. See text for discussion.

TABLE 2.—Sensitivity of a one-dimensional (1D) PetroMod® basin and petroleum system model to error in corrected BHT (±8° C or 14° F) tested using data from a pseudowell in the upper Cook Inlet, Alaska. Several identical models of the pseudowell were calibrated by adjusting heat flow to fit BHT in the range 92.4 ± 8° C (198.3 ± 14° F) at 3382 m (11,096 ft). Present-day depth to top of the Tuxedni Group source rock is 10,144 m (33,281 ft; Fig. 8).

BHT (°C/°F)	Error (°C/°F)	Adjusted heat flow (mW/m²)	Tuxedni Group source rock = 0.6% R_o Age (Ma)	Tuxedni Group source rock = 0.6% R_o Depth (m/ft)	Difference from measured BHT (million years)	Difference from measured BHT (m/ft)	Tuxedni Group source rock = 1.2% R_o Age (Ma)	Tuxedni Group source rock = 1.2% R_o Depth (m/ft)	Difference from measured BHT (million years)	Difference from measured BHT (m/ft)
102.4/216.3	+10/18.0	75.30	89.86	2048/6719	7.36	361/1184	24.77	3825/12,549	2.04	362/1188
100.4/212.7	+8/14.4	73.83	88.73	2104/6903	6.23	305/999	24.48	3875/12,713	1.75	312/1024
97.4/207.3	+5/9.0	71.50	86.58	2211/7254	4.08	198/650	23.91	3970/13,025	1.18	217/712
92.4/198.3	0/0	67.65	82.50	2409/7904	0	0/0	22.73	4187/13,669	0	0/0
91.4/196.5	−1/−1.8	66.80	81.72	2455/8054	0.78	46/151	22.42	4217/13,835	0.31	30/98
90.4/194.7	−2/−3.6	66.10	81.13	2479/8133	1.37	70/230	22.18	4258/13,970	0.55	71/233
89.4/192.9	−3/−5.4	65.38	80.55	2507/8225	1.95	98/323	21.93	4298/14,101	0.80	111/364
87.4/189.3	−5/−9.0	63.80	78.88	2592/8504	3.62	183/600	21.42	4383/14,380	1.31	196/643
84.4/183.9	−8/−14.4	63.04	77.96	2640/8661	4.54	231/758	21.11	4436/14,554	1.62	249/817
82.4/180.3	−10/−18.0	59.92	73.47	2859/9380	9.03	450/1476	19.53	4698/15,413	3.20	511/1677

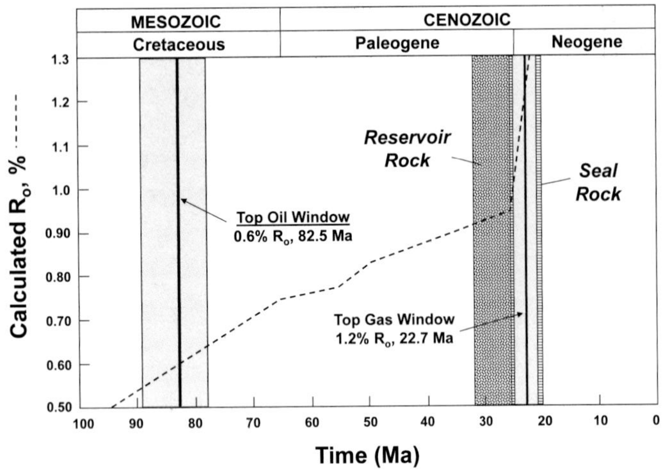

FIG. 10.—Calculated vitrinite reflectance (R_o, %; dashed line) for Jurassic Tuxedni Group source rock in a pseudowell from the upper Cook Inlet, Alaska, based on a PetroMod® 1D model that used the EASY%R_o method of Sweeney and Burnham (1990). Top of oil and gas windows occurred at 0.6% and 1.2% R_o (solid vertical lines at 82.5 and 22.7 Ma, respectively; Table 2). Shaded areas define uncertainty of the calibration if one assumes a BHT of 92.4 ± 8° C (198.3 ±14° F) as discussed in the text. Reservoir rock (Hemlock Conglomerate; stippled) and seal rock (lower portion of Tyonek Formation; horizontal lines) were deposited approximately 32–26 Ma and 26–20 Ma, respectively (Magoon 1994), thus allowing accumulation of late-generated petroleum from the source rock.

1.6 million years later and 249 m deeper than that calculated using the 92.4° C BHT, respectively. The oil window is the range of maturity where source rock generates both oil and hydrocarbon gas, whereas the gas window refers to higher maturity where only hydrocarbon gas is generated, either from the source rock or from thermal cracking of crude oil.

As an example of interpretation, Figure 10 shows the range of ages to the top of the oil (0.6% R_o) and gas windows (1.2% R_o) based on the plausible error in BHT (shaded areas). The best-fit scenario (bold vertical lines) results in an interpretation where most petroleum from the Tuxedni Group source rock was lost due to generation and migration prior to formation of the reservoir rock (Hemlock Conglomerate; stippled) and seal rock (lower part of the Tyonek Formation; horizontal lines). In this simple case, the interpretation from the model is not seriously affected by the error associated with the BHT measurement (Fig. 10).

An older model (Magoon 1994) required that petroleum generated from the Tuxedni Group source rock in Paleocene time migrate updip to the basin flanks where it was trapped stratigraphically until structural traps formed during Miocene deformation. Unlike the older model, our new model suggests that petroleum generation began in Upper Cretaceous rather than Paleocene time. However, both models confirm that the generation–migration–accumulation process continues today and that petroleum generated prior to deposition of the Hemlock Conglomerate reservoir rock (ca. 26–32 Ma) and the overlying seal rock in the lower portion of the Tyonek Formation seal rock (ca. 20–26 Ma) could be lost. Both models suggest that only a highly mature fraction of the generated petroleum consisting mainly of gas with some condensate could be trapped because of the relative timing of generation and formation of reservoir and seal rock.

CONCLUSIONS

In general, the most reliable formation temperatures are from production tests and drill-stem tests, provided that several criteria are met. For example, the temperature gauge must be positioned close to the perforations. Fluid production must be of sufficient volume and duration (>100 bbl) to allow the temperature to equilibrate with the producing formation. Care must be taken to avoid temperature disturbances caused by gas production and the Joule-Thompson effect.

Bottom-hole temperatures are ubiquitous, but they are generally biased much lower than true formation temperature due to lack of thermal equilibration. BHT data should not be used without quality assessment and Horner or other suitable corrections. BHT data are usually manually recorded by the logger. Incorrectly recorded temperatures and times are common, especially when a long work shift runs past midnight. A common shortcut is to read the thermometers for only one temperature run and record the same number for the other runs.

Acquiring just a few accurate formation temperatures for calibration of basin and petroleum system models is difficult and time-consuming because the vast majority of temperature data are unsuitable. Despite these difficulties, investigators must recognize the need to rigorously screen and discard poor-quality or questionable temperature data. Uncorrected BHT data can be 20 to 30° C (36–54° F) below true formation temperature. The best Horner-corrected temperature data are rarely more accurate than ±8° C or ±14° F, and systematic error may also be present. Because most wells target shallow reservoir rocks and do not penetrate deep, thermally mature source rock, errors associated with shallow BHT data can propagate to the greater depth of the source rock where temperature and thermal maturity are to be estimated. Interpreters should address this error using sensitivity analysis where several models are run through the range of possible BHT values.

A PetroMod® 1D model for a pseudowell in the upper Cook Inlet shows the way in which error in formation temperature inferred from BHT measurements impacts the predicted thermal maturity of deep source rock. For example, calibration using one BHT of 92.4° C at 3382 m extrapolated to the pseudowell from the COST-1 well in the lower Cook Inlet results in a modeled top oil window for the Jurassic Tuxedni Group source rock at 82.50 Ma and 2409 m paleodepth. BHT values 8° C above or below 92.4° C (one standard deviation based on published data) result in top oil window 6.2 million years (7.6%) earlier and 305 m (12.7%) shallower or 4.5 million years (5.5%) later and 231 m (9.6%) deeper than that calculated using the 92.4° C BHT.

Finally, temperature data to be used in basin and petroleum system modeling should be mapped and examined in the context of basin-scale geology. A simple technique is to map and contour all temperature data from a formation or formations in the form of temperature–depth ratios. One may then decide not to use corrected BHT data from anomalous areas. Geological impacts on present-day temperature include rock units with unusually high or low thermal conductivity, convective heat flow, heat refraction associated with salt and mud diapirs, and abnormally high temperature gradients in overpressured zones. If these geological factors are not included in the basin and petroleum system model, then such temperature data should not be used to constrain model input.

ACKNOWLEDGMENTS

We thank Ken Bird, Rick Stanley, Paul Hsieh (US Geological Survey), Les Magoon (USGS, retired), Wolf Rottke (Schlumberger), Nick Harris (University of Alberta), Jeffrey Nunn (Louisiana State University), and an anonymous reviewer for useful comments. Use of trade, product, or firm names is for descriptive purposes only and does not imply endorsement by the US Geological Survey.

REFERENCES

Barker CE, Pawlewicz MJ. 1994. Calculation of vitrinite reflectance from thermal histories and peak temperatures: a comparison of methods. *In* Mukhopadhay PK, Dow W (Editors). *Vitrinite Reflectance as a Maturity Parameter: Applications and Limitations*, Symposium Series 570: American Chemical Society, Washington, DC. p. 216–229.

Bird KJ, Burrus RC, Pawlewicz MJ. 1999. Chapter VR thermal maturity. *In* ANWR Assessment Team (Editors). *The Oil and Gas Resource Potential of the 1002 Area, Arctic National Wildlife Refuge, Alaska*: US Geological Survey. Open File Report 98-34, Washington, DC. 64 p.

Blanchard DC, Tailleur IL. 1982. Temperatures and Interval Geothermal-Gradient Determinations from Wells in National Petroleum Reserve in Alaska: US Geological Survey. Open File Report 82-391, Washington, DC. 79 p.

Bowen R. 1986. *Groundwater*, 2nd ed.: Springer, New York. 444 p.

Bullard EC. 1947. The time necessary for a bore hole to attain temperature equilibrium. *Monthly Notices, Royal Astronomical Society, Geophysical Supplement* 5(5):127–130.

Carter LS, Kelley SA, Blackwell DD, Naeser ND. 1998. Heat flow and thermal history of the Anadarko Basin, Oklahoma. *American Association of Petroleum Geologists Bulletin* 82:291–316.

Conaway JG, Beck AE. 1977. Fine-scale correlation between temperature gradient logs and lithology. *Geophysics* 42:1401–1410.

Deming D. 1989. Application of bottom-hole temperature corrections in geothermal studies. *Geothermics* 18:775–786.

Dowdle WL, Cobb WM. 1975. Static formation temperature from well logs—an empirical method. *Journal of Petroleum Technology* 27:1322–1330.

Goutorbe B, Lucazeau F, Bonneville A. 2007. Comparison of several BHT correction methods: a case study on an Australian data set. *Geophysics Journal International* 170:913–922.

Hantschel T, Kauerauf A. 2009. *Fundamentals of Basin and Petroleum Systems Modeling*: Springer Verlag, Heidelberg, Germany. 476 p.

Hearst JR, Nelson PH, Paillet FL. 2000. *Well Logging for Physical Properties*: Wiley, Chichester, UK. 483 p.

Hermanrud C, Cao S, Lerche I. 1990. Estimates of virgin rock temperature derived from BHT measurements: bias and error. *Geophysics* 55:924–931.

Horner DR. 1951. Pressure build-up in wells. *In* Proceedings of the Third World Petroleum Congress; The Hague, Netherlands, 30 May–5 June 1951, Section II: E.J. Brill, Leiden, Netherlands. p. 503–521.

Hunt JM. 1996. *Petroleum Geochemistry and Geology*: W.H. Freeman and Company, New York. 743 p.

Jessop A. 1990. Comparison of industrial and high-resolution thermal data in a sedimentary basin. *Pure and Applied Geophysics* 133:251–267.

Johnsson MJ, Howell DG (Editors). 1996. Thermal Evolution of Sedimentary Basins in Alaska. US Geological Survey. Bulletin 2142, Washington, DC. 138 p.

Johnsson MJ, Howell DG, Bird KJ. 1993. Thermal maturity patterns in Alaska: implications for tectonic evolution and hydrocarbon potential. *American Association of Petroleum Geologists Bulletin* 77:1874–1903.

Kehle RO. 1972. *Geothermal Survey of North America, 1971 Annual Progress Report to the American Association of Petroleum Geologists (January 15, 1972)*: University of Texas, Austin, Texas. 31 p.

Lachenbruch AH, Brewer MC. 1959. Dissipation of the Temperature Effect of Drilling a Well in Arctic Alaska: US Geological Survey. Bulletin 1083-C, Washington, DC. p. 73–109.

Lachenbruch AH, Sass JH, Lawver LA, Brewer MC, Marshall BV, Munroe RJ, Kennelly JP Jr, Galanis SP Jr, Moses TH Jr. 1988. Temperature and depth of permafrost on the Arctic slope of Alaska. *In* Gryc G (Editor). *Geology and Exploration of the National Petroleum Reserve in Alaska, 1974 to 1982*: US Geological Survey Professional Paper 1399, Washington, DC. p. 645–656.

Magoon LB. 1986. Geologic Studies of the Lower Cook Inlet COST No. 1 Well, Alaska Outer Continental Shelf: US Geological Survey. Bulletin 1596, Washington, DC. 99 p.

Magoon LB. 1994. Tuxedni-Hemlock(!) petroleum system in Cook Inlet, Alaska, USA. *In* Magoon LB, Dow WG (Editors). *The Petroleum System—From Source to Trap*, Memoir 60: American Association of Petroleum Geologists, Tulsa, Oklahoma. p. 359–370.

Magoon LB, Dow WG. 1994. The petroleum system. *In* Magoon LB, Dow WG (Editors). *The Petroleum System—From Source to Trap*, Memoir 60: American Association of Petroleum Geologists, Tulsa, Oklahoma. p. 3–24.

Majorowicz JA, Judge A, Jones FW. 1990. Deep subpermafrost thermal regime in the Mackenzie Delta Basin, Northern Canada—Analysis from petroleum bottom-hole temperature data. *Geophysics* 55:362–371.

Nelson PH. 2003. Enhanced temperature gradient in the deep Tuscaloosa trend—is it really there? *In* Transactions of the 53rd Annual Convention of the Gulf Coast Association of Geological Societies: Gulf Coast Association of Geological Societies, Houston, TX, 10 p.

O'Sullivan PB, Green PF, Bergman SC, Decker J, Duddy IR, Gleadow AJW, Turner DL. 1993. Multiple phases of Tertiary uplift and erosion in the Arctic National Wildlife Refuge, Alaska, revealed by apatite fission track analysis. *American Association of Petroleum Geologists Bulletin* 77:359–385.

Paul LE, Choromanski DR, Turner RF, Flett TO. 1994. Geological, Geochemical, and Operational Summary, Aurora Well, Outer Continental Shelf Y-0943-1, Beaufort Sea, Alaska: Minerals Management Service OCS. Report MMS 94-0001, Anchorage, AK. 71 p.

Peters KE, Magoon LB, Hudson T. 2006. Origin and generation timing of Cook Inlet petroleum: circum-Arctic chemometrics and 1-D geohistory modeling. *Geological Society of America Abstract with Programs* 38(5)71.

Peters KE, Magoon LB, Lampe C, Hosford Scheirer A, Lillis PG, Gautier DL. 2008. A four-dimensional petroleum systems model for the San Joaquin Basin Province, California. *In* Hosford Scheirer A (Editor). *Petroleum Systems and Geological Assessment of Oil and Gas in the San Joaquin Basin Province, California*: US Geological Survey. Professional Paper 1713, Chapter 12, Washington, DC. 35 p. (http://pubs.usgs.gov/pp/pp1713/).

Sweeney JJ, Burnham AK. 1990. Evaluation of a simple model of vitrinite reflectance based on chemical kinetics. *American Association of Petroleum Geologists Bulletin* 74:1559–1570.

Timko DJ, Fertl WH. 1972. How down-hole temperatures, pressures affect drilling. Part 5, predicting hydrocarbon environments with wireline data. *World Oil* 174(10):73–88.

Waples DW, Pacheco J, Vera A. 2004. A method for correcting log-derived temperatures in deep wells, calibrated in the Gulf of Mexico. *Petroleum Geoscience* 10:239–245.

Waples DW, Pedersen MR. 2004. Evaluation of Horner plot–corrected log-derived temperatures in the Danish Central Graben, North Sea. *Natural Resources Research* 13:223–227.

Waples CW, Ramly M. 2001. A statistical method for correcting log-derived temperatures. *Petroleum Geoscience* 7:231–240.

Welte DH, Horsfield B, Baker DR (Editors). 1997. *Petroleum and Basin Evolution*: Springer-Verlag, New York. 535 p.

ZetaWare, Inc., 2007. ZetaWare, Inc. http://zetaware.com/utilities/bht/horner.html. Accessed May 21, 2007.

MOLECULAR INDICATORS OF GEOTHERMAL HISTORY

CLIFFORD C. WALTERS
ExxonMobil Corporate Strategic Research, 1545 US Highway 22E, Annandale, New Jersey 08801-0998 USA
e-mail: clifford.c.walters@exxonmobil.com

PAUL G. LILLIS
US Geological Survey, Denver Federal Center, Box 25046, MS 977, Denver, Colorado 80225-0046 USA

AND

KENNETH E. PETERS
Schlumberger Information Solutions, 18 Manzanita Place, Mill Valley, California 94941-1018 USA

ABSTRACT: Many basin simulators include the option to use calculated biomarker ratios from published first-order kinetic parameters as a means to calibrate thermal history. Published kinetic parameters for many biomarker reactions are based on simplistic and unfounded assumptions of reaction pathways, lack the needed precision, or have insufficient documentation to evaluate their effectiveness. We believe that calibrations of thermal history based on biomarker ratios are less reliable than those based on more common methods, such as vitrinite reflectance and corrected bottom-hole temperatures, because of uncertainty in the reaction mechanisms and kinetic parameters. Some research indicates that certain reactions, such as cyclization of *ortho*-substituted biphenyls and thermal decomposition of alkyl esters, could provide useful kinetics with which to calibrate basin models, but these efforts have not been pursued by the geochemical community.

KEY WORDS: biomarker, thermal maturity, stereoisomerization, kinetics

INTRODUCTION

The chemical composition of crude oil and rock bitumen changes with increasing thermal maturation. Systematic alteration of molecular composition occurs during cracking, condensation, aromatization, and rearrangement reactions in response to increased time and temperature. Geochemists have developed a host of molecular parameters and routinely use them to characterize the thermal maturity of oil and source rock. These molecular indicators of thermal maturity could supplement more traditional approaches to calibrate basin models, such as vitrinite reflectance (R_o), corrected bottom-hole temperature, apatite fission track temperature, and fluid inclusion homogenization temperature. However, in practice, molecular maturity parameters are rarely used to constrain geothermal history.

The purpose of this review is to examine the practical application of molecular indictors of thermal maturity to calibrate or constrain basin heat flow or temperature history. We discuss thermal maturity parameters based on the extent of biomarker isomerization and aromatization and show why the inferred reactions are too simplistic to allow reliable calibration of basin models. We then discuss thermal maturity parameters based on alkylaromatic distributions and their limitations. Finally, we discuss some alternative molecular reactions, such as decomposition of alkyl esters, which have potential application in basin modeling but have not yet been adequately documented.

Biomarkers are molecular fossils in sediment, rock, and crude oil that consist of complex organic compounds having chemical structures readily linked to their parent compounds in living organisms. We stress that biomarker and other geochemical analyses provide critical information needed to validate basin model results by comparing the predicted oil and gas yields, timing, migration pathways, and oil quality against measurements made on actual surface and subsurface samples. The reader is directed to Peters et al. (2005) for further information on these techniques.

DISCUSSION

Biomarker Maturation Parameters

Early geochemical studies revealed that ratios of certain biomarkers in source rock extracts systematically changed with thermal maturity (Mackenzie et al. 1980, 1981, 1982; Seifert and Moldowan 1980). Some of these ratios involved stereoisomers, which are compounds having the same molecular formula and basic chemical structures but different spatial arrangements of atoms around one or more asymmetric (chiral) carbon atoms. To be asymmetric, a carbon atom must be linked to four different substituents. When a substituent group changes orientation at one or more of these chiral carbons, the molecule is said to isomerize. Stereoisomers containing more than one asymmetric center are called epimers. In a molecule containing more than one asymmetric center, an inversion of all the chiral centers leads to its enantiomer (mirror image). An inversion of less than all the asymmetric centers yields an epimer or diastereomer.

The thermal processes that convert biological precursors to their geologic products are thought to include intermolecular reactions that shift the distribution toward endpoint mixtures (sometimes called equilibrium mixtures) of stable and less stable stereoisomers. For example, bacteriohopanetetrol has several asymmetric carbon atoms (Fig. 1). Three key asymmetric centers occur at C-17 and C-21 in the terminal ring and at C-22 in the alkyl side chain. The alcohol functional groups are lost during burial, but the resulting geolipid (compound ii in Fig. 1) retains the stereochemistry [17β,21β(H)-22R] of the initial biomolecule (i) This configuration is thermally unstable and isomerizes to other configurations, such as 17α,21β(H)-22R (iv) and 17α,21β(H)-22S (v). Epimers with 22S and 22R configurations reach an endpoint ratio for the 22S/(22S+22R) ratio at ~60% in mature oils and rock extracts. Similarly, most sterols have a double bond at C-5 (compound i in Fig. 2), and hydrogenation during diagenesis results in 5β- and 5α-steranes. The latter are more stable

Analyzing the Thermal History of Sedimentary Basins: Methods and Case Studies
SEPM Special Publication No. 103, Copyright © 2012
SEPM (Society for Sedimentary Geology), Print ISBN 978-1-56576-315-9, CD/DVD ISBN 978-1-56576-317-3, p. 17–28.

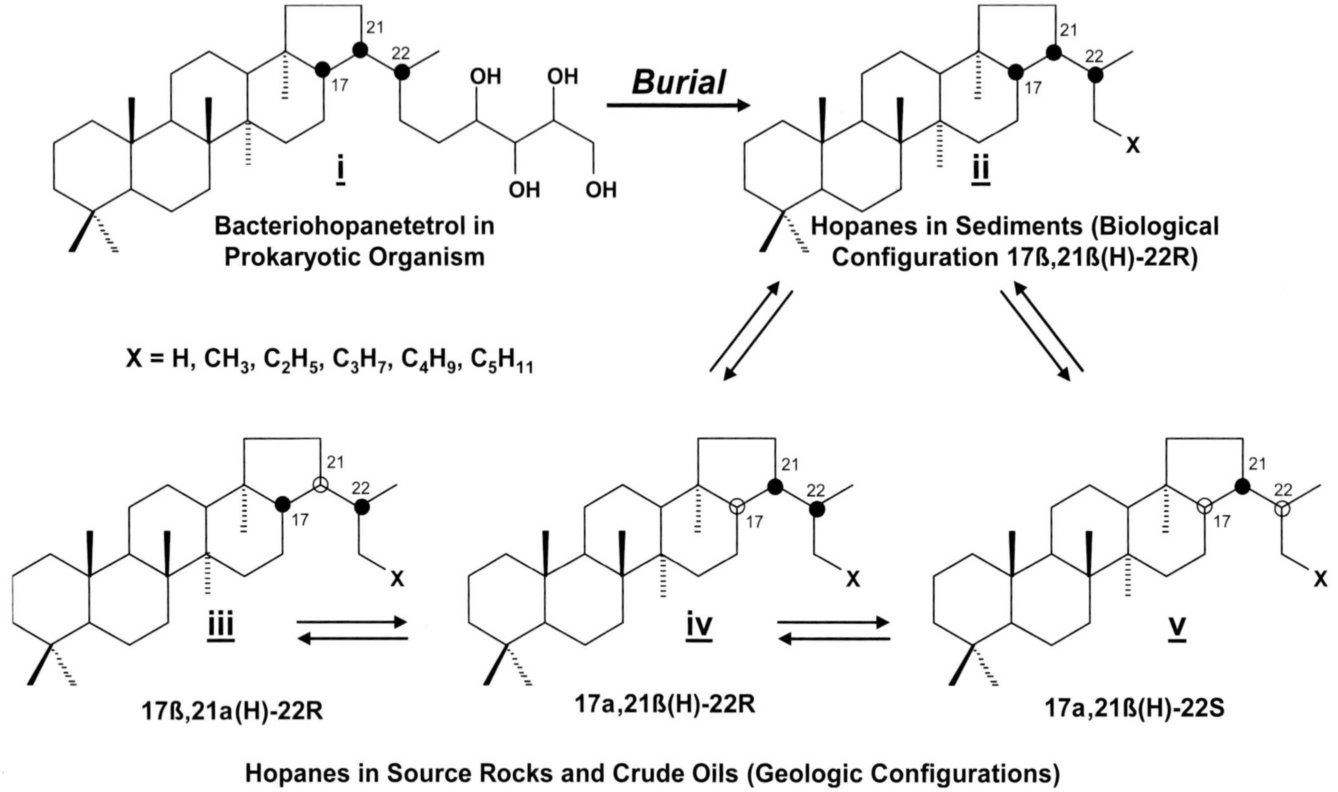

FIG. 1.—Formation of hopanes in petroleum from bacteriohopanetetrol (i) in the lipid membranes of prokaryotic organisms and examples of subsequent stereoisomerization. Stereochemistry is indicated by open (α) and solid (β) dots where the hydrogen is directed into and out of the page, respectively. The biological configuration (17β,21β(H)-22R) imposed on bacteriohopanetetrol and its immediate saturated product (ii) by enzymes in living organisms is unstable during thermal maturation and undergoes isomerization to geological configurations (e.g., iii, iv, v). The 17β,21α(H)-hopanes (e.g., iii) are called moretanes, while all others are hopanes (e.g., ii, iv, v). Hopane isomerization in basin models refers to the ratio 22S/(22S+22R) for 17α,21β(H)-hopanes. Physically, 17β,21β isomerization to 17α,21β and 17β,21α configurations result in a shift from a relatively flattened (all chair) to a bent (boat) structure at the five-carbon ring. Isomerization at C22 orients the alkyl side chain from an elongated to a more compact structure.

and are the dominant geological epimers. Sterols have three other asymmetric centers at C-14, C-17, and C-20. During early diagenesis, biological precursors give rise to steranes with only the 20R configuration, which isomerize under geologic conditions to mixtures dominated by the 20S configuration. Biological sterols have the configuration 14α,17α(H)-20R (ii), which converts during burial to 14β,17β(H)-20S (iii), 14β,17β(H)-20R (iv), and 14α,17α(H)-20S (v) epimers. The ratio 5α,14α,17α 20S/(20S+20R) achieves equilibrium at ~0.52 to 0.55. Other thermal maturity parameters include biomarkers that are not stereoisomers. For example, C-ring monoaromatic steroid hydrocarbons (MA) were considered to fully aromatize (i.e., the formation of aromatic rings from saturated naphthene rings) to their triaromatic equivalents (TA) by an irreversible reaction (Fig. 3). A maturity parameter based on these compounds, TA/(MA+TA), ranges from ~0 in immature sediment to 1 in mature oil and bitumen.

As discussed in the next section, it was once thought that the sterane and hopane isomerizations and the sterane aromatization could be expressed as simple first-order reactions. In contrast, various other biomarker parameters thought to be effective indictors of thermal maturity from early diagenesis through the oil window (Fig. 4) are empirical relationships involving multiple reactions that would be difficult to resolve and express as a single first-order reaction.

Aromatization–Isomerization Diagrams

Mackenzie and McKenzie (1983) and Mackenzie et al. (1984) first advocated the use of biomarker ratios to assess paleotemperature for basin models. They compared the extent of the isomerization and aromatization between extensional basins with different thermal histories. In young basins with rapid heating rates, aromatization quickly reached completion at about the same time that the hopane stereoisomers achieved the endpoint mixture at the C-22 position and before the sterane stereoisomers reached the endpoint at the C-20 position. Conversely, in old basins with slow heating rates, the hopanes rapidly isomerized, while the slower sterane isomerization and aromatization reactions progressed at about the same rate. Using measurements from basins with known temperature histories and the Arrhenius equation for first-order reactions, $k = Ae^{-Ea/RT}$, they derived the apparent activation energies (E_a) and frequency factors (A) that control the rate constant (k) for each biomarker reaction (Fig. 5). From these kinetic models, Mackenzie et al. (1984) prepared a series of aromatization–isomerization (AI) diagrams that could be used as guides for predicting average heating rate in other basins (Fig. 6). By plotting measured biomarker ratios for a given source unit at different levels of maturity within a basin (hence, fixing time and varying temperature), a comparison of the extent of hopane or sterane

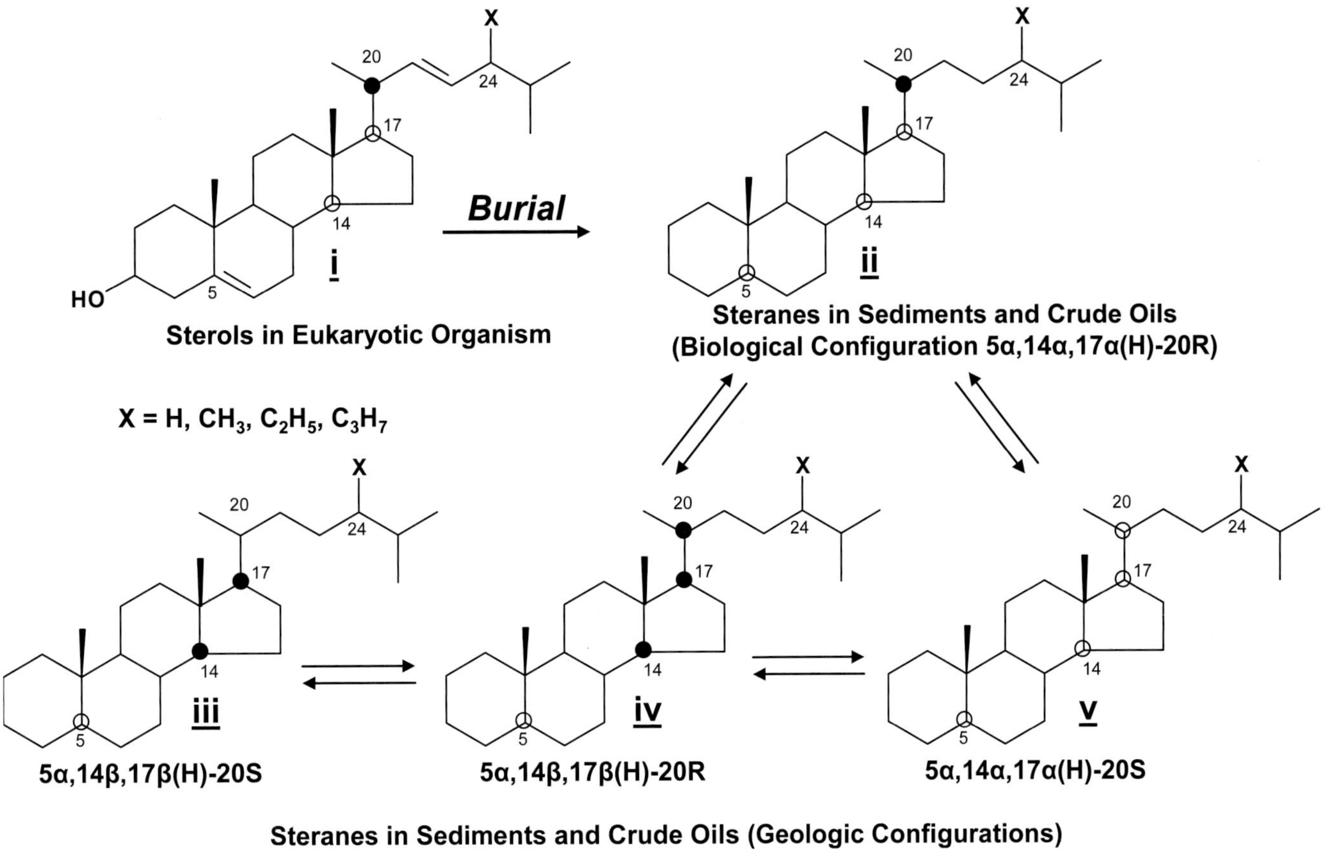

FIG. 2.—Isomerization of sterols to form geological epimers during burial. Stereochemistry is indicated by open (○) and solid (•) dots (hydrogen directed into and out of the page, respectively). Sterane isomerization in basin models typically refers to the 20S/(20S+20R) ratio for 14α,17α(H)-steranes. The 5α,14α,17α- 20R configuration has a relatively flat (all chair) structure that bends upon isomerization to a bent (boat) structure at the five-carbon ring. Isomerization at the C-20 carbon orients the alkyl side chain from an elongated structure to one in which the alkyl side chain folds back across the naphthenic rings.

isomerization vs. steroidal aromatization should define the effective heating rate experienced by that source unit. By applying the first-order kinetic parameters, AI diagrams can be constructed for a basin with a more complex geothermal history or to test scenarios in which the geothemal history is uncertain. The AI technique was applied by others (e.g., Mackenzie et al. 1984, 1985; Beaumont et al. 1985; Hong et al. 1986; Armagnac et al. 1988; King and Lillis 1990; Marzi et al. 1990; Walters and Kotra 1990; Marzi and Rullkötter 1992; Hermansen 1993; Ritter et al. 1993c), and its apparent success led to the inclusion of biomarker kinetics in basin simulators (Welte and Yalçin 1988). Lewis (1993) provides a thorough review of this approach and summarizes several case studies.

The ability to model thermal history using molecular parameters depends on the accuracy of the biomarker-reaction rate constants. After

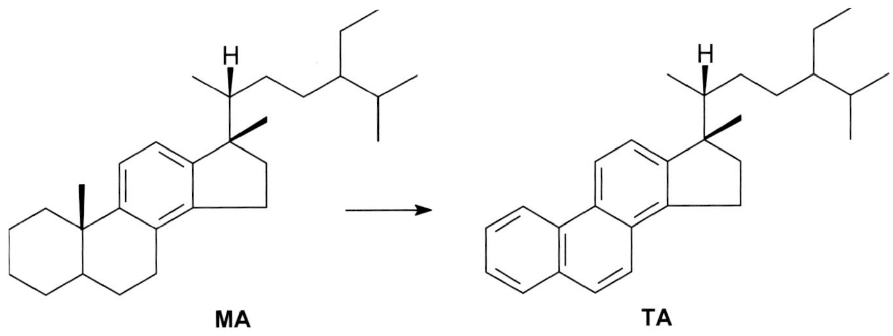

FIG. 3.—Proposed aromatization reaction of C_{29} MA to C_{28} TA (Mackenzie et al. 1981). The loss of hydrogen from naphthenic rings produces the more stable aromatic form. The aromatization ratio used in basin modeling refers to TA/(MA+TA).

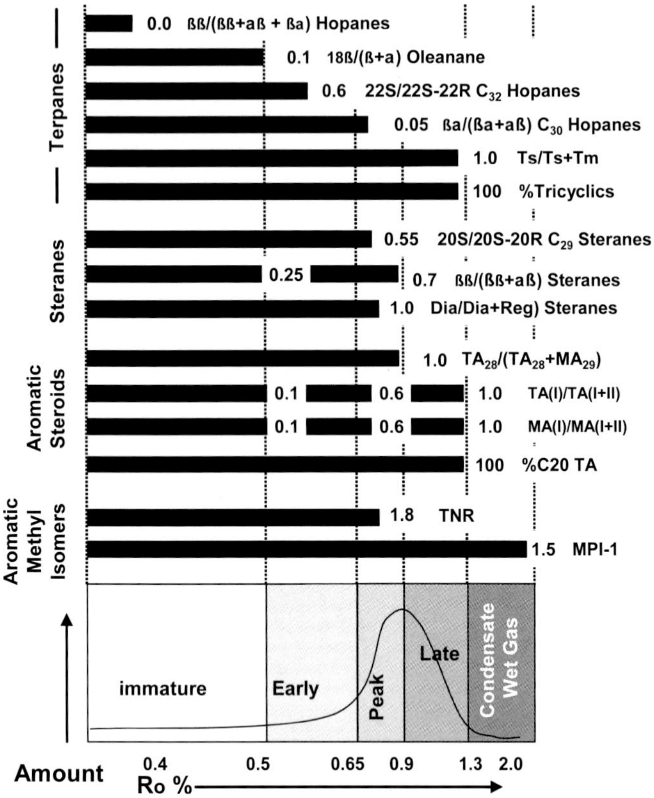

FIG. 4.—Biomarker maturation parameters respond in different ranges of maturity, allowing estimates of the extent of maturity for crude oil and source rock extracts relative to the oil window (modified from Peters et al. [2005]). Estimates are approximate because most parameters depend on temperature and time (thermal stress) rather than on the amount of petroleum generated. Approximate ranges of biomarker maturity parameters are shown vs. R_o and a generalized oil generation curve. Variations of $\pm 0.1\%$ reflectance for biomarker ratios are common, and even greater variations can occur. All values and ranges with respect to oil generation and R_o are approximate and may vary with heating rate, lithofacies, and organic facies of the source rock. The solid bars indicate the range for each ratio with respect to the stages of oil generation. The ratio reaches a constant value indicated by the number at the end of the bar. This number is a maximum value for most of the parameters, except 18β/(18α+18β)-oleanane, βα/(αβ+βα)-hopanes (C_{30}), and ββ/(ββ+αβ+βα)-hopanes, for which 0.1, 0.05, and 0.0, respectively, are minimum values. For a few ratios, intermediate values are shown within the bars to indicate earlier stages of maturity. The solid bars are not calibrated scales and do not indicate a linear change in each parameter up to its maximum value (e.g., the reader is cautioned against inferring R_o from a given biomarker ratio).

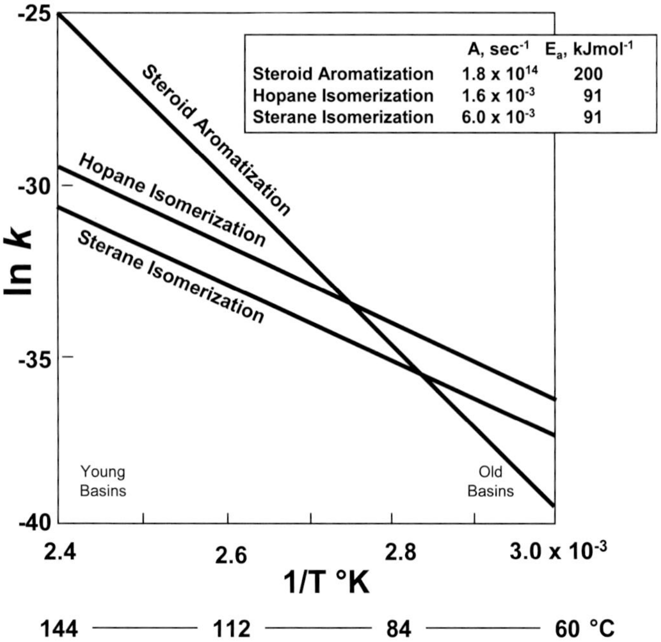

FIG. 5.—Arrhenius plots show the temperature dependence of biomarker reaction rate constants. Modified from Mackenzie et al. (1984).

the work of Mackenzie and McKenzie (1983), others attempted to measure rates (Fig. 7) from laboratory simulations (Suzuki 1984), natural sample suites (Alexander et al. 1986b, Strachan et al. 1989), or both (Rullkötter and Marzi 1988; Marzi et al. 1990; Marzi and Rullkötter 1992; Ritter et al. 1993a, 1993b, 1993c). Differences in the resulting rate constants were attributed to different mechanisms between laboratory and natural systems, uncertainties in the thermal histories, organofacies (different types of source kerogens), mineral matrix effects, and the mathematical methods used to derive the rate constants. Lewis (1993) reviews the various experimental methods used to determine Arrhenius parameters from field and laboratory data as well as their limitations.

Inherent uncertainties in both laboratory and field data allow the derived kinetic models to span a range of possible A and E_a values; that is, the same approximate results are produced by models with lower A and higher E_a as are produced with models with higher A and lower E_a. Thermal conditions associated with laboratory heating experiments and geological heating rates constrain the range of possible A and E_a values that allow the reaction to occur (Fig. 8). Assuming that the same reactions occur under both laboratory and natural conditions, then the allowable Arrhenius rate constants are constrained to a narrow range where the two intersect. Hence, when Marzi and Rullkötter (1992) and Ritter et al. (1993a, 1993b, 1993c) combined natural and laboratory data, their derived rate constants were required to fall within this constrained range. Marzi and Rullkötter (1992) concluded that since the rate constants for the isomerization and aromatization reactions do not appear to differ significantly—and, hence, different heating rates would not yield distinguishably different AI-curves—the AI diagrams of Mackenzie et al. (1984) are invalid. Indeed, only a few studies making use of AI diagrams and kinetics have been published since these concerns were raised. In a field study of four wells offshore of East Kalimantan, Indonesia, Mudford (1996) found that published biomarker kinetics did not agree with the known heat flow history.

Laboratory and natural studies conducted after the initial publication of the Mackenzie and McKenzie model undermine one of the basic assumptions of that model, that the sterane and hopane isomerization ratios result from single-step equilibrium reactions at asymmetric carbon centers. Hydrous pyrolysis simulations by Lewan et al. (1986) and Peters et al. (1990) revealed unexpected reversals in sterane 20S/(20S+20R) at high temperatures (>330° C). Pyrolysis experiments indicate that the thermal release of kerogen-bound biomarkers and their subsequent degradation are far more significant than is isomerization

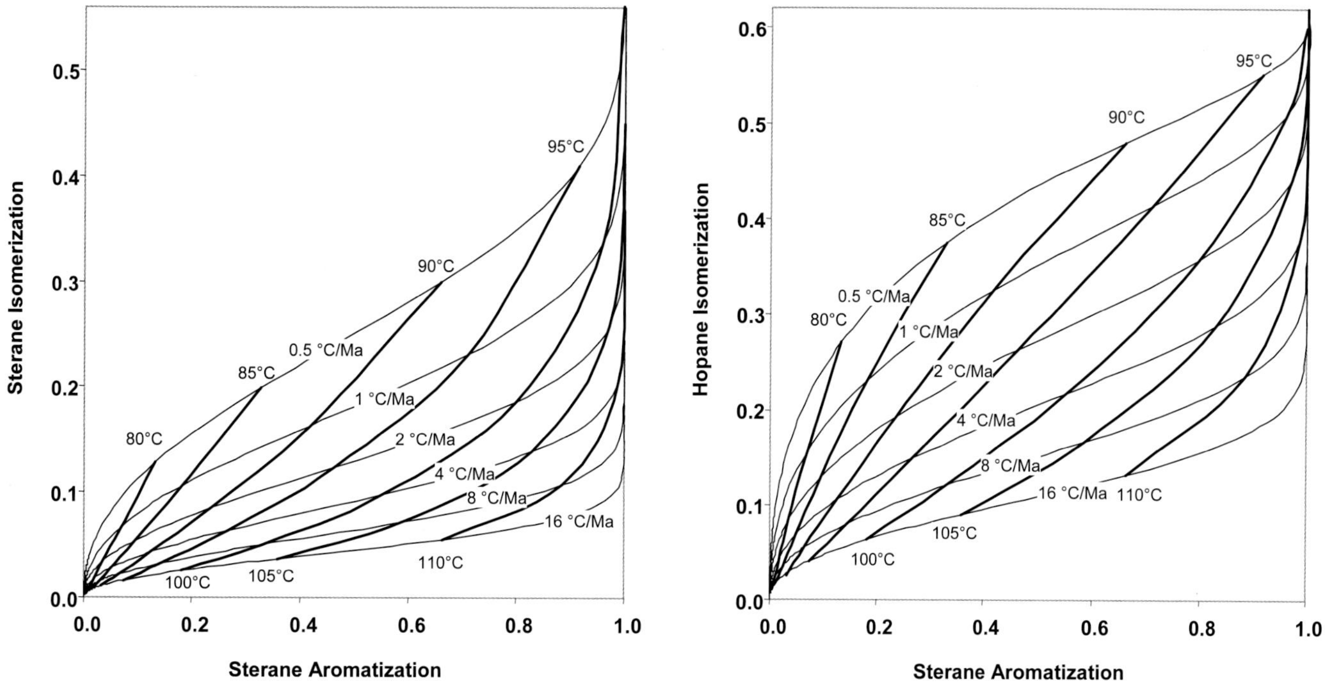

FIG. 6.—Aromatization–sterane isomerization (left) and aromatization–hopane isomerization (right) diagrams based on Arrhenius constants defined by Mackenzie and McKenzie (1983). Thin lines indicate burial at a constant heating rate; thick lines join points having the same temperature. Ma = million years. Modified from Mackenzie et al. (1984).

(Lu et al. 1989, Wang 1990). Other experiments showed that 5α-cholestane does not isomerize in the presence of organic-rich shale under laboratory simulations, indicating that once liberated from kerogen, steranes do not isomerize (Abbott et al. 1990, Abbott and Stott 1999). Van Duin et al. (1996, 1997) showed that isomerization could occur only through a complex series of metastable intermediates. Furthermore, the apparent rate of isomerization is strongly influenced by the rock matrix (Strachan et al. 1989).

It is now clear from field studies using quantitative analysis of biomarker abundance that sterane and hopane isomerization ratios are controlled by a complex balance between relative thermal stabilities of individual epimers, by the rates of release of the stereoisomers from kerogen during generation and expulsion, and by intramolecular rearrangements (Bishop and Abbott 1993; Requejo 1994; Dzou et al. 1995; Farrimond et al. 1996, 1998). Abbott et al. (1990) and Abbott and Stott (1999) attempted to determine the individual rate constants for sterane 22R and 22S epimers as they are released from kerogen and thermally destroyed during hydrous pyrolysis experiments (Fig. 9). Collectively, these experiments and observations show that the net effect of sterane and hopane isomerization cannot be expressed or approximated by a simple first-order equilibrium reaction.

The sterane aromatization parameter exhibits similar problems. Abbott et al. (1985) and Abbott and Maxwell (1988) conducted laboratory experiments proving that MA can be converted to their TA equivalents. However, the experiments used sulfur to generate free radicals, and the derived kinetic parameters do not apply to the geologic realm. Hydrous pyrolysis experiments by Wang (1990) showed increasing concentrations of triaromatic steroids over time at reaction temperatures >330° C, but no such trends were observed at lower temperatures. Bishop and Abbott (1993) studied extracts from rocks heated by an igneous intrusion and concluded that changes in TA/(MA+TA) are not governed by aromatization but by rates of generation and thermal destruction. Hence, it is likely that the TA/(MA+TA) ratio also cannot be described by a single first-order reaction.

We conclude that the biomarker maturity method of Mackenzie and McKenzie (1983) does not reliably constrain basin geothermal history. The method fails because it attempts to represent multiple processes, including generation, expulsion, conversion, and destruction, using a single first-order rate expression. More complex kinetic models might be sufficient to make the method useful, but this would require knowledge of the kinetics of all reactions as well as their dependence on the source rock setting (e.g., organic matter type or mineralogy).

Alkylaromatic Parameters

Various ratios of C_0-C_3 akylated naphthalenes, phenanthrenes, and dibenzothiophenes are molecular indicators of thermal maturity. The first described and most commonly employed of these indictors is the methylphenanthrene index (MPI-1), defined as $[1.5 \times (2MP + 3MP)]/[P + 1MP + 9MP]$, where MP indicates methylphenanthrene and P indicates phenanthrene (Radke and Welte 1983). The MPI-1 attempts to capture both the effects of greater stability of α-methylphenanthrene (2MP and 3MP) over the β-methylphenanthrene (1MP and 9MP) and cleavage of the methyl group to form phenanthrene. The net effect on the ratio is that it initially increases within increasing maturation as 2MP and 3MP increase over 1MP and 9MP and then decreases as phenanthrene formed by demethylation becomes dominant. Because MPI-1 does not measure a single reaction, simple first-order kinetics cannot be derived and applied to the calibration of thermal history models.

Many field studies (Fig. 10) show that MPI-1 exhibits a positive linear correlation for R_o within the oil window and a negative linear correlation at higher maturity (Radke and Welte 1983, Boreham et al. 1988, Chandra et al. 1994, Fabiańska and Kruszewska 2003). Therefore, it might appear that MPI-1 could serve as a molecular

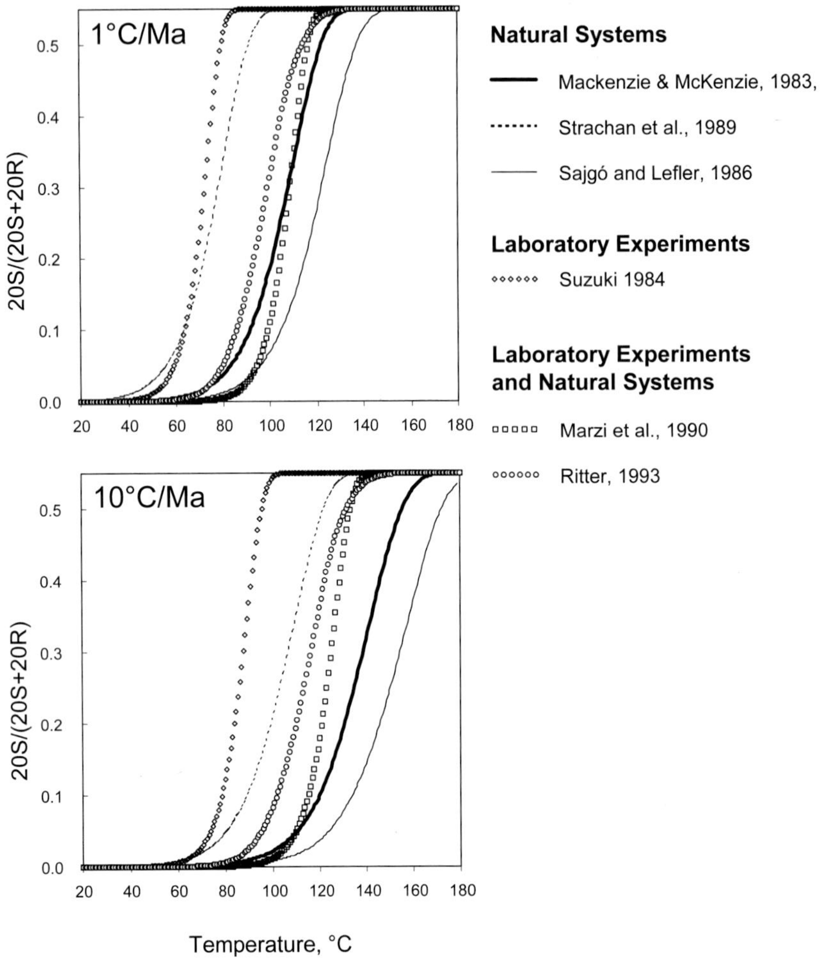

FIG. 7.—Comparison of published sterane isomerization kinetic models at different heating rates. The kinetic parameters for these models were determined from geologic data collected in basins with known thermal histories (Mackenzie and McKenzie 1983, Sajgó and Lefler 1986, Strachan et al. 1989), laboratory experiments (Suzuki 1984), or a combination of both approaches (Marzi et al. 1990, Ritter 1993a). These models differ substantially from each other and yield different results at varying heating rates.

proxy for R_o in strata in which vitrinite is absent. However, methylphenanthrene distributions strongly depend on source rock organofacies. The MPI-1 only correlates with R_o when measured in rocks with terrigenous Type III kerogen, and the ratio exhibits poor correlation with R_o and other molecular indicators from oil-prone source rocks (Cassani et al. 1988, Radke 1988, Budzinski et al. 1995). Furthermore, while the field calibration studies of MPI-1 to R_o are internally consistent, they do not agree (Fig. 10). In fact, it is unreasonable to expect MPI-1 to have a universal linear correlation with R_o because neither parameter measures a single first-order reaction.

Thermal maturity parameters based on the distribution of methyl-, ethyl-, dimethyl-, and trimethyl-naphthalenes, phenanthrenes, and thiophenes and dibenzothiophenes have limited use in basin modeling because the reactions that determine the values of these alkylaromatic parameters are influenced by organofacies and/or the mineral matrix. This was recognized by Alexander et al. (1986b), who determined first-order kinetic results for three of the alkylaromatic parameters: DP-1 (1,8-DMN/2,7-DMN), TP-1 [(1,4,6-TMN+1,3,5-TMN)/2,3,6-TMN], and PP-1 [1-MP/(2-MP + 3M)], where DMN indicates dimethylnaphthalene and TMN indicates trimethylnaphthalene (Fig. 4). Jurassic Dingo Claystone shales and Triassic Mungaroo Formation coals were examined from two Carnarvon Basin wells having different heating rates. The MP-1 and DP-1 exhibit nearly equal first-order kinetic parameters only for the Mungaroo Formation coals. The kinetic parameters for TP-1 not only differed between the shales and coals but were substantially different from the kinetic parameters derived for MP-1 and DP-1.

Recognizing that common molecular parameters of thermal maturity are not useful to constrain geothermal history, Alexander et al. (1986a, 1988) and Cumbers et al. (1987) examined several molecular reactions that could be free of organofacies or mineral effects. One such reaction emerged from their observations: *ortho*-substituted alkylbiphenyls become depleted relative to *meta*- and *para*-substituted biphenyls with increasing thermal maturity (Fig. 11). Because field data could not provide the needed accuracy, kinetic models were determined from laboratory experiments. The cyclization of *ortho*-substituted 2,3-dimethylbiphenyl to 1-methylfluorene yields first-order kinetic parameters ($A = 8.2 \times 10^7$ sec^{-1} and $E_a = 181$ kJ mol^{-1}) that were determined using pure compounds heated isothermally at 460 to 500° C (Alexander et al. 1988). However, the direct application of this kinetic model to geologic samples proved impractical because no systematic

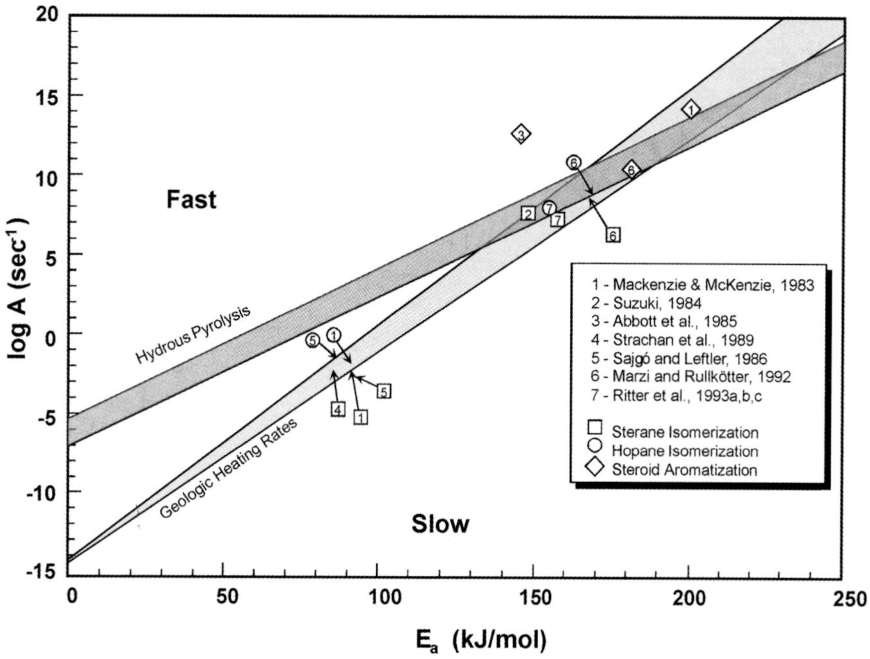

FIG. 8.—Published Arrhenius constants for biomarker kinetic reactions compared to fields of allowable frequency factor (A) and activation energy (E_a) pairs that yield reasonable results corresponding to geologic heating rates (1–10° C/my) and temperatures typical of hydrous pyrolysis experiments (280–350° C). Approaches that combined data from both geologic basins and laboratory experiments, such as those of Marzi and Rullkötter (1992) and Ritter (1993a), are constrained to the range where the two shaded areas overlap.

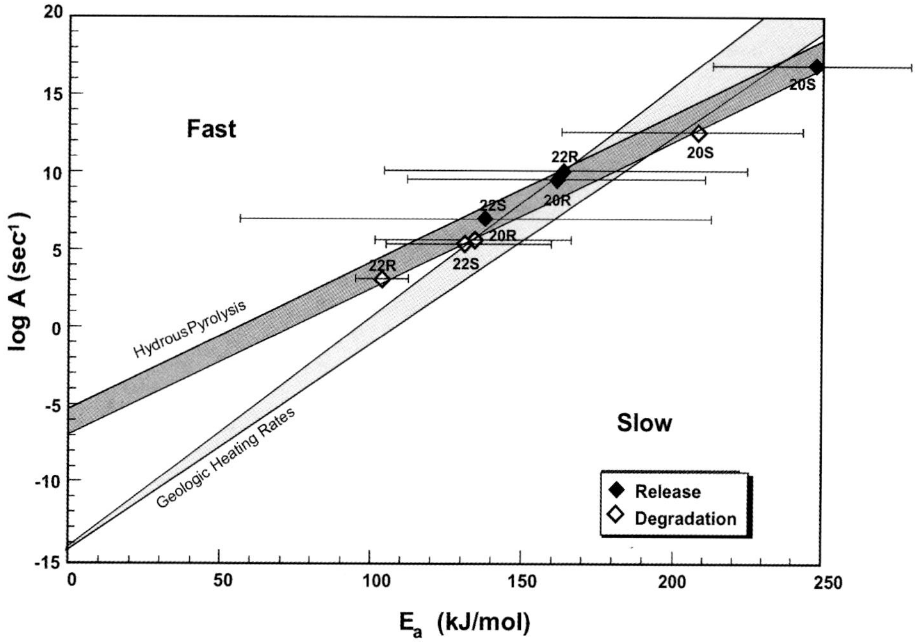

FIG. 9.—Arrhenius constants derived from hydrous pyrolysis experiments for the release (generation and expulsion from the source rock or kerogen matrix) and degradation (thermal destruction of unbound epimers) of the 20S+R sterane and 22S+R hopane isomers (Abbott and Stott 1999). The kinetic rates are mostly consistent with the E_a and A pairing needed to work for both laboratory conditions and natural settings, but the experimental error is large, preventing use of these kinetics to effectively constrain thermal history.

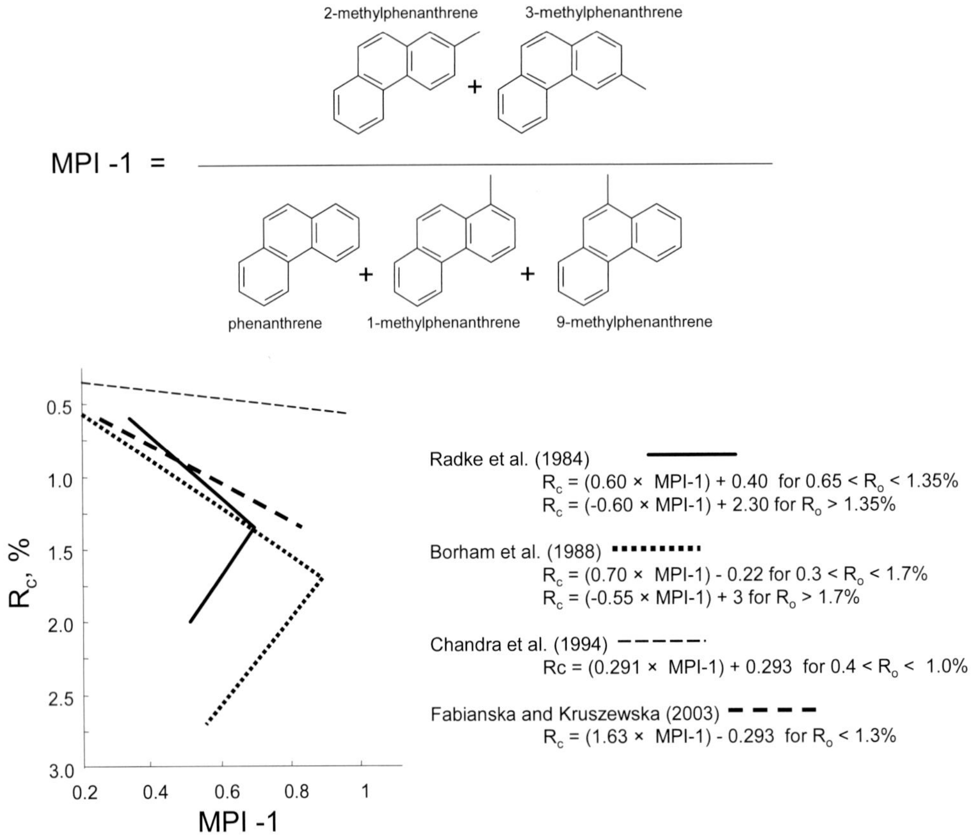

FIG. 10.—MPI-1 is a maturity indicator based on the ratio of methylphenanthrene isomers and phenanthrene. The 2- and 3-methylphenanthrene isomers are more thermally stable than the 1- and 9-methylphenanthrene isomers, and, hence, the MPI-1 value increases with increasing thermal maturation. The ratio reverses and decreases with increasing maturation as phenanthrene is formed from the thermal cracking of the methyl groups. Published comparisons of vitrinite reflectance (R_c %) and the methylphenanthrene index (MPI-1) for coals and coaly shales generally show a good correlation; however, these correlations differ from basin to basin.

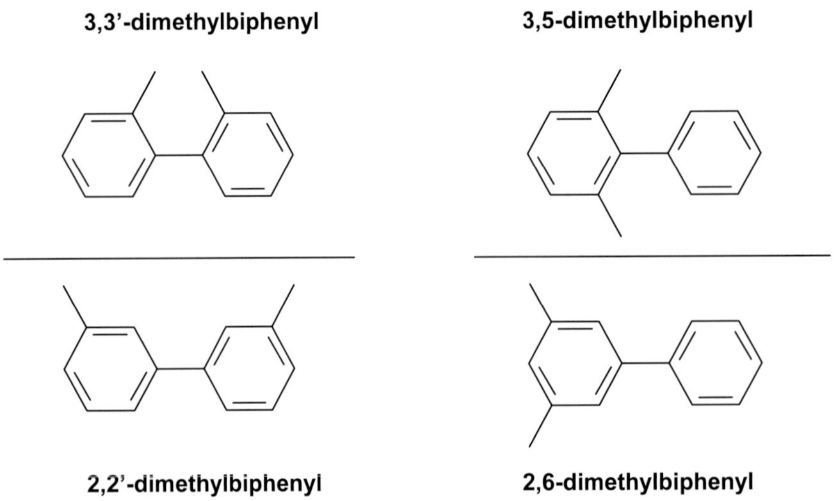

FIG. 11.—Dimethylbiphenyls with *ortho*-substitutions (e.g., 3,3' and 3,5) increase with depth relative to species with *meta*-substitutions (e.g., 2,2' and 2,6). Ratios of these compounds (3,3'/2,2' and 3,5/2,6) can change over several orders of magnitude in sediments that span the window of oil generation (Cumbers et al. 1987).

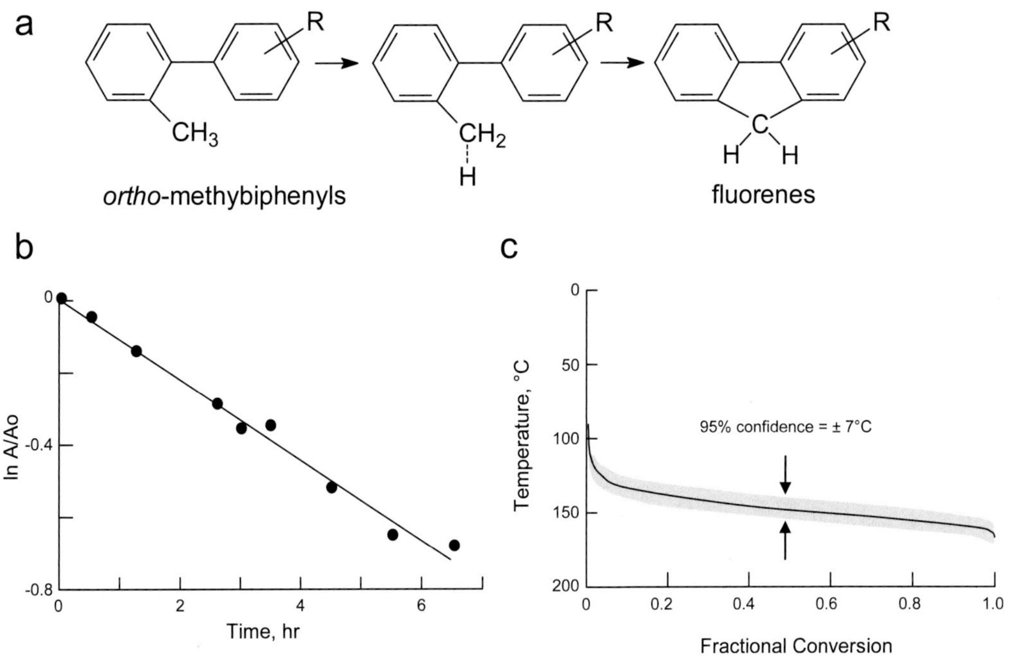

FIG. 12.—Cyclization reaction of *ortho*-methylbiphenyls to fluorenes (Kagi et al. 1990). (a) The rate-limiting step is the initial cleavage of the benzylic C-H bond. This was shown by comparing the rates of cyclization for 2-methylbiphenyl and its deuterated methyl equivalent. (b) Relative concentration of 2-methylbiphenyl when heated in a Dexsil solution at 450° C. Perdeuterobiphenyl was used as internal standard. (c) Fraction reacted of 2-methylbiphenyl relative to 3-methylbiphenyl vs. temperature calculated by applying a 1.5° C Ma^{-1} heating rate to the laboratory-derived kinetic model.

change in the ratio of 1-methylfluorene to 2,3-dimethylbiphenyl occurs in rock extracts of increasing maturity. This was attributed to additional, unspecified reactions occurring in source rocks involving the formation of 1-methylfluorene from other precursors. To use the 2,3-dimethylbiphenyl cyclization kinetics, the concentration of 3,5-dimethylbiphenyl was compared to 2,3-dimethylbiphenyl. Laboratory simulations showed that 3,5-dimethylbiphenyl, which has no methyl group adjacent to the other phenyl ring, is inert under experimental conditions. The ratio of 3,5- to 2,3-dimethylbiphenyl therefore would follow the kinetics determined for the decrease of the 2,3-isomer due to cyclization, assuming no other reaction removes the 2,3-isomer. The derived kinetic model was consistent with maturity data from two Australian wells.

Cyclization of other *ortho*-substituted biphenyls was further investigated by Kagi et al. (1990). Comparative rate measurements involving the elimination of 2-methybiphenyl, its deuterated methyl equivalent, 2,3-dimethylbiphenyl, and 2,3,5-dimethylbiphenyl provided strong evidence that the rate-limiting step in the cyclization reaction is cleavage of the benzylic C-H bond (Fig. 12a). Addition of aluminum montmorillonite did not promote the cyclization of 2-methylbiphenyl to fluorene, but it generated products resulting from dealkylation, isomerization, and transalkylation reactions. These results lead Kagi et al. (1990) to conclude that kinetics based on the depletion rate of *ortho*-alkylbiphenyls, rather than the formation rate of fluorenes, are suitable for reconstruction of thermal histories and that these reactions are not substantially influenced by acid clay catalysis (Fig. 12b). The laboratory results indicate that the kinetics of this reaction would be applicable to constraining basinal temperatures in the range associated with late oil/condensate generation (>130° C) with an accuracy of ±7° C. Unfortunately, to our knowledge, this assertion has never been tested. Kagi et al. (1990) stated that they determined first-order kinetic parameters for six different compounds, but except for saying that their activation energies ranged from 184 to 204 kJ mol^{-1}, further discussion was deemed beyond the scope of the article, and the kinetic parameters for the *ortho*-substituted alkylbiphenyls were not published.

Alkyl Ester Parameters

Alexander et al. (1992) also proposed that thermal decomposition of alkyl esters could be an ideal reaction with which to provide kinetics for use in basin modeling. In this reaction, alkyl or acyl esters decompose into alkenes and carboxylic acids by a concerted reaction (Fig. 13). Concerted reactions are those in which all bond breaking and bond making occur in a single step. These reactions are not affected by free radical inhibitors and are insensitive to catalytic effects. Therefore, the kinetics of these reactions could be useful to constrain geothermal histories. First-order kinetic parameters were first determined for alkyl esters (Alexander et al. 1992) and then for stanyl and triterpanyl esters (Alexander et al. 1997). The release of n-C_{26} alkane+alkene from brown coal, Δ^2 and Δ^3 sterenes from an oil shale, and the depletion of triterpenyl (oleanyl) esters in a sedimentary sequence from the Gippsland Basin were shown to be consistent with the kinetic models derived from pure compounds. The thermal decomposition of stanyl esters shows kinetics similar to steroid aromatization. Alexander et al. (1997) proposed that these two processes may be linked because triaromatic steroids could form by reactions of C-ring monoaromatic steroids that still contain an ester function at the C3 position in the A-ring. We are aware of no subsequent published studies that have used the kinetic parameters for ester decomposition to calibrate basin geothermal history.

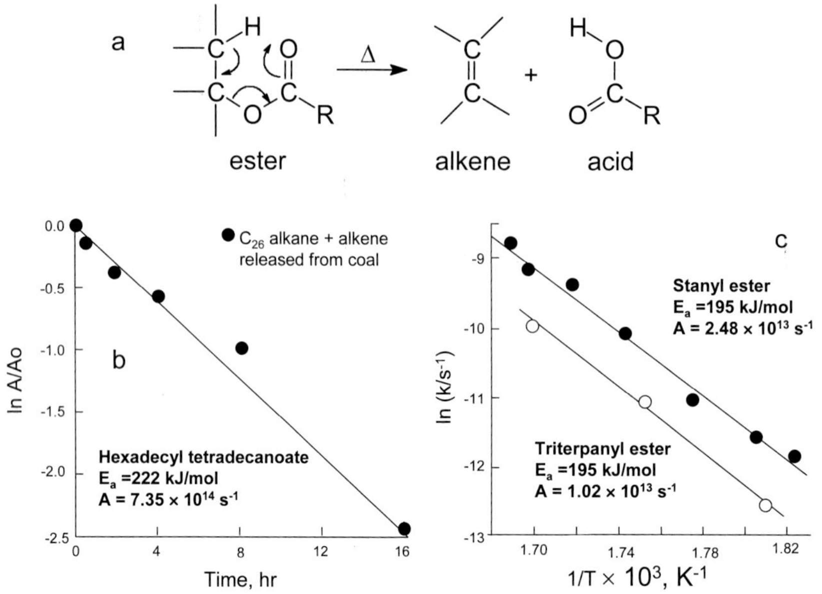

FIG. 13.—(a) Alkyl esters thermally decompose into an alkene and a carboxylic acid via a concerted reaction that is insensitive to free radical inhibitors. (b) The solid line, calculated from the first-order kinetic parameters determined for the thermal decomposition of hexadecyl tetradecanoate, is compared to the release of *n*-C_{26} alkane+1-alkene from immature brown coal. These results led Alexander et al. (1992) to conclude that the kinetic model developed for the pure compound is a good approximation for the thermal decomposition of kerogen-bound esters. (c) Arrhenius plots for the thermal decomposition of stanyl and triterpenyl esters (Alexander et al. 1997).

CONCLUSIONS

Molecular indicators of thermal maturity were developed to supplement other maturity parameters, such as vitrinite reflectance. Mackenzie and McKenzie (1983) first described how biomarker maturity parameters might be used to constrain geothermal history. The technique appeared to be promising, and several case studies were published through the early 1990s. Many basin simulators include the option to display biomarker ratios calculated based on published first-order kinetic parameters to calibrate the predicted thermal history. Nonetheless, early attempts to use various apparent biomarker reactions for calibration, such as hopane and sterane isomerizations and the conversion of monoaromatic to triaromatic steroids, resulted in only limited success. The use of biomarker kinetics fell out of favor, and only a few additional publications using biomarker kinetics in basin models have appeared since (e.g., Pearson et al. 1996, Mazurek et al. 2006).

There are several reasons why molecular maturity parameters have not been widely used to calibrate thermal basin models.

1. Molecular maturity parameters may not accurately represent the thermal history of rock samples. Unlike kerogen (insoluble particulate organic matter) in the rock matrix, petroleum is mobile. Thus, extracts from rock may contain indigenous material, migrated oil, contamination, or mixtures of these components.
2. Most molecular maturity parameters are influenced by nonthermal factors. For example, organic matter input and mineralogy of the source rock can influence certain maturity parameters. Secondary alteration processes, such as biodegradation and thermochemical sulfate reduction, can also influence these parameters.
3. Most molecular maturity parameters do not arise from simple reversible isomerization reactions or irreversible transformations of reactants to products that obey first-order kinetics. Rather, molecular changes during thermal alteration result from complex processes. For example, multiple reactants may yield the same product or one reactant may yield different products. Certain products may originate in part from reactants in the bitumen but may also be supplemented by products released from kerogen in the source rock during maturation.
4. Molecular maturity parameters require advanced separation techniques and interpretations. Although biomarker and other hydrocarbon analyses are routinely performed by industrial, academic, and service laboratories, the methods are not standardized and may be difficult to compare. Expert interpretation of gas chromatography–mass spectrometry data may be needed for peak identification and evaluation of the precision and accuracy of the reported compounds, particularly in cases involving co-elution.

It is our opinion that published kinetics for many of these reactions are based on simplistic and unfounded assumptions of reaction pathways, lack the needed precision, and/or are characterized by insufficient documentation. Many of the promising molecular kinetic reactions, such as alkylbiphenyl (Alexander et al. 1988, Kagi et al. 1990) and ester (Alexander et al. 1992, 1997) decompositions, have not been tested or applied by others since their publication. Without convincing evidence that molecular thermal parameters provide useful calibration of modeled thermal history, basin modelers have little motivation to use them.

Nonetheless, molecular maturity parameters have the potential to improve calibration of basin models. Apatite fission tracks have an upper temperature limit of ~120° C. Vitrinite originates from land plants and is absent in pre-Devonian sediments and rare in younger sediments deposited far from land plant input (e.g., distal marine shales). Fluid inclusions may be absent in low temperature strata, record only the formation temperature when they formed, and are difficult, if not impossible, to date. Molecular thermal maturity parameters could overcome these limitations and would be most effective if used in conjunction with nonmolecular maturity measurements. Furthermore, models of complex reactions involving intermediate steps (e.g., hopane isomerization; Van Duin et al. 1997) may overcome some of the limitations.

ACKNOWLEDGMENTS

We thank Joseph Curiale, Daniel Dawson, and Nicholas Harris for their useful reviews and our respective organizations for permission to publish.

REFERENCES

Abbott GD, Lewis CA, Maxwell JR. 1985. Laboratory models for aromatization and isomerization of hydrocarbons in sedimentary basins. *Nature* 318:651–653.

Abbott GD, Maxwell JR. 1988. Kinetics of the aromatisation of rearranged ring-C monoaromatic steroid hydrocarbons. *Organic Geochemistry* 13:881–885.

Abbott GD, Stott AW. 1999. Biomarker reaction kinetics during kerogen microscale pyrolysis. *In* Pályi G, Zucchi C, Caglioti L (Editors). *Advances in Biochirality*: Elsevier, Amsterdam. p. 231–246.

Abbott GD, Wang GY, Eglinton TI, Home AK, Petch GS. 1990. The kinetics of sterane biological marker release and degradation processes during hydrous pyrolysis of vitrinite kerogen. *Geochimica et Cosmochimica Acta* 54:2451–2461.

Alexander R, Cumbers KM, Kagi RI. 1986a. Alkylbiphenyls in ancient sediments and petroleums. *Organic Geochemistry* 10:841–845.

Alexander R, Fisher SJ, Kagi RI. 1988. 2,3-Dimethylbiphenyl; Kinetics of its cyclisation reaction and effects of maturation upon its relative concentration in sediments. *Organic Geochemistry* 13:833–837.

Alexander R, Kralert PG, Kagi RI. 1992. Kinetics and mechanism of the thermal decomposition of esters in sediments. *Organic Geochemistry* 19:133–140.

Alexander R, Kralert PG, Sosrowidjojo IB, Kagi RI. 1997. Kinetics and mechanism of the thermal elimination of alkenes from secondary stanyl and triterpenyl esters: Implications for sedimentary processes. *Organic Geochemistry* 26:391–398.

Alexander RR, Strachan MG, Kagi RI, van Bronswijk W. 1986b. Heating rate effects on aromatic maturity indicators. *Organic Geochemistry* 10:997-1003.

Armagnac C, Kendall CGSC, Kuo C, Lerche I, Pantano J. 1988. Determination of paleoheat flux from vitrinite reflectance data, and from sterane and hopane isomer data. *Journal of Geochemical Exploration* 30:1–28.

Beaumont C, Boutitier R, Mackenzie AS, Rullkötter J. 1985. Isomerization and aromatization of hydrocarbons and the paleothermometry and burial history of Alberta Foreland Basin. *American Association of Petroleum Geologists Bulletin* 69:546–566.

Bishop AN, Abbott GD. 1993. The interrelationship of biological marker maturity parameters and molecular yields during contact metamorphism. *Geochimica et Cosmochimica Acta* 57:3661–3668.

Boreham CJ, Crick IH, Powell TG. 1988. Alternative calibration of the methylphenanthrene index against vitrinite reflectance: Application to maturity measurements on oils and sediments. *Organic Geochemistry* 12:289–294.

Budzinski H, Garrigues Ph, Connan J, Devillers J, Domine D, Radke M, Oudins JL. 1995. Alkylated phenanthrene distributions as maturity and origin indicators in crude oils and rock extracts. *Geochimica et Cosmochimica Acta* 59:2043–2056.

Cassani F, Gallango O, Talukdar S, Vallejos C, Ehrmann U. 1988. Methylphenanthrene maturity index of marine source rock extracts and crude oils from the Maracaibo Basin. *Organic Geochemistry* 13:73–80.

Chandra K, Mishra CS, Samanta U, Anita Gupta K, Mehrotra KL. 1994. Correlation of different maturity parameters in the Ahmedabad–Mehsana block of the Cambay Basin. *Organic Geochemistry* 21:313–321.

Cumbers KM, Alexander R, Kagi RI. 1987. Methylbiphenyl, ethylbiphenyl and dimethylbiphenyl isomer distribution in some sediments and crude oils. *Geochimica et Cosmochimica Acta* 51:3105–3112.

Dzou LIP, Noble RA, Sentfle JT. 1995. Maturation effects on absolute biomarker concentration in a suite of coals and associated vitrinite concentrates. *Organic Geochemistry* 23:681–697.

Fabiańska MJ, Kruszewska KJK. 2003. Relationship between petrographic and geochemical characterisation of selected South African coals. *International Journal of Coal Geology* 54:95–114.

Farrimond P, Bevan JC, Bishop AN. 1996. Hopanoid hydrocarbon maturation by an igneous intrusion. *Organic Geochemistry* 25:149–164.

Farrimond P, Taylor T, Telnæs N. 1998. Biomarker maturity parameters: The role of generation and thermal degradation. *Organic Geochemistry* 29:1181–1197.

Hermansen D. 1993. Optimization of temperature history—aspects of vitrinite reflectance and sterane isomerization. *In* Dore AG, Augustson JH, Hermanrud C, Stewart DJ, Sylta O (Editors). *Basin Modeling: Advances and Applications*, Norwegian Petroleum Society Special Publication 3: Elsevier, Amsterdam. p. 119–126.

Hong Z-H, Li H-X, Rullkötter J, Mackenzie A. 1986. Geochemical application of sterane and triterpane biological marker compounds in the Linyi Basin. *Organic Geochemistry* 10:433–439.

Kagi RI, Alexander R, Toh E. 1990. Kinetics and mechanism of the cyclisation reaction of *ortho*-methylbiphenyls. *Organic Geochemistry* 16:161–166.

King JD, Lillis PG. 1990. Thermal modeling using biomarkers in the Santa Maria Basin, California. *American Association of Petroleum Geologists Bulletin* 74(5):695.

Lewan MD, Bjorøy M, Dolcater DL. 1986. Effects of thermal maturation on steroid hydrocarbons as determined by hydrous pyrolysis of Phosphoria Retort Shale. *Geochimica et Cosmochimica Acta* 50:1977–1987.

Lewis CA. 1993. The kinetics of biomarker reactions. Implications for the assessment of the thermal maturity of organic matter in sedimentary basins. *In* Engel MH, Macko SA (Editors). *Organic Geochemistry: Principles and Applications*: Plenum Press, New York. p. 491–510.

Lu ST, Ruth E, Kaplan IR. 1989. Pyrolysis of kerogens in the absence and presence of montmorillonite; I. The generation, degradation and isomerization of steranes and triterpanes at 200 and 300° C. *Organic Geochemistry* 14:491–499.

Mackenzie AS, Beaumont C, Boutitier R, Rullkötter J. 1985. The aromatization and isomerization of hydrocarbons and the thermal and subsidence history of the Nova Scotia margin. *Philosophical Transactions of the Royal Society London, Series A* 315:203–232.

Mackenzie AS, Beaumont C, McKenzie DP. 1984. Estimation of the kinetics of geochemical reactions with geophysical models of sedimentary basins and applications. *Organic Geochemistry* 6:875–884.

Mackenzie AS, Hoffmann CF, Maxwell JR. 1981. Molecular parameters of maturation in the Toarcian shales, Paris Basin, France—III. Changes in aromatic steroid hydrocarbons. *Geochimica et Cosmochimica Acta* 45:1345–1355.

Mackenzie AS, Lamb NA, Maxwell JR. 1982. Steroid hydrocarbons and the thermal history of sediments. *Nature* 295:223–226.

Mackenzie AS, McKenzie D. 1983. Isomerization and aromatization of hydrocarbons in sedimentary basins formed by extension. *Geology Magazine* 120:417–470.

Mackenzie AS, Patience RL, Maxwell JR, Vandenbroucke M, Durand B. 1980. Molecular parameters of maturation in the Toarcian shales, Paris Basin, France—I. Changes in the configuration of acyclic isoprenoid alkanes, steranes, and triterpanes. *Geochimica et Cosmochimica Acta* 44:1709–1721.

Marzi R, Rullkötter J. 1992. Qualitative and quantitative evolution and kinetics of biological marker transformations; Laboratory experiments and application to the Michigan Basin. *In* Moldowan JM, Albrecht P, Philp RP (Editors). *Biological Markers in Sediments and Petroleum*: Prentice-Hall, Englewood Cliffs, New Jersey. p. 18–41.

Marzi R, Rullkötter J, Perriman WS. 1990. Application of the change of sterane isomer ratios to the reconstruction of geothermal histories: Implications of the results of hydrous pyrolysis experiments. *Organic Geochemistry* 16:91–102.

Mazurek M, Hurford AJ, Leu W. 2006. Unraveling the multi-stage burial history of the Swiss Molasse Basin: Integration of apatite fission track, vitrinite reflectance and biomarker isomerisation analysis. *Basin Research* 18:27–50.

Mudford BS. 1996. Can biomarker ratio data be used quantitatively to constrain thermal histories? A case study from East Kalimantan, Indonesia. *Energy Exploration & Exploitation* 14:607–625.

Pearson MJ, Duncan AD. 1996. Biomarker maturity profiles in the Inner Moray Firth Basin and implications for inversion estimates. *In* Hurst A, Johnson HD,

Burley SD, Mackertich DS (Editors). *Geology of Humber Group: Central Graben and Moray Firth, UKCS*, Special Publication 114: Geological Society, London. p. 287–298.

Peters KE, Moldowan JM, Sundararaman P. 1990. Effects of hydrous pyrolysis on biomarker thermal maturity parameters: Monterey phosphatic and siliceous members. *Organic Geochemistry* 15:249–265.

Peters KE, Walters CC, Moldowan JM. 2005. *The Biomarker Guide: Biomarkers and Isotopes in Petroleum Exploration and Earth History*: Cambridge University Press, Cambridge, UK. 1155 p.

Radke M. 1988. Application of aromatic compounds as maturity indicators in source rocks and crude oils. *Marine and Petroleum Geology* 5:224–236.

Radke M, Welte DH. 1983. The methylphenanthrene index (MPI). A maturity parameter based on aromatic hydrocarbons. *In* Bjorøy M, Albrecht P, Cornford C, de Groot K, Eglinton G, Galimov E, Leythaeuser D, Pelet R, Rullkötter J, Speer G (Editors). *Advances in Organic Geochemistry 1981*: John Wiley and Sons, New York. p. 504–512.

Requejo AG. 1994. Maturation of petroleum source rocks—II. Quantitative changes in extractable hydrocarbon content and composition associated with hydrocarbon generation. *Organic Geochemistry* 21:91–105.

Ritter U, Aareskjold K, Schou L. 1993a. Distributed activation energy models of isomerization reactions from hydrous pyrolysis. *Organic Geochemistry* 20:511–520.

Ritter U, Myhr MB, Aareskjold K, Ganz HH, Mostafa A, Schou L. 1993b. Distributed activation energy models from hydrous pyrolysis: Towards a second-generation tool of thermal histories. *In* Dore AG, Augustson JH, Hermanrud C, Stewart DJ, Sylta O (Editors). *Basin Modeling: Advances and Applications*, Norwegian Petroleum Society Special Publication 3: Elsevier, Amsterdam. p. 201–208.

Ritter U, Myhr MB, Aareskjold K, Schou L. 1993c. Validation and first application of paleotemperature models of isomerization and NMR aromatization parameters. *In* Dore AG, Augustson JH, Hermanrud C, Stewart DJ, Sylta O (Editors). *Basin Modeling: Advances and Applications*, Norwegian Petroleum Society Special Publication 3: Elsevier, Amsterdam. p. 185–199.

Rullkötter J, Marzi R. 1988. Natural and artificial maturation of biological markers in a Toarcian shale from northern Germany. *Organic Geochemistry* 13:639–645.

Sajgó C, Lefler J. 1986. A reaction kinetic approach to the time-temperature history of sedimentary basins. *In* Bunterbath G, Stegena L (Editors). *Paleogeothermics*, Lecture Notes in Earth Science 5: Springer, Berlin. p. 119–151.

Seifert WK, Moldowan JM. 1980. The effect of thermal stress on source-rock quality as measured by hopane stereochemistry. *Physics and Chemistry of the Earth* 12:229–237.

Strachan MG, Alexander R, Subroto EA, Kagi RI. 1989. Constraints upon the use of 24-ethylcholestane diastereomer ratios as indicators of the maturity of petroleum. *Organic Geochemistry* 14:423–432.

Suzuki N. 1984. Estimation of maximum temperature of mudstone by two kinetic parameters; Epimerization of sterane and hopane. *Geochimica et Cosmochimica Acta* 8:2273–2282.

Van Duin ACT, Bass JMA, van de Graaf B. 1996. A molecular mechanics force field for tertiary carbocations. *Journal of the Chemical Society of Faraday Transactions* 92:353–362.

Van Duin ACT, Sinninghe Damsté JS, Koopmans MP, van de Graaf B, de Leeuw JW. 1997. A kinetic calculation method of homohopanoid maturation: Applications in the reconstruction of burial histories of sedimentary basins. *Geochimica et Cosmochimica Acta* 61:2409–2429.

Walters CC, Kotra RK. 1990. Thermal maturity of Jurassic shales from the Newark Basin, U.S.A.: Influence of hydrothermal fluids and implications to basin modeling. *Applied Geochemistry* 5:211–225.

Wang G-Y. 1990. A quantitative and qualitative investigation of hydrocarbon release from kerogens during hydrous pyrolysis [unpublished PhD dissertation]: The University of Newcastle-Upon-Tyne, UK. 326 p.

Welte DH, Yalçin MN. 1988. Basin modeling—a new comprehensive method in petroleum geology. *Organic Geochemistry* 13:141–151.

STABLE HYDROGEN ISOTOPES OF ISOPRENOIDS AND *N*-ALKANES AS A PROXY FOR ESTIMATING THE THERMAL HISTORY OF SEDIMENTS THROUGH GEOLOGICAL TIME

ERCIN MASLEN, KLITI GRICE, DANIEL DAWSON, SUE WANG

WA Organic and Isotope Geochemistry Centre, The Institute for Geoscience Research, Department of Applied Chemistry, Curtin University of Technology, GPO Box U1987, Perth, WA 6845, Australia

e-mail: ercin.maslen@gmail.com

AND

BRIAN HORSFIELD

GeoForschungsZentrum Potsdam, Section 4.3, Telegrafenberg, B425, D-14473, Potsdam, Germany

Abstract: The effect of thermal maturation on the δD values of individual petroleum hydrocarbons (*n*-alkanes and regular isoprenoids) from sedimentary organic matter over geological timescales has been explored in six different sedimentary sequences covering a wide range of maturities; i.e., 0.53%–1.6% vitrinite reflectance (R_o or equivalent; i.e., R_e, R_c). These include new data and recently reported literature data on formations ranging in age from the Early Cretaceous to the Permian. The application of Deuterium/Hydrogen (D/H) of biomarkers as a maturity proxy for Devonian source rocks from the Western Canada Sedimentary Basin is also presented here, extending this application to much older sediments than previously studied. In each case, pristane and phytane are predominantly derived from the lipid side chain of chlorophyll *a* present in most photosynthetic organisms, with an additional (but minor) contribution to pristane from tocopherol of land plants in selected cases. The *n*-alkanes represent contributions of algae, bacteria, and in certain cases higher plants. In general, the *n*-alkanes, pristane, and phytane from relatively immature sediments have δD values that retain the isotopic signature of their natural product precursors; i.e., biosynthesized lipid components made up of acetyl and isoprene subunits, respectively. With increasing maturity, pristane and phytane become more enriched in deuterium (D), while the *n*-alkanes generally remain at a constant isotopic composition until an overmature level is reached, at which point there is a significant enrichment of D in *n*-alkanes. The enrichment of D in pristane and phytane with increasing maturity correlates strongly with changes in traditional maturity parameters, including vitrinite reflectance, T_{max}, and molecular parameters, providing evidence that D enrichment is associated with thermal maturation.

Key words: hydrogen isotopes, isoprenoids, *n*-alkanes, Devonian, maturation

INTRODUCTION

Hydrogen is the lightest and most abundant chemical element in the universe and is present in water and all organic compounds. Hydrogen has two naturally occurring stable isotopes, hydrogen (^1H) and deuterium (^2H, also abbreviated as D). Hydrogen in hydrocarbons can be either bound to (1) primary, secondary, or tertiary carbon atoms in alkyl moieties (Fig. 1), (2) carbon atoms in aromatics (aromatic C-bound hydrogen), or (3) nitrogen, oxygen, or sulfur atoms in polar compounds (N,O,S–bound hydrogen).

The strength of the hydrogen bond differs considerably among different chemicals, and this is a critical factor in assessing the exchangeability and therefore the rate of hydrogen exchange at different molecular positions in organic components. For example, aliphatic C-bound hydrogen is linked via a strong covalent bond and is often referred to as "nonexchangeable hydrogen" (e.g., Epstein et al. 1976, Schimmelmann 1991). In contrast, aromatic C-bound hydrogen is more susceptible to exchange due to the enhanced stability of aromatic systems and their ability to stabilize charged transition states. N,O,S–bound hydrogen is linked via weaker hydrogen bonds, making exchange very favorable.

The rate of exchange of hydrogen bound in alkyl moieties (i.e., part or portion of a molecule) depends on the extent of substitution of carbon atoms within the compound. For instance, compounds containing tertiary carbon atoms exchange their hydrogen atoms more easily and more rapidly than compounds containing secondary or primary carbon atoms (Fig. 1). Although aliphatic C-bound hydrogen is often considered as "nonexchangeable" hydrogen, a few reports describe aliphatic C-bound hydrogen exchange in model compounds under laboratory conditions. Alexander et al. (1984) heated *meso*-pristane with a deuterated clay (montmorillonite) catalyst at 160° C (670 hours) and reported that approximately 60% of the total pristane (based on gas chromatography–mass spectrometry [GC-MS] analysis) was deuterated.

Based on ^1H Nuclear Magnetic Resonance (NMR) spectroscopic analysis, the same workers also reported that 40% of the methyl hydrogen, and 40% of methylene (secondary carbon) and methine (tertiary carbon) hydrogen were replaced by D. Estimated exchange half-times were 2.1 and 1.5 years for primary and secondary hydrogen, respectively, which are much slower than those determined for naphthalene under comparable conditions (Alexander et al. 1982, 1983; Sessions et al. 2004). Sessions et al. (2004) used compound-specific isotope-ratio mass spectrometry to examine *n*-icosane (n-C_{20}) incubated with heavy water (D_2O) on montmorillonite or silica under a range of temperatures. They reported exchange half-times ranging from ~10,000 years (at 60° C) to ~100,000 years (at 7° C), which are shorter than those reported for *n*-hexane (n-C_6) without a catalyst (Koepp 1978). The faster exchange rate was attributed to the presence of a mineral catalyst. Larcher et al. (1986) studied alkyl hydrogen exchange in a series of acyclic isoprenoidal acids when heated at 160° C in the presence of deuterated montmorillonite. GC-MS analysis of the methyl ester derivatives indicated that all of the isoprenoidal acids had undergone hydrogen exchange with water adsorbed on the clay surface, and exchange was most rapid at the carbon adjacent to the carboxyl group (Larcher et al. 1986).

These studies show that (1) aliphatic C-bound hydrogen can indeed

Analyzing the Thermal History of Sedimentary Basins: Methods and Case Studies
SEPM Special Publication No. 103, Copyright © 2012
SEPM (Society for Sedimentary Geology), Print ISBN 978-1-56576-315-9, CD/DVD ISBN 978-1-56576-317-3, p. 29–43.

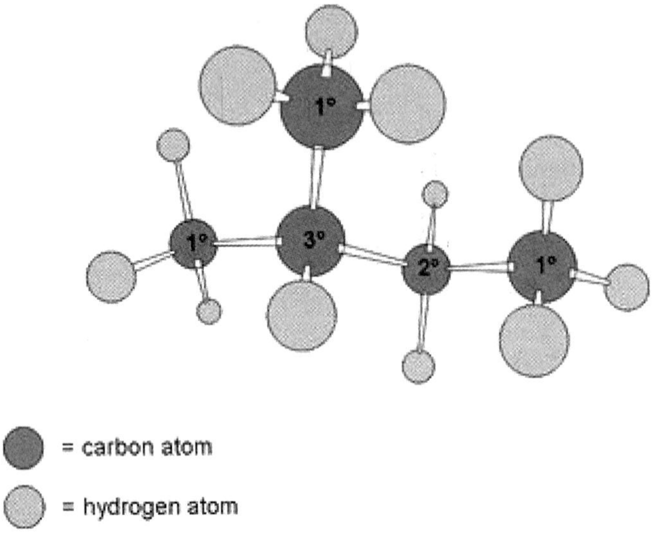

FIG. 1.—A structural diagram to illustrate the differences among primary, secondary, and tertiary carbon positions.

undergo hydrogen exchange, although at significantly slower rates than aromatic C-bound and N-, O-, S-bound hydrogen, and (2) the laboratory conditions under which alkyl hydrogen exchange has been observed in model compounds suggest that similar processes could occur in sedimentary organic matter during petroleum generation. The slower rate of exchange of C-bound hydrogen in aliphatic compounds is favorable to the measurement of δD values of hydrocarbons formed over geologic timescales, because the δD signature of their precursors is either preserved or only slightly altered, therefore allowing an assessment of the nature of their depositional environment.

A range of techniques has been used to artificially mature sedimentary organic matter. The techniques involve heating a sample in either an open or closed pyrolysis system. Artificial maturation experiments can also be performed under anhydrous or hydrous conditions. Hydrous experiments appear to more accurately reproduce the natural maturation process, and thus have been used more widely in D/H studies (Koepp 1978, Hoering 1984, Lewan 1997, Schimmelmann et al. 1999, Leif and Simoneit 2000, Sessions et al. 2004), and it has been shown that water-derived hydrogen can exchange readily with aliphatic C-bound hydrogen during artificial maturation of sedimentary organic matter.

Published work on maturation in natural systems provides evidence of gradual D enrichment of organic matter with increasing thermal maturity (Hoering 1977, Rigby et al. 1981, Smith et al. 1982, Li et al. 2001). The effect of thermal maturation processes on the δD value of individual hydrocarbons (regular isoprenoids and *n*-alkanes) has been studied recently (Dawson et al. 2005, 2007; Radke et al. 2005; Pedentchouk et al. 2006). A large (~115‰) biologically derived offset between the δD values of *n*-alkanes and regular isoprenoids from immature sediments has been shown to gradually decrease with rising maturity, because the isoprenoids become enriched in D more readily than the *n*-alkanes (Dawson et al. 2005). The D enrichment in isoprenoids has been shown to correlate strongly with measured, calculated, and equivalent (modeled) vitrinite reflectance (R_o, R_c, R_e, respectively), as well as other traditional molecular maturity parameters such as those based on biomarker ratios. This suggests that D/H exchange and the associated D enrichment during maturation occur possibly via a mechanism involving carbocation-like intermediates (i.e., any cation containing an excess positive charge on one or more carbon atoms), which proceeds more rapidly with compounds containing tertiary carbon centers.

The potential of using δD values of individual hydrocarbons (isoprenoids and *n*-alkanes) to estimate the thermal maturity of sedimentary organic matter warrants further investigation. A maturity indicator based on compound-specific δD values may prove useful in cases where traditional biomarker maturity parameters are ineffective, for example, at high maturity levels (i.e., $\%R_o > 1.0$) or where their associated reactants and products either equilibrate or are thermally degraded. In addition, such a maturity measurement could be applicable to pre-Devonian sediments, where vitrinite reflectance measurements cannot be made because the higher-plant precursors of vitrinite had not yet evolved. This paper reviews previous work that demonstrates systematic variation in D/H of individual compounds in sediments as a function of thermal maturity and extends the application of D/H of biomarkers to Devonian samples from the Duvernay Formation of the Western Canada Sedimentary Basin (WCSB).

MATERIALS AND METHODS

Sample Description and Preparation

The Upper Devonian Duvernay Formation is an organic matter–rich basinal carbonate succession and one of the most prolific marine source rocks in the central Alberta region of the WCSB (Stoakes and Creaney 1984, Creaney and Allan 1990, Chow et al. 1995). The Duvernay Formation, in the central Alberta basin of the WCSB, exhibits a wide maturity range, from very immature in the east to overmature in the west (Stoakes and Creaney 1984). Samples in this study were taken from a number of different exploration wells

TABLE 1.—Sample description, organic carbon (TOC %), and Rock-Eval pyrolysis parameters of source rocks from the Duvernay Formation, WCSB. Samples are listed in order of increasing maturity (see Maslen et al. [2008] for further details).

Locality	Depth (m)	TOC (wt%)[a]	T_{max} (°C)[a]	HI (mg HC/g TOC)[a]	S2 (mg/g)
Redwater	1147.3	5.4	419	591	31.7
Imperial Kingman	1402.7	2.7	426	410	11.06
Imperial Kingman	1404.5	3.2	426	432	13.88
Leduc	1766.9	3.6	434	465	16.5
Ferrybank	2239.5	4.0	437	273	0.99
Ferrybank	2248.7	3.3	437	247	8.17
Imperial Cynthia	2971.5	1.5	440	55	0.82
Imperial Cynthia	2973.7	4.2	441	303	12.7

[a] TOC = total organic carbon; HI = hydrogen index; T_{max} = determined temperature of maximum hydrocarbon generation during Rock-Eval pyrolysis.

covering a broad maturity range of the Duvernay Formation (Maslen et al. 2008). Table 1 shows the investigated source rock samples with their location, depth, and Rock-Eval screening data. No vitrinite has been found in the studied sample set. The eight Duvernay Formation sediment samples used in this study have total organic carbon (TOC) values ranging from 1.5 to 5.4% and Hydrogen Index (HI) values ranging from 55 to 591 mg HC/g TOC (Table 1). The thermal maturity of the source rock increases with depth, with T_{max} values varying between 419 and 441° C. The sediment samples were surface-extracted using a solution of dichloromethane and methanol (9:1), and samples were dried before grinding. The sediment samples were ground to a particle size of ≤150 μm using a ring-mill and were then extracted as described next.

Solvent Extractions

The ground sediments were accurately weighed into pre-extracted stainless-steel extraction automated accelerated solvent extractor (ASE) cells containing prerinsed glass-fiber filters. The samples were extracted using the ASE (ASE 200 Dionex, Sunnyvale, California, USA) cells using the following method. The cells were preheated at 100° C (2 minutes). A mixture of dichloromethane and methanol (9:1) was pumped into the cells up to a pressure of 6895 kPa (1000 psi) for 5 minutes. The solvent residue was then purged with high-purity compressed nitrogen, flushing the solvent-extract from the cell into a collection vial. The entire procedure was repeated until the solvent in the collection vial became colorless. Finally, the solvent extracts were combined, and the excess solvent was carefully removed under a nitrogen purge.

Fractionation of Sediment Extracts Using Column Chromatography

The extracts were fractionated using a large-scale liquid chromatographic method. The sediment extract was applied to the top of a large column (20 cm by 0.9 cm internal diameter) of activated silica gel (120° C, 8 hours). The saturated fraction was eluted with *n*-hexane (35 mL), the aromatic hydrocarbon fraction with a solution of dichloromethane in *n*-hexane (40 mL, 30%), and the polar fraction with a solution of equal parts of dichloromethane and methanol (40 mL). Finally, the excess solvent was carefully removed under a nitrogen purge.

5A Molecular Sieving

Straight-chain hydrocarbons were separated from branched and cyclic hydrocarbons by treating the saturated fractions with activated (250° C, 8 hours) 5A molecular sieves (Murphy 1969, Dawson et al. 2005, Grice et al. 2008) in cyclohexane. In a typical 5A molecular sieving separation, a portion of the saturate fraction in cyclohexane was added to a 2-mL vial that was three-quarters full of activated 5A molecular sieves. The vial was capped and placed into a preheated aluminum block (85° C, overnight). The resulting solution was then cooled and filtered through a small column of silica (prerinsed with cyclohexane), and the sieves were rinsed thoroughly with cyclohexane, yielding the branched/cyclic fraction (5A excluded). The *n*-alkanes were obtained by dissolving the sieve with HF acid as described by Grice et al. (2008).

Gas Chromatography–Mass Spectrometry (GC-MS)

GC-MS analyses were performed on a Hewlett Packard (HP) 5973 interfaced to a HP6890 gas chromatograph containing a 60 m × 0.25 mm internal diameter Wall Coated Open Tubular (WCOT) fused silica capillary column coated with a 0.25-μm-thick film (DB-5, J&W Scientific). The GC oven was programmed from 40 to 310° C at 3° C/min with initial and final hold times of 1 and 30 minutes, respectively. Samples (in *n*-hexane) were injected (split/splitless, pulsed splitless mode) using a HP6890 series autosampler. The carrier gas used was ultrahigh-purity helium at a constant flow of 1 mL/min.

Gas Chromatography–Isotope Ratio–Mass Spectrometry (GC-IR-MS)

Compound-specific isotope analyses were carried out on a Micromass IsoPrime isotope ratio–mass spectrometer coupled to a HP6890 gas chromatograph (GC-IR-MS), utilizing a column identical to that used for GC-MS analysis (see previous paragraph). For the external organic reference standards, the GC oven was programmed from 50 to 310° C at 10° C/min with initial and final hold times of 1 and 10 minutes, respectively. The δD value was calculated by integration of the *m/z* 2 and 3 ion currents of the hydrogen peaks produced by pyrolysis of the GC-separated compounds using chromium powder catalyst maintained at 1050° C. Contributions to *m/z* 3 from H_3^+ ions produced in the ion source were corrected (see Dawson [2007] for details). The δD values of individual compounds in the samples were calculated relative to that of reference gas pulses produced by allowing H_2 gas of a known D/H ratio into the IR-MS. Reported values are the average and standard deviations of three to five analyses. The δD values are given in the delta notation in per mil (‰) relative to Vienna Standard Mean Ocean Water (VSMOW). Precision and accuracy of the instrument were determined daily via the analysis of a series of reference compounds.

REVIEW OF PREVIOUS STUDIES

It is generally accepted that maturation of organic matter during burial affects its stable carbon and hydrogen isotopic composition (e.g., Peters and Moldowan 1993, and references therein). Other than source effects, temperature and time, in other words thermal maturity, are the primary factors driving these isotopic shifts (Seewald 2001). D enrichment with increasing thermal maturity is thought to be a result of hydrogen exchange with formation waters, which in most cases are significantly enriched in D relative to the organic hydrogen in sediments (Schoell 1984).

Carbon isotope studies preceded hydrogen isotope studies for two pertinent reasons. First, analytical techniques in the past experienced difficulties in measuring stable hydrogen isotopes. Second, conceptual difficulties exist in interpreting the δD results due to the fact that hydrogen in certain positions may undergo rapid exchange with a variety of hydrogen sources in the surrounding environment. However, the recent developments of (1) a method for compound-specific isotope analyses of hydrogen and (2) laboratory standards for controlling the isotopic composition of easily exchangeable hydrogen in bulk organic samples have changed the situation, enabling geochemists to better understand hydrogen isotopic changes accompanying the thermal maturation history of sediments through geological time.

In the present paper, we provide a comprehensive literature review, focusing on variations in δD values of sediment extracts, crude oils (including bulk organic matter and hydrocarbon fractions as well as individual *n*-alkanes and isoprenoids), and kerogen. Table 2 presents a summary of the sample details and their key parameters such as age, source, thermal maturity, and corresponding range of δD values for each study. In a following section, we test the application of D/H of hydrocarbons as a maturity parameter with new data from Devonian source rocks in the WCSB, extending the application of D/H measurements to much older sediments (i.e., Devonian) than previously studied.

TABLE 2.—A summary of δD measurements made on sediments and oils of varying maturity and age (modified after Schimmelmann et al. 2006).

Sample	Age	Thermal maturity	Summary	Reference
South African coals	Permian–Carboniferous	Not available	δD from −127‰ to −147‰, different climate conditions during deposition	Schiegl and Vogel (1970)
German and Netherland coals	Middle Carboniferous	Not available	δD from −92‰ to −112‰, different climate conditions during deposition	Schiegl and Vogel (1970)
Coals from different origins and ranks	Unknown	R_o: 0.25–4.8%	δD from −65.2‰ to −153.8‰	Redding (1978)
Crude oils	Ordovician–Late Tertiary	Not available	δD from −90‰ to −160‰, increase in δD with age and increase through the series, saturate, aromatics, and NOS compounds	Schoell and Redding (1978)
Crude oils and fractions	Precambrian–Pliocene	Not available	Older samples tend to be enriched in δD	Hoering (1977)
Kerogens from marine sediments	Unknown	R_o: 0.25–2.5%	Kerogen δD increases by 70–150‰ with maturity	Simoneit et al. (1978)
Oils and alkane extracts, Australian coals	Cretaceous–Late Tertiary	Not available	δD values increasing with depth	Rigby et al. (1981)
Marine kerogens	Unknown	Not available	Kerogens near a dike intrusion are D enriched	Simoneit and Mazurek (1981)
Australian coals, macrolithotypes	Permian–Miocene	R_o: 0.7–1.04%	Increase in δD values with increasing maturity	Smith et al. (1982)
Coal lignite seam, Germany	Tertiary	R_o: 0.29–2%	δD values of coals increases from −120‰ to −94‰ by contact metamorphism	Redding et al. (1980), Schoell (1984)
Type-I, -II, and -III kerogens	Unknown	R_o: I 0.7–1.39%, II 2.1–2.5%, III 0.35–2.81%	All kerogen types show D enrichment with increasing maturity	Redding et al. (1980), Schoell (1984)
Coals occurring worldwide	Unknown	Not available	Gondwana coals with δD −135‰ to −170‰, American and European coals with δD −90‰ to −140‰, large hydrogen isotopic variations—primary, source controlled	Redding et al. (198)
Crude oils and fractions from Viking Formation, WCSB	Cretaceous	Not available	δD values increasing with depth for all fractions except NOS fractions	Schoell (1984)
Condensates	Unknown	R_o: 0.5–1.5%	δD values increasing with maturity	Schoell (1984)
Type-III kerogen, saturate, aromatic, and NOS fractions, Mahakam Delta	Miocene	T_{max} 414–458° C	δD values increasing both in kerogen and in all the fractions with maturity	Schoell et al. (1983)
Type-II and -III kerogen, saturate, aromatic, NOS fractions, North Sea	Jurassic	Not available	δD values increasing in all the fractions with depth	Schou et al. (1985)
Jurassic oils and saturate fractions	Jurassic	Increasing maturity w/ biomarker data	δD values vary around −78‰ ± 6‰ in oils and in sat fractions	Baker (1987)
Bitumen, Kupferschiefer Formation, Germany	Permian	Not available	Increase in δD values in bitumen near the intrusive body	Bechtel and Puttmann (1992)

TABLE 2.—*Continued.*

Sample	Age	Thermal maturity	Summary	Reference
Oils, condensates, saturate, aromatic fractions, NE Japan	Unknown	Oil window by biomarker data	δD values increasing with maturity except aromatic fractions	Waseda (1993)
Oils from lacustrine, marine–evaporitic, and mixed source, Potiguar Basin, Brazil	Cretaceous	Oil window by biomarker data	At low levels of thermal maturity, three different paleoenvironments distinguished by their δD values (lacustrine −90‰, mixed −110‰, marine–evaporitic −120‰ to −135‰) with increasing thermal maturity, δD values converge around −110‰; unique study showing decrease in δD values with increasing maturity	Dos Santos Neto and Hayes (1999)
Oils, condensates, bitumens, asphaltenes, maltenes, Ukraine	Unknown	Oil window	δD values increasing with maturity in older samples	Lukin (1999)
Jurassic oils and saturate fractions	Jurassic	Increasing maturity w/ biomarker data	δD values vary around −78‰ ± 6‰ in oils and in sat fractions	Baker (1987)
Bitumen, Kupferschiefer Formation, Germany	Permian	Not available	Increase in δD values in bitumen near the intrusive body	Bechtel and Puttmann (1992)
Oils, condensates, saturate, aromatic fractions, NE Japan	Unknown	Oil window by biomarker data	δD values increasing with maturity except aromatic fractions	Waseda (1993)
Oils from lacustrine, marine–evaporitic, and mixed source, Potiguar Basin, Brazil	Cretaceous	Oil window by biomarker data	At low levels of thermal maturity, three different paleoenvironments distinguished by their δD values (lacustrine −90‰, mixed −110‰, marine–evaporitic −120‰ to −135‰) with increasing thermal maturity, δD values converge around −110‰; unique study showing decrease in δD values by increasing maturity	Dos Santos Neto and Hayes (1999)
Oils, condensates, bitumens, asphaltenes, maltenes, Ukraine	Unknown	Oil window	δD values increasing with maturity in older samples	Lukin (1999)
Type-I, -II, -IIS, and -III kerogens (I—Green River Formation, Utah; II—New Albany Shale, Indiana; IIS—Ghareb limestone, Jordan; III—lignite in Wilcox Group, Texas)	(I—Eocene, II—Devonian–Mississippian, IIS—Senonian, III—Paleocene)	Thermal maturation simulation	Hydrogen in kerogen, bitumen, and expelled oil from most to least isotopically influenced by water-derived hydrogen in the order IIS > II ≈ III > I	Schimmelmann et al. (1999)

TABLE 2.—*Continued.*

Sample	Age	Thermal maturity	Summary	Reference
n-alkanes and acyclic isoprenoids by compound-specific isotope analysis (CSIA) in oils, WCSB	Ordovician–Cretaceous	Increasing maturity with biomarker data	δD values increasing for all *n*-alkanes with increasing maturity of the oils	Li et al. (2001)
Type-II-S kerogens, Monterey Formation, Canada	Miocene	R_o: 0.23–0.33% and 0.29–0.42%	More-mature kerogens are enriched in δD by ~40‰ relative to less-mature kerogens	Schimmelmann et al. (2001)
Oils, saturate, aromatic, NOS fractions, *n*-alkanes (CSIA), kerogen, terrestrially sourced oils in Australia	Early Permian–Late Eocene	Not available	Isotopic exchange in δD values is limited to the aromatic fractions, not aliphatic fractions. In oils derived from a single source, whole oil, fractions and *n*-alkanes have similar δD values. δD increasing with carbon number in *n*-alkanes	Schimmelmann et al. (2004)
n-alkanes and isoprenoids (CSIA) from bitumen and oils, Perth Basin, Australia	Early Triassic	R_o: 0.53–1.13% based on biomarkers	δD values of isoprenoids increase by ~150‰, *n*-alkanes by ~42‰ with increasing maturity	Dawson et al. (2005)
n-alkane and isoprenoids (CSIA), Kupferschiefer and Posidonia Formation, Germany	Late Permian and Early Jurassic	R_o: 0.48–1.3% based on methylphenanthrene index	δD values increase with maturity, more for isoprenoids	Radke et al. (2005)
n-alkanes (CSIA), terrestrial source rocks, China	Permian–Oligocene	Not available	δD values from terrestrial source rocks vary between −140‰ and −250‰, and more negative than marine *n*-alkanes; depletion in D values for *n*-alkanes in the order of saline lacustrine to freshwater lacustrine and to swamp	Xiong et al. (2005)
Type-II and -III kerogens, New Albany Shale and Exshaw Formation	Devonian–Mississippian	R_o: 0.3–3%	δD values increase with maturity up to R_o ~1.8%	Lis et al. (2006)
n-alkanes, isoprenoids, kerogen, Gabon Basin, West Africa	Early Cretaceous	R_o: 0.55–0.7%	δD values of isoprenoids increase dramatically to similar values of *n*-alkanes at R_o 0.7%; kerogen and *n*-alkanes change slightly	Pedentchouk et al. (2006)
n-alkanes and isoprenoids (CSIA) in bitumen and oils Vulcan Subbasin, northern Australia	Late Jurassic	R_o: 0.7–1.6%	δD values increase with maturity, both for *n*-alkanes and more for isoprenoids	Dawson et al. (2007)

TABLE 3.—δD values of *n*-alkanes, pristane (Pr), and phytane (Ph), and corresponding ΔδD values (see footnote for definition) for the Perth Basin sediment extracts (Dawson et al. 2005).

Well and depth (m) (ft in parentheses)	R_e[a] (%)	$Ts/(Ts + Tm)$[b]	*n*-alkanes δD range (‰)[c]	δD Pr (‰)[c]	δD Ph (‰)[c]	ΔδD[d] (‰)
BMR 10 973–976 (3193–3203)	0.53	0.30	−183 to −128 $(13)^4$	$-267\ (7)^5$	$-249\ (4)^5$	−115
BMR 10 989–991 (3245–3250)	0.53	0.25	−188 to −130 $(8)^5$	$-278\ (3)^3$	$-266\ (1)^3$	−116
Dongara 4 1674 (5491)	0.60	0.32	−175 to −140 $(8)^5$	$-217\ (3)^5$	$-185\ (0)^3$	−43
Dongara 4 1675 (5495.5)	0.60	0.34	−157 to −112 $(6)^5$	$-208\ (6)^5$	$-166\ (7)^5$	−55
Dongara 4 1678 (5505)	0.60	0.38	−165 to −116 $(5)^4$	$-175\ (6)^5$	$-168\ (2)^3$	−29
Yardarino 2 2289 (7509)	0.78	0.76	−146 to −109 $(8)^3$	$-153\ (3)^3$	$-144\ (1)^3$	−23
Yardarino 2 2290 (7512.5)	0.78	0.79	−166 to −123 $(7)^3$	$-159\ (7)^3$	$-141\ (3)^3$	−6
Arrowsmith 1 2494 (8781)	1.13	0.89	−145 to −80 $(4)^3$	$-102\ (16)^3$	$-97\ (16)^3$	3 (0)
Arrowsmith 2678 (8787)	1.13	0.80	−123 to −73 $(3)^3$	$-121\ (4)^3$	$-108\ (4)^3$	−19

[a] R_e = calculated vitrinite reflectance values (Thomas and Barber 2004).
[b] $Ts/(Ts + Tm)$ = a maturity parameter based on the 18α[*H*]- and 17α[*H*]-trisnorhopanes.
[c] Numbers in parentheses are standard deviations (average is shown for *n*-alkanes); superscript numbers are number of replicate analyses.
[d] ΔδD = difference between the average δD value of Pr and Ph and the average δD value of the *n*-alkanes (relative to VSMOW).

Earlier Studies

Early studies on the distribution of D/H in coals and crude oils began with Schiegl and Vogel (1970), Redding (1978), and Schoell and Redding (1978). Schiegl and Vogel (1970) reported δD values from −140‰ to −127‰ for 19 samples of South African coals and from −112‰ to −92‰ for six German coals, although no thermal maturity data were provided (Table 2). Redding (1978) recorded a wider range of δD values, from −154‰ to −65‰, for 24 coals of different origins and for a wide interval of thermal maturity, from 0.25% to 4.8% R_o. Schoell and Redding (1978) examined 26 crude oils of Ordovician to Late Tertiary age that yielded δD values from −160‰ to −90‰. They found enrichment in D with age in the bulk oils and within each series of saturated, aromatic, and NOS hydrocarbon fractions. Hoering (1977) studied 11 whole oils and oil fractions with source rock ages from the Pliocene to the Precambrian, and they found that organic matter in older samples had a tendency to be enriched in D. Simoneit et al. (1978) demonstrated the effect of maturation on the δD values of kerogens from marine sediments, presenting δD values of kerogens that increase from −150‰ to −70‰ with increasing maturity over an R_o range of 0.25% to 2.5%. Schoell et al. (1983) observed that type-III kerogen, and associated saturated, aromatic, and NOS fractions all trend toward more positive values in Miocene sediment samples from the Mahakam Delta (Indonesia) with increasing burial depth. The δD values of kerogen and extractable fractions from the North Sea (Schou et al. 1985) and extracts from Australian coals (Rigby et al. 1981) show similar patterns.

These patterns are repeated in the δD values of specific organic compounds. Li et al. (2001) measured the δD values of individual *n*-alkanes and regular isoprenoids from a number of crude oils from the WCSB. The *n*-alkanes from a marginally mature crude oil displayed a range from −155‰ to −120‰ (averaging −134‰), while pristane and phytane gave δD values of −188‰ and −168‰, respectively (averaging −178‰). The *n*-alkanes, pristane, and phytane were found to be more enriched in D in the more mature crude oil relative to respective δD values from the less mature crude oil. The *n*-alkane δD values range from −144‰ to −136‰, respectively (averaging −140‰) in the less mature crude oil, and pristane and phytane δD values range from −131‰ to −97‰ (averaging −110‰) in the more mature crude oil. These data show trends consistent with the results obtained in this study and other reviewed literature.

Other studies, however, have documented a lack of isotopic trends in organic matter with thermal maturity. For example, Baker (1987) investigated a suite of Jurassic oils and their saturate fractions, where δD values varied around −78‰ ± 6‰ regardless of maturity. Schimmelmann et al. (1999) later attributed these values to the organic matter being relatively enriched in D prior to maturation (δD values > −100‰).

Studies of δD in n-Alkanes and Isoprenoids—Six Studies

The effect of thermal maturation sedimentary processes on the δD values of individual petroleum hydrocarbons (*n*-alkanes and isoprenoids) over geological timescales has only recently been explored. Studies of six different sedimentary sequences have been published to date: Perth Basin (marine, Western Australia; Dawson et al. 2005); Vulcan Subbasin (two data sets, marine, northern Australia; Dawson et al. 2007); Kissenda Formation (lacustrine, Gabon Basin; Pedentchouk et al. 2006); Poland (marine, Kupferschiefer; Radke et al. 2005); and Germany (marine, Posidonia Shale; Radke et al. 2005). These six data sets cover a range of thermal maturities from 0.53 to 1.6% R_o and ages from the Early Cretaceous to the Permian.

Perth Basin, Australia: Comprehensive organic geochemical studies of sediments from the Lower Triassic marine Kockatea Shale in the Perth Basin have been carried out by Thomas and Barber (2004) and Grice et al. (2005a, 2005b), including Rock-Eval pyrolysis, distributions of biomarkers and thermal maturity estimates. The Hovea Member of the Kockatea Shale is classified as a type-II, organic matter–rich source rock (Thomas and Barber 2004). Due to the problems associated with the direct measurement of vitrinite reflectance (% R_o) of sediments from the Kockatea Shale (Kantsler and Cook 1979), Thomas and Barber (2004) calculated equivalent vitrinite reflectance values (% R_e) for the samples using basinwide burial history modeling. The Perth Basin sediments span thermal maturities from immature to late mature (% R_e = 0.53 to 1.13) and have $Ts/(Ts + Tm)$ values (a maturity parameter based on 18α[*H*]- and 17α[*H*]-trisnorhopanes; Seifert and Moldowan 1981) ranging from 0.25 to 0.89 (Table 3).

The range of δD values of *n*-alkanes (from C_{12} to C_{32}), and the δD values of pristane and phytane from extracts of the Perth Basin sediments were reported by Dawson et al. (2005) and are summarized in Table 3. The ΔδD values were determined by subtracting the average δD value of the *n*-alkanes from the average δD value of pristane and phytane. Figure 2 shows average δD values of the *n*-alkanes, and the δD values of pristane and phytane vs. depth in sediments obtained from

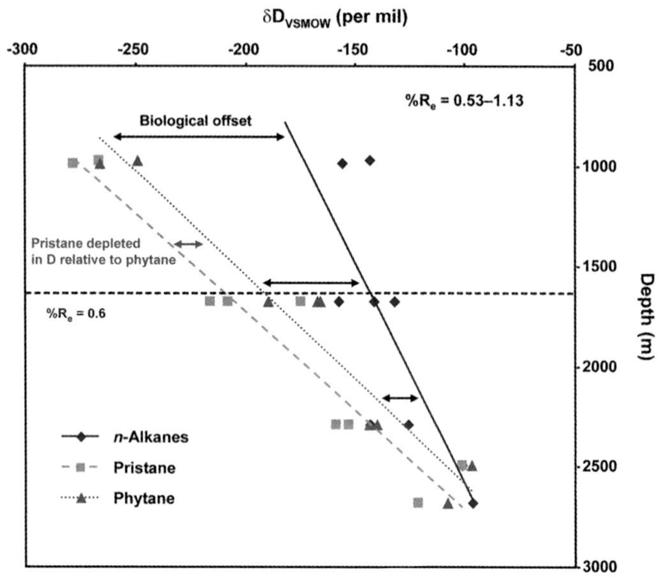

FIG. 2.—Average δD value of *n*-alkanes, and the δD values of pristane and phytane vs. burial depth in extracts from Perth Basin sediments (Dawson et al. 2005).

the Perth Basin. Both *n*-alkanes and isoprenoids show clear trends of D enrichment with depth, with the rate of enrichment more rapid in isoprenoids than in *n*-alkanes. The δD values of *n*-alkanes from extant plants vary from −187‰ to −152‰, representing the different photosynthetic assimilation processes (C3, C4, and Crassulacean Acid Metabolism (CAM)) and plant classes (e.g., Chikaraishi and Naraoka 2001). The δD values of phytol in aquatic plants and flagellates range from −278‰ to −357‰, respectively (Sessions et al. 1999). A large (~115‰) biologically derived offset between the δD values of *n*-alkanes from extant plants and δD values of phytol (i.e., lipid side chain of chlorophyll *a* is a precursor of phytane) has been observed between the δD values of *n*-alkanes and pristane and phytane from immature sediments (Dawson et al. 2005, 2007). This large offset gradually decreases with increasing maturity, because the isoprenoids become enriched in D more rapidly than the *n*-alkanes. The D enrichment in isoprenoids correlates strongly with R_e and traditional molecular maturity parameters. Figure 2 also clearly shows that the change in δD values of pristane and phytane with depth, and their similar rates of D enrichment suggest that they probably exchange hydrogen at similar rates. It is possible (although unlikely) that the *n*-alkanes and isoprenoids exchange hydrogen at similar rates, and that the observed changes in their δD values with maturation are a result of the extent to which the compounds are out of isotopic equilibrium with formation water, which is typically enriched in D. In relatively immature samples, pristane and phytane are on average much further away from the point of isotopic equilibrium than *n*-alkanes, because the isoprenoids' biological precursors are significantly more depleted in D compared to the precursors of the straight-chain hydrocarbons. Therefore, for a given increment in hydrogen exchange, there would be a larger change in the δD value of the isoprenoids compared to that of the *n*-alkanes, even though the two compound classes are hypothetically exchanging hydrogen at the same rate (Dawson 2007). This concept is illustrated in Figure 3.

Vulcan Subbasin, Australia: Comprehensive organic geochemical studies on sediment samples from the Paqualin-1 and Vulcan-1B wells (Upper Jurassic, Lower Vulcan Formation) in the Vulcan Subbasin (Timor Sea, northern Australia) have been carried out by Smith and Sutherland (1991), Van Aarssen et al. (1998a, 1998b), and Edwards et al. (2004). Smith and Sutherland (1991) and Edwards et al. (2004) reported Rock-Eval pyrolysis and R_o results. The Lower Vulcan Formation, intersected by the Paqualin-1 and Vulcan-1B wells, is classified as a type-II/III source rock. The Paqualin-1 sediments range in maturity from early mature to overmature, with measured R_o from 0.62% to 1.6% and T_{max} values from 441 to 455° C. The Vulcan-1B sediments cover a maturity range from early mature to postmature (% R_o = 0.69–1.3).

These sediments were analyzed to study the effect of maturation on the δD values of individual hydrocarbons from sedimentary sequences, extending the original work of Dawson et al. (2005) on the Perth Basin

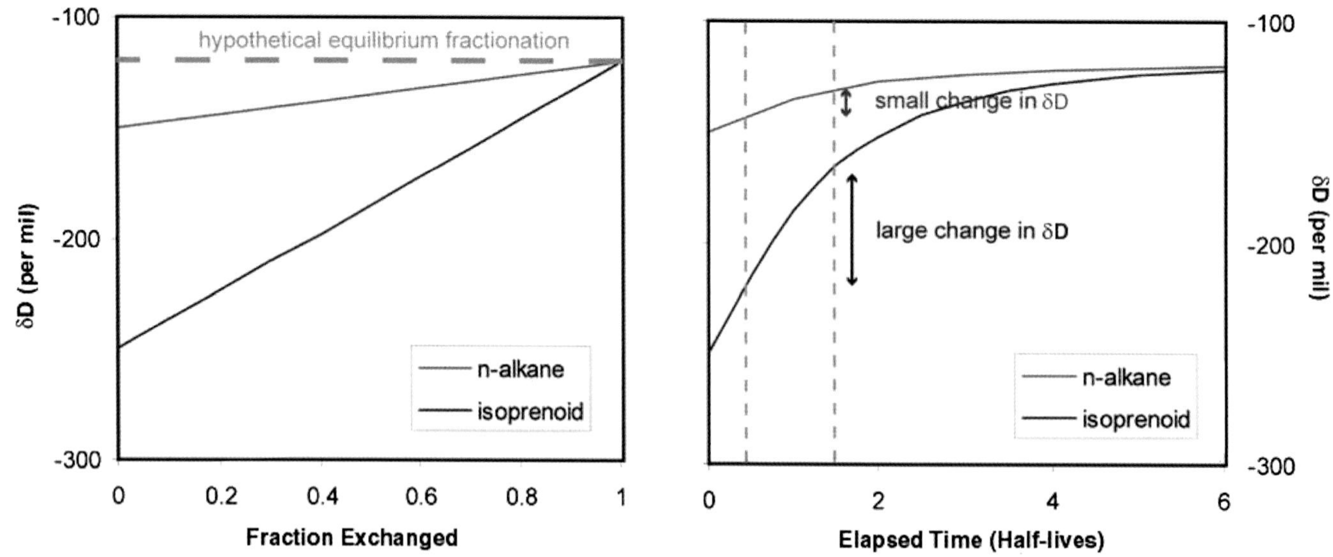

FIG. 3.—Modeled trends for (a) δD value vs. fraction of total hydrogen exchanged, and (b) δD value vs. elapsed time (half-lives) for an *n*-alkane and an isoprenoid (Dawson et al. 2007; AL Sessions, personal communication, January 2006).

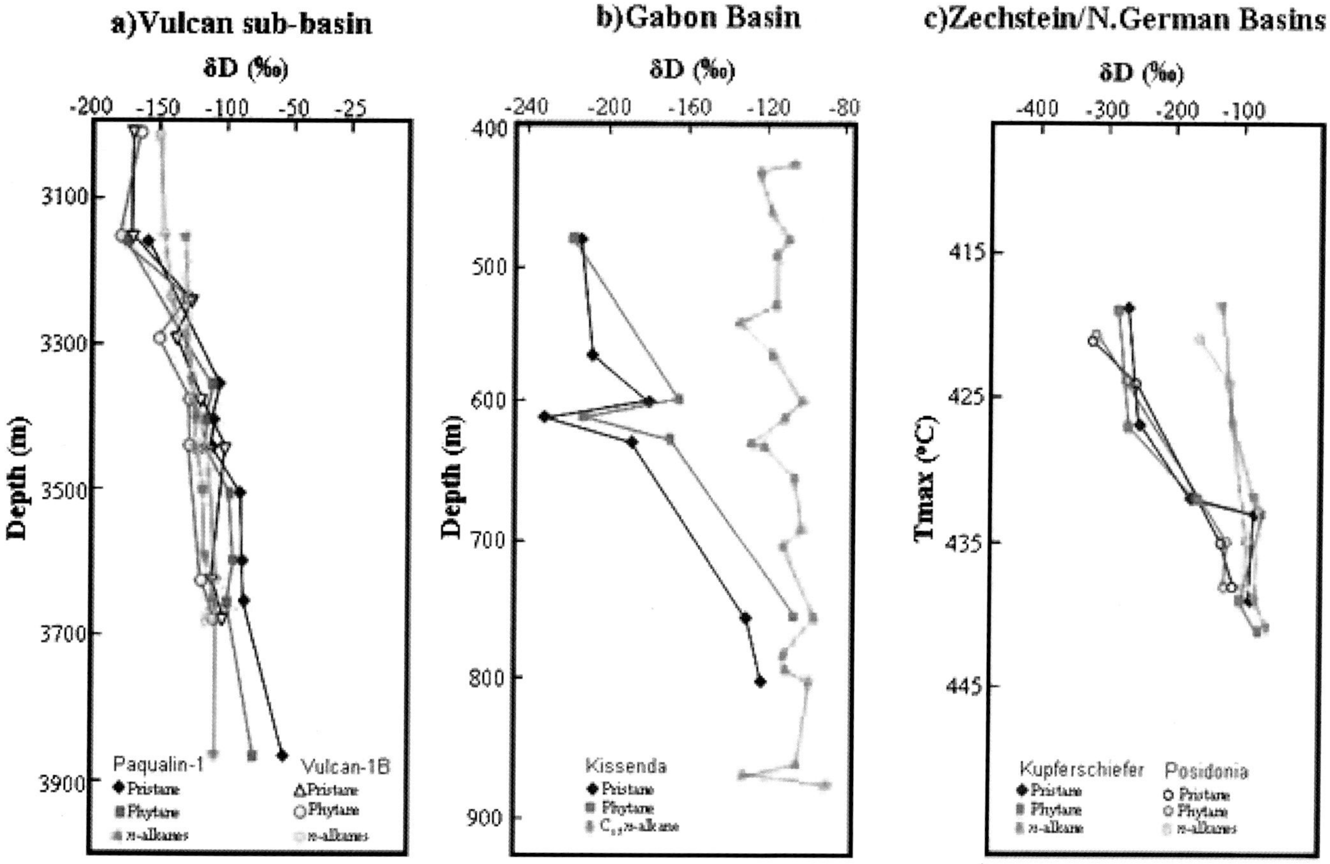

FIG. 4.—δD values of isoprenoids and *n*-alkanes with depth and T_{max}. Data have been replotted from (a) Dawson et al. (2007), where R_o varied from 0.62% to 1.6% and T_{max} varied from 441 to 455° C in the Paqualin-1 well, and R_o varied from 0.69% to 1.3% and T_{max} varied from 430 to 441° C in the Vulcan-1B well; (b) Pedentchouk et al. (2006); (c) Radke et al. (2005); no depth data were available from the Zechstein and North German Basins.

to a higher level of maturity. The δD values of *n*-alkanes in extracts from the Paqualin-1 sediments average between −136‰ and −108‰, and the *n*-alkanes in extracts from the Vulcan-1B sediments average between −146‰ and −110‰ (Fig. 4a). The δD values of *n*-alkanes extracted from the sediments of lowest maturity are consistent with a source from the marine mudstones of the Lower Vulcan Formation, while the samples of higher maturity contain *n*-alkanes more enriched in D. The δD values of pristane and phytane in the Paqualin-1 samples average between −165‰ and −68‰, and values in Vulcan-1B are between −166‰ and −105‰ (Fig. 4). It was reported that the Paqualin-1 sediments below ~3159 m seem to be relatively more mature than is reflected by their R_o values, based on a significant enrichment of D (~21‰) in pristane and phytane relative to *n*-alkanes, signifying that the available R_o data do not accurately reflect maturity in this particular case (Dawson et al. 2007). Pristane and phytane are more rapidly enriched in D with ongoing maturation, similar to observations in the Perth Basin sediments of Dawson et al. (2005). Pristane and phytane from postmature Paqualin-1 sediments (% R_o = 1.6) are significantly enriched in D (~40‰) relative to the *n*-alkanes.

This result indicates that D enrichment continues at extremely high maturity and continues to be more rapid for regular isoprenoids than *n*-alkanes. Published equilibrium fractionation factors between C-bound hydrogen and water for primary, secondary, and tertiary C-H (Sessions et al. 2004, and references therein) do not allow for isoprenoids to be enriched in D relative to *n*-alkanes. However, the data presented by Dawson et al. (2007) indicate that the δD values of isoprenoids can, in fact, be more positive than *n*-alkanes.

Gabon Basin, Africa: Samples from well ONEZ-1, Lower Cretaceous, Gabon Basin, represent a 450-m-thick section in the Kissenda Formation (Neocomian), covering about 7 million years of lacustrine deposition (Pedentchouk et al. 2006). The kerogen from most of the samples is classified as type I, based on Rock-Eval pyrolysis. Both bulk and biomarker indicators show an increase in organic matter thermal maturity, based upon $22S/(22S + 22R)C_{31}$ homohopanes (the value of 0.6 for this ratio is equivalent to $R_o = 0.6$; Peters and Moldowan 1993), $20S/(20S + 20R)$ C_{29} steranes (0.4, R_o = 0.6), and $Ts/(Ts + Tm)$ (0.5, $R_o = 0.6$) ratios, with depth from immature to early mature. The onset of oil generation was estimated to occur at a depth of 750 m to 800 m in this section (Pedentchouk et al. 2006).

The distribution of δD values of *n*-C_{17}, pristane, and phytane from the Kissenda Formation, from a depth interval covering a thermal maturity range from immature to early mature, is shown in Figure 4b. This study reported changes in δD values of *n*-alkanes and isoprenoids against δD of kerogen through the transition from thermally immature to early mature organic matter. Substantial differences in δD values (up to 40‰ for *n*-C_{17}, up to 30‰ for *n*-C_{29}, and up to 10‰ for bulk kerogen) were evident from one sample to the next throughout the ONEZ-1 well. However, from the top of the well to the bottom, in general, δD values of individual *n*-alkanes showed a slight D

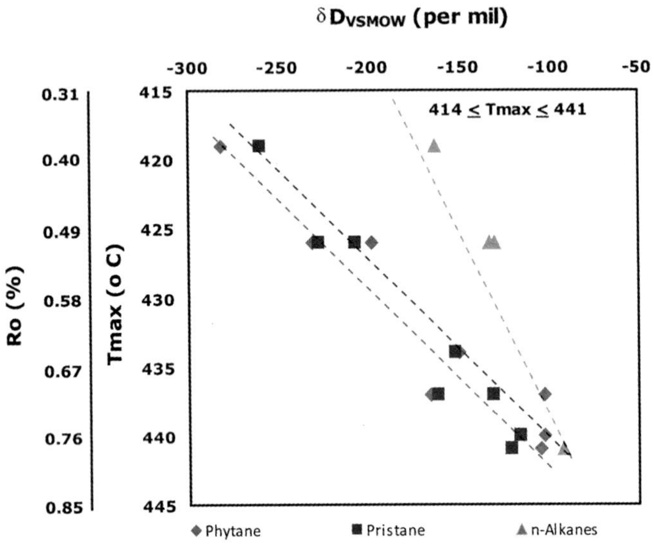

FIG. 5.—Average δD values of pristane, phytane, and *n*-alkanes vs. thermal maturity (T_{max}) in samples from the Duvernay Formation, WCSB. (The formula for deriving R_o, vitrinite reflectance, from T_{max} is $R_o[\%] = [0.0180 \times T_{max}] - 7.16$ and is taken from Jarvie et al. 2001.)

enrichment with depth and/or maturity. A general trend of increasing δD values with increasing *n*-alkane chain length was an interesting feature observed for all the samples. The δD values of kerogen samples were found to overlap with the δD values of *n*-alkanes; this was attributed to the kerogen being preferentially enriched in D due to its overall higher susceptibility to H transfer in comparison with saturated hydrocarbons (Schimmelmann et al. 1999). The largest δD offset (100‰) between the *n*-alkanes and isoprenoids was observed in the upper section of the core, due to the δD values of the isoprenoids being more greatly affected by hydrogen exchange than the *n*-alkanes due to the greater extent of preservation of δD values at the shallower burial depth resulting from a lower thermal maturity.

Zechstein and Northern German Basins, Europe: Two stratigraphic sections from northern Europe with different thermal histories were investigated by Radke et al. (2005): the Permian Kupferschiefer unit from the Zechstein Basin; and the Lower Jurassic Posidonia Shale, from the Vlotho Massif, Northern German Basins. Kupfererschiefer samples come from the Lower Permian basal Zechstein "Kupferschiefer" horizon of the barren zone, Zechstein Basin, NE Poland (Bechtel et al. 2000). Both sedimentary sequences contain type-II algal kerogen. Samples from the two sequences represent a range of thermal maturity from 0.48 to 1.3% R_o, determined by vitrinite reflectance measurements and equivalent reflectance values calculated from the methylphenanthrene index. Thermal maturation of the Kupferschiefer horizon from the lower Zechstein Basin of Poland was associated with burial, while the Posidonia Shale experienced contact metamorphic thermal maturation related to a magmatic intrusion during the Lower Cretaceous.

The δD values of individual alkanes and isoprenoids in the Kupferschiefer samples range from −198‰ to −53‰ and −281‰ to −84‰, respectively. The δD values of the *n*-alkanes and isoprenoids show a positive correlation with thermal maturity (Fig. 4c). D enrichment in isoprenoids, in particular, correlates strongly with traditional molecular maturity parameters, for example, the methylphenanthrene index (0.4–1.5 MPI-1). The mean δD values of phytane and pristane range from −275‰ in the immature to −147‰ in the mature Kupferschiefer samples (Fig. 4). The D enrichment in the alkanes is +45‰ from the immature to the mature sample. Consistent with the earlier work of Dawson et al. (2005), again the rate of enrichment of δD with increasing maturity is more rapid for the isoprenoids than for the *n*-alkanes.

The δD values of alkanes from Posidonia sediments vary between −204‰ and −76‰. The D enrichment in the alkanes is +67‰ from the immature to the mature sample (Fig. 4). In the Posidonia Shale samples, the δD values of isoprenoids are between −317‰ and −123‰ in immature and mature samples, respectively. Different timescales and temperature regimes during burial of sediments in the Kupferschiefer and Posidonia Shale probably led to different hydrogen isotope fractionations within the same compound class.

RESULTS FROM THE DUVERNAY FORMATION, WESTERN CANADA SEDIMENTARY BASIN

Devonian strata include extremely important sources of hydrocarbons in the WCSB, accounting for 55% of oil reserves and 27% of gas reserves in that basin (Fowler et al. 2001). The Upper Devonian Duvernay Formation, part of the Late Devonian Woodbend Group in the WCSB of central Alberta, has excellent hydrocarbon source potential and is a major source for oil and gas in the basin (Stoakes and Creaney 1984, 1985; Creaney and Allan 1990; Chow et al. 1995). Comprehensive organic geochemical characteristics of sediments from the Duvernay Formation are reported by Creaney and Allan (1992) and Requejo et al. (1992). The Duvernay Formation maturity increases from very immature in the east to overmature in the west of the WCSB in central Alberta (Stoakes and Creaney 1985).

The range of δD values of *n*-alkanes and the δD values of pristane and phytane from extracts of the Duvernay Formation sediments are shown in Figure 5 and summarized in Table 4 (including the number of individual analyses, range, and standard deviation for each sample). The δD values of *n*-alkanes (n-C_{10} to n-C_{32}) in extracts from the Duvernay Formation sediments average between −163‰ and −92‰. The δD values of pristane and phytane in the Duvernay Formation samples range between −260‰ and −116‰ and between −282‰ and −103‰, respectively. Figure 5 shows a clear linear relationship of D enrichment in isoprenoids and *n*-alkanes with increasing thermal maturity, represented by the T_{max} parameter. The δD values of pristane and phytane change at similar rates with increasing maturation, suggesting they exchange hydrogen at similar rates. The difference between the δD values of *n*-alkanes and isoprenoids gradually decreases with thermal maturity (Table 4). While pristane and phytane become more rapidly enriched in D with increasing maturity, *n*-alkanes are less rapidly enriched in D (if at all) until a relatively more mature stage is reached. These results, and the correlation between D/H of biomarkers and maturity parameters for Devonian source rocks from the WCSB, show a strong link to the previous work.

DISCUSSION

The δD values of individual hydrocarbons (*n*-alkanes and isoprenoids) can be used to estimate the thermal maturity of sedimentary organic matter from source rocks. The application of D/H of individual hydrocarbons as a maturity parameter for Devonian source rocks from the WCSB has been presented here, extending the application of D/H measurements to much older sediments than previously studied. Figure 6 shows a plot of T_{max} against the ΔδD values (determined by subtracting the average δD value of the *n*-alkanes from the average δD value of pristane and phytane) for the several sedimentary sequences discussed in this review (Vulcan Subbasin, Zechstein Basin, Northern German Basin, and WCSB). A good correlation is evident (R^2 = 0.74), even though this data set

TABLE 4.—δD values of pristane (Pr), phytane (Ph), and *n*-alkanes for the WCSB (Duvernay Formation) sediment extracts.

Sample	Depth (m)	T_{max} (°C)	δD Pr[a] (‰)	δD Ph[a] (‰)	Average δD *n*-alkanes (‰)	ΔδD[b] (‰)
G000542	1147.3	419	−260 (7.3)[3]	−282 (5.2)[3]	−163	−108
G000584	1402.7	426	−228 (12.4)[3]	−231 (28.9)[3]	−130	−100
G000587	1404.5	426	−207 (4.6)[2]	−198 (0.2)[2]	−133	−75
G000553	1766.9	434	−152 (0.9)[3]	−150 (0.1)[3]	n.d.	n.d.
G000595	2239.5	437	−162 (6.5)[3]	−166 (10.8)[3]	n.d.	n.d.
G000599	2248.7	437	−131 (7.3)[3]	−103 (4.2)[3]	n.d.	n.d.
G000601	2971.5	440	−116 (0.5)[2]	−103 (2.0)[2]	n.d.	n.d.
G000603	2973.7	441	−121 (5.5)[3]	−105 (7.6)[3]	−92	−21

[a] Superscript numbers are number of replicate analyses; numbers in parentheses are standard errors.
[b] ΔδD = difference between the average δD value of Pr and Ph and the average δD value of the *n*-alkanes (relative to VSMOW).

includes sediments of different age, facies, and source-rock/kerogen type. It is clear that the ΔδD values provide a good approximation of thermal maturity of sediments. These systematic hydrogen isotopic trends with thermal maturity can provide an estimate of thermal maturity in cases where none of the traditional parameters can be used.

Investigating Mechanism(s) of Hydrogen Exchange

Fundamental chemical principles, together with observations of stereochemical changes occurring during thermal maturation, provide insight into possible thermally induced mechanisms by which organic

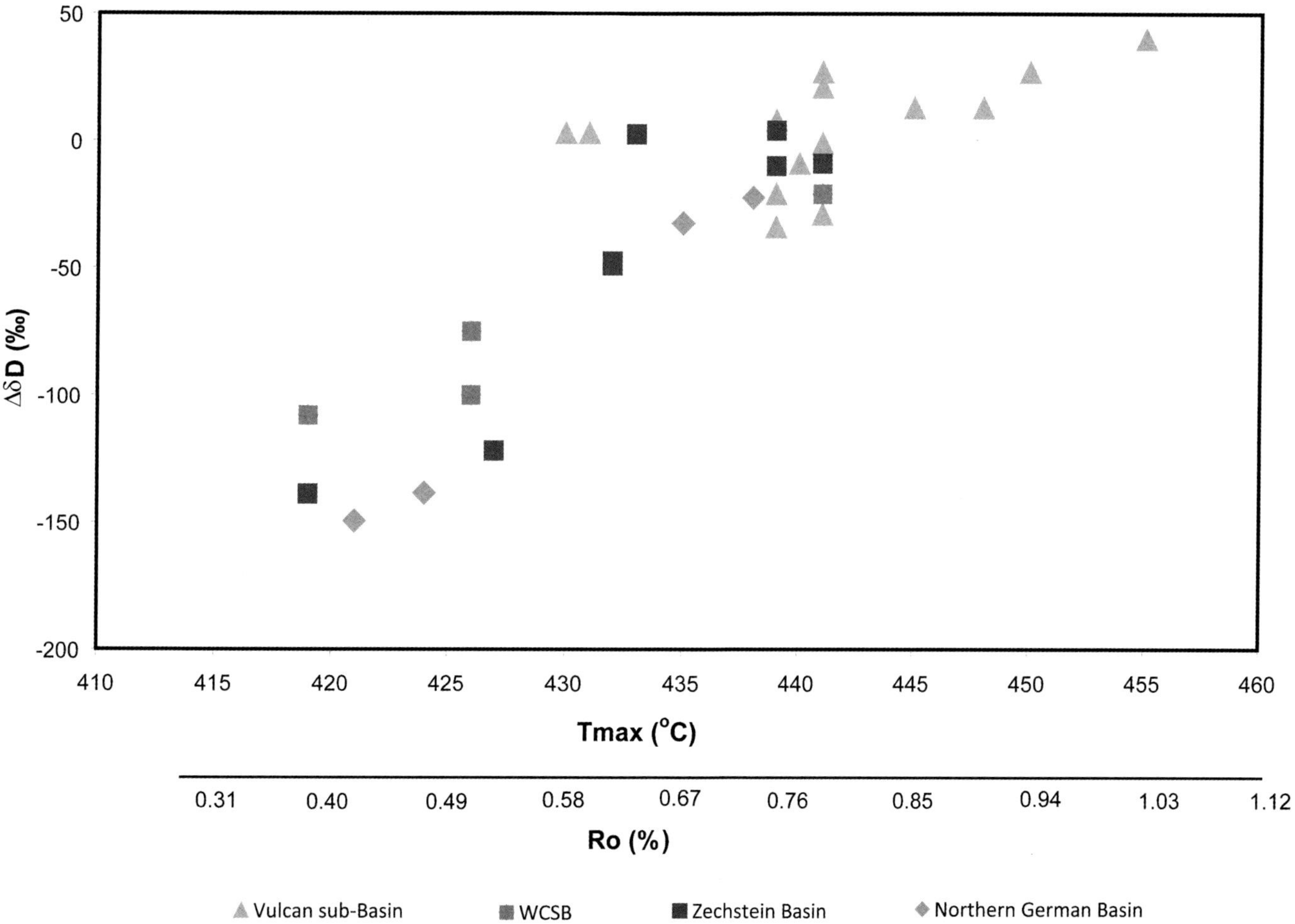

FIG. 6.—Combined results of ΔδD vs. T_{max} for several sedimentary basins. (ΔδD is the difference between the average δD value of pristane and phytane and the average δD value of *n*-alkanes.) The formula for deriving R_o, vitrinite reflectance, from T_{max} is R_o (%) = (0.0180 × T_{max}) − 7.16, and it is taken from Jarvie et al. (2001).

matter in the sedimentary environment might exchange hydrogen with its surrounding water, resulting in D enrichment. The *n*-alkanes and the isoprenoids pristane and phytane are quantitatively the most significant hydrocarbon components of crude oils. While they belong to a similar compound class (alkanes), *n*-alkanes and regular isoprenoids exhibit some fundamental chemical differences that result in significantly different behaviors during various types of chemical reactions. The *n*-alkanes are straight-chain hydrocarbons containing only primary (methyl) and secondary (methylene) carbon atoms (Fig. 1). Regular isoprenoids, however, also contain tertiary (methine) carbon atoms. Some chemical processes, for example, carbon–carbon bond cleavage, occur more readily at tertiary carbon atoms due to the ability of additional alkyl substituents to stabilize charged transition states (e.g., carbocations or radicals) through inductive and resonance effects. For this reason, hydrogen exchange readily occurs at tertiary carbons, and even more readily at carbon atoms adjacent to the tertiary centers (e.g., Alexander et al. 1984).

Structural changes of molecules, such as stereochemical inversion or rearrangement of chiral centers, provide insight into the mechanisms of hydrogen exchange. Pristane and phytane both contain chiral carbons, and thus are ideal candidates for stereochemical studies. The pristane molecule is symmetrical, with two chiral centers at C-6 and C-10 resulting in three diastereomers, 6(*R*),10(*R*)-pristane (*RR*); 6(*S*),10(*S*)-pristane (*SS*); and 6(*R*),10(*S*)-pristane (*RS*, *meso*). All three have been discovered in sedimentary organic matter, but only *meso*-pristane retains the configuration of the precursor, while the *RR* and *SS* enantiomers form from the isomerization of the *meso* compound during maturation (Patience et al. 1978). For example, the immature Messel Shale (Eocene, Germany) contains only *meso*-pristane, while the thermally mature Irati Shale (Permian, Brazil) contains a 1:1 mixture of the *meso* and *RR*/*SS* isomers, respectively (Patience et al. 1978). Phytane has three chiral centers at C-6, C-10, and C-14, resulting in eight diastereomers, with the eight possible configurations at the chiral centers being *RRR*, *SSS*, *RRS*, *SSR*, *RSR*, *SRS*, *SRR*, and *RSS*, respectively. All eight occur in sedimentary organic matter (Patience et al. 1980).

Dawson et al. (2005, 2007) used pristane and phytane diastereomer ratios (PrDR′ and PhDR, respectively; see also Hansen et al. 2003) to monitor the extent of hydrogen exchange of pristane and phytane in Perth Basin and Vulcan Subbasin sediments. The plots in Figure 7 show (1) PrDR′ vs. the δD value of pristane, and (2) PhDR vs. the δD value of phytane, for the Perth Basin and Vulcan Subbasin sediments (Dawson et al. 2007). Dawson et al. (2005) showed that PrDR′ and PhDR derived from the Perth Basin sediments increase and approach a value of 1 from initial values of 0.71 and 0.77, respectively, with the increase concurring with enrichment in D. The values increased linearly with a moderate to strong relationship up to PrDR′ and PhDR values of 0.93 and 0.95, respectively, for two late-mature samples. On the other hand, the Vulcan Subbasin sediments start at PrDR′ and PhDR values of approximately 0.85 (corresponding to 85% epimerization) at their lowest maturity level (early mature, R_o = 0.62%). Significant epimerization (~85–90%) of Pr and Ph has occurred in all the Vulcan Subbasin sediments analyzed by Dawson et al. (2007). This is consistent with results obtained from the Perth Basin sediments, where 85–90% epimerization had also occurred at an equivalent level of thermal maturity (Dawson et al. 2007).

Patience et al. (1978) suggested that an equilibrium mixture of approximately 1:1 of the *RS* and *RR*/*SS* diastereomers of pristane would be reached as a result of epimerization during maturation. Patience et al. (1980) separated the diastereomers of phytane in a crude oil into two partially resolved peaks with a 1:1 ratio, and Mackenzie et al. (1980) found that the extent of isomerization of phytane with maturation appeared to be virtually identical to that of pristane. The values of PrDR′ and PhDR obtained from the immature Perth Basin sediments (0.71 and 0.77, respectively) suggested that significant epimerization

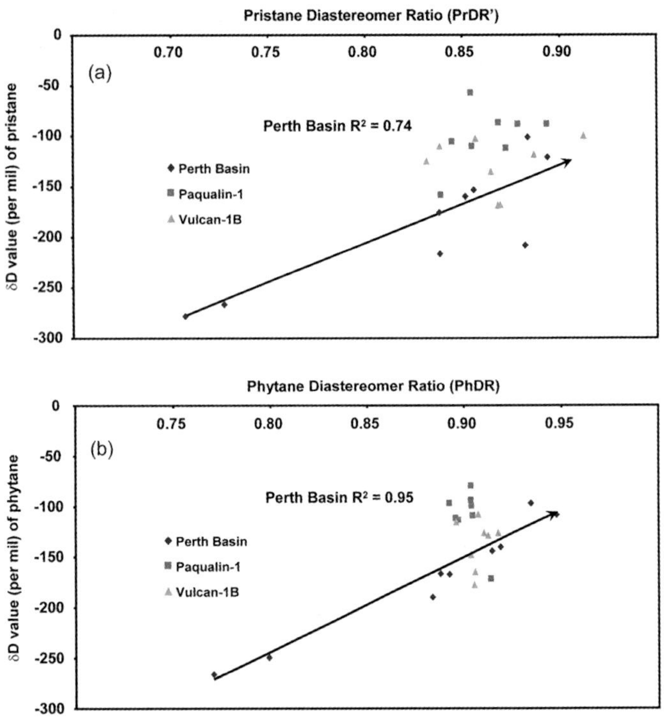

FIG. 7.—Plot of (a) the δD value of pristane vs. the pristane diastereomer ratio (PrDR′) and (b) the δD value of phytane vs. the phytane diastereomer ratio (PhDR) for the Vulcan Subbasin sediment extracts, superimposed on the data obtained from the Perth Basin sediment extracts (Dawson et al. 2005, 2007). The direction of the arrow indicates increasing maturity. Both PrDr′ and PhDR increase as a function of thermal maturity.

had already occurred at a relatively low maturity level (R_e = 0.53%; Dawson et al. 2007). For example, a PrDR′ of 0.71 indicates that there has been approximately 70% conversion of *meso*-pristane to *RR*- and *SS*-pristane in these immature samples, even though pristane appears to have essentially retained its precursor's isotopic composition (δD = −278‰; Dawson et al. 2007). The conversion approaches 85–90% at an early mature level, equivalent to an R_e value of 0.78%.

The extent of epimerization in the immature and early mature Perth Basin samples would suggest significant hydrogen exchange and thus significant modification of indigenous δD signatures; however, this was clearly not the case (Dawson et al. 2007). Indeed, pristane and phytane have largely retained the isotopic composition of their precursors (see previous).

A mechanism has been proposed that can account for both D/H exchange and epimerization of pristane and phytane in the sedimentary environment (Fig. 8; Dawson et al. 2007). Upon interaction with a catalytic surface, a compound containing tertiary carbon centers may form an intermittent alkene structure via a tertiary carbocation-like species. The hydrogen attached to the carbon adjacent to the tertiary carbon may then separate and exchange with other ^{1}H or D atoms in the vicinity. Then, the resulting sp^2 hybridized tertiary carbocation undergoes ^{1}H (or D) reattachment, resulting in additional hydrogen exchange along with epimerization of the chiral center.

CONCLUSIONS

The earlier studies described here and the new study of the Duvernay Formation demonstrate that the δD values of the *n*-alkanes, pristane,

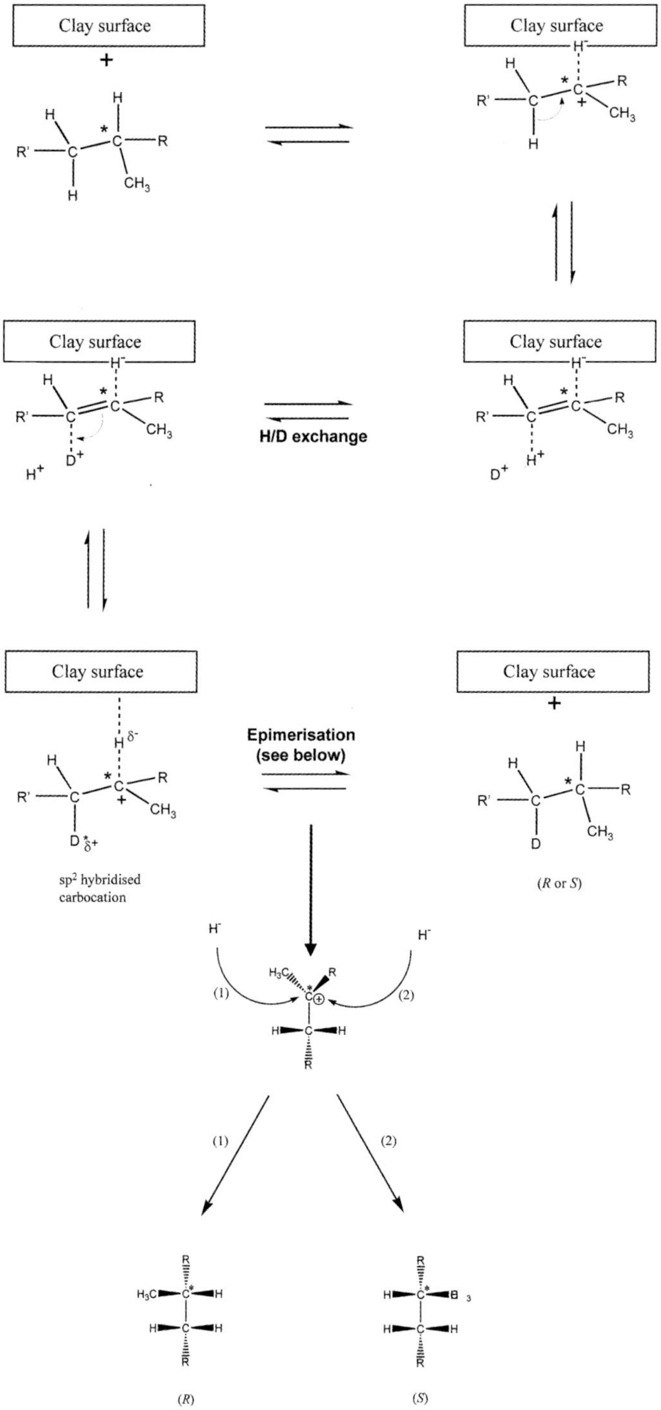

FIG. 8.—Mechanism of hydrogen exchange (after Alexander et al. 1984), and epimerization of a chiral center, where $R' \geq R$ in terms of substituent priority, and $R \neq CH_3$. * denotes a chiral center (Dawson et al. 2007).

and phytane are affected by thermal maturation. In general, the *n*-alkanes, pristane, and phytane of immature sediments have initial δD values that represent the accepted isotopic composition of their natural product precursors. With increasing maturity, pristane and phytane become more enriched in D, while the *n*-alkanes generally remain at a constant isotopic composition until a mature–late mature–overmature level is reached, after which there is a significant enrichment of D in *n*-alkanes. With increasing maturity, the hydrocarbon compounds become enriched in D through hydrogen exchange, with the isoprenoids becoming enriched more quickly than the *n*-alkanes. Significant epimerization of pristane and phytane during early maturation suggests that hydrogen exchange occurs at the tertiary carbons of these hydrocarbons; however, it does not result in a significant alteration of precursor-derived δD values. Dawson et al. (2007) suggested that perhaps hydrogen exchange and epimerization proceed via different mechanisms during the early stages of maturation. These results provide an important insight into the chemical processes that may occur in the subsurface during petroleum formation and generation.

It still remains uncertain whether the D/H of individual hydrocarbons can be used as a precise maturity tool for pre-Devonian sedimentary rocks (i.e., predating the evolution of higher plants and those containing no vitrinite). The results also emphasize that care is needed when interpreting δD values in the context of the source(s) of hydrocarbons in sediments of high thermal maturity, particularly those hydrocarbons with a high reactivity toward hydrogen exchange. The effect of maturation on indigenous δD values may result in an erroneous interpretation.

ACKNOWLEDGMENTS

The authors thank Mr. Geoff Chidlow for his assistance with GC-MS analysis and maintenance. K.G. acknowledges the ARC for a QEII fellowship (DP0211875, DP0877167). E.M. gratefully acknowledges the Department of Education, Science and Training for receipt of a Curtin University Postgraduate Award research scholarship and support from an ARC Discovery grant. The reviewers are thanked for their constructive comments, including Associate Professor Nick Harris, Dr. Barry Katz, and an anonymous reviewer. Dr. Ken Peters and Associate Professor Nick Harris are thanked for acting as editors of the special edition.

REFERENCES

Alexander R, Kagi RI, Larcher AV. 1982. Clay catalysis of aromatic hydrogen exchange reactions. *Geochimica et Cosmochimica Acta* 46:219–222.

Alexander R, Kagi RI, Larcher AV. 1984. Clay catalysis of alkyl hydrogen exchange reactions—reaction mechanisms. *Organic Geochemistry* 6:755–760.

Alexander R, Kagi RI, Larcher AV, Woodhouse GW. 1983. Aromatic hydrogen exchange in petroleum source rocks. *In* Bjorøy M (Editor). *Advances in Organic Geochemistry, 1981*: John Wiley & Sons Ltd., New York. p. 69–71.

Baker KL. 1987. An investigation of ^2H and ^{13}C abundance in a sequence of related oils of increasing maturity [MSc thesis]: Indiana University, Bloomington, Indiana. 138 p.

Bechtel A, Puttmann W. 1992. Combined isotopic and biomarker investigations of temperature- and facies-dependent variations in the Kupferschiefer of the Lower Rhine Basin, northwestern Germany. *Chemical Geology* 102:23–40.

Bechtel A, Shieh YN, Elliott WC, Oszczepaski S, Hoernes S. 2000. Mineralogy, crystallinity and stable isotopic composition of illitic clays within the Polish Zechstein Basin: implications for the genesis of Kupferschiefer mineralization. *Chemical Geology* 163:189–205.

Chikaraishi Y, Naraoka H. 2001. Organic hydrogen-carbon isotope signatures of terrestrial higher plants during biosynthesis for distinctive photosynthetic pathways. *Geochemical Journal* 35:451–458.

Chow N, Wendte J, Stasiuk LD. 1995. Productivity vs. preservation controls on two organic-rich carbonate facies in the Devonian of Alberta: sedimentological and organic petrological evidence. *Bulletin of Canadian Petroleum Geology* 43:433–460.

Creaney S, Allan J. 1990. Hydrocarbon generation and migration in the Western Canada Sedimentary Basin. *In* Brooks J (Editor). *Classic Petroleum*

Provinces: Geological Society, London, UK. Special Publication 50, p. 189–202.

Creaney S, Allan J. 1992. Petroleum systems in the foreland basin of western Canada. *In* Macqueen RW, Leckie DA (Editors). *Foreland Basins and Fold Belts*: American Association Petroleum Geologist, Tulsa, Oklahoma. Memoir 55, p. 279–308.

Dawson D. 2007. Stable hydrogen isotope ratios of individual hydrocarbons in sediments and petroleum [PhD thesis]: Curtin University of Technology, Perth, Western Australia. 175 p.

Dawson D, Grice K, Alexander R. 2005. Effect of maturation on the indigenous δD signatures of individual hydrocarbons in sediments and crude oils from the Perth Basin (Western Australia). *Organic Geochemistry* 36:95–104.

Dawson D, Grice K, Alexander R, Edwards DS. 2007. The effect of source and maturity on the stable isotopic compositions of individual hydrocarbons in sediments and crude oils from the Vulcan Sub-basin, Timor Sea, northern Australia. *Organic Geochemistry* 38:1015–1038.

Dos Santos Neto EV, Hayes JM. 1999. Use of hydrogen and carbon stable isotopes characterizing oils from the Potiguar Basin (onshore), northeastern Brazil. *American Association of Petroleum Geologists Bulletin* 83:496–518.

Edwards DS, Preston JC, Kennard JM, Boreham CJ, Van Aarssen BGK, Summons RE, Zumberge JE. 2004. Geochemical characteristics of hydrocarbons from the Vulcan Sub-basin, western Bonaparte Basin, Australia. *In* Ellis GK, Baillie PW, Munson TJ (Editors). *Timor Sea Petroleum Geoscience*: Proceedings of the Timor Sea Symposium, Darwin, Northern Territory, June 19–20, 2003; Northern Territory Geological Survey, Darwin. p. 169–201.

Epstein S, Yapp CJ, Hall JH. 1976. The determination of the D/H ratio of non-exchangeable hydrogen in cellulose extracted from aquatic and land plants. *Earth and Planetary Science Letters* 30:241–251.

Fowler MG, Stasiuk LD, Hearn M, Obernajer M. 2001. Devonian hydrocarbon source rocks and their derived oils in the Western Canada Sedimentary Basin. *Bulletin of Canadian Petroleum Geology* 49:117–148.

Grice K, Cao C, Love GD, Böttcher ME, Twitchett RJ, Grosjean E, Summons RE, Turgeon SC, Dunning W, Jin Y. 2005a. Photic zone euxinia during the Permian–Triassic superanoxic event. *Science* 307:706–709.

Grice K, De Mesmay R, Glucina A, Wang S. 2008. An improved and rapid 5A molecular sieve method for gas chromatography isotope ratio mass spectrometry of *n*-alkanes (C_8-C_{30+}). *Organic Geochemistry* 39:284–288.

Grice K, Summons RE, Grosjean E, Twitchett RJ, Dunning W, Wang SX, Böttcher ME. 2005b. Depositional conditions of the northern onshore Perth Basin (Basal Triassic). *Australian Petroleum Production and Exploration Association Journal* 45:263–274.

Hansen D, Bastow TP, Van Aarssen BGK, Alexander R, Kagi RI. 2003. Maturation and biodegradation effects on the diastereomers of pristane and phytane. *In* Book of Abstracts, Part II, 21st International Meeting on Organic Geochemistry, 8–12 September 2003; Krakòw, Poland. 355 p.

Hoering TC. 1977. The stable isotopes of hydrogen in Precambrian organic-matter. *In* Ponnamperuma C (Editor). *Chemical Evolution of the Early Precambrian*: Academic Press, New York. p. 81–86.

Hoering TC. 1984. Thermal reactions of kerogen with added water, heavy water and pure organic substances. *Organic Geochemistry* 5:267–278.

Jarvie DM, Claxton BL, Henk F, Breyer JT. 2001. Oil and shale gas from the Barnett Shale, Fort Worth Basin, Texas. *American Association of Petroleum Geologists, Annual Meeting Program,* v. 10, p. A100.

Kantsler AJ, Cook AC. 1979. Maturation patterns in the Perth Basin. *Australian Petroleum Exploration Association Journal* 19:94–107.

Koepp M. 1978. D/H isotope exchange reaction between petroleum and water: a contributory determinant for D/H-isotope ratios in crude oils? *In* Zartman RE (Editor). *Short Papers of the Fourth International Conference, Geochronology, Cosmochronology, Isotope Geology*: US Geological Survey. Open-File Report 78-701, Washington, DC. p. 221–222.

Larcher AV, Alexander R, Rowland SJ, Kagi RI. 1986. Acid catalysis of alkyl hydrogen exchange and configurational isomerization reactions: acyclic isoprenoid acids. *Organic Geochemistry* 10:1015–1021.

Leif RN, Simoneit RT. 2000. The role of alkenes produced during hydrous pyrolysis of a shale. *Organic Geochemistry* 31:1189–1208.

Lewan MD. 1997. Experiments on the role of water in petroleum formation. *Geochimica et Cosmochimica Acta* 61:3691–3723.

Li M, Huang Y, Obermajer M, Jiang C, Snowdon LR, Fowler MG. 2001. Hydrogen isotopic compositions of individual alkanes as a new approach to petroleum correlation: case studies from the Western Canada Sedimentary Basin. *Organic Geochemistry* 32:1387–1399.

Lis G, Schimmelmann A, Mastalerz M, Stankiewicz BA. 2006. D/H ratios and hydrogen exchangeability of type-II kerogens with increasing maturity. *Organic Geochemistry* 37:342–353.

Lukin AE. 1999. Geodynamics-related differences in the isotopic composition of the hydrogen of oils and condensates from oil- and gas-bearing regions of Ukraine. *Earth Science* 369A:1214–1216.

Mackenzie AS, Patience RL, Maxwell JR. 1980. Molecular parameters of maturation in the Toarcian shales, Paris Basin, France—I. Changes in the configurations of acyclic isoprenoid alkanes, steranes and triterpanes. *Geochimica et Cosmochimica Acta* 44:1709–1721.

Maslen E, Grice K, Gale J, Horsfield B, Hallmann C. 2008. The origin and significance of crocetane in Devonian source rocks and crude oils from the Western Canada Sedimentary Basin. *Organic Geochemistry* 40:1–11.

Murphy MTJ. 1969. Analytical methods. *In* Eglinton G, Murphy MTJ (Editors). *Organic Geochemistry: Methods and Results*: Springer-Verlag, New York. p. 576–598.

Patience RL, Rowland SJ, Maxwell JR. 1978. The effect of maturation on the configuration of pristane in sediments and petroleum. *Geochimica et Cosmochimica Acta* 42:1871–1875.

Patience RL, Yon DA, Ryback G, Maxwell JR. 1980. Acyclic isoprenoid alkanes and geochemical maturation. *In* Douglas AG, Maxwell JR (Editors). *Advances in Organic Geochemistry, 1979*: Pergamon Press, Oxford, UK. p. 287–294.

Pedentchouk N, Freeman KH, Harris NB. 2006. Different response of δD values of *n*-alkanes, isoprenoids, and kerogen during thermal maturation. *Geochimica et Cosmochimica Acta* 70:2063–2072.

Peters KE, Moldowan JM. 1993. *The Biomarker Guide. Interpreting Molecular Fossils in Petroleum and Ancient Sediments*: Prentice Hall, Englewood Cliffs, New Jersey. 1155 p.

Radke J, Bechtel A, Gaupp R, Püttmann W, Schwark L, Sachse D, Gleixner G. 2005. Correlation between hydrogen isotope ratios of lipid biomarkers and sediment maturity. *Geochimica et Cosmochimica Acta* 69:5517–5530.

Redding C. 1978. Hydrogen and carbon isotopes in coals and kerogens. *In* Zartman RE (Editor). *Short Papers of the Fourth International Conference, Geochronology, Cosmochronology, Isotope Geology*: US Geological Survey. Open-File Report 78-701, Washington, DC. p. 348–349.

Redding CE, Schoell M, Monin JC, Durand B. 1980. Hydrogen and carbon isotopic composition of coals and kerogens. *In* Douglas AG, Maxwell JR (Editors). *Advances in Organic Geochemistry, 1979*: Pergamon, Oxford, UK. p. 711–723.

Requejo AG, Gray NR, Freund H, Thomann H, Melchior MT, Gebhard LA, Bernardo M, Pictroski CF, Hsu CS. 1992. Maturation of petroleum source rocks—1. Changes in kerogen structure and composition associated with hydrocarbon generation. *Energy Fuels* 6:203–214.

Rigby D, Batts BD, Smith JW. 1981. The effect of maturation on the isotopic composition of fossil fuels. *Organic Geochemistry* 3:29–36.

Schiegl WE, Vogel JC. 1970. Deuterium content of organic matter. *Earth and Planetary Science Letters* 7:307–313.

Schimmelmann A. 1991. Determination of the concentration and stable isotopic composition of non-exchangeable hydrogen in organic matter. *Analytical Chemistry* 63:2456–2459.

Schimmelmann A, Boudou J-P, Lewan MD, Wintsch RP. 2001. Experimental controls on D/H and $^{13}C/^{12}C$ ratios of kerogen, bitumen and oil during hydrous pyrolysis. *Organic Geochemistry* 32:1009–1018.

Schimmelmann A, Lewan MD, Wintsch RP. 1999. D/H isotope ratios of kerogen, bitumen, oil, and water in hydrous pyrolysis of source rocks containing kerogen types I, II, IIS, and III. *Geochimica et Cosmochimica Acta* 63:3751–3766.

Schimmelmann A, Sessions AL, Boreham CJ, Edwards DS, Logan GA, Summons RE. 2004. D/H ratios in terrestrially sourced petroleum systems. *Organic Geochemistry* 35:1169–1195.

Schimmelmann A, Sessions AL, Mastalerz M. 2006. Hydrogen isotopic (D/H) composition of organic matter during diagenesis and thermal maturation. *Annual Review of Earth and Planetary Sciences* 34:501–533.

Schoell M. 1984. Recent advances in petroleum isotope geochemistry. *Organic Geochemistry* 6:645–663.

Schoell M, Redding C. 1978. Hydrogen isotopic composition of selected crude oils and their fractions. *In* Zartman RE (Editor). *Short Papers of the Fourth International Conference, Geochronology, Cosmochronology, Isotope Geology*: US Geological Survey. Open-File Report 78-701, Washington, DC. p. 384–385.

Schoell M, Teschner M, Wehner H, Durand B, Oudin JL. 1983. Maturity related biomarker and stable isotope variations and their application to oil/source rock correlation in the Mahakam Delta Kalimantan. *In* Bjorøy M (Editor). *Advances in Organic Geochemistry, 1981*: John Wiley & Sons Ltd., New York. p. 156–163.

Schou L, Eggen S, Schoell M. 1985. Oil–oil and oil–source rock correlation, northern North Sea. *In* Thomas BM, Dore AG, Eggen SS, Home PC, Larsen RM (Editors). *Petroleum Geochemistry in Exploration of the Norwegian Shelf*: Graham and Trotman Ltd. (for Norwegian Petroleum Society), London. p. 101–117.

Seewald JS. 2001. Aqueous geochemistry of low molecular weight hydrocarbons at elevated temperatures and pressures: constraints from mineral buffered laboratory experiments. *Geochimica et Cosmochimica Acta* 65:1641–1664.

Seifert WK, Moldowan JM. 1981. Paleoreconstruction by biological markers. *Geochimica et Cosmochimica Acta* 45:783–794.

Sessions AL, Burgoyne TW, Schimmelmann A, Hayes JM. 1999. Fractionation of hydrogen isotopes in lipid biosynthesis. *Organic Geochemistry* 30:1193–1200.

Sessions AL, Sylva SP, Summons RE, Hayes JM. 2004. Isotopic exchange of carbon-bound hydrogen over geologic timescales. *Geochimica et Cosmochimica Acta* 68:1545–1559.

Simoneit BRT, Brenner S, Peters KE, Kaplan IR. 1978. Thermal alteration of Cretaceous black shale by basaltic intrusions in the eastern Atlantic. *Nature* 273:501–504.

Simoneit BRT, Mazurek MA. 1981. Organic geochemistry of sediments from the Southern California Borderland, Deep Sea Drilling Project Leg 63. *Initial Reports of the Deep Sea Drilling Project* 63:837–853.

Smith JW, Gould KW, Rigby D. 1982. The stable isotope geochemistry of Australian coals. *Organic Geochemistry* 3:111–131.

Smith PM, Sutherland ND. 1991. Discovery of salt in the Vulcan Graben: a geophysical and geological evaluation. *Australian Petroleum Exploration Association Journal* 31:229–243.

Stoakes FA, Creaney S. 1984. Sedimentology of a carbonate source rock: the Duvernay Formation of central Alberta. *In* Eliuk L, Kaldi J, Watts N, Harrison G (Editors). *Carbonates in Subsurface and Outcrop*; Canadian Society of Petroleum Geologists Core Conference; October 18–19, 1984; Calgary, Alberta. p. 132–147.

Stoakes FA, Creaney S. 1985. Sedimentology of a carbonate source rock: the Duvernay Formation of central Alberta. *In* Longman MW, Shanley KW, Lindsay RF, Eby DE (Editors). *Rocky Mountain Carbonate Reservoirs*: Society for Sedimentary Geology (SEPM), Tulsa, Oklahoma. Core Workshop Notes 7, p. 343–375.

Thomas BM, Barber CJ. 2004. A re-evaluation of the hydrocarbon habitat of the northern Perth Basin. *Australian Petroleum Production and Exploration Association Journal* 44:59–92.

Van Aarssen BGK, Alexander R, Kagi RI. 1998a. Higher plant biomarkers on the North West Shelf: application in stratigraphic correlation and palaeoclimate reconstruction. *In* Purcell PG, Purcell RR (Editors). *The Sedimentary Basins of Western Australia 2, Symposium Proceedings*: Petroleum Exploration Society of Australia, Perth. p. 123–128.

Van Aarssen BGK, Alexander R, Kagi RI. 1998b. Molecular indicators for palaeoenvironmental changes. *Petroleum Exploration Society of Australia Journal* 26:98–105.

Waseda A. 1993. Effect of maturity on carbon and hydrogen isotopes of crude oils in northeast Japan. *Journal of Japanese Association of Petroleum Technology* 58:199–208.

Xiong Y, Geng A, Pan C, Liu D, Peng P. 2005. Characterization of the hydrogen isotopic composition of individual *n*-alkanes in terrestrial source rocks. *Applied Geochemistry* 20:455–464.

FLUID INCLUSION GEOTHERMOMETRY IN SEDIMENTARY SYSTEMS: FROM PALEOCLIMATE TO HYDROTHERMAL

ROBERT H. GOLDSTEIN

Department of Geology, University of Kansas, 1475 Jayhawk Boulevard, 120 Lindley Hall, Lawrence, Kansas 66045, USA

e-mail: gold@ku.edu

ABSTRACT: Fluid inclusion geothermometry is useful for establishing detailed thermal histories of sedimentary rocks that cannot be gleaned from other techniques. The fluid inclusion technique requires careful attention to petrography and evaluation of thermal reequilibration. The fluid inclusion assemblage approach (FIA) is most important in evaluating the extent to which fluid inclusions have been altered.

Fluid inclusion geothermometry can be accomplished in both low- and high-temperature sedimentary systems. At low temperature, it can be used in paleoclimate work by generating bubbles from metastable inclusions, using cooling and femtosecond laser techniques. At elevated paleotemperature, homogenization temperatures (Th) can be measured from inclusions composed of high-temperature aqueous liquid without dissolved gas, high-temperature aqueous liquid with dissolved gas, gas, and petroleum. Given consistent FIAs and an understanding of the pressure–volume–temperature (PVT) relations for the inclusions, fluid inclusion Th data can be pressure-corrected given certain constraints.

Inclusions can be used to evaluate the detailed spatial, temporal, tectonic, and fluid composition history of a system that cannot be determined in other ways. Fluid inclusions are used as a tool for determining the maximum temperature achieved in a sample, but this commonly would be an underutilization of the data generated. Unlike other techniques, fluid inclusions can link thermal history to fluid flow, as a means of evaluating whether heating was associated with normal burial conditions, hydrothermal systems, or cool fluids flowing into warmer rocks. Hydrothermal systems can be identified when geothermometry indicates: (1) repeated increases and decreases in temperature inconsistent with the burial and unroofing history; (2) paleotemperatures higher than what the most liberal burial history analysis will allow; (3) paleogeothermal gradients or pressure–temperature data inconsistent with normal burial; (4) evidence for locally increased temperature at the same depth within a region; and (5) geothermometric evidence that higher temperatures are focused in fracture, fault, or stratigraphic conduits for paleofluid flow.

KEY WORDS: geothermometry, fluid inclusions, paleoclimate, basin analysis, hydrothermal, diagenesis

INTRODUCTION

Introductory geology students learn that sedimentary rocks owe their origin to fluids. A fluid is always present during the history of a sediment particle or sedimentary rock. The sediments themselves are formed by fluid-controlled weathering reactions or are precipitated directly from aqueous fluids. Sediments are transported and deposited in the presence of a fluid phase, water or air; and after deposition, all sediments are continually bathed in fluids. If a record of those fluids is preserved with their original composition and density intact, then any number of important aspects of geologic history could be recorded, including a history of the temperature, pressure, and composition of fluids that existed during the entire history of a sediment particle and sedimentary rock.

Fluid inclusions are tiny fluid-filled cavities sealed within mineral phases. They can be formed during mineral growth, as irregularities develop during passive precipitation of minerals from a fluid. They also can be trapped as microcracks and other deformation features heal. Fluid inclusions can leak and refill with various fluids when minerals fracture, recrystallize, or when internal overpressure causes fluid inclusions to burst. With so many mechanisms for entrapment of fluids in inclusions, there is potential for fluid inclusions to be trapped during much of the history of a sediment or sedimentary rock. If those fluid inclusions remain intact over geologic time, sealed within their mineral bottles, then fluid inclusions have the potential to provide a detailed geologic record.

This paper concerns the use of fluid inclusions for paleotemperature analyses in sediments and sedimentary rocks. It is not meant to be a full-blown analysis of the subject, but it addresses many of the more important points in applying the fluid inclusion technique in sedimentary systems. All aspects of the sedimentary realm are discussed, including near-surface systems useful in paleoclimate analysis, shallow diagenetic systems, and burial diagenetic systems. In particular, fluid inclusions are useful for discriminating hydrothermal systems from other thermal regimes. An expanded discussion of criteria for recognition of ancient hydrothermal systems is included. For a more detailed analysis of applications of the fluid inclusion technique, the reader is referred to Roedder (1984), Goldstein and Reynolds (1994), Goldstein (2001a), and Samson et al. (2003).

History of Fluid Inclusion Geothermometry in Sedimentary Systems

Fluid inclusion geothermometry has a long history of application in sedimentary systems. Sorby (1858) originated the use of fluid inclusions as geothermometers, observing that fluid inclusions commonly contained a liquid phase and a small gas bubble. Sorby reasoned that the bubbles were caused by differential thermal contraction of a liquid during natural cooling of the inclusion fluid from a higher temperature earlier in geologic history. He further reasoned that reheating an inclusion in the laboratory would lead to disappearance of the bubble, and that this temperature of disappearance could be used to estimate the temperature of mineral formation. Since that time, this basic concept has been shown to be the most useful technique in fluid inclusion geothermometry. The temperature of disappearance of a bubble in a liquid-rich fluid inclusion during heating in the laboratory is called the homogenization temperature (Th).

Prior to the 1980s, fluid inclusion applications to geothermometry in sedimentary systems mostly concentrated on evaporites and on Mississippi Valley–type ore deposits. During the early 1980s, oil industry research laboratories were actively pursuing fluid inclusion

Analyzing the Thermal History of Sedimentary Basins: Methods and Case Studies
SEPM Special Publication No. 103, Copyright © 2012
SEPM (Society for Sedimentary Geology), Print ISBN 978-1-56576-315-9, CD/DVD ISBN 978-1-56576-317-3, p. 45–63.

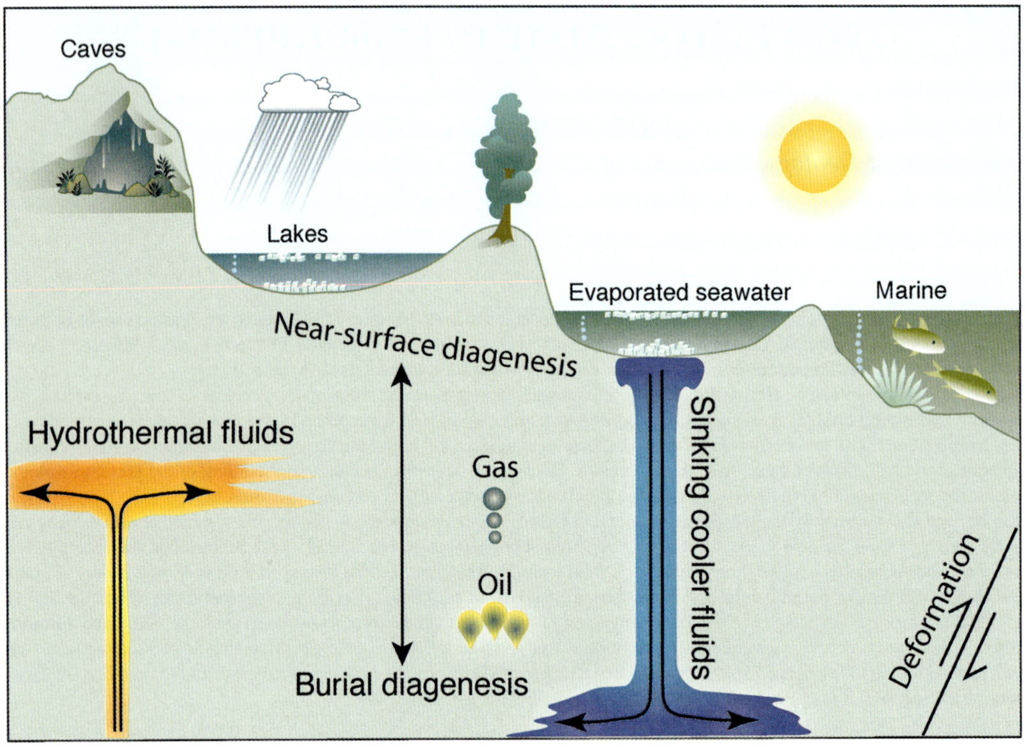

FIG. 1.—Illustration of the wide range of environments in which the fluid inclusion geothermometer is applied. Surface and near-surface paleoenvironments such as caves, lakes, evaporative lagoons, and seas hold great potential for paleotemperature and paleoclimate analysis. In the shallow subsurface, low-temperature diagenetic systems hold potential for refining our understanding of low-temperature diagenetic processes. In the subsurface, fluid inclusion geothermometry is useful for a wide range of applications. Going from right to left on the diagram, fluid inclusion geothermometry can be used: to determine the temperature of fracturing and faulting; resolve the downward flow of cool fluids deeper into the warm subsurface; measure the temperature of oil and gas migration and accumulation; determine subsurface temperatures during normal burial; and identify hydrothermal systems, where warm subsurface fluids flow into cooler rocks. Figure is modified from Goldstein (2003).

studies. Robert Burruss at Gulf Research and Roger McLimans at Conoco Research were the industry leaders in working with petroleum fluid inclusions and finding new applications to sedimentary systems.

Fluid inclusion research took a giant leap forward when Edwin Roedder published a monumental work in 1984 (Roedder 1984). This encyclopedic volume is still the basic reference on fluid inclusion work and is required reading for all fluid inclusionists working in any system. At the time of its publication, the unique problems and prospects of sedimentary systems had not yet been worked out. At the same time as its publication, researchers were becoming aware of particular issues faced by fluid inclusionists in sedimentary systems.

In the 1970s and early 1980s, fluid inclusion geothermometric applications (e.g., Nelson 1973, Klosterman 1981, Moore and Druckman 1981, Wagner and Matthews 1982) led to quite a bit of controversy. It was found that, when analyzed in the same way as ore deposits, fluid inclusion homogenization temperatures did not consistently give the expected results. Soon after this discovery was made, thermal reequilibration of fluid inclusions was found to be the main culprit leading to complexities in interpreting fluid inclusion data from sedimentary systems (Goldstein 1986, Prezbindowski and Larese 1987).

Since then, new methodologies have been developed for doing fluid inclusion geothermometry in sedimentary rocks. These methodologies concentrate on techniques that allow researchers to recognize homogenization temperature data that are representative of the original conditions of entrapment as well as conditions of later reequilibration. The new methodologies require painstaking petrographic analysis of fluid inclusions, and identification of fluid inclusions that appear to have been trapped at the same time as one another (as much as the limits of petrography allow). The approach, termed the fluid inclusion assemblage (FIA) approach, is a methodology for working on fluid inclusions in sedimentary systems in general, and it was described by Goldstein and Reynolds (1994). This book is an important reference for any researcher interested in fluid inclusion geothermometry. After originating from particular problems faced by fluid inclusion researchers in sedimentary geology, the FIA approach has now become the required approach applied in fluid inclusion studies from most subdisciplines in geology (e.g., Bodnar 2003a, Goldstein 2003, Samson et al. 2003).

Applications in Sedimentary Geology

Fluid inclusion geothermometry has a wide range of applications in sedimentary systems. As a technique, it is unusual in that there is potential for evaluating a detailed history of temperature conditions from a single sample and placing it in the framework of mineral paragenesis and deformation. Fluid inclusions are typically regarded as having been entrapped instantaneously, and therefore they have potential to provide temporal resolution higher than other techniques. Other techniques require assumptions about duration at a particular temperature, or must be modeled as one of an infinite number of

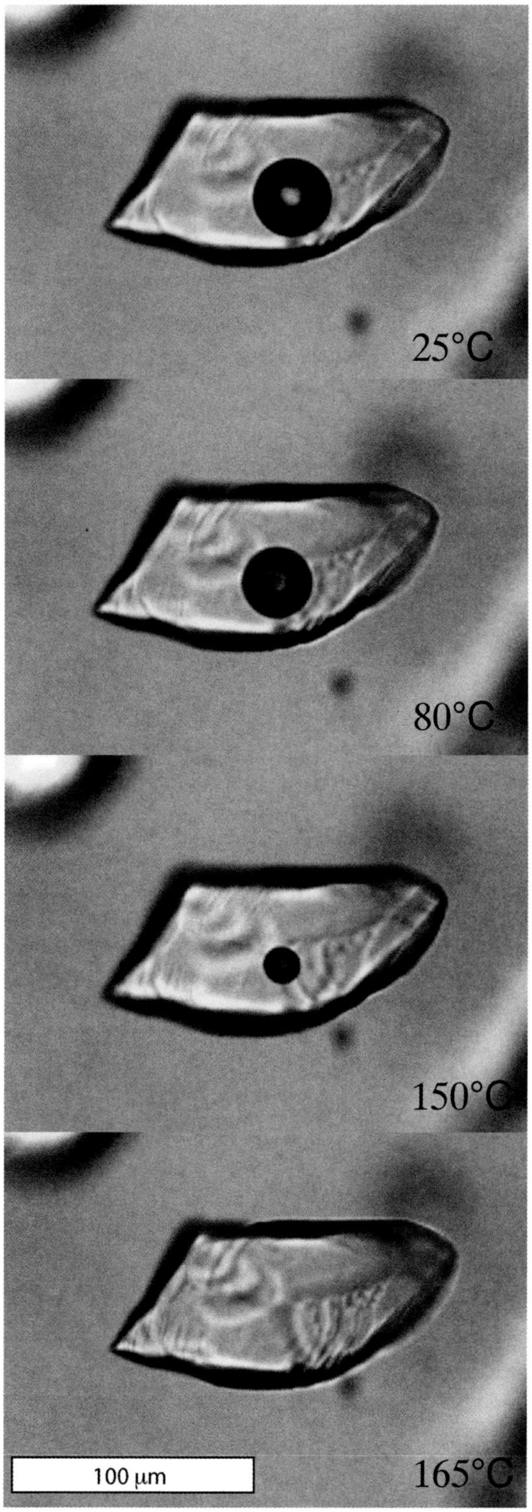

FIG. 2.—Sequence of transmitted light photomicrographs of an aqueous fluid inclusion during a heating run on a microscope heating stage. Notice that as temperature increases, the bubble shrinks and eventually disappears at 165° C. The temperature of disappearance of the bubble is the homogenization temperature (Th). Illinois fluorite.

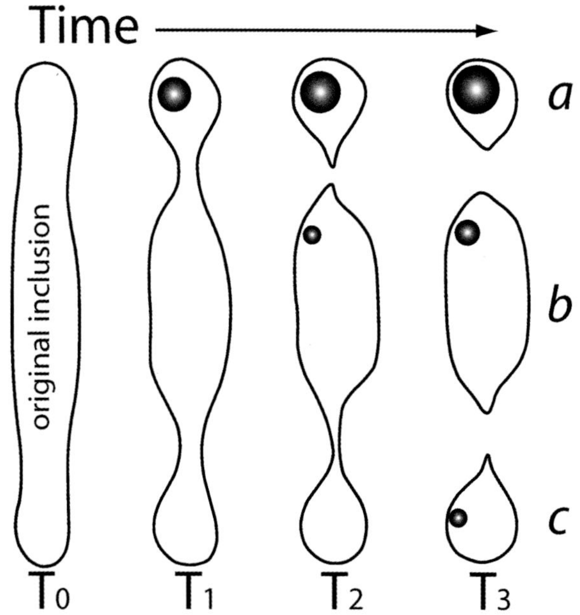

FIG. 3.—Sketch of the way in which necking down of a single fluid inclusion after a phase change can lead to multiple fluid inclusions, each with densities different from the original. Inclusion starts as a single liquid phase at high temperature (T_0). As inclusion cools (T_1), inclusion changes shape, necks begin forming, and gas bubble nucleates. As inclusion continues to cool (T_2), upper part of inclusion necks off into separate inclusion, trapping a large gas bubble. At the same time, a bubble nucleates in the lower inclusion. As cooling continues (T_3), the lower inclusion necks off, trapping a bubble. At the same time, a new bubble nucleates in the lower inclusion. Inclusion *a* has a very low density, and it is unlikely to homogenize during laboratory heating; a bubble would remain even at a temperature higher than that expected for homogenization for the original inclusion had it not necked down. Inclusion *c* would homogenize at a very low temperature, and inclusion *b* would homogenize at some geologically meaningless temperature. The summed volumes in inclusions *a*, *b*, and *c* are equal to the volume of the original single inclusion. Figure is modified from Roedder (1962).

possible solutions. The fluid inclusion geothermometer does not require those sorts of assumptions.

Most other geothermometry techniques only provide a single datum from the thermal history of a system, but fluid inclusions have the potential to provide multiple records of temperatures linked to paragenesis, by virtue of trapping fluids in inclusions at multiple stages. When this paragenesis is tied to stratigraphic, structural, or fluid-flow events, there is great potential for interpreting geologic history. When it is tied additionally to absolute ages through dating of thermal history (e.g., Wagner and Van den Haute 1992, Farley and Stockli 2002), fluid inclusion entrapment (e.g., Pettke and Diamond 1995, 1996), or mineral precipitation (e.g., Luczaj and Goldstein 2000), the potential record of geologic history is even more extensive.

Thus, fluid inclusion geothermometry is useful to a broad range of research topics in sedimentary systems, as illustrated in Figure 1, ranging from paleoclimate analysis, and records of surface or near-surface environments, to high-temperature systems in deep burial regimes.

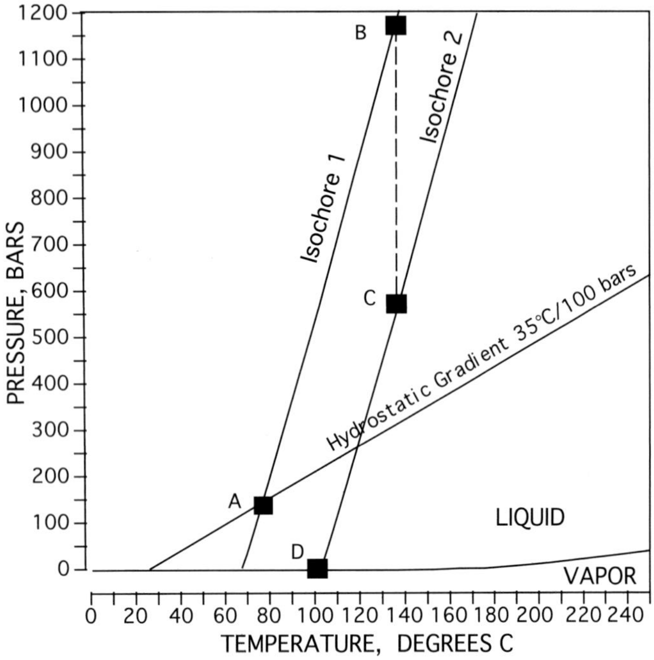

FIG. 4.—Pressure–temperature (P-T) phase diagram for pure water that shows how stretching occurs when a fluid inclusion is heated beyond its temperature of entrapment. If the fluid inclusion were trapped at point A (75° C) and were then heated during burial, the inclusion's internal pressure would follow isochore 1 toward point B, and its internal pressure would increase greatly compared to the surrounding hydrostatic pressure. If the inclusion were to stretch from this internal overpressure, its volume would increase and density and pressure would decrease. This is represented by a drop along a line of constant temperature to point C, and it places the inclusion on the new isochore (isochore 2). Cooling of the inclusion would result in P-T conditions within the inclusion intersecting the liquid–vapor curve at about 100° C, and the inclusion would then have a density that would produce Th measurements of 100° C. If the inclusion had never stretched, it would have yielded a Th at 65° C. Figure is after Goldstein and Reynolds (1994).

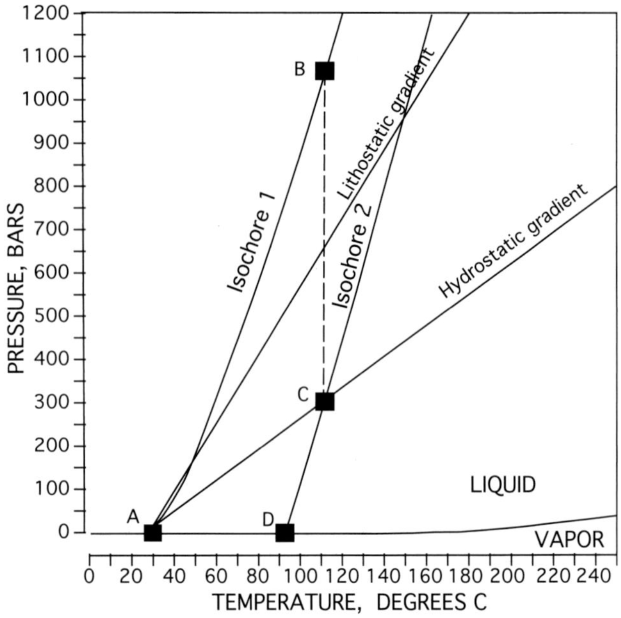

FIG. 5.—Pressure–temperature (P-T) phase diagram for pure water with hydrostatic thermobaric gradient of 28° C/100 bars and lithostatic thermobaric gradient of 28° C/226 bars superimposed. If a fluid inclusion were trapped as liquid at the surface (point A) and were then heated during burial, the fluid inclusion's internal pressure would rise along isochore 1 toward point B. At point B, the inclusion is internally overpressured relative to the surrounding hydrostatic thermobaric gradient. If the inclusion were to decrepitate (burst) from internal overpressure, it would immediately equilibrate with the conditions in the surrounding pore fluid and jump down in pressure to point C on the hydrostatic thermobaric gradient. Because the inclusion has decrepitated, it can refill with whatever pore fluids happen to be present at the time of opening of the inclusion. Rehealing of the inclusion followed by cooling would result in fluid inclusion P-T conditions following a new isochore (isochore 2; different density and possibly different composition) that intersects the liquid–vapor curve at a higher temperature (point D). This thermal reequilibration causes an increase in homogenization temperature of fluid inclusions and may result in their refilling with new fluids. Figure is after Goldstein and Reynolds (1994).

HOW IS THE FLUID INCLUSION GEOTHERMOMETER APPLIED?

Requirements for Geothermometry

Geothermometry is accomplished by observing fluid inclusion phase changes on a microscope heating stage. The most important phase change observable for fluid inclusion geothermometry is the temperature of homogenization (Th). Th typically is measured in the laboratory when two-phase fluid inclusions are found in polished thick sections at room temperature. Most commonly, these contain a small gas bubble floating in a surrounding liquid phase (Fig. 2). As the temperature is increased during laboratory heating, the liquid phase expands, and the bubble shrinks. The temperature at which the bubble finally disappears (Fig. 2) is recorded as Th and typically is reproducible and accurate to within 1° C. If various requirements have not been violated, then Th is an exceptionally high-precision measure of the minimum entrapment temperature of the fluid inclusion.

Certain requirements are necessary for reliable application of fluid inclusions as geothermometers (Roedder 1984). For sedimentary systems, these requirements have been scrutinized extensively (e.g., Goldstein 1993, Goldstein and Reynolds 1994, Bodnar 2003b). All fluid inclusion research requires careful evaluation of each of these requirements.

- The first requirement is that inclusions must have remained chemically closed (isoplethic) systems since entrapment. As an approximation, the volume of an inclusion must be constant (isochoric) since entrapment.
- The inclusion must be trapped as a single fluid phase, not a mixture of phases. The fluid inclusions must follow equilibrium predictions of phase changes rather than be kinetically controlled.

- Necking down of one fluid inclusion into multiple fluid inclusions cannot have occurred in the presence of multiple phases (Fig. 3); or, if it did, the fluid inclusionist recognizes the problem and refrains from measuring Th.
- The chemical composition and phase equilibria of the fluid inclusion must be largely known.
- There can be no irreversible phase or chemical changes to the inclusion fluid after it is entrapped.

Most of these requirements are relatively simple to evaluate through petrographic or other approaches (e.g., Goldstein and Reynolds 1994; Goldstein 1993, 2003). The most critical requirement for geothermometry in sedimentary systems, however, is that individual inclusions are closed systems, with volumes that have not changed since the time of entrapment. In the sedimentary realm, it is clear that fracturing and deformation can lead to violation of this requirement, leading to resetting of inclusions. More importantly, it appears that if fluid inclusions are heated beyond their entrapment temperature, there is potential for some reequilibration. Progressive heating during burial, and hydrothermal heating, are well known in the sedimentary realm, and therefore, thermal reequilibration must be evaluated carefully in each fluid inclusion study.

Thermal Reequilibration

When working with any group of fluid inclusions from the sedimentary realm, one must carefully evaluate the possibility that fluid inclusions have been altered through thermal reequilibration. Thermal reequilibration is the general term for alteration due to internal overpressure associated with heating above entrapment temperature. It includes stretching, which is permanent plastic deformation around a fluid inclusion in response to internal overpressure, and leakage and refilling, which is brittle deformation around an internally overpressured fluid inclusion followed by resealing. Both forms of thermal reequilibration are well known from the sedimentary realm (Goldstein 1986).

Natural stretching of fluid inclusions due to overheating can be caused by progressive burial and heating after a fluid inclusion is trapped, or by hydrothermal overheating. If an aqueous liquid fluid inclusion were trapped during normal burial conditions, its initial density would be the density of the fluid present at the time the inclusion formed (Fig. 4, point A). If the mineral and its inclusion were then heated further, the mineral enclosing the inclusion would essentially maintain a constant volume and, thus, so would the inclusion cavity. Because of this, a fluid inclusion trapped in a mineral is constrained to follow a constant density (isochoric) pressure–temperature (P-T) path. During an increase in temperature, that isochoric path shows that pressure would increase significantly in a fluid inclusion with only a small increment of burial or hydrothermal heating (Fig. 4, point B). If the mineral enclosing the inclusion were to relieve that increased pressure through permanent plastic expansion, the volume of the inclusion cavity would increase and the density of the inclusion fluid would decrease (Fig. 4, point C). A decrease in density would lead to an increase in the inclusion's Th value in the laboratory (Fig. 4, point D). Stretched aqueous fluid inclusions may yield homogenization temperatures that in some cases approach the temperature at which stretching occurred, but typically these Th values are significantly lower. Stretched fluid inclusions do not yield Th values above the temperature of reequilibration.

Natural leakage and refilling of fluid inclusions due to overheating also can be caused by progressive burial after a fluid inclusion is trapped, or by hydrothermal overheating. If an aqueous liquid fluid inclusion were formed, its initial density would be controlled by the density of the fluid present when the inclusion formed (Fig. 5, point A). If the mineral and inclusion were then heated further, pressure within the inclusion would be constrained to follow an isochoric P-T path. This would result in internal overpressure in the inclusion (Fig. 5, point B). If the mineral enclosing the inclusion were to relieve that pressure by fracturing, fluid could leak out of the inclusion and into the fracture. If the fracture were to intersect the mineral surface, the inclusion fluid would equilibrate, chemically and with respect to pressure, with the pore fluid present (Fig. 5, point C). Healing of the fracture would result in the original inclusion cavity now filled by a later fluid of a different density, yielding a higher Th than the original inclusion (Fig. 5, point D). Leaked-and-refilled fluid inclusions may yield homogenization temperatures that, in some cases, approach the temperature at which reequilibration occurred, but typically Th would be lower. Assuming no immiscibility during refilling, leaked-and-refilled fluid inclusions should not yield Th values above the temperature of reequilibration.

Whereas the strength of the enclosing mineral appears to be the primary control on whether or not a fluid inclusion thermally reequilibrates (Turgarinov and Vernadsky 1970), the inclusion size also appears to be important. In a given mineral, it appears that large fluid inclusions reequilibrate more easily, and to a greater extent, than small fluid inclusions (Leroy 1979, Swanenberg 1980, Bodnar and Bethke 1984, Goldstein 1986, Prezbindowski and Larese 1987, Bodnar et al. 1989). Inclusion shape also seems to be a control, with irregularly shaped inclusions with pointy protrusions more likely to reequilibrate than spherical fluid inclusions (Bodnar and Bethke 1984, Goldstein 1986, Bodnar et al. 1989, Prezbindowski and Tapp 1991).

If several aqueous fluid inclusions of variable size and shape were all trapped under the same conditions, and then cooled, there would be no thermal reequilibration because the inclusions had not been heated above their entrapment temperature. When heated in the laboratory to homogenization, each inclusion would yield essentially the same Th. If that same group of inclusions of variable size and shape were subjected to natural overheating, it is likely that some of the inclusions would survive with their original density intact without reequilibrating. Others of less favorable size and shape would reequilibrate either by stretching or leakage and refilling. With enough variation in size and shape, the various inclusions would likely reequilibrate to different degrees. This is supported theoretically and by experimental data (Fig. 6). Thus, when returned to the surface after natural overheating, each inclusion might have a different density, and inclusions likely would homogenize with a wide range of temperatures.

The tendency for some fluid inclusions to be altered by thermal reequilibration is the most important concern in fluid inclusion work in sedimentary rocks. Because of it, fluid inclusion workers must always evaluate the degree to which fluid inclusions have reequilibrated and the degree to which the original inclusions are preserved. This requires fluid inclusion work in sedimentary systems to have especially careful petrographic study.

Fluid Inclusion Petrography

Petrography is the single most important step in any fluid inclusion study because it allows the researcher to establish a paragenesis of fluid inclusion entrapment relative to other deformational, depositional, and diagenetic fabrics. It is also the only technique that allows a researcher to evaluate variability in Th data to determine if thermal reequilibration has occurred. The methodology for classifying the origin of fluid inclusions was summarized by Roedder (1979, 1984) and updated with currently available imaging approaches by Goldstein (2003). Fluid inclusions are classified as primary, secondary, pseudosecondary, and indeterminable (Fig. 7). Fluid inclusion origin must be determined petrographically using transmitted-light petrography, but this microscopy commonly is aided by backscattered electron imaging, optical cathodoluminescence, scanning electron microscope (SEM)–cathodo-

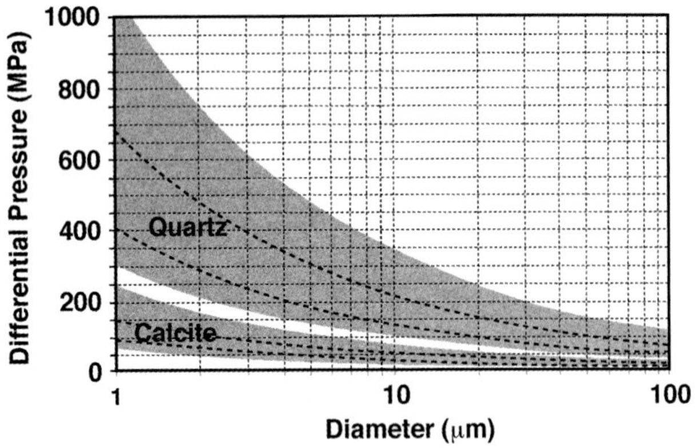

FIG. 6.—Differential pressure (internal inclusion pressure minus external pressure) in MPa (bars/10) required for decrepitation of fluid inclusions of various shapes and sizes in calcite and quartz. Base of each shaded area is the differential pressure at which the weakest shape modeled should decrepitate. Top of each shaded area is the differential pressure at which the strongest modeled inclusion shape should decrepitate. Dashed lines within each shaded area define the differential pressure at which decrepitation should occur for shapes of intermediate strength. No inclusions should decrepitate at conditions below each shaded area. All inclusions should decrepitate above each shaded field. Curves are defined using shape factors in Lacazette (1990) and illustrate the relationship between size and differential pressure of decrepitation for inclusions of a given shape within each mineral. Notice the important controls of both size and shape on whether a fluid inclusion decrepitates. Figure is modified from Lacazette (1990).

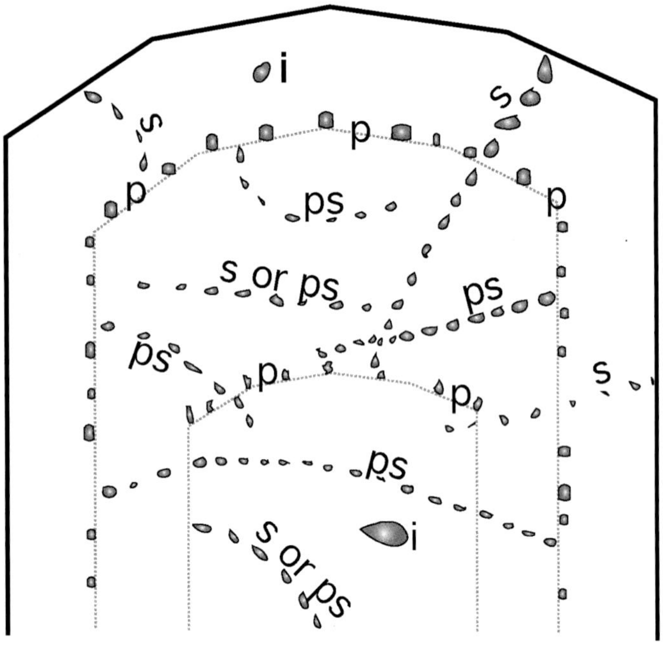

FIG. 7.—Composite sketch of primary (p), secondary (s), pseudosecondary (ps), and inclusions of indeterminable origin (i) in a hypothetical view within a single crystal. Note the complex history of fluid inclusion entrapment that can be determined from superpositional and crosscutting relationships. Figure is after Goldstein (2003).

luminescence, and energy dispersive/wavelength dispersive element mapping.

The term *primary* normally is used to refer to inclusions with petrographic evidence of entrapment during and as a direct result of crystal growth (Fig. 7). The applicable criterion, in identifying fluid inclusions as primary, is petrographic evidence for a relationship to growth of the crystal (see Goldstein [2003] for details). In sedimentary systems, many primary fluid inclusions are trapped during recrystallization of mineral phases (Fig. 8). *Secondary* fluid inclusions are trapped when fractures or deformation features heal and preserve petrographic evidence of entrapment after crystal growth is complete (Fig. 7). Crosscutting relationships can be used to put thermal events in sequence (Fig. 9). *Pseudosecondary* fluid inclusions preserve petrographic evidence of entrapment before crystal growth is complete, but not as a direct result of crystal growth. Typically, such inclusions are trapped along microfractures or other deformation features, and then the crystal is overgrown by further precipitation (Fig. 7). *Indeterminable* fluid inclusions are those for which the origin of the inclusion cannot be determined by petrographic means. With the tremendous opportunities for preservation of multiple fluid inclusion records of thermal history in each sample, it should be clear that the fluid inclusion geothermometer stands alone in the detailed history that can be gleaned from it.

Because the fourfold classification of inclusion origin bears no implications as to whether inclusions have leaked and refilled or undergone reequilibration, another approach must be brought to bear to determine how well fluid inclusions preserve geothermometric information. That approach requires the definition of fluid inclusion assemblages.

The Fluid Inclusion Assemblage (FIA)

Finely discriminated groups of petrographically associated fluid inclusions are termed *fluid inclusion assemblages* (Goldstein and Reynolds 1994). Each fluid inclusion assemblage (FIA) is defined on the basis of a petrographic association in which all inclusion vacuoles of each FIA represent the finest temporal resolution possible as to timing of initial closure of the inclusion cavities. Later reequilibration of the fluid in the inclusion cavity is not considered in this definition unless there is petrographic evidence for it. The FIA represents the fluid inclusionist's finest petrographic ability to identify those inclusions sealed at approximately the same time, even though one can never expect to be able to define truly instantaneous events of entrapment. Examples of FIAs are all those fluid inclusions trapped along a single healed microfracture or along a single growth zone (Fig. 10).

The FIA approach is essential for identifying alteration of fluid inclusions from necking down after a phase change, stretching, or leakage and refilling (Goldstein 1993, Goldstein and Reynolds 1994, Bodnar 2003a, 2003b). If all fluid inclusions in an FIA have about the same Th, and inclusions have variable sizes and shapes, it is unlikely that thermal reequilibration has occurred. Alternatively, if inclusions in an FIA yield variable Th, with 90% of the inclusions ranging more than 10–15° C, then thermal reequilibration may have affected some of the fluid inclusions.

Geothermometry using fluid inclusions in the sedimentary realm typically requires use of the FIA approach because of problems with thermal reequilibration and because of the complexity of fluid inclusion records preserved in diagenetic minerals.

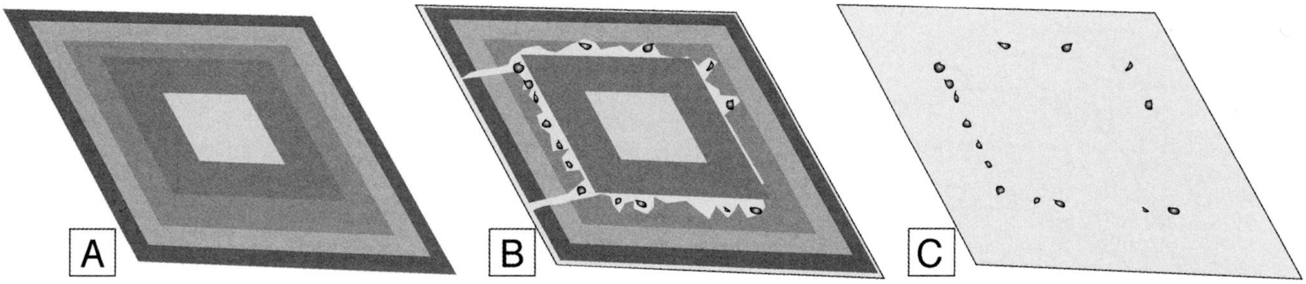

FIG. 8.—Mechanism for entrapment of primary fluid inclusions from internal recrystallization in a crystal of dolomite. Scale is microscopic view through a single crystal. **A)** Illustration of compositional growth zoning in a crystal as viewed in cathodoluminescence or backscattered electron imaging. **B)** Texture observable with cathodoluminescence or backscattered electron imaging after internal growth zone has been dissolved and refilled with later phase of mineral. Fluid inclusions are superimposed to illustrate relationship between recrystallized internal growth zone and entrapment of fluid inclusions. **C)** Same crystal as (B) but viewed in transmitted light, which does not effectively highlight zones of dissolution and reprecipitation. Note that fluid inclusions are concentrated in a particular internal growth zone and would be interpreted as primary. Using transmitted light petrography alone, the inclusionist would incorrectly identify the inclusions as trapped during growth of an early zone of the crystal, whereas in reality, the inclusions were trapped very late in the paragenesis during recrystallization of the early growth zone. Figure is after Goldstein (2003).

GEOTHERMOMETRY SCENARIOS

Fluid inclusions in geothermometry of sedimentary systems can be applied in a wide variety of environments and fluid systems (Fig. 1), ranging from determining paleotemperatures near the surface to determining high temperatures in the subsurface. Geothermometry requires a careful analysis of FIAs followed by measurement of Th. Thorough analysis of the data requires careful attention to the requirements for fluid inclusion geothermometry as outlined in the previous sections. The following discussion presents various scenarios of fluid inclusion FIAs that could be encountered in the sedimentary realm. All aspects of possible scenarios are not discussed; a more comprehensive discussion can be found in Goldstein and Reynolds (1994) and Goldstein (2001a). In each scenario, it is assumed that the inclusionist has evaluated any problems with necking down after a phase change, heterogeneous entrapment of multiple fluid phases, or reequilibration.

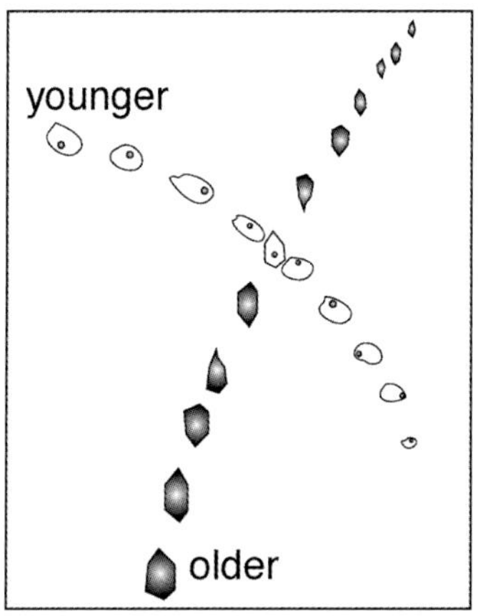

FIG. 9.—Methodology for using crosscutting relationships involving planar arrays of secondary fluid inclusions (e.g., Touret 1981; after Goldstein 2003).

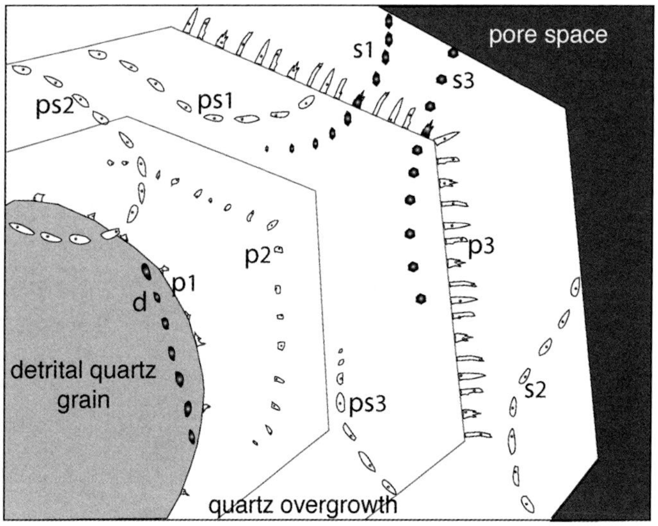

FIG. 10.—Ten fluid inclusion assemblages identifiable in this sketch of a quartz overgrowth on a detrital quartz grain illustrate the variety of geothermometric data that can be gathered for a single sample. The sample contains an FIA transported with the detrital quartz grain (d). That is followed by formation of two FIAs of primary fluid inclusions formed during growth of the quartz overgrowth (p1 and p2). After further precipitation of quartz overgrowth, there are one or more events of fracturing and healing, trapping three FIAs of pseudosecondary fluid inclusions (ps1, ps2, and ps3). After further growth of quartz, an FIA of primary fluid inclusions is formed (p3). Then, after quartz growth is complete, multiple events of fracturing and healing trap three FIAs of secondary fluid inclusions with at least two different compositions (s1, s2, and s3). Figure is modified from Goldstein (2003).

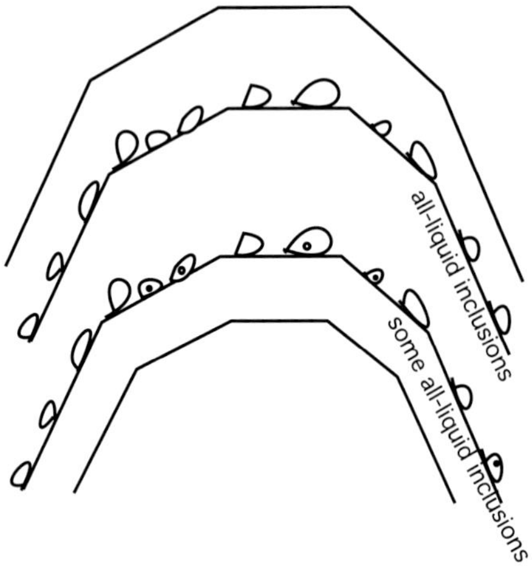

FIG. 11.—Diagram illustrating FIAs containing all-liquid aqueous fluid inclusions. The first (inner growth band) FIA contains all-liquid fluid inclusions alongside two-phase fluid inclusions with small bubbles. As large inclusions are also all-liquid in the FIA, it is unlikely that the all-liquid inclusions are high-temperature inclusions that have not nucleated a bubble. It is likely that they represent low temperature of entrapment. Two-phase inclusions are likely those that have thermally reequilibrated. In the second FIA, only all-liquid inclusions are present. If they are not significantly metastable, then they are typical of low temperature of entrapment.

Low Temperature, Near Surface

Fluid inclusions can be used to reconstruct paleotemperature in surface and near-surface environments. Typically, such inclusions remain entirely liquid at laboratory conditions (Fig. 11). After burial heating, it is common for some of these high-density inclusions to thermally reequilibrate. The presence of all liquid inclusions in an FIA, however, is evidence of a low temperature of entrapment, assuming that there is not significant metastability in nucleating a vapor bubble. If abundant all-liquid inclusions are present in an FIA, it typically suggests entrapment of the fluid inclusions below about 50° C (Goldstein 1993, Goldstein and Reynolds 1994), given that certain conditions are met. Inclusions trapped above about 50° C typically nucleate small bubbles upon cooling to surface temperature, given geologic time and sufficient size.

Even with high temperature of entrapment, some inclusions may not nucleate bubbles. It is well known that some fluid inclusions, especially small ones, can remain metastable over long periods of time at the surface, even though the inclusions should contain bubbles at laboratory temperature and should homogenize at temperatures even higher than 100° C. Because of this, fluid inclusionists always must attempt to nucleate bubbles in all-liquid inclusions by keeping the samples at low temperature as a means of evaluating if inclusions are significantly metastable (Goldstein and Reynolds 1994). Thus, the less-than-about-50° C interpretation always should be considered a very approximate boundary; it requires the presence of large fluid inclusions and geologic time under cool conditions to promote nucleation of a bubble.

If metastability is overcome and liquid inclusions are cooled below their entrapment temperature, they generate a tiny bubble. Reheating of the inclusion to measure Th reveals paleotemperature data that provide a direct measure useful in climate and low-temperature paleoenvironment studies (Goldstein 2001a, 2001b). Geologic materials useful for such analyses would include speleothems, evaporites, marine cements, and any near-surface diagenetic phase. Until recently, the preferred mechanism for generating bubbles in all-liquid fluid inclusions has been to cool the fluid inclusions in the laboratory until a small percentage of the inclusions generates bubbles (Roberts and Spencer 1995). These inclusions are then heated in the laboratory to measure Th, yielding direct measures of paleotemperature.

The most successful work has been done on halite, where abundant fluid inclusion Th data yield the water temperature of ancient systems (e.g., Lowenstein et al. 1998, Benison and Goldstein 1999, Satterfield et al. 2005). The temporal resolution detectable from these records is astounding. Cloudy-clear growth zones in halites tend to form rapidly, in some cases, on the time scale of day–night cycles. Fluid inclusion data are capable of resolving similar day–night temperature variation in ancient systems (Fig. 12; Benison and Goldstein 1999) as well as longer-term paleoclimate data.

Although analysis of geothermometry of such low-temperature samples has yielded excellent results, wide application has suffered because of difficulty in nucleating bubbles in all-liquid fluid inclusions. In many sample suites, cooling in the laboratory over a period of weeks has proven insufficient to overcome nucleation metastability. Recently, however, a new technique was developed to overcome the kinetic problems of nucleating bubbles (Gilg et al. 2007, Krüger et al. 2007) that has great potential for improving the application of low-temperature fluid inclusion geothermometry. The approach involves cooling all-liquid fluid inclusions to 4° C. The chilled inclusions are then hit with femtosecond laser pulses, which nucleate bubbles. The system uses an amplified Ti:sapphire laser (800 nm) with a nominal maximum output pulse energy of 4 μJ and a full-width at half-maximum pulse duration of around 330 fs. The laser is focused through a conventional 100×/0.80 long-working-distance objective. This treatment is reported to generate bubbles in all-liquid fluid inclusions reliably. Once a bubble has been generated, the sample can be heated on the stage to measure the homogenization temperature. Th values as low as 6.1° C have been measured with this technique, and it holds great potential for expanding the fluid inclusion geothermometer for routine use in paleoclimate and paleoenvironmental studies.

High-Temperature Aqueous Liquid

During deeper burial conditions, aqueous liquids can be trapped at high temperature. Such fluids tend to yield FIAs consisting mostly of liquids with small bubbles at room temperature (Fig. 13). The phase equilibria of homogenization in such systems depend on the salts in solution (Fig. 14A, B). In purely aqueous systems at room temperature, the bubbles consist of water vapor at low pressure. Irrespective of the salts in solution, the measured Th values represent minimum temperatures of entrapment.

To reconstruct actual entrapment temperature from Th, the composition of the fluid inclusion must be known. Concentration of salts in solution typically can be determined by measuring freezing point depressions in fluid inclusions on a microscope freezing stage (e.g., Goldstein and Reynolds 1994). More detailed compositional data can be extracted from other low-temperature phase behavior (see Goldstein and Reynolds 1994) and a variety of destructive and nondestructive analytical techniques (e.g., Anderson and Mayanovic 2003, Burruss 2003a, Gagnon et al. 2003, Gleeson 2003, Salvi and Williams-Jones 2003). The composition is used to model the pressure–volume–temperature (PVT) properties of the fluid (Bakker and Brown 2003, Bodnar 2003c), providing the appropriate vapor:liquid field boundary and isochore in PT space (an isochore is a curve in pressure–temperature space denoting a fixed molar volume or bulk density). Then, the pressure or thermobaric gradient at the time of entrapment must be assumed. Its intersection with the isochore yields the actual entrapment temperature (Fig. 14). This pressure correction of the

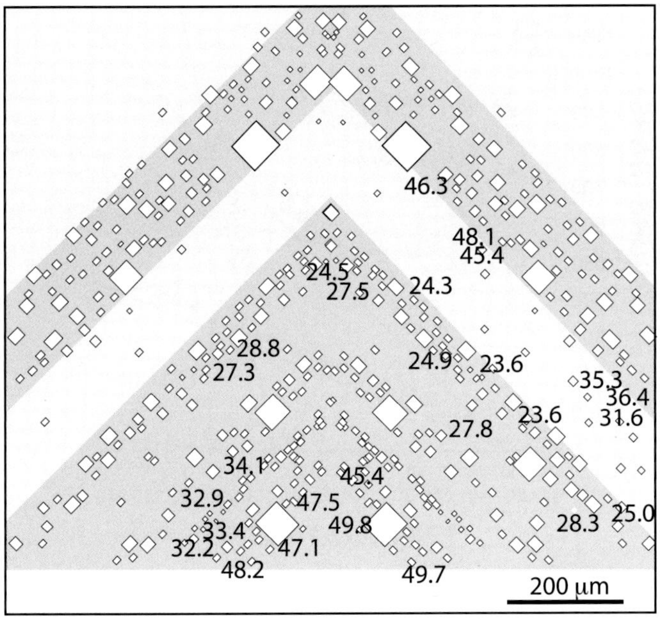

FIG. 12.—Illustration of distribution of fluid inclusions in cloudy–clear growth zones in halite from the Permian Nippewalla Group, Kansas. Spatial distribution of Th data is illustrated. Note that Th values rise and fall as measurements are followed outward through successive growth zones. These values may be showing day–night temperature variation or temperature variation at some other short time scale in an evaporitic lake. Figure is modified from Benison and Goldstein (1999).

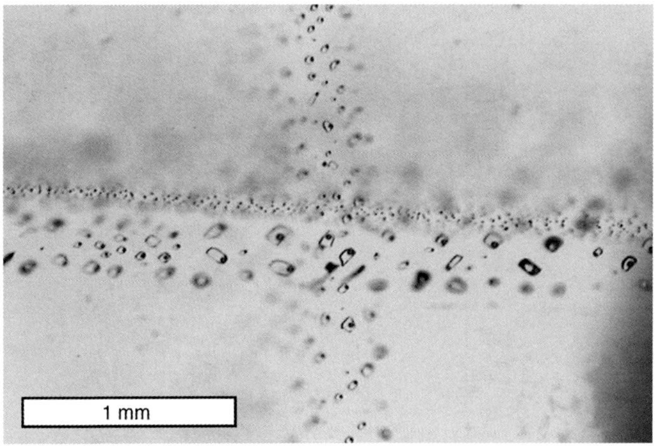

FIG. 13.—Aqueous inclusions with small bubbles and consistent vapor:liquid ratio. When these inclusions are heated in the laboratory, they yield consistent Th values. Several planar arrays of secondary fluid inclusions have formed in healed fractures. Fluorite, southern Illinois. Figure is after Goldstein (2003).

homogenization temperature can be large in aqueous systems lacking dissolved gas (see Fig. 14 for an example).

High-Temperature Aqueous Liquid and Dissolved Gas

During burial conditions, aqueous liquids with dissolved gases can be trapped at high temperatures, in systems in which the gas component (commonly methane, carbon dioxide, or other component) is undersaturated with respect to the aqueous solution. Such fluids yield FIAs consisting mostly of liquid with small bubbles. In transmitted light microscopy, they appear identical to the simple high-temperature aqueous liquid inclusions (Fig. 13). The bubbles consist of gas under high pressure, rather than low-pressure water vapor. The phase equilibria of homogenization in such systems depend on the salts in solution, the composition of the dissolved gases, and their amounts. Irrespective of the specific fluid composition, the measured Th values represent minimum entrapment temperatures.

To reconstruct actual entrapment temperature from Th, the composition of the fluid inclusion must be known. That composition is used to model the PVT properties of the fluid, providing the appropriate bubble point curve (solubility curve) and isochore in pressure–temperature space. Then, the pressure or thermobaric gradient at the time of entrapment must be assumed. Its intersection with the isochore yields the actual entrapment temperature (Fig. 15). This pressure correction of the homogenization temperature is typically quite small in aqueous systems containing dissolved gas (see Fig. 15 for an example).

High-Temperature Aqueous Liquid and Immiscible Gas

During burial conditions, aqueous liquids with dissolved gases can be trapped at high temperatures, in systems in which the gas phase is saturated with respect to the aqueous solution. Such fluids yield FIAs consisting of liquid with small bubbles, but they also trap inclusions with either all gas or mixtures of gas and liquid (Fig. 16). In such systems, the bubbles consist of gas under high pressure. The liquid-rich end members in FIAs are those fluids trapped with just the aqueous liquid and no gas phase; measurement of their homogenization temperatures is equal to the entrapment temperature. No pressure correction is required (Fig. 17).

Gas Fluid Inclusions

Fluid inclusions consisting of natural gas are common in sedimentary systems. These inclusions typically contain complex mixtures of gases with large amounts of methane and carbon dioxide. The methane-rich inclusions typically appear as a single dark fluid phase at room temperature (Fig. 18). All fluid inclusions in this system are above their homogenization temperature at laboratory temperature. By using a cooling stage, inclusions can be cooled until they separate into two fluid phases; they can then be warmed to measure homogenization temperature. Homogenization temperatures in such systems are far below the actual entrapment temperature. A huge pressure correction is required to obtain the actual entrapment temperature.

If fluid composition is known, the PVT properties can be modeled to construct a vapor:liquid field boundary and isochore for the inclusions in the FIA (Fig. 19). The conditions of fluid inclusion entrapment would lie somewhere along the isochore. One way of determining the actual entrapment temperature requires assuming the pressure at the time of entrapment and evaluating its intersection with the isochore.

Petroleum Fluid Inclusions

The temperature of entrapment of petroleum fluid inclusions has received significant attention in recent years as a means of documenting the timing, temperature, and pressure during petroleum migration and accumulation. Petroleum fluid inclusions are common in

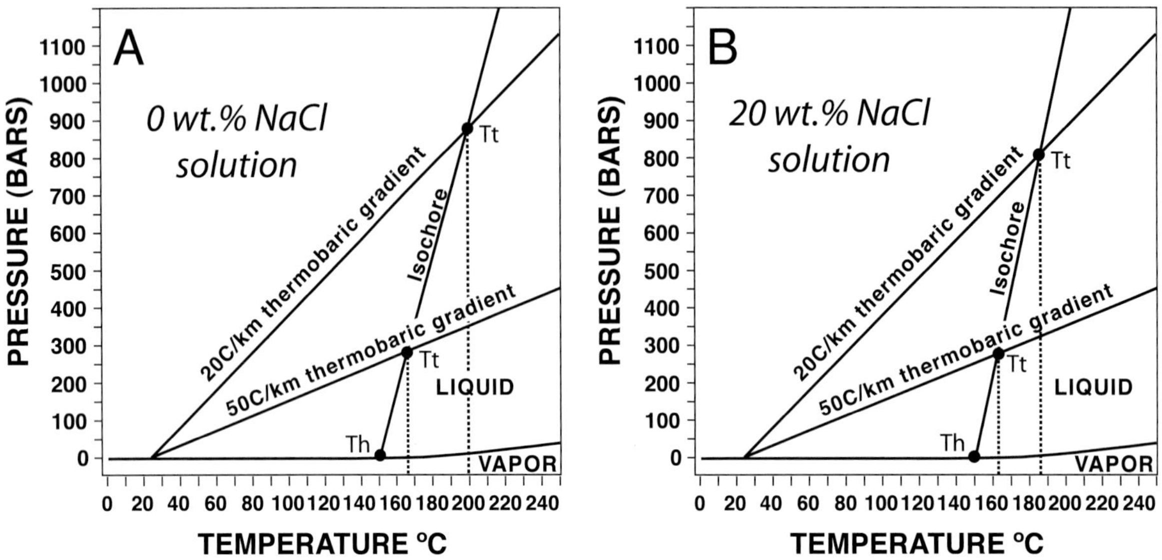

FIG. 14.—Phase diagrams for **A)** pure water system and **B)** 20 wt% NaCl solution, each with an isochore drawn to intersect the liquid–vapor field boundary at 150° C. Note that the salt in solution has an effect on the slope of the isochore. If inclusions in an FIA homogenize at 150° C, homogenization takes place at a low pressure on the liquid–vapor field boundary at 150° C (Th). This value is well below the actual entrapment temperature. The Th can be pressure-corrected to the actual entrapment temperature by drawing the isochore for the appropriate composition and Th, and assuming a thermobaric gradient or pressure of entrapment. **A)** For an inclusion containing pure water and homogenizing at 150° C, the inclusion would have been entrapped along the isochore shown. If one were to apply a pressure correction using a thermobaric gradient of 20° C/100 bars, one would interpret entrapment temperature to be just below 200° C on the basis of the intersection between the isochore and thermobaric gradient (Tt). If one were to apply a pressure correction using a thermobaric gradient during inclusion entrapment of 50° C/ 100 bars, one would interpret an entrapment temperature just below 170° C on the basis of the intersection between the isochore and thermobaric gradient (Tt). **B)** For an inclusion containing a 20 wt% NaCl solution and homogenizing at 150° C, the inclusion would have been entrapped along the isochore shown. If one were to apply a pressure correction using a thermobaric gradient of 20° C/100 bars, one would interpret entrapment temperature to be just below 190° C on the basis of the intersection between the isochore and thermobaric gradient (Tt). If one were to apply a pressure correction using a thermobaric gradient during inclusion entrapment of 50° C/100 bars, one would interpret an entrapment temperature just above 160° C on the basis of the intersection between the isochore and thermobaric gradient (Tt).

sedimentary systems and are easily identifiable by their fluorescence under ultraviolet (UV) illumination (Burruss 1981, 1991, 2003b). The most common petroleum fluid inclusions consist of a liquid phase with a small gas bubble. When the inclusions are heated in the laboratory, they homogenize by disappearance of the bubble. Typically, Th of petroleum fluid inclusions is lower than Th of coeval entrapped aqueous inclusions. The explanation for this is that most petroleum fluid inclusion Th values require large corrections for pressure compared to aqueous inclusions.

Challenges: Significant efforts have been devoted to correcting petroleum fluid inclusion Th for pressure in order to obtain trapping temperature. Although results have been highly encouraging, there are two important aspects of this endeavor that make it challenging. First, there are a number of assumptions specific to petroleum inclusion systems that must be evaluated. These assumptions are in addition to the typical requirements for fluid inclusion geothermometry presented in the foregoing sections, and they are discussed briefly next. Second, because petroleum is a chemically complex fluid, modeling the PVT properties of the inclusions to make the pressure correction is not trivial.

Some petroleum fluid inclusions contain bitumen-like solids that do not dissolve upon heating. If these solids formed from irreversible chemical changes in fluid inclusions after entrapment, then such inclusions should not be used for geothermometry. It also remains possible that those petroleum inclusions lacking such solid phases may have undergone similar, although potentially undetectable, irreversible chemical changes. Although many petroleum fluid inclusion studies seem to indicate that there is no general problem with irreversible chemical changes to the included fluid, the extent of the problem is still unknown.

Many petroleum inclusions have an obvious to barely visible rim of aqueous liquid that was trapped heterogeneously along with liquid oil at the time of entrapment. Entrapment of a separate water phase is reasonable because most oil fluid inclusions are probably trapped in water-wet rather than oil-wet systems. If the amount of trapped water varies from inclusion to inclusion, then there will be an effect on Th, potentially making the temperature of disappearance of the bubble difficult to interpret.

Pressure Correction and Pressure–Volume–Temperature–Composition (PVTX): Interpretation of petroleum fluid inclusion Th, by disappearance of the bubble, should be possible if the PVT relations of the complex petroleum system can be determined. Observed Th values are partial homogenization temperatures, however, because homogenization of a thin rim of aqueous liquid in such inclusions cannot be resolved. Interpretation of the observed Th assumes that the inclusion: (1) has not trapped a separate water phase at the time of inclusion entrapment, (2) has not undergone irreversible chemical changes, and (3) has not reequilibrated.

PVT relations for the inclusion fluid can be modeled by several means (Karlsen et al. 1993; Pironon et al. 1998; Aplin et al. 1999; Thiéry et al. 2000, 2002; Tseng and Pottorf 2002; Burruss 2003b), with

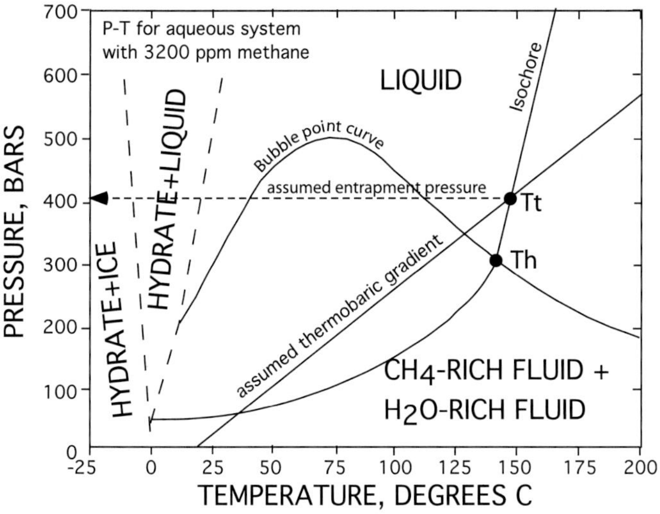

FIG. 15.—Pressure–temperature (P-T) phase diagram for H_2O-CH_4 system with 3200 ppm CH_4. Homogenization temperature for the FIA is 140° C. Homogenization is at a high pressure, where the isochore for the inclusions intersect (Th) the bubble point curve (solubility curve). Actual entrapment took place at a higher temperature and pressure in the liquid field. If a thermobaric gradient or entrapment pressure is assumed, then the actual entrapment temperature can be interpreted from the intersection with the isochore (Tt). In this case, entrapment temperature is just below 150° C. This small pressure correction is typical for inclusions containing large amounts of dissolved gas. Figure is modified from Hanor (1980).

the ultimate goal of constructing bubble point curves and isochores for the fluid inclusion Th values measured. The accuracy of these models is still in need of further testing. For some applications, the PVT relations for the petroleum in the inclusion can be assumed to be similar to a produced petroleum, for which PVT properties are known.

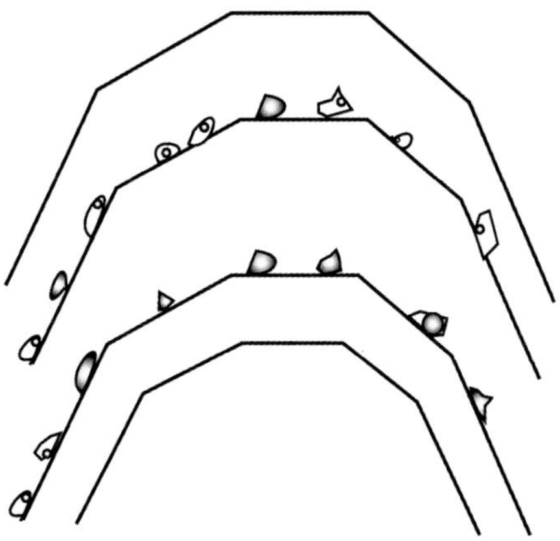

FIG. 16.—Typical petrographic associations of FIAs in a high-temperature aqueous system with immiscible gas. Two FIAs are represented. Notice the variability in ratio of liquid-to-gas among inclusions. Some inclusions are dark and appear to be all gas. Others have a large amount of aqueous liquid and a small gas bubble.

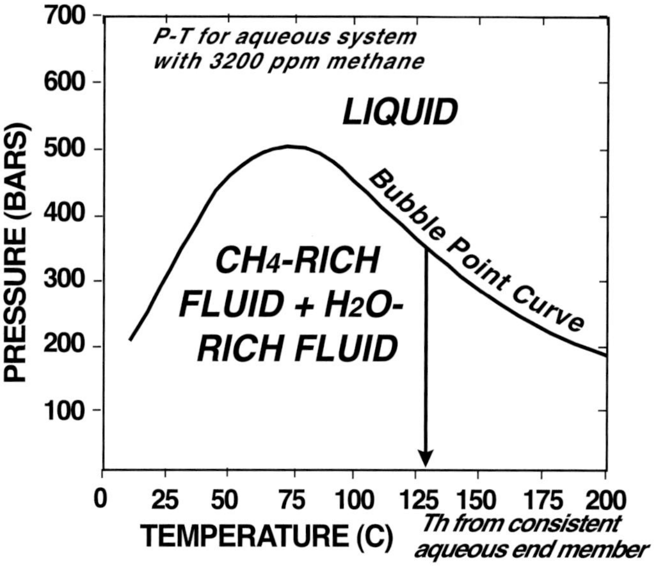

FIG. 17.—Pressure–temperature (P-T) phase diagram for H_2O-CH_4 system with 3200 ppm CH_4. If there is petrographic evidence of immiscibility, and if a consistent aqueous-rich end-member FIA with 3200 ppm CH_4 yields consistent Th values of about 130° C, the temperature of fluid inclusion entrapment would have been 130° C. No pressure correction would be necessary. Figure is modified from Hanor (1980).

In other cases, these PVT relations can be modeled from the fluid inclusions themselves. Fluids in inclusions can be extracted and analyzed chemically, or otherwise characterized in situ, to model the PVT relations.

As one of several alternative examples, the PIT approach can be applied. PIT refers to *petroleum inclusion thermodynamic*, used for modeling PVTX relations of petroleum in inclusions (Thiéry et al. 2000, 2002). In this and other approaches for modeling petroleum inclusion systems, it is acknowledged that small amounts of water would be dissolved in the petroleum-rich fluid at the time of

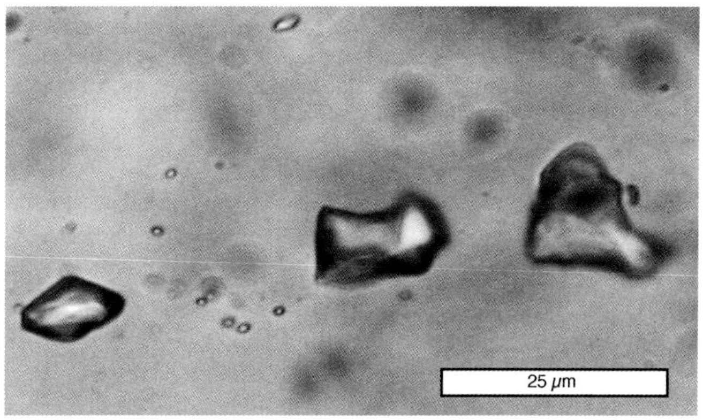

FIG. 18.—Transmitted light photomicrograph of three dark fluid inclusions in quartz, all filled with a single "gas" phase. Gas inclusions commonly (but not in all cases) have a dark appearance in transmitted light.

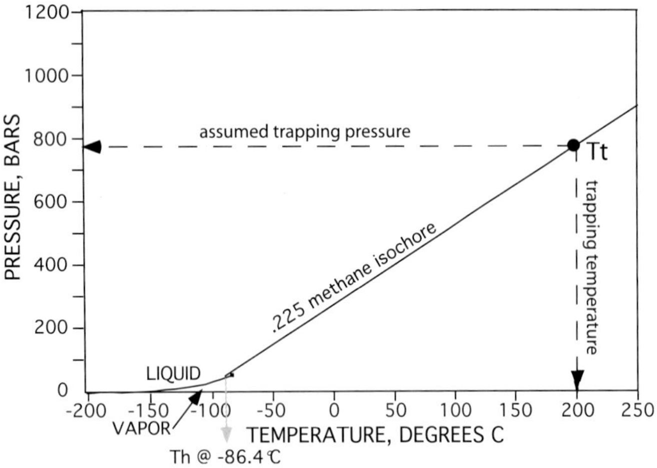

FIG. 19.—Pressure–temperature (P-T) phase diagram for pure CH_4. If a methane FIA homogenizes to liquid at −86.4° C, then the inclusions must have been entrapped along the .225 methane isochore. If the pressure at the time of entrapment is assumed to have been 775 bars, then the temperature of entrapment would be interpreted as 200° C, where the assumed pressure intersects the isochore (Tt).

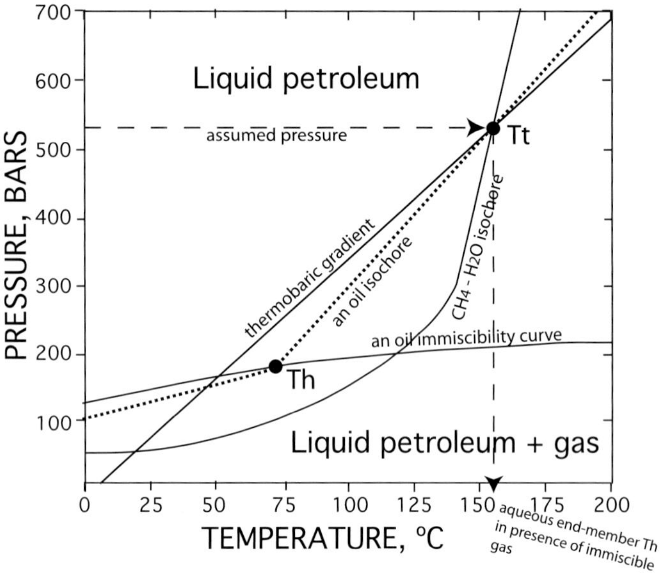

FIG. 20.—Pressure–temperature (P-T) phase diagram illustrating determination of trapping temperature from petroleum fluid inclusion Th. Immiscibility curve is for a chosen composition of petroleum. Below the curve, both oil and gas coexist in the fluid inclusion. Above the curve, only a single liquid phase is present. In this example, oil inclusions are observed to homogenize at about 75° C (Th). The inclusion follows the oil isochore (dotted line) into the one-phase field. Inclusion trapping conditions were somewhere along that isochore. Intersection of an assumed thermobaric gradient with the isochore (Tt) can yield the trapping temperature, in this case 155° C. Alternatively, a pressure can be assumed at the time of trapping. Its intersection with the inclusion's isochore yields the trapping temperature. Or, if aqueous fluid inclusions are trapped along with the oil inclusions in the same FIA, an isochore can be calculated for the aqueous inclusions given known composition and measured Th. In this case, the isochore is drawn for an aqueous inclusion containing 3200 ppm methane that homogenized at approximately 140° C. Its intersection with the oil isochore yields the trapping temperature. Or, if both aqueous inclusions and immiscible gas were trapped in the same FIA with the oil inclusion, the end-member aqueous inclusion (yielding the lowest Th of all liquid-rich water–gas inclusions) would yield the trapping temperature.

entrapment. This aqueous liquid would come out of solution upon cooling and would generate a thin rim of aqueous liquid. It would be difficult or impossible to observe the homogenization of this aqueous liquid into the oil in the inclusion. Thus, it is typically assumed that this water phase has little effect on the interpretation of the phase equilibrium for temperature of bubble disappearance (Pironon et al. 2000). The PIT approach uses confocal laser scanning microscopy to evaluate volumes of oil and gas phases at various temperatures to obtain volumetric properties. Compositional fields can be constrained by techniques such as Fourier transform infrared spectroscopy (FTIR) on the oil inclusions and related to a two-parameter compositional model that describes a wide range of petroleum compositions.

If consistent Th values are produced from an FIA of petroleum inclusions, and the bubble point curve and isochore are accurately modeled, the Th can be corrected to trapping temperature in one of four ways (Fig. 20): (1) Intersection of an assumed thermobaric gradient with the oil isochore can yield the trapping temperature; (2) intersection of an assumed pressure with the isochore would yield trapping temperature; (3) if aqueous fluid inclusions are trapped along with the oil inclusions in the same FIA, an isochore can be calculated for the aqueous inclusions if their composition is known and if they yield consistent Th values—that isochore's intersection with the oil isochore yields the trapping temperature; and (4) in situations in which both aqueous inclusions and immiscible gas were trapped in the same FIA with the oil inclusion, the end-member aqueous inclusion (yielding the lowest Th of all aqueous liquid-gas inclusions) would yield the trapping temperature. No pressure correction is needed for these aqueous inclusions.

DETERMINING MAXIMUM OR PEAK TEMPERATURE

In sedimentary systems, fluid inclusions are one of the many techniques useful in determining the maximum or peak temperature experienced by the material. Compared to other techniques in geothermometry, there are several advantages to fluid inclusions in the analysis of peak temperatures. For example, fluid inclusions can record instantaneous events, and no temporal assumptions are necessary. The entrapment of FIAs also can be related to mineral paragenesis, making it possible to place fluid inclusion maximum paleotemperatures in the context of a particular phase of deformation or mineral precipitation.

Highest Th FIA

Given the many FIAs typically preserved in a given system, it is likely that some will be trapped at the maximum temperature. The highest consistent aqueous Th FIA in many systems turns out to have been trapped at or near the maximum temperature (Goldstein and Reynolds 1994). After evaluating the fluid inclusion composition and determining a pressure correction, this FIA can be used as an estimate of maximum temperature, but there will always be some risk that FIAs were not trapped at the maximum temperature. Because of this possibility, the maximum pressure-corrected Th from a consistent FIA should be considered a minimum estimate of the maximum temperature. In other words, the maximum temperature experienced by the rock was at least as high as the pressure-corrected FIA values.

Statistical Approaches

Over the years, there have been statistical calibrations of fluid inclusion Th to the maximum temperature experienced in a sedimentary basin or geothermal system. Most of these approaches have concentrated on using fluid inclusions in soft minerals like calcite. Although fluid inclusionists admit that we can do better than this approach, there are cases in which it is useful, especially where the careful FIA approach is not possible; e.g.; where time constraints or lack of training do not permit a detailed fluid inclusion study. All researchers would admit that, for a statistical correlation between Th values and maximum temperature, there is a probability that measured Th values are accurately calibrated to maximum temperature, and a corresponding statistical probability that they are not. This is the greatest weakness in all approaches of this sort.

Barker and Goldstein (1990) published an extensive data set from fluid inclusions in calcite, concentrating on subsurface oil field and geothermal systems at peak temperature. A remarkably good correlation was obtained between the uppermost mode in the Th data (Th mode) and the present peak temperature (Tpeak; Fig. 21). This correlation appeared despite the fact that some of the Th data came from calcite formed at low temperature. The logic to explain this was that some fluid inclusions in soft minerals, like calcite, are expected to thermally reequilibrate during natural heating. It was found that modal values in populations of Th data provided a better correlation to Tpeak than the mean values (Tpeak = 12.1 + 0.965Th mode; r = 0.96). Th mode is the single mode in unimodal populations of Th data, or the highest mode in polymodal populations of Th data. Most of the Th mode values were below the present peak temperature; this was expected, considering the need for pressure corrections and the overpressure necessary for thermal reequilibration. A small proportion of the Th mode data was above the present peak temperature; this was interpreted to result from expected errors in correcting subsurface data to actual temperature, as well as the possibility that the wells had been hotter than the temperatures measured downhole in the wells.

More recently, Tobin and Claxton (2000) published a new data set and new correlation between fluid inclusion Th data in calcite and interpreted maximum temperature. Their correlation with peak temperature was established only roughly, however. Peak temperature was calculated using the EASY%R_o function in BasinMod and assuming a constant 100-million-year heating duration for all of the systems studied (Fig. 22). They concluded that, as others had shown previously (Barker and Goldstein 1990, 1991; Bodnar and Burruss 1991; Goldstein and Reynolds 1994), fluid inclusion Th data should not be regarded as *equivalent* to peak temperature, except in the case where inclusions can be shown to have been trapped at peak temperature and where immiscibility in the system precluded the need for a pressure correction.

Fluid inclusion Th data compare favorably with other measures of maximum temperature. For example, correlations between fluid inclusion Th data in calcite have been correlated to vitrinite reflectance data. Barker and Goldstein (1990) and Barker and Pawlewicz (1994) published useful correlations incorporating burial and geothermal systems. Tobin and Claxton (2000) published a revised correlation using an internally consistent data set and high-grading the data set, by only using data from burial systems and avoiding samples from geothermal areas or areas with intrusions. They attempted to make the data worldwide in scope and from various ages, and they instituted criteria for acceptance or rejection of fluid inclusion data deemed inappropriate to the method. The high-graded fluid inclusion data in calcite are highly correlated to vitrinite reflectance values (Fig. 23).

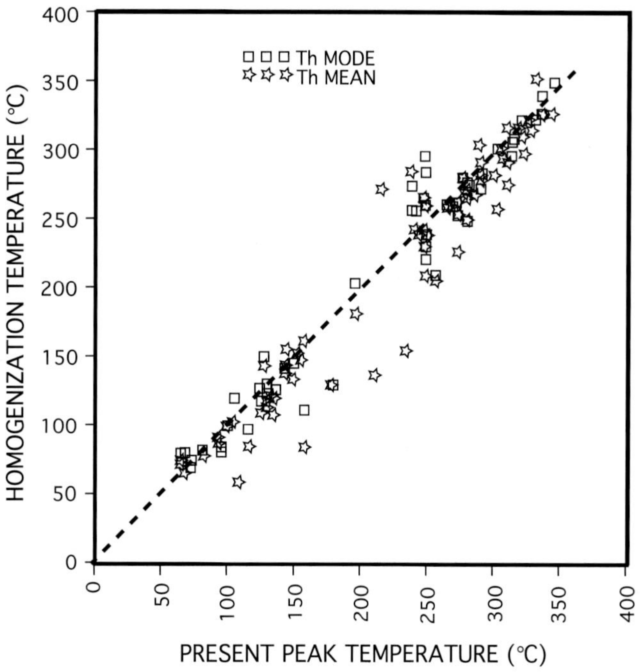

FIG. 21.—Plot of aqueous Th data from calcite (means and upper mode) in comparison to present peak temperature in the subsurface. Dashed line indicates what would be a 1:1 correlation. As expected, most fluid inclusion Th data lie below present peak temperature. The best correlation to present peak temperature is the uppermost mode in the data. These data have not been pressure corrected or studied carefully using the FIA technique. Although a correlation exists, that correlation does not mean that lower-temperature FIAs are not preserved. On the contrary, in many of the included data sets, lower-temperature FIAs are well preserved. As this approach to interpreting peak temperature is a statistical one, there is a real possibility for significant error (Barker and Goldstein 1990). Researchers should only resort to this technique when the FIA approach is not possible and detailed analysis of PVT relations cannot be accomplished. Figure is modified from Barker and Goldstein (1990).

DISTINGUISHING BETWEEN HYDROTHERMAL AND OTHER PALEOTEMPERATURE REGIMES

Commonly, subsurface paleotemperatures are associated with normal burial thermal regimes, where conductive heat flow rather than fluid movement controls temperatures. If vertical fluid movement is rapid enough, however, subsurface temperatures also are modified by warmer fluids flowing into cooler rocks or cooler fluids flowing into warmer rocks (Fig. 1). Each of these three scenarios should be evaluated in any study of paleotemperature in a sedimentary system.

Normal burial thermal regimes, with relatively stagnant fluid conditions, are typically the ones modeled in sedimentary basins. Researchers now, however, increasingly recognize the importance of hydrothermal systems where warm fluids flow into cooler rocks. There is an unfortunately wide variety of definitions used for the term

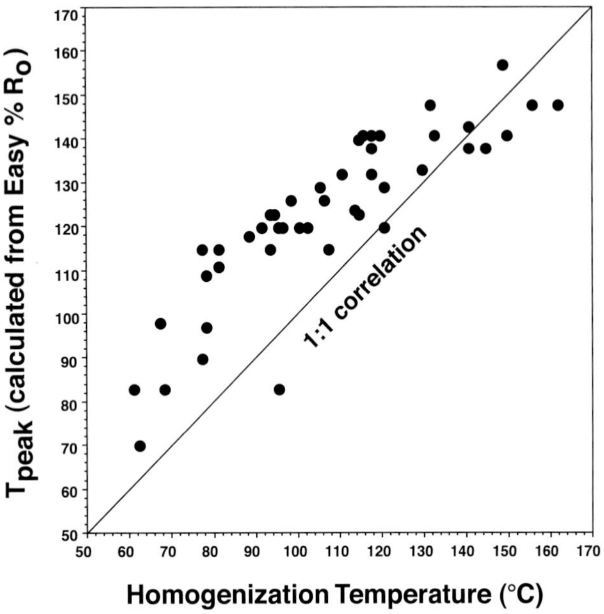

FIG. 22.—Tobin and Claxton's (2000) cross plot of median Th from calcite compared to peak temperature roughly calculated from the EASY%R_o function in BasinMod. These rough interpretations were not meant to illustrate a calibration between median Th and peak temperature by Tobin and Claxton. Figure is modified from Tobin and Claxton (2000).

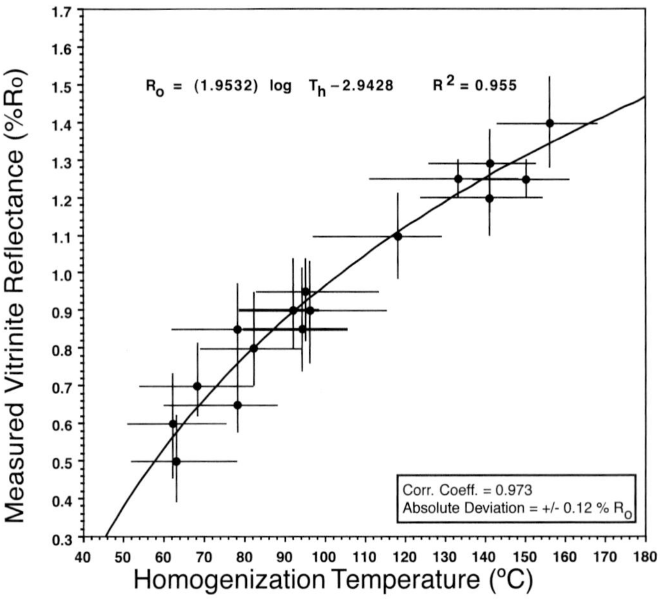

FIG. 23.—Cross plot of vitrinite reflectance data (%R_o) and median fluid inclusion Th from high-graded calcite samples. The logarithmic best fit yields an R^2 of 0.96. Figure is modified from Tobin and Claxton (2000).

hydrothermal (see discussion in Machel and Lonnee 2002), but for the purposes of this paper, if the flow of warm fluids into cooler rocks affects the temperature regime significantly, the system will be considered hydrothermal (Stearns et al. 1935, White 1957).

Hydrothermal alteration is clearly important in the localization of sandstone and carbonate oil and gas reservoirs as well as Mississippi Valley-Type (MVT) ore deposits (e.g., Leach and Sangster 1993, Wojcik et al. 1997, Rossi et al. 2002, Cantrell et al. 2004, Davies and Smith 2006, Smith and Davies 2006). Fluid inclusions and other geothermometers are essential in recognizing ancient hydrothermal fluid flow.

Fluid inclusions are also useful in recognizing where cooler fluids flow into warmer rocks (termed hydrofrigid by Machel and Lonnee 2002). An example of a "hydrofrigid" system is one in which brines sink downward into sedimentary basins by density inversion or reflux. Reflux diagenesis is important in its effect on sandstone or carbonate oil and gas reservoir rocks (e.g., Prather 1992, Ceriani et al. 2002, Luczaj and Goldstein 2000, Jones et al. 2003). Fluid inclusions are one of the few tools capable of distinguishing among these three thermal regimes in sedimentary basins.

Geothermometers, such as fluid inclusions, are useful in identifying when and where the normal burial system has been perturbed by hydrothermal fluid flow and flow of cool waters into warmer rocks. Hydrothermal systems have been of great interest of late because of recent discoveries in the Trenton–Black River reservoirs (e.g., Davies and Smith 2006).

Hydrothermal systems typically require a mechanism to move fluids upward from deeper parts of the basin, a setting in which warmer fluids exist and are capable of being transmitted into cooler rocks, and conduits such as fractures or permeable horizons that provide a focus for rapid fluid flow. There has been much discussion of the criteria necessary to identify ancient hydrothermal heating (e.g., Machel and Lonnee 2002, Esteban and Taberner 2003, Davies and Smith 2006). It has been suggested that a paleotemperature measure higher than predicted from burial history modeling is an appropriate indication. On the other hand, it must be pointed out that the many assumptions of such modeling and the nature of the thermal maturity data typically available (i.e., vitrinite reflectance, pyrolysis data from Rockeval) normally make it impossible to unequivocally distinguish between a hydrothermal system, higher heat flow than modeled, or deeper burial than interpreted. Here, I present additional criteria incorporating fluid inclusions and other geothermometers.

Fluctuating Paleotemperature

Fluid inclusions may provide very strong evidence, in the form of Th data that indicate repeated increases and decreases in temperature, that are not consistent with the burial and unroofing history. This typically results from fault- or fracture-controlled episodic injection of warm fluids followed by cooling (see an example in Fig. 24; e.g., Rossi et al. 2002). Fluid inclusion Th values in such systems typically show repeated rises and falls during precipitation of diagenetic phases. Similar rises and falls could also be observed in systems where cool waters are injected into warm rocks ("hydrofrigid"). Criteria described next can be used to differentiate between hydrothermal and hydrofrigid systems.

Geothermometers Higher than Possible from Burial History Models

In ancient hydrothermal systems, it is common to find that fluid inclusion Th and pressure-corrected Th values are higher than expected from burial history analysis. In cases in which this pressure-corrected Th value exceeds the maximum possible paleotemperature from basin modeling, based on highest possible geothermal gradient and the deepest possible burial, this is a useful criterion for identifying hydrothermal systems (Fig. 25).

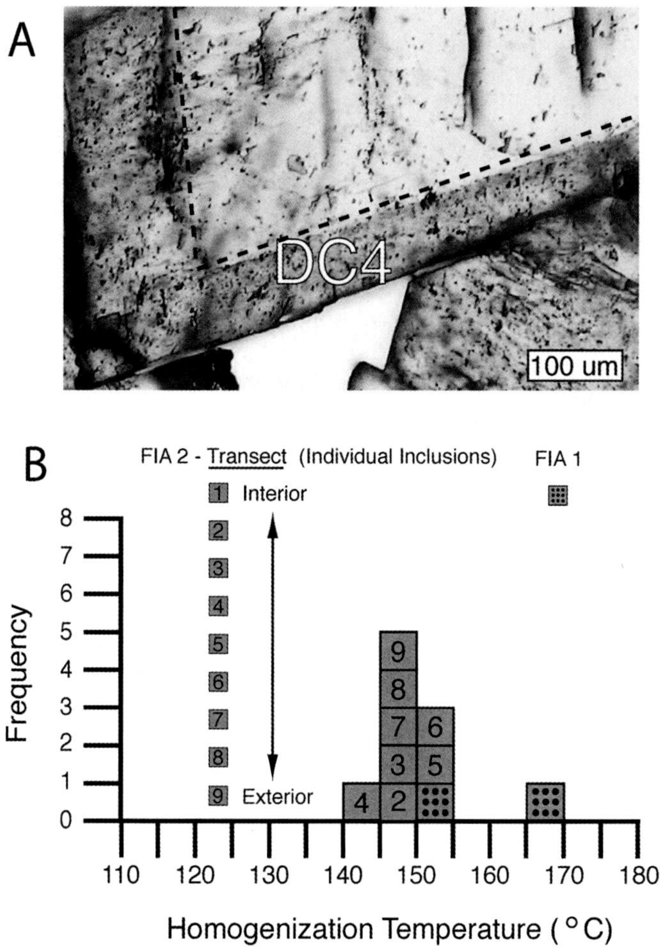

FIG. 24.—Fluid inclusion Th data from two FIAs in a single sample of dolomite from the Indian Basin field of New Mexico (Hiemstra and Goldstein 2004, 2005). **A)** Photomicrograph of dolomite, with cathodoluminescence cement stratigraphy zone DC4 labeled. Notice primary fluid inclusions in zone beyond dashed line. **B)** Frequency histogram illustrating that as Th measurements progress outward in the crystal from FIA1 to FIA2, there is a temperature drop. In FIA2, values on histogram indicate a fall, rise, and fall in Th from a transect going progressively outward. Such fluid inclusion evidence for rises and falls in temperature is consistent with fluid flow perturbing temperature associated with episodic injection of fluids of a temperature different from the normal burial temperature.

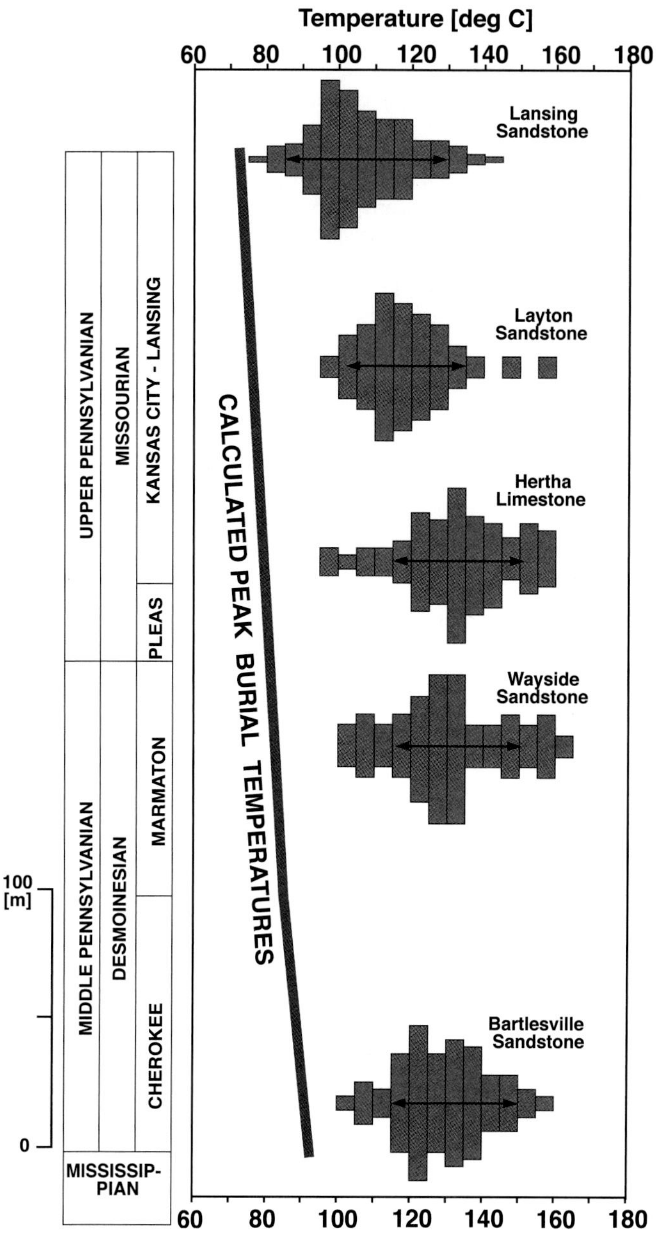

FIG. 25.—Fluid inclusion Th histograms from time-equivalent ankerite in sandstone and Fe-dolomite in limestone compared to maximum possible burial temperatures from southeastern Kansas. Double arrows within histograms illustrate data with consistent Th in FIAs. Data are plotted relative to depth, and stratigraphic units are noted. Calculated peak temperatures are based on high geothermal gradients and maximum burial depth possible for this region. Note that fluid inclusion Th values are well above the maximum possible burial temperature, and this is consistent with a hydrothermal system. Also note that the fluid inclusions are useful in reconstructing the paleogeothermal gradient. Note that the lower three stratigraphic horizons show no geothermal gradient, and the upper three indicate a steep gradient. These results are as expected in a system dominated by hydrothermal fluid flow. Figure is modified from Wojcik et al. (1994).

Gradients and Pressure Data Inconsistent with Normal Thermal Regime

Maximum temperature calculations from a burial history analysis represent interpretations requiring many assumptions, and it is always possible that the assumptions of the burial history modeling are in error. Thus, paleotemperatures from burial history modeling must be looked upon with healthy skepticism. On the other hand, the assumptions in the burial history analysis can be evaluated.

Fluid inclusions and other geothermometers can be used to help in evaluating the assumptions of the burial history model. Primary fluid inclusions trapped in a time-equivalent diagenetic phase, or trapped in association with other time-equivalent geologic features can be used to reconstruct a geothermal gradient to evaluate assumptions in the burial history model (Fig. 25). Other geothermometers also may be useful in reconstructing the paleogeothermal gradient.

Either exceptionally high or exceptionally low paleogeothermal gradients, inconsistent with heat flow in a stagnant system, are strong indications of fluid flow in hydrothermal or hydrofrigid systems. The location and distribution of these anomalies, relative to hypothesized conduits, are the keys in identifying the distribution of subsurface fluid flow and whether the system was hydrothermal or hydrofrigid (see Fig. 25).

Fluid inclusions in sedimentary rocks are used to reconstruct pressure as well as temperature (Goldstein and Reynolds 1994). Because pressure can be equated to depth, this allows for further evaluation of the burial history model assumptions, providing both depth and temperature from the same datum. In addition, it can be used to distinguish normal ancient thermal regimes from hydrothermal or hydrofrigid systems. If paleotemperature is too high for a given paleodepth, a hydrothermal system is indicated. If paleotemperature is too low for a given paleodepth, a hydrofrigid system is indicated.

Variation in Paleotemperature at Same Depth

In a hydrothermal system, warm fluids commonly are focused by faults and fractures on a local or regional scale. This results in lateral variation in geothermometers at the same depth, given the same stratigraphy. Such evidence for warm spots, with increased thermal maturity not associated with intrusives or stratigraphic controls, normally is considered good evidence for hydrothermal fluid flow (Fig. 26).

Higher Paleotemperatures in Conduits

Hydrothermal systems typically require warm fluids to flow along permeable beds or through faults or fractures. If such conduits can be identified, then one can compare paleotemperatures inside the conduit and those at various distances away from it. In some cases, the paleofluid flow may have been focused enough to result in higher temperatures in the conduit in comparison to temperatures away from it.

In other cases, the flow may have been so dispersed or long-lived that differences between the hypothesized conduit and surrounding rock are not apparent. Such scenarios are common in ancient systems. In highly permeable and fractured systems, flow may be somewhat dispersed. Where flow has been long-lived, the fluid may have transferred significant heat to the surrounding rock conductively, making differences between conduit and surrounding rock not obvious.

Thermal maturity in proposed conduits can be compared to areas outside of conduits using vitrinite reflectance, Rockeval pyrolysis, fluid inclusions, or some other technique. Paleotemperature data from a time-equivalent diagenetic phase or set of fluid inclusions also can be used. If there is evidence for higher paleotemperatures near the conduit, then a case for hydrothermal fluid flow can be made (Fig. 27). If lower paleotemperatures are found in the conduit, then the system must have been hydrofrigid rather than hydrothermal.

CONCLUSIONS

Fluid inclusion geothermometry is an important tool in basin analysis and in understanding thermal history of sedimentary rocks. Fluid inclusions provide opportunities for establishing detailed thermal histories of sedimentary rocks that cannot be gleaned using other techniques. They are useful in a wide variety of systems in the sedimentary realm, but to be properly applied, the geologist must first pay careful attention to the petrography of the studied samples. The most important concern in using fluid inclusions is alteration of inclusions from thermal reequilibration. The fluid inclusion assemblage approach is most important in evaluating the extent to which fluid inclusions have been altered.

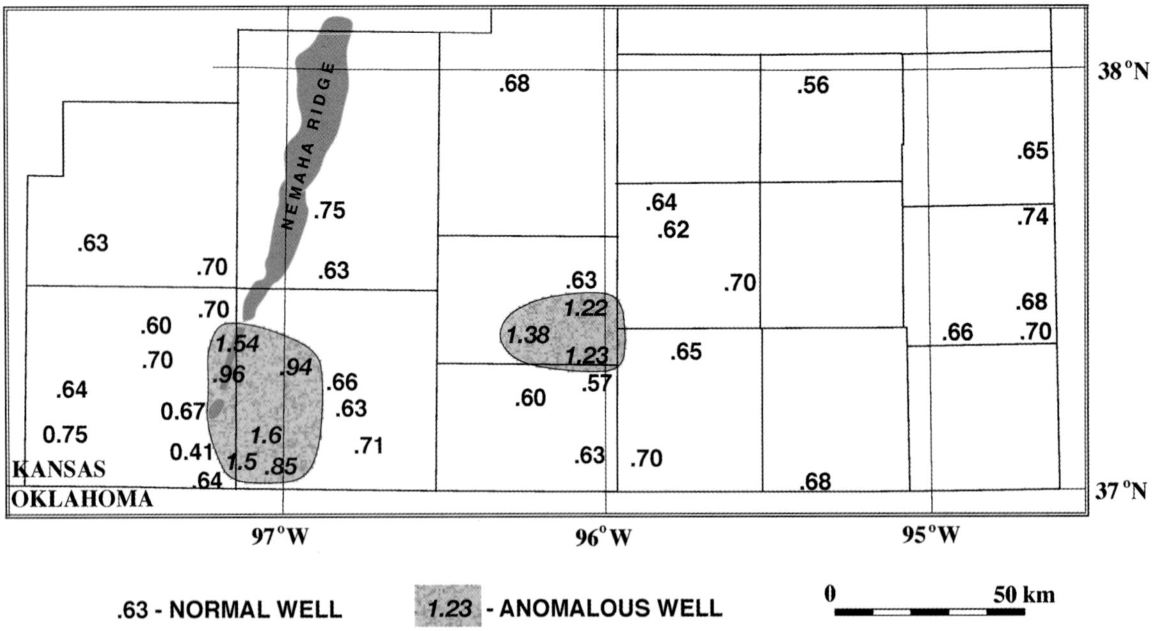

FIG. 26.—Plot of regional variation in mean random vitrinite reflectance data from Pennsylvanian strata of southeastern Kansas. High reflectance values come from wells with anomalously variable vitrinite reflectance. Note that several "warm spots" exist with elevated vitrinite reflectance. These patterns indicate warm fluids preferentially moving up section in the warm spots. Figure is modified from Barker et al. (1992) and Walton et al. (1995).

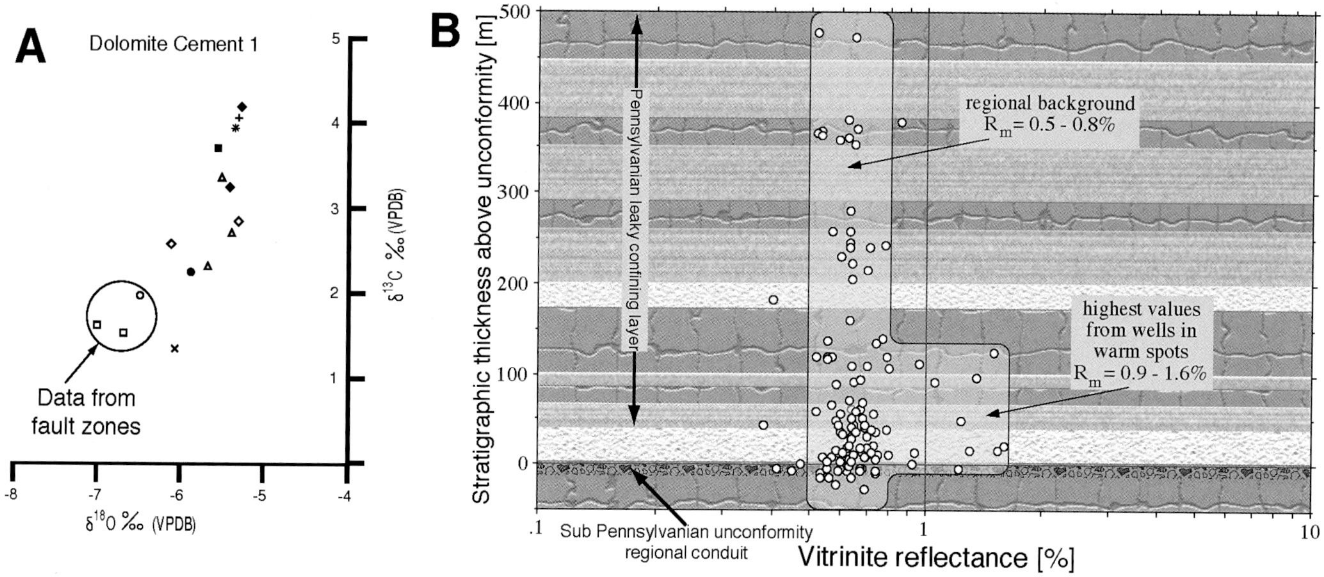

FIG. 27.—**A)** Carbon and oxygen stable isotope data from Indian Basin field of New Mexico. Data come from dolomite cement #1, which has been correlated with cement stratigraphy and shown to be approximately time equivalent throughout the area. Circled data points are from fault zones, and other data come from the same dolomite phase away from fault zones. The more negative oxygen isotopic data indicate that the dolomite from the fault zones formed from warmer fluids than the dolomite outside of the fault zones. This confirms a hydrothermal origin in which the fault zones were preferred conduits for fluid flow (Hiemstra and Goldstein 2004, 2005). **B)** Stratigraphic variation in vitrinite reflectance values from Pennsylvanian strata of southeastern Kansas. Locally, anomalously high vitrinite values are located in close stratigraphic proximity to the sub-Pennsylvanian unconformity. The diagram shows schematically that the sub-Pennsylvanian unconformity is a regional paleokarst, and that it acts as a regional stratigraphic conduit for hydrothermal fluid flow. The Pennsylvanian section, however, acts as a leaky confining unit. This pattern confirms hydrothermal fluid flow for the system. Data are from Barker et al. (1992). Figure is modified from Walton et al. (1995).

Fluid inclusion geothermometry can now be applied to low-temperature systems and used for paleoclimate work using femtosecond laser techniques. Fluid inclusions are well suited for geothermometry at elevated temperatures. Paleotemperature data can be generated from inclusions composed of high-temperature aqueous liquid, high-temperature aqueous liquid with dissolved gas, gas, and petroleum. Inclusions can be used to evaluate the spatial, temporal, tectonic, and fluid composition history of a system that cannot be determined in other ways. Like other techniques, fluid inclusions can be used as a tool for determining the maximum temperature achieved in a sample. Unlike other techniques, fluid inclusions can link thermal history to fluid flow as a means of evaluating whether heating was associated with normal burial conditions, hydrothermal systems, or cool fluids flowing into warmer rocks.

ACKNOWLEDGMENTS

The manuscript benefited greatly by constructive reviews from Nicholas B. Harris, T. James Reynolds, and Kelly Dempster. Cynthia Keeffe assisted with edits at an early stage. Research was aided by interactions with K.M. Wojcik, A.W. Walton, C.E. Barker, E.J. Hiemstra, and K.C. Benison. Many of the ideas for this were generated through productive interactions with T.J. Reynolds over the years. Partial funding was provided by Kansas Interdisciplinary Carbonates Consortium (KICC).

REFERENCES

Anderson AJ, Mayanovic RA. 2003. Electron, nuclear and X-ray probe microanalysis of fluid inclusions. *In* Samson I, Anderson A, Marshall D (Editors). *Fluid Inclusions: Analysis and Interpretation*, Short Course 32: Mineralogical Association of Canada, Quebec, Canada. p. 323–351.

Aplin AC, Macleod G, Larter SR, Pedersen KS, Sorensen H, Booth T. 1999. Combined use of confocal laser scanning microscopy and PVT simulation for estimating the composition and physical properties of petroleum in fluid inclusions. *Marine and Petroleum Geology* 16:97–110.

Bakker RJ, Brown PE. 2003. Computer modelling in fluid inclusion research. *In* Samson I, Anderson A, Marshall D (Editors). *Fluid Inclusions: Analysis and Interpretation*, Short Course 32: Mineralogical Association of Canada, Quebec, Canada. p. 175–212.

Barker CE, Goldstein RH. 1990. Fluid inclusion technique for determining maximum temperature and its comparison to the vitrinite reflectance geothermometer. *Geology* 18:1003–1006.

Barker CE, Goldstein RH. 1991. Fluid inclusion technique for determining maximum temperature and its comparison to the vitrinite reflectance geothermometer—reply: *Geology* 19:1242.

Barker CE, Goldstein RH, Hatch JR, Walton AW, Wojcik KM. 1992. Burial history and thermal maturation of Pennsylvanian rocks, Cherokee basin, southeastern Kansas. *In* Johnson KS Cardott BJ (Editors). *Source Rocks in the Southern Midcontinent*, Circular 93: Oklahoma Geological Survey, Norman, Oklahoma. p. 299–310.

Barker CE, Pawlewicz MJ. 1994. Calculation of vitrinite reflectance from thermal histories and peak temperatures. *In* Mukhopadhyay PK, Dow W G (Editors). *Vitrinite Reflectance as a Maturity Parameter; Applications and Limitations*, Symposium Series 570: American Chemical Society, Washington, DC. p. 216–229.

Benison KC, Goldstein RH. 1999. Permian paleoclimate data from fluid inclusions in halite. *Chemical Geology* 154:113–132.

Bodnar RJ. 2003a. Introduction to fluid inclusions. *In* Samson I, Anderson A,

Marshall D (Editors). *Fluid Inclusions: Analysis and Interpretation*, Short Course 32: Mineralogical Association of Canada, Quebec, Canada. p. 1–8.

Bodnar RJ. 2003b. Re-equilibration of fluid inclusions. *In* Samson I, Anderson A, Marshall D (Editors). *Fluid Inclusions: Analysis and Interpretation*, Short Course 32: Mineralogical Association of Canada, Quebec, Canada. p. 213–230.

Bodnar RJ. 2003c. Interpretation of data from aqueous-electrolyte fluid inclusions. *In* Samson I, Anderson A, Marshall D (Editors). *Fluid Inclusions: Analysis and Interpretation*, Short Course 32: Mineralogical Association of Canada, Quebec, Canada. p. 81–100.

Bodnar RJ, Bethke PM. 1984. Systematics of stretching of fluid inclusions; fluorite and sphalerite at 1 atmosphere confining pressure. *Economic Geology* 79:141–161.

Bodnar RJ, Binns PR, Hall DL. 1989. Synthetic fluid inclusions VI. Quantitative evaluation of the decrepitation behavior of fluid inclusions in quartz at one atmosphere confining pressure. *Journal of Metamorphic Geology* 7:229–242.

Bodnar RJ, Burruss RC. 1991. Fluid inclusion technique for determining maximum temperature and its comparison to the vitrinite reflectance geothermometer—comment. *Geology* 19:1241–1242.

Burruss RC. 1981. Hydrocarbon fluid inclusions in studies of sedimentary diagenesis. *In* Hollister LS, Crawford ML (Editors). *Short Course in Fluid Inclusions: Application to Petrology*, Short Course Handbook 6: Mineralogical Association of Canada, Quebec, Canada. p. 138–156.

Burruss RC. 1991. Practical aspects of fluorescence microscopy of petroleum fluid inclusions. *In* Barker CE, Kopp OC (Editors). *Luminescence Microscopy and Spectroscopy*, Short Course 25: Society of Economic Paleontologists and Mineralogists, Tulsa, Oklahoma. p. 1–7.

Burruss RC. 2003a. Raman spectroscopy of fluid inclusions: an introduction. *In* Samson I, Anderson A, Marshall D (Editors). *Fluid Inclusions: Analysis and Interpretation*, Short Course 32: Mineralogical Association of Canada, Quebec, Canada. p. 179–189.

Burruss RC. 2003b. Petroleum fluid inclusions: an introduction. *In* Samson I, Anderson A, Marshall D (Editors). *Fluid Inclusions: Analysis and Interpretation*, Short Course 32: Mineralogical Association of Canada, Quebec, Canada. p. 159–174.

Cantrell DL, Swart PK, Hagerty RM. 2004. Genesis and characterization of dolomite, Arab-D reservoir, Ghawar field, Saudi Arabia. *GeoArabia* 9(2):1–26.

Ceriani A, Di Guilio A, Goldstein RH, Rossi C. 2002. Diagenesis associated with cooling during burial: an example from Lower Cretaceous reservoir sandstones (Sirt basin, Libya). *American Association of Petroleum Geologists Bulletin* 86:1573–1591.

Davies GR, Smith T. 2006. Structurally controlled hydrothermal dolomite reservoir facies: an overview. *American Association of Petroleum Geologists Bulletin* 90:1641–1690.

Esteban M, Taberner C. 2003. Secondary porosity development during late burial in carbonate reservoirs as a result of mixing and/or cooling brines. *Journal of Geochemical Exploration* 78–79:355–359.

Farley KA, Stockli DF. 2002. (U-Th)/He dating of phosphates: apatite, monazite, and xenotime. *In* Kohn ML, Rakovan J, Hughes JM (Editors). *Phosphates—Geochemical, Geobiological, and Materials Importance*, Reviews in Mineralogy 48: Mineralogical Society of America, Chantilly, Virginia. p. 559–577.

Gagnon JE, Samson IM, Fryer BJ. 2003. Laser ablation ICP-MS analysis of fluid inclusions. *In* Samson I, Anderson A, Marshall D (Editors). *Fluid Inclusions: Analysis and Interpretation*, Short Course 32: Mineralogical Association of Canada, Quebec, Canada. p. 291–322.

Gilg HA, Krüger Y, Stoller P, Frenz M, Boni M, Ziemann M. 2007. Temperatures in sulfide oxidation zones: microthermometry of monophase aqueous fluid inclusions and stable isotope systematic. *In* European Current Research on Fluid Inclusions (ECROFI) XIX Abstract Volume: University of Bern, Switzerland, p. 130.

Gleeson SA. 2003. Bulk analysis of electrolytes in fluid inclusions. *In* Samson I, Anderson A, Marshall D (Editors). *Fluid Inclusions: Analysis and Interpretation*, Short Course 32: Mineralogical Association of Canada, Quebec, Canada. p. 233–246.

Goldstein RH. 1986. Reequilibration of fluid inclusions in low-temperature calcium-carbonate cement. *Geology* 14:792–795.

Goldstein RH. 1993. Fluid inclusions as microfabrics: a petrographic method to determine diagenetic history. *In* Rezak R, Lavoi D (Editors). *Carbonate Microfabrics; Frontiers in Sedimentary Geology*: Springer-Verlag, New York. p. 279–290.

Goldstein RH. 2001a. Fluid inclusions in sedimentary and diagenetic systems. *Lithos* 55:159–193.

Goldstein RH. 2001b. Perspectives: paleoenvironment, clues from fluid inclusions. *Science* 294:1009–1010.

Goldstein, R.H., 2003, Petrographic analysis of fluid inclusions. *In* Samson I, Anderson A, Marshall D (Editors). *Fluid Inclusions: Analysis and Interpretation*, Short Course 32: Mineralogical Association of Canada, Quebec, Canada. p. 9–53.

Goldstein RH, Reynolds TJ. 1994. *Systematics of Fluid Inclusions in Diagenetic Minerals*, Short Course 31: Society of Economic Paleontologists and Mineralogists, Tulsa, Oklahoma. 199 p.

Hanor JS. 1980. Dissolved methane in sedimentary brines—potential effect on the PVT properties of fluid inclusions. *Economic Geology* 75:603–609.

Hiemstra EJ, Goldstein RH. 2004. The Indian Basin, New Mexico: a tectonically valved hydrothermal dolomite reservoir. *In* American Association of Petroleum Geologists Annual Meeting Expanded Abstracts, Vol. 13: American Association of Petroleum Geologists, Tulsa, Oklahoma. p. 62–63.

Hiemstra EJ, Goldstein RH. 2005. The diagenesis and fluid migration history of the Indian Basin Field, Eddy County, New Mexico. *In* Lufholm P, Cox D (Editors). *Unconventional Reservoirs, Technologies, and Strategies; New Perspectives for the Permian Basin*, Publication 05-115: West Texas Geological Society, Midland, Texas. p. 197–206.

Jones GD, Smart PL, Whitaker FF, Rostron BJ, Machel HG. 2003. Numerical modeling of reflux dolomitization in the Grosmont platform complex (Upper Devonian), Western Canada sedimentary basin. *American Association of Petroleum Geologists Bulletin* 87:1273–1298.

Karlsen DA, Nedkvitne T, Larter SR, Bjørlykke K. 1993. Hydrocarbon composition of authigenic inclusions—applications to elucidation of petroleum reservoir filling history. *Geochimica et Cosmochimica Acta* 57:3641–3659.

Klosterman MJ. 1981. Applications of fluid inclusion techniques to burial diagenesis in carbonate rock sequences. *In Applied Carbonate Research Program Technical Series Contributions*, Vol. 7: Louisiana State University, Baton Rouge, Louisiana. 102 p.

Krüger Y, Stoller P, Ricka J, Frenz M. 2007. Femtosecond lasers in fluid inclusion analysis: overcoming metastable phase states. *In* European Current Research on Fluid Inclusions (ECROFI) XIX Abstract Volume: University of Bern, Switzerland, p. 184.

Lacazette A. 1990. Application of linear elastic fracture mechanics to the quantitative evaluation of fluid-inclusion decrepitation. *Geology* 18:782–785.

Leach DL, Sangster DF. 1993. Mississippi Valley–type lead-zinc deposits. *In* Kirkham RV, Sinclair WD, Thorp RI, Duke JM (Editors). *Mineral Deposit Modeling*, Special Paper 40: Geological Association of Canada, St. John's, Newfoundland. p. 289–314.

Leroy J. 1979. Contribution a l'etalonnage de la pression interne des inclusions fluides lors de leur décrepitation. *Bulletin de Minéralogie* 120:584–593.

Lowenstein TK, Li J, Brown CB. 1998. Paleotemperatures from fluid inclusions in halite: method verification and a 100,000 year paleotemperature record, Death Valley, California. *Chemical Geology* 150:233–245.

Luczaj JA, Goldstein RH. 2000. Diagenesis of the lower Permian Krider Member, southwest Kansas: fluid inclusion, U-Pb, and fission-track evidence for reflux dolomitization during latest Permian time. *Journal of Sedimentary Research* 70:762–773.

Machel HG, Lonnee J. 2002. Hydrothermal dolomite—a product of poor definition and imagination. *Sedimentary Geology* 152:163–171.

Moore CH, Druckman Y. 1981. Burial diagenesis and porosity evolution, Upper Jurassic Smackover, Arkansas and Louisiana. *American Association of Petroleum Geologists Bulletin* 65:597–628.

Nelson RC. 1973. Fluid inclusions as a clue to diagenesis of carbonate rocks (Abstract). *Geological Society of America Abstracts with Programs* 5:748.

Pettke T, Diamond LW. 1995. Rb-Sr isotopic analysis of fluid inclusions in quartz: evaluation of bulk extraction procedures and geochronometer systematics using synthetic fluid inclusions. *Geochimica et Cosmochimica Acta* 59:4009–4027.

Pettke T, Diamond LW. 1996. Rb-Sr dating of sphalerite based on fluid inclusion–host mineral isochrons; a clarification of why it works. *Economic Geology and the Bulletin of the Society of Economic Geologists* 91:951–956.

Pironon J, Canals M, Dubessy J, Walgenwitz F, Laplace-Buihle C. 1998. Volumetric reconstruction of individual oil inclusion by confocal scanning laser microscopy. *European Journal of Mineralogy* 10:1143–1150.

Pironon J, Thiéry R, Teinturier S, Walgenwitz F. 2000. Water in petroleum inclusions. Evidence from Raman and FT-IR measurements, PVT consequences. *Journal of Geochemical Exploration* 69–70:663–668.

Prather BE. 1992. Origin of dolostone reservoir rocks; Smackover Formation (Oxfordian), northeastern Gulf Coast, USA. *American Association of Petroleum Geologists Bulletin* 76:133–163.

Prezbindowski DR, Larese RE. 1987. Experimental stretching of fluid inclusions in calcite—implications for diagenetic studies. *Geology* 15:333–336.

Prezbindowski DR, Tapp JB. 1991. Dynamics of fluid inclusion alteration in sedimentary rocks: a review and discussion. *Organic Geochemistry* 17:131–142.

Roberts SM, Spencer RJ. 1995. Paleotemperatures preserved in fluid inclusions in halite. *Geochimica et Cosmochimica Acta* 59:3929–3942.

Roedder E. 1962. Ancient fluids in crystals. *Scientific American* 207:38–47.

Roedder E. 1979. Fluid inclusion evidence on the environment of sedimentary diagenesis, review. *In* Scholle PA, Schlunger PR (Editors). *Aspects of Diagenesis*, Special Publication 26: Society of Economic Paleontologists and Mineralogists, Tulsa, Oklahoma. p. 89–107.

Roedder E. 1984. *Fluid Inclusions*, Reviews in Mineralogy 12: Mineralogical Society of America, Chantilly, Virginia. 644 p.

Rossi C, Goldstein RH, Ceriani A, Marfil R. 2002. Fluid inclusions record thermal and fluid evolution in reservoir sandstones, Khatatba Formation, Western Desert, Egypt: a case for fluid injection. *American Association of Petroleum Geologists Bulletin* 86:1773–1799.

Salvi S, Williams-Jones AE. 2003. Bulk analysis of volatiles in fluid inclusions. *In* Samson I, Anderson A, Marshall D (Editors). *Fluid Inclusions: Analysis and Interpretation*, Short Course 32: Mineralogical Association of Canada, Quebec, Canada. p. 247–278.

Samson I, Anderson A, Marshall D (Editors). 2003. *Fluid Inclusions: Analysis and Interpretation*, Short Course 32: Mineralogical Association of Canada, Quebec, Canada. 374 p.

Satterfield CL, Lowenstein TK, Vreeland RH, Rosenzweig WD. 2005. Paleobrine temperatures, chemistries, and paleoenvironments of Silurian Salina Formation F-1 salt, Michigan Basin, USA, from petrography and fluid inclusions in halite. *Journal of Sedimentary Research* 75:534–546.

Smith LB Jr, Davies GD. 2006. Structurally controlled hydrothermal alteration of carbonate reservoirs: introduction. *American Association of Petroleum Geologists Bulletin* 90:1635–1640.

Sorby HC. 1858. On the microscopic structure of crystals, indicating the origin of minerals and rocks. *Geological Society of London Quarterly Journal* 14:453–500.

Stearns ND, Stearns HT, Waring GA. 1935. Thermal Springs in the United States. U.S. Geological Survey, Water Supply Paper 679-B, p. 59–191.

Swanenberg HEC. 1980. *Fluid Inclusions in High-Grade Metamorphic Rocks from SW Norway*, Vol. 20: Geologica Ultraiectina, University of Utrecht, Netherlands. 147 p.

Thiéry R, Pironon J, Walgenwitz F, Montel F. 2000. PIT (petroleum inclusion thermodynamic); a new modeling tool for the characterization of hydrocarbon fluid inclusions from volumetric and microthermometric measurements. *Journal of Geochemical Exploration* 69–70:701–704.

Thiéry R, Pironon J, Walgenwitz F, Montel F. 2002. Individual characterization of petroleum fluid inclusions (composition and P–T trapping conditions) by microthermometry and confocal laser scanning microscopy: inferences from applied thermodynamics of oils. *Marine and Petroleum Geology* 19:847–859.

Tobin RC, Claxton BL. 2000. Multidisciplinary thermal maturity studies using vitrinite reflectance and fluid inclusion microthermometry: a new calibration of old techniques. *American Association of Petroleum Geologists Bulletin* 84:1647–1665.

Touret JLR. 1981. Fluid inclusions in high-grade metamorphic rocks. *In* Hollister LS, Crawford ML (Editors). *Short Course in Fluid Inclusions: Application to Petrology*, Short Course Handbook 6: Mineralogical Association of Canada, Quebec, Canada. p. 182–208.

Tseng HY, Pottorf RJ. 2002. Fluid inclusion constraints on petroleum PVT and compositional history of the Greater Alwyn–South Brent petroleum system, northern North Sea. *Marine and Petroleum Geology* 19:797–809.

Turgarinov AI, Vernadsky VI. 1970. Dependence of the decrepitation temperature of minerals on their gas-liquid inclusions and hardness. *Akademiya Nauk SSSR Doklady* 195:112–114.

Wagner G, Van den Haute P. 1992. *Fission Track Dating*: Dordrecht, Netherlands, Kluwer Academic Publishers. 285 p.

Wagner PD, Matthews RK. 1982. Porosity preservation in the Upper Smackover (Jurassic) carbonate grainstone, Walker Creek Field, Arkansas: response of paleophreatic lenses to burial processes—reply. *Journal of Sedimentary Petrology* 52:24–25.

Walton AW, Wojcik KM, Goldstein RH, Barker CE. 1995. Diagenesis of Upper Carboniferous rocks in the Ouachita foreland shelf in the Mid-Continent USA: an overview of widespread effects of a Variscan-equivalent orogeny. *Geologische Rundschau* 84:535–551.

White DE. 1957. Thermal waters of volcanic origin. *Geological Society of America Bulletin* 68:1637–1658.

Wojcik KM, Goldstein RH, Walton AW. 1994. History of diagenetic fluids in a distant foreland area, Middle and Upper Pennsylvanian, Cherokee basin, Kansas: fluid inclusion evidence. *Geochimica et Cosmochimica Acta* 58:1175–1192.

Wojcik KM, Goldstein RH, Walton AW. 1997. Regional and local controls of diagenesis driven by basin-wide flow system: Pennsylvanian sandstones and limestones, Cherokee basin, southeastern Kansas. *In* Montañez I, Gregg J, Shelton K (Editors). *Basin-Wide Diagenetic Patterns: Integrated Petrologic, Geochemical, and Hydrologic Considerations*, Special Publication 57: Society for Sedimentary Geology, Tulsa, Oklahoma. p. 235–252.

THERMAL HISTORY RECONSTRUCTION IN SEDIMENTARY BASINS USING APATITE FISSION-TRACK ANALYSIS AND RELATED TECHNIQUES

PAUL F. GREEN AND IAN R. DUDDY

Geotrack International Pty Ltd., 37 Melville Road, Brunswick West, Victoria 3055, Australia
e-mail: mail@geotrack.com.au

ABSTRACT: While several methods have been developed for assessing the magnitude of postdepositional heating of sedimentary rocks, apatite fission track analysis (AFTA®) can also define the time at which a sedimentary rock cooled from its maximum postdepositional temperature, up to ~110° C. This information is particularly important in hydrocarbon exploration, e.g., in defining the timing of hydrocarbon generation and identifying regions where the main phase of generation postdates formation of potential trapping structures.

Based on analysis of naturally occurring radiation damage features (fission tracks) in detrital apatite grains, the foundation of the technique is a detailed understanding of the kinetics of fission-track "annealing," incorporating observations from both laboratory and geological field conditions and making explicit allowance for apatite composition (chlorine content), which exerts a crucial influence over fission-track annealing kinetics. The thermal response of fission tracks is dominated by the maximum postdepositional paleotemperature, and this fundamental aspect of the technique imposes strict limitations on the information that can be obtained. In particular, no information can be obtained on the thermal history prior to the onset of cooling from the maximum postdepositional temperature. However, one or possibly two additional episodes of heating and cooling can often be resolved following the paleo–thermal maximum. Integration of AFTA data with paleotemperature estimates from other methods, in particular vitrinite reflectance (VR), provides additional support for thermal history interpretations and can often help to refine solutions from AFTA.

Most importantly, the combined use of AFTA and VR in a vertical sequence allows construction of profiles of paleotemperature with depth (or elevation), enabling quantitative determination of paleogeothermal gradient, which in turn allows identification of the mechanisms of heating and cooling. Heating due to deeper burial produces a linear paleotemperature profile with a similar gradient to the present temperature profile, whereas heating due primarily to increased basal heat flow will produce a profile with a higher gradient than the present temperature profile. Extrapolation of such profiles above the appropriate unconformity identified from AFTA to a suitable paleo-surface temperature allows determination of the magnitude of additional burial responsible for the observed heating. Estimations of additional burial in this way depend on assumptions concerning the lithology (i.e., thermal conductivity) of the eroded sequence and wherever possible should be combined with estimates of burial based on nonthermal processes, such as sonic velocity, in order to provide consistent constraints on the burial history. Nonlinear profiles are produced by contact heating around intrusive bodies and by passage of hot fluids within confined aquifer horizons. More complex situations that involve nonlinear profiles resulting from thick sequences with extreme thermal conductivities (e.g., coal or salt) can also be assessed by inspection of the variation of paleotemperature with depth.

AFTA has been applied to hydrocarbon exploration in a wide variety of settings. Some of the most important outcomes of AFTA analysis, in terms of events that affect regional hydrocarbon prospectivity, are: (1) the recognition of regional kilometer-scale exhumation (implying earlier deeper burial), often in areas that have traditionally been considered stable; (2) definition of major Phanerozoic paleo-thermal events in Proterozoic basins; and (3) revelation of the importance of hot fluids in transporting heat in sedimentary basins. In basins with complex histories, for example, exhumed basins or those in which heat flow was higher in the past, AFTA can provide unique constraints on the timing of hydrocarbon generation, which can significantly reduce exploration risk in such areas.

KEY WORDS: apatite, fission track, thermochronology, sedimentary rocks, hydrocarbon exploration

INTRODUCTION

The degree to which a sedimentary unit has been heated after deposition is important from a number of practical viewpoints, including the maturation of hydrocarbon source rocks and the possible extent of diagenetic effects in potential reservoir units, and in the determination of the extent of eroded sections in basins that have undergone some degree of exhumation. A wide variety of methods may be applied to assessing the severity of postdepositional heating in sedimentary units (e.g., Naeser and McCulloh 1989), but while many techniques can provide a relative or qualitative measure of the magnitude of postdepositional heating, quantitative treatment is only possible using those methods for which the kinetics are understood in sufficient detail, such as vitrinite reflectance (Burnham and Sweeney 1989). Even then, such techniques provide only the magnitude of the maximum paleotemperature.

Application of thermochronological methods (i.e., methods for defining temperature–time histories of rock samples) to sedimentary rocks provides the possibility to determine the timing of the thermal maximum, as well as the magnitude. This information is particularly important in hydrocarbon exploration, by defining the timing of hydrocarbon generation and identifying regions where the main phase of generation postdates formation of potential trapping structures.

Numerous applications of conventional geochronological methods (Ar-Ar, Rb-Sr, U-Pb) to metasediments utilize the differing thermal sensitivities of these techniques in the temperature range 200 to 500° C and above. But here we restrict discussion to temperatures below ~200° C, as encountered in the upper few kilometers of the crust in typical sedimentary basin settings. For defining thermal histories of sedimentary units in this range, apatite fission-track analysis (AFTA®) constitutes the only practical method currently available. Since initial development in the early 1980s, as an adaptation of a technique previously applied to basement terrains, this method has found wide application to sedimentary sequences in a wide variety of settings, and routine commercial implementation in the form of AFTA as a tool for hydrocarbon exploration reflects the widespread acceptance of the method. While other techniques have been proposed as potentially useful, such as ^{40}Ar-^{39}Ar in K-feldspars (Harrison and Burke 1989) and (U-Th)/He in apatites (Crowhurst et al. 2002), these have not achieved routine application and provide useful information only in highly specific conditions.

Here, we focus on apatite fission-track analysis, and specifically on

Analyzing the Thermal History of Sedimentary Basins: Methods and Case Studies
SEPM Special Publication No. 103, Copyright © 2012
SEPM (Society for Sedimentary Geology), Print ISBN 978-1-56576-315-9, CD/DVD ISBN 978-1-56576-317-3, p. 65–104.

AFTA, which describes the system of methods and principles embodied in our own approach. We first review the historical development and the technical basis of the technique, then discuss various key aspects that control the type of information that can and cannot be obtained, and briefly contrast our own approach with those adopted in other laboratories, discussing various strengths and weaknesses. We stress the importance of integrating information from AFTA with other types of information. In this context, we briefly describe the newer technique of apatite (U-Th)/He dating, highlighting the initial success and later problems that have beset this technique. Finally, we describe a number of case studies that illustrate the way in which AFTA has been applied in a variety of different settings.

HISTORICAL BACKGROUND AND PHYSICAL BASIS

Naturally occurring fission tracks are radiation damage trails produced by the spontaneous fission of ^{238}U atoms, in which a uranium atom splits into two highly energetic fragments, stripped of electrons. Due to electrostatic repulsion between the fragments, they travel rapidly in opposite directions, creating a linear zone of intense disruption through the crystal lattice, known as a fission track. Apatite grains typically contain around 10 ppm uranium (range 1–100 ppm), so over geological time, fission tracks accumulate in the crystal lattice. If a collection of apatite grains is mounted and polished, tracks can be revealed where they intersect the polished surface by a simple etching treatment (Fig. 1). Since spontaneous fission is a form of radioactive decay, in principle, the number of tracks in the surface is controlled by uranium content and time, through standard decay laws, so by counting the number of tracks and measuring the uranium content, a "fission-track age" can be measured, which, in the absence of other factors, should indicate the time over which tracks have accumulated. Reviews of the basics of fission-track dating are provided by Fleischer et al. (1975), Wagner and Van den Haute (1992), and Galbraith (2006).

Early Insights into the Thermal Sensitivity of Fission Tracks in Apatite

Early applications of fission-track dating to accessory apatites from crystalline basement rocks revealed that the technique was extremely thermally sensitive, suggesting that fission-track ages could be reset at relatively low temperatures around 100° C over geological timescales (e.g., Wagner and Reimer 1972). This was supported by early laboratory annealing studies (Wagner 1968, Naeser and Faul 1969), and subsequently confirmed by direct measurement of fission-track ages in subsurface samples (Naeser and Forbes 1976). Integration of fission-track ages with confined track length measurements (Fig. 2), first reported by Bhandari et al. (1971), led to deeper understanding of the method. Early measurements showed that even in volcanic rocks that have experienced only very low temperatures after initial posteruption cooling, mean confined track lengths (around 14 to 15 μm) were shorter than induced tracks (~16 μm) in the same apatites. Green (1980) showed that this can be understood in terms of thermal annealing of these tracks at low temperatures (<50° C) over geological timescales, highlighting the sensitivity of the technique.

Results from boreholes in the Otway Basin of SE Australia (Gleadow and Duddy 1981) provided the first quantitative insights into the thermal stability of tracks in geological conditions, revealing the progressive reduction in fission-track age with depth and temperature, and showing that this was complemented by a corresponding decrease in track length (Fig. 3). Compilation of confined track length data in a large number of apatite samples showed that the form of the track length distribution was a sensitive indicator of the style of thermal history (Gleadow et al. 1986). Laboratory studies (Green et al. 1985, 1986; Green 1988), together with detailed mathematical analysis (Laslett et al. 1982; Galbraith and Laslett 1988, 1990), established that the reduction in track length (which is, in turn, a manifestation of the reduction in the degree of damage within the track region) causes the reduction in fission-track age by reducing

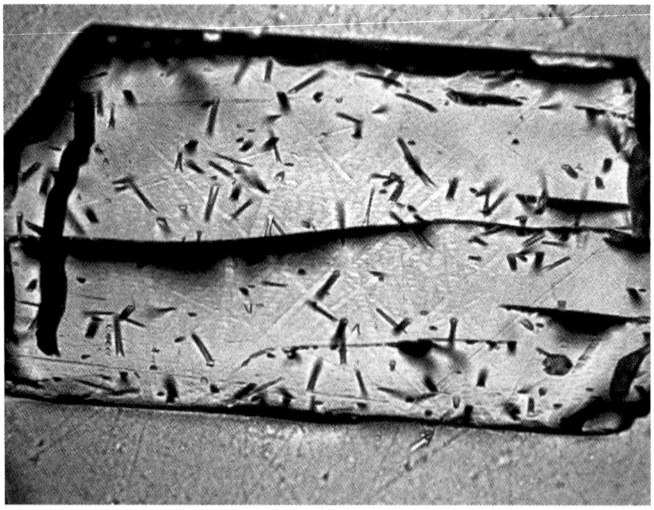

FIG. 1.—Spontaneous fission tracks in a detrital apatite grain ~0.25 mm in length. Fission tracks were revealed by etching in 5 M HNO_3 for 20 seconds at 20° C. The majority of linear features in the image are fission tracks. They are randomly orientated in three dimensions, and the number of tracks revealed at the surface is controlled by the uranium content of the grain, the time over which tracks have accumulated, and the distribution of track lengths, which in turn is a function of the thermal history.

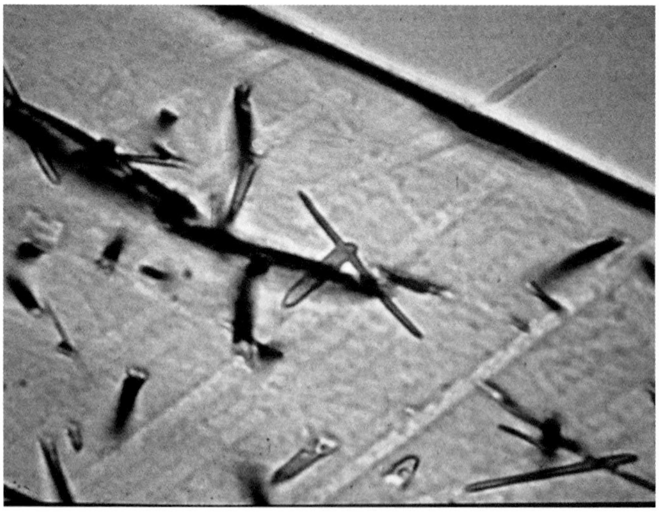

FIG. 2.—Confined fission-track lengths in apatite. In order to determine the distribution of track lengths, measurements are made of the full length of horizontal tracks that are totally enclosed within the body of the crystal, which have been etched via fractures or other tracks that intersect the surface of the grain. Note that difference in width of the tracks is due to a difference in etch rate with crystallographic orientation, with a higher etch rate parallel to the *c*-axis.

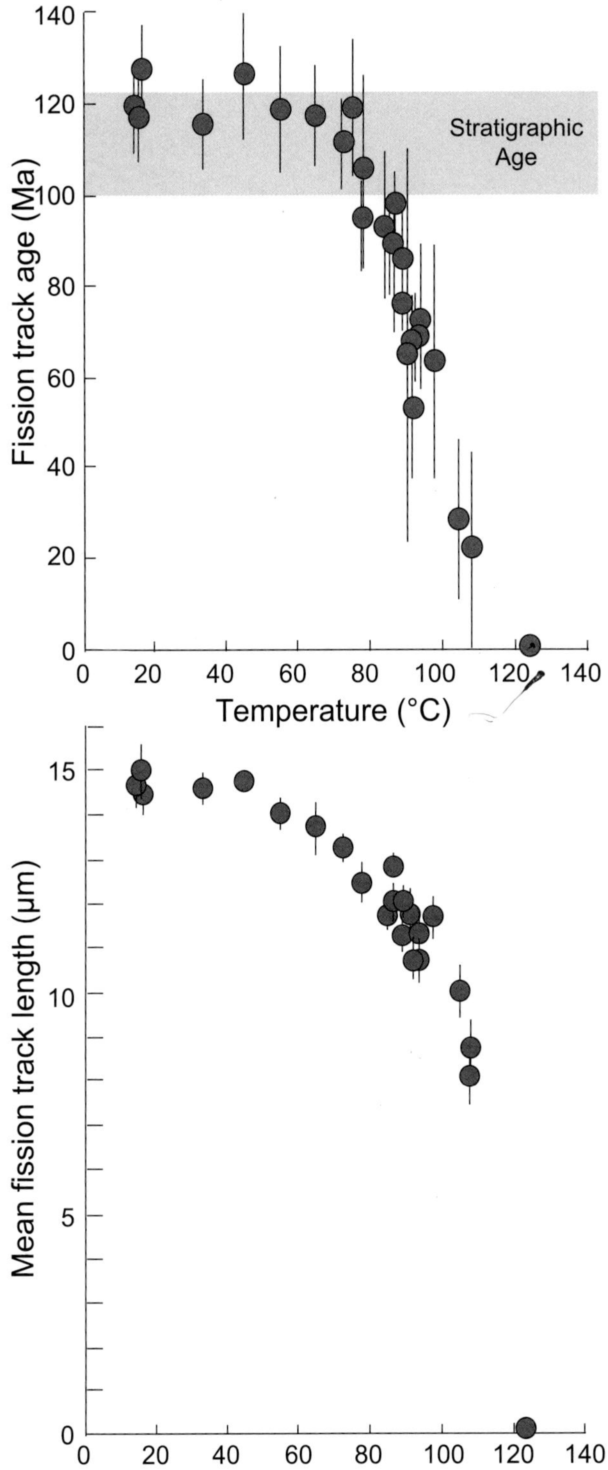

FIG. 3.—Variation of measured fission-track age and mean track length with present-day temperature in samples from Otway Basin (Australia) exploration wells, which are at maximum temperatures at the present day. With increasing maximum temperature, there is a progressive reduction in fission-track age to zero at ≈125° C, which is the result of the progressive reduction in mean confined track length. These observations provide a direct expression of the thermal stability of fission tracks in geological conditions (after Gleadow and Duddy 1981).

the proportion of tracks that can intersect a polished grain surface. This realization underpins all subsequent studies involving quantitative prediction of apatite fission-track (AFT) parameters and extraction of thermal history information from such data.

The Laslett et al. (1987) Model

Early laboratory annealing studies, based on measurement of track density as an indicator of the degree of annealing, in accessory apatites of varying character led to a wide variety of kinetic models of conflicting styles (as reviewed by Green et al. 1986, 1988). Based on the recognition of the key role of track length in the annealing process, coupled with the fact that track lengths can be measured with greater precision than track densities, Green et al. (1986) carried out a series of detailed annealing experiments on a well-characterized apatite of uniform composition (Durango apatite; Young et al. 1969), in which mean track length was used to indicate the degree of annealing. This study demonstrated that as a result of heating, the mean track length is progressively reduced, and the tracks effectively "shrink" from each end, until in the final stages, individual tracks may break up into several segments. This behavior can be understood in terms of a progressive reduction in the degree of radiation damage within the track region, as displaced atoms return to their original lattice sites by thermally activated diffusion, until in the final stages, discrete zones of etchability are separated by regions that are fully "healed." Laslett et al. (1987) subsequently showed that the variation of mean track length with temperature and time was well described by a "fanning Arrhenius plot" model, in which contours of equal track length reduction form straight lines in a plot of time against inverse absolute temperature, with the slope of these lines (reflecting an "activation energy") increasing as the degree of annealing increases (Fig. 4).

Variable Temperature Behavior

The improved definition of the kinetics of fission-track annealing provided by using mean confined track length as the fundamental parameter (Laslett et al. 1987), combined with a detailed understanding of the way in which reduction in track length is manifested in the fission-track age (Green 1988), provided the basis for making realistic predictions of apatite fission-track parameters in geological situations. A key step in this process is the transition from isothermal annealing models to variable temperature behavior as encountered in geological settings. Duddy et al. (1988) provided a way forward by adopting the principle of "equivalent time" (originally postulated by Goswami et al. 1984), in which the rate of annealing of a track at any given time depends only on the length to which the track has already been reduced, and the prevailing temperature, and not on the history of the mechanism by which the track reached that length. The validity of this assumption is by no means straightforward, and Duddy et al. (1988) provided experimental verification through a series of variable-temperature annealing experiments, laying the basis for the successful extension of isothermal annealing models to variable-temperature geological histories.

Quantitative Modeling of AFTA Parameters

Green et al. (1989b) built on these advances to develop methods for quantitatively modeling the response of fission tracks in apatite to various styles of thermal history. The basis of the approach is the

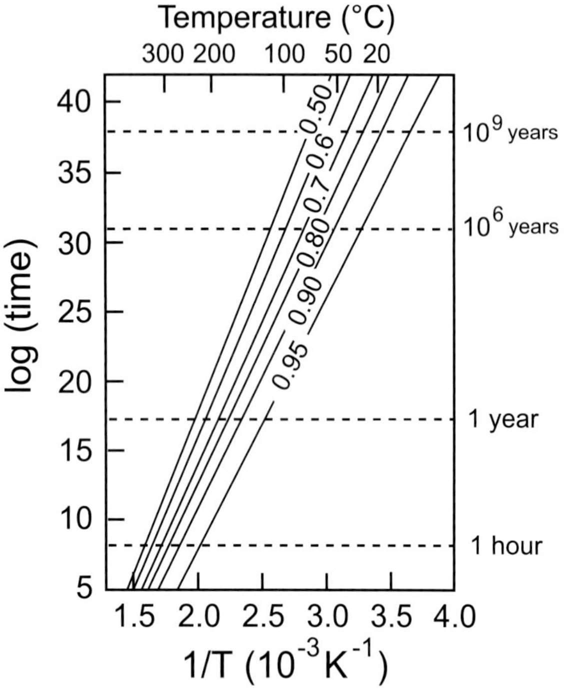

FIG. 4.—"Fanning Arrhenius plot" adopted by Laslett et al. (1987) to describe the variation of mean track length with temperature and time for Durango apatite. In this plot, contours of equal track length reduction form straight lines in a plot of log (time) against inverse absolute temperature (contoured values are the ratio of predicted length to initial length, L/L_o).

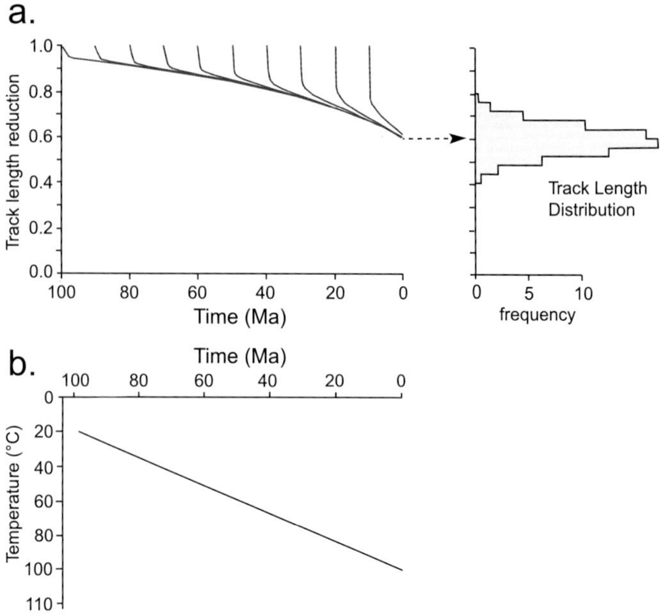

FIG. 5.—(a) Predicted pattern of track length reduction with time (left) and final track length distribution (right) for populations of tracks formed at different times through the notional thermal history shown in (b), involving progressive heating from 20 to 100° C over 100 million years. The final track length distribution is calculated by applying the principle of equivalent time to each population of tracks using the fanning Arrhenius model of Laslett et al. (1987), and summing the track lengths distributions for all populations of tracks in the appropriate length-biased proportions (Laslett et al. 1982). After Green et al. (1989a).

recognition that track length reduction is the controlling process in determining the AFTA parameters that result from any particular history. New fission tracks are continually generated as time passes, such that different tracks sample different proportions of the whole history. If a specified thermal history is broken down into discrete intervals, the pattern of track length reduction with time for populations of tracks formed at different times through the history can be calculated by applying the principle of equivalent time (above) to the thermal history appropriate to each population, resulting in a predicted mean length for each population of tracks at the present day (Fig. 5).

Fission tracks show an inherent spread in track length, reflecting the range of energies and masses of the fission fragments produced by spontaneous fission. This spread, measured by the standard deviation of the length distribution, increases as the mean length is reduced, largely as a result of an increasing anisotropy of the annealing process (Green et al. 1986). So, to calculate the final distribution of track lengths predicted for a sample at the present day, the component distributions of track length resulting from populations of tracks produced at different times throughout the history must be summed, employing the appropriate spread of lengths for each population. The component populations must also be added in the appropriate proportions, allowing for the biases involved in revelation of tracks of different length (Laslett et al. 1982). By summing components of fission-track age appropriate to the length of each time interval, moderated by the effect of length reduction on the reduction of age (Green 1988), the final fission-track age for the sample can also be predicted.

Influence of Apatite Composition on Annealing Rates

Studies of apatite fission-track parameters in subsurface samples from the Otway Basin (Gleadow and Duddy 1981; Green et al. 1985, 1986) showed that chlorine content exerts a systematic influence on annealing rates (Fig. 6). This has subsequently been confirmed in laboratory studies by Carlson et al. (1999) and Barbarand et al. (2003a). In both these studies, the authors tend to downplay the influence of chlorine in favor of other factors. However, results from both studies clearly illustrate the first-order control exerted by chlorine (Fig. 7), while also suggesting that other, possibly more exotic, factors may exert second-order control in some cases. The importance of differential annealing within individual samples related to wt% Cl has now been demonstrated in a number of geological studies (e.g., Argent et al. 2002, Crowhurst et al. 2002, Green et al. 2002, Green 2005).

Several studies have suggested that etch pit diameters can be used as an indicator of differential annealing rates between various apatite species (Burtner et al. 1994, Ketcham et al. 1999, Barbarand et al. 2003a). However, Green et al. (2005a) demonstrated that annealing rates in the Barbarand et al. (2003a) data set show a much stronger correlation with wt% Cl than to etch pit size (Fig. 8).

Evidence for systematic differences in annealing rates in natural geological samples due to any element other than Cl has yet to be demonstrated. In practical application of AFTA, variation in fission-track age and track length with wt% Cl allows identification of any anomalous grains that might represent unusual annealing properties (e.g., Crowhurst et al. 2002). This approach also allows detection of contaminant grains, which can be important in working with ditch

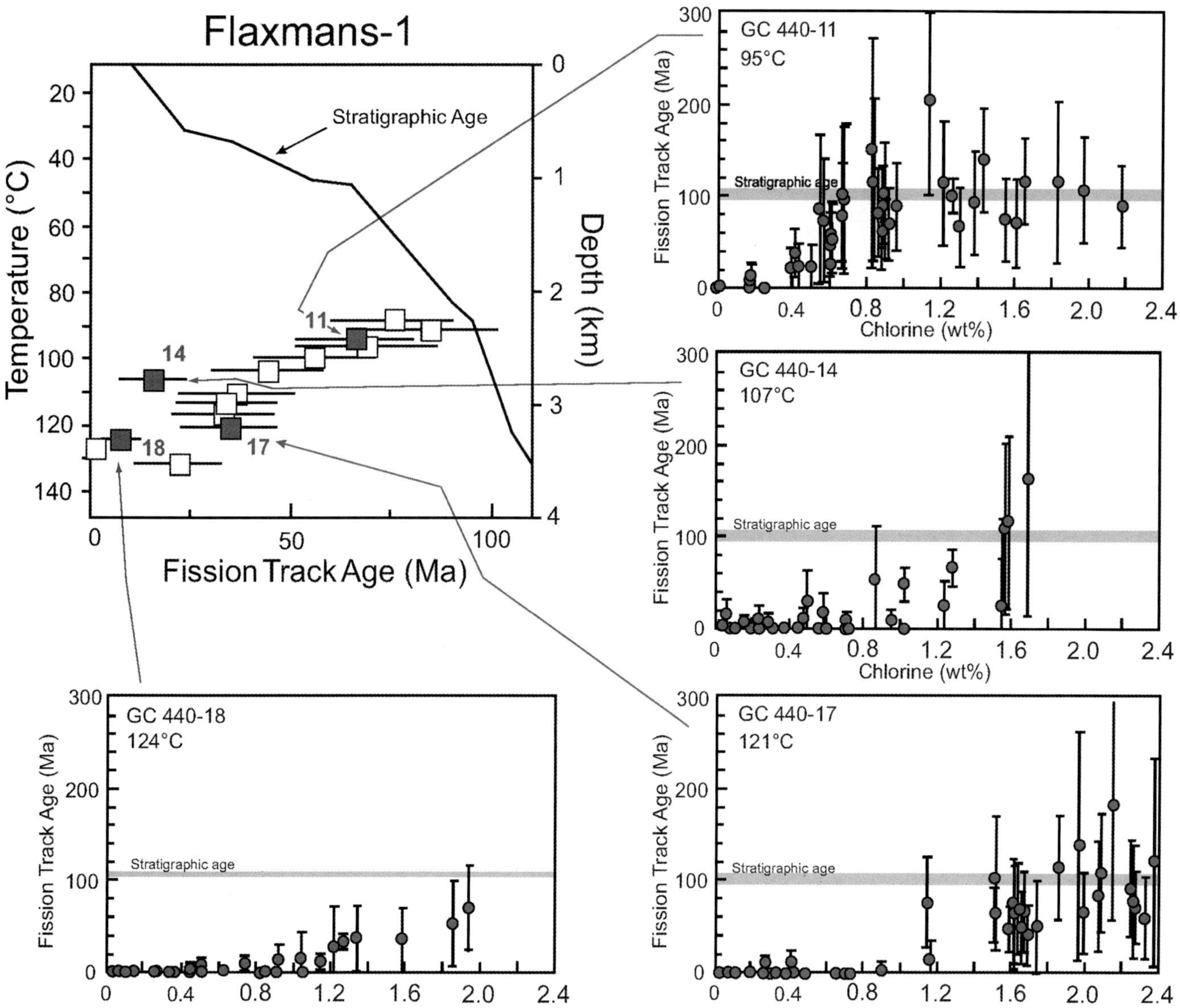

FIG. 6.—Variation of central fission-track age with depth for samples from the Flaxmans-1 well, Otway Basin, SE Australia, together with the variation of fission-track age with chlorine content for individual apatite grains from four selected samples. In each of these samples, the most sensitive (i.e., low wt% chorine) grains are totally annealed (i.e., zero fission-track age), while grains with more retentive (higher Cl) grains give ages up to the depositional age and above. With increasing present-day downhole temperature (the maximum postdepositional temperature in these samples), the transition to total annealing shifts to progressively higher Cl contents, demonstrating the systematic influence of chlorine content on annealing (Gleadow and Duddy 1981; Green et al. 1985, 1986). Note that while most of the central fission-track ages define a generally smooth decrease with increasing temperature, the age for sample GC440-14 is off trend. This is due to the absence of grains with chlorine >1.6 wt% Cl compared to adjacent samples, which are dominated by more retentive grains and therefore give higher central ages. This major effect of chlorine on apatite fission-track age must be taken into account in order to extract meaningful geological constraints from apatite fission-track data.

cuttings in subsurface samples, where "caving" of material from shallower levels can be recognized in this way (Japsen et al. 2005). Such anomalous grains can be eliminated from the data set prior to extraction of thermal history solutions.

In geological conditions, differential annealing effects within individual samples are maximized in rocks that have been heated into the critical temperature range (typically 90–120° C), where the most sensitive (i.e., low Cl) apatites are totally annealed, while more resistant apatites (high Cl) are unaffected (Fig. 6). In such cases, the systematic dispersion in fission-track age, correlating with wt% Cl, provides added precision to a thermal history solution (Crowhurst et al. 2002) (Fig. 9).

Numerous examples exist in the literature of apparently anomalous observations that can probably be simply explained in terms of

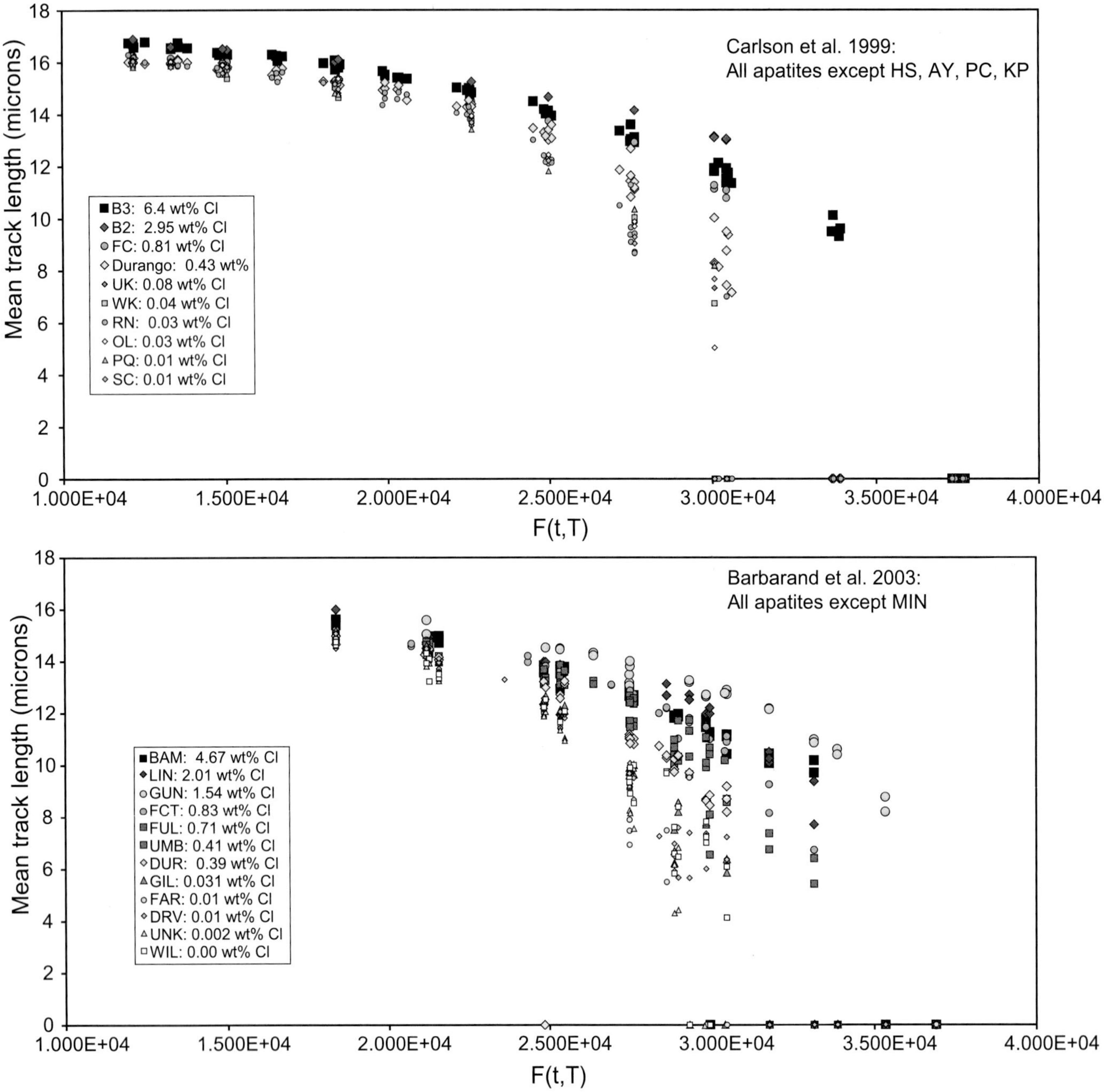

FIG. 7.—Mean track lengths from laboratory annealing experiments reported by Carlson et al. (1999) and Barbarand et al. (2003a), plotted against a unifying function of temperature and time, which reduces all data to a common scale. This function is of the form $F(t,T) = [\log t - \log t_o]/[(1/T) - (1/T_o)]$, where $\log t_o = -10$, and $1/T_o = 0.001$. Apatites of different Cl content are coded to illustrate this variation, with high Cl contents (>1 wt% Cl) shown in blue colors and large symbols, apatites low in Cl (<0.1 wt % Cl) shown in pale colors and small symbols, and intermediate compositions shown in yellow, red, and green colors. The Durango apatite (yellow diamonds) and Fish Canyon Tuff apatite (green circles) are common to both data sets. These results clearly illustrate the first-order control on annealing rates exerted by Cl content, with apatites high in chlorine giving longer lengths for any given heat treatment than those low in Cl. While the first-order control from chlorine is clear, other elements produce additional variation, and several apatites have been omitted from these plots as they are not consistent with the main body of data. See the CD/DVD version to view this figure in color.

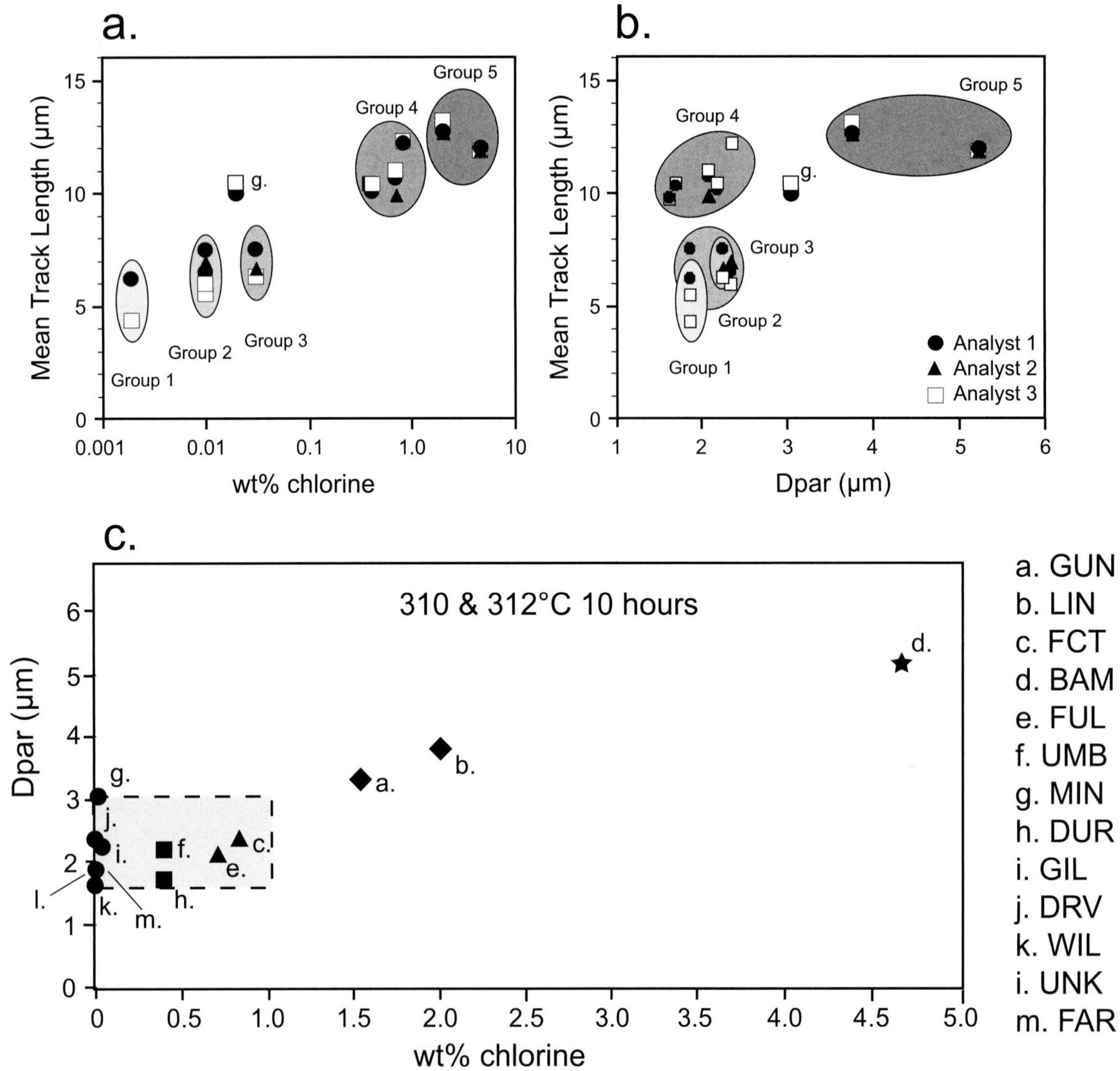

FIG. 8.—Comparison of the variation in mean track length with chlorine content (a) and with the width of track openings parallel to the *c*-axis (Dpar) (b) from the laboratory annealing experiments of Barbarand et al. (2003a). Track size shows only a poor correlation with mean fission track length, whereas chlorine contents show a very strong correlation, illustrating that Dpar is only a poor proxy for fission-track annealing rates. Similar behavior is seen in results from other annealing conditions reported in the data set of Barbarand et al. (2003a), as discussed in detail by Green et al. (2005a). The track-size parameter Dpar shows only a poor correlation with chlorine contents between 0.and 1 wt% (c), and it should therefore be expected that Dpar would provide only a limited indication of differential annealing sensitivities between different apatite species.

differences in wt% Cl. One example is the apparent difference in resistance to erosion attributed to adjacent gneissic and charnockitic terrains in India, reported by Gunnell (2000), which is more likely to be due to a difference in annealing rates in apatites from the two rock types (with charnockitic apatites likely to be richer in Cl, and therefore giving older ages). In such cases, measurement of Cl contents in the analyzed grains can easily resolve such effects.

Later Kinetic Models

Subsequent to publication of the Laslett et al. (1987) kinetic model of fission-track annealing, various refinements or alternative forms of the basic fanning Arrhenius plot model have been published. Carlson (1990, p. 1120) suggested a model based on the laboratory annealing data from Green et al. (1986) and other (then unpublished) data, and claimed that because this model was "based on explicit physical mechanisms, extrapolations of annealing rates to the lower temperatures and longer timescales required for the interpretation of natural fission track length distributions can be made with greater confidence than is the case for purely empirical relationships fitted to the experimental annealing data." However, as explained by Green et al. (1993), all aspects of Carlson's model are in fact purely empirical, and this model is inherently no more reliable than any other model. In addition, as reported by Crowley (1993), detailed inspection shows that Carlson's model does not fit the laboratory data set at all well.

Crowley et al. (1991) published new annealing data for three apatites

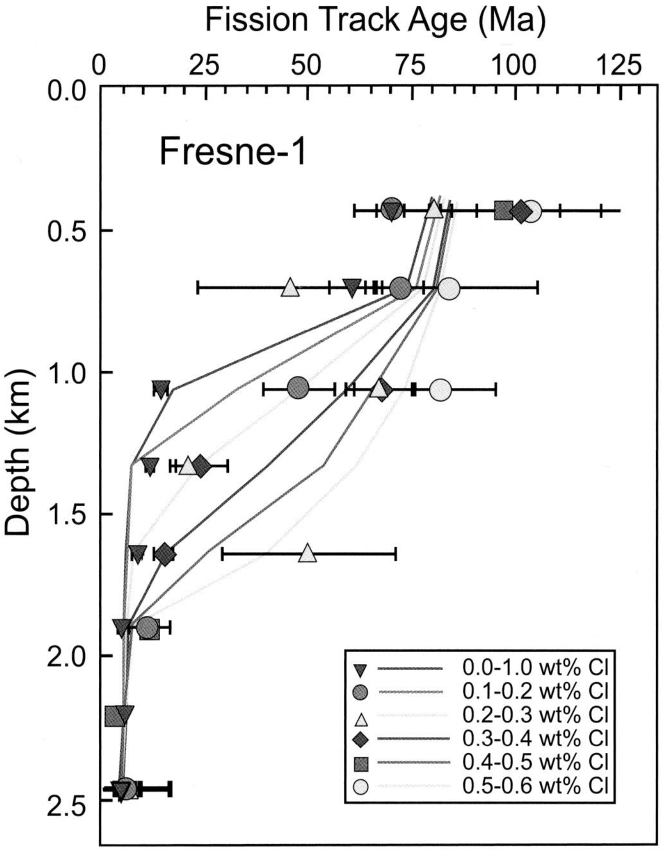

FIG. 9.—Variation of fission-track age with depth for apatites of different chlorine content in samples from the Fresne-1 well, Taranaki Basin, New Zealand (after Crowhurst et al. 2002). Apatites with lower chlorine content achieve a given degree of annealing (as expressed by the reduction in fission-track age) at shallower levels, corresponding to lower maximum paleotemperatures, compared to apatites with higher chlorine. In particular, note the progressive shift in the transition from partial to total annealing (defined by the inflection in the age–depth trend), with apatites containing 0.5–0.6 wt% Cl becoming totally annealed almost 1 km deeper than samples containing between 0.0 and 0.1 wt% Cl. Curves show the variation of fission-track age for each compositional group predicted from the final reconstructed history for the Fresne-1 well, as reported in Crowhurst et al. (2002).

of different composition, including the Durango apatite from which the Laslett et al. (1987) model was derived, and fitted kinetic models that were similar to the Laslett et al. (1987) model except for the use of a nonzero inverse temperature intercept ($1/T_o$). It should be noted that Laslett et al. (1987) considered such models but showed that the best estimate of $1/T_o$ in their analysis was not significantly different from zero, and because the assumption of $1/T_o = 0$ simplifies the mathematical description, they preferred the simpler model. Despite an apparently superior experimental design compared to the Green et al. (1986) study, the kinetic models offered by Crowley et al. (1991), as well as that of Carlson (1990), give predictions in geological conditions that are not consistent with observations (as reviewed in more detail later), and these models have not achieved widespread use. Crowley et al. (1991) investigated Arrhenius plot models in which contours of equal annealing are curved, although linear models were favored. Crowley et al. (1991) also fitted a revised model to the annealing data for Durango apatite published by Green et al. (1986). Predictions of the revised model are not very much different to those from the Laslett et al. (1987) model (below), and this model has also not found widespread use.

Subsequently, Laslett and Galbraith (1996) showed that rather than using the reduced track length L/L_o (where L_o is the mean length of unannealed tracks) as the controlling parameter, improved models could be achieved by fitting to the measured mean track lengths, L, instead of L/L_o, while L_o becomes simply an additional empirical parameter to be estimated. This is of particular importance because Donelick et al. (1990) showed that tracks produced by thermal neutron irradiation begin to anneal (shorten) over very short timescales (minute to hours), even at room temperature. For this reason, mean lengths of unannealed induced confined tracks that are measured weeks to months after irradiation have already undergone some degree of length reduction and therefore have no fundamental significance.

In our own work, we use a series of fanning Arrhenius plot models of the form advocated by Laslett and Galbraith (1996), with finite $1/T_o$ intercept, fitted to measured track lengths, employing coefficients that vary smoothly with wt% Cl, such that the overall annealing rate decreases as wt% Cl increases. This model is based on a combination of laboratory annealing data and geological constraints, using data from a series of exploration wells in which the thermal histories can be reconstructed with confidence. This avoids uncertainties associated with extrapolation of models from laboratory to geological timescales (discussed later herein). Effectively, the relative behavior of different apatite species in laboratory conditions (similar to that shown in Fig. 7) is mapped onto the variation within the geological data set, because there is insufficient overlap over the full range of apatite compositions in the geological data to provide a satisfactory basis for model fitting based on these data alone.

In similar fashion, Ketcham et al. (1999) fitted a number of models to laboratory annealing data (Fig. 7) in a variety of apatite species reported by Carlson et al. (1999) and showed how these could be combined into a multikinetic model that predicts parameters in samples where ranges of kinetic species are present. While their laboratory data favor linear fanning Arrhenius plot models, Ketcham et al. (1999) found that the predictions of curvilinear models provide a closer match to their chosen geological constraints than the linear models.

Precision and Accuracy of Model Predictions in Geological Conditions

Application of any empirically constructed kinetic model based on laboratory data to derive meaningful constraints from measured data in geological conditions depends critically on the accuracy and precision involved in the extrapolation of the model over many orders of magnitude in time, and validation of such models is an important step in reliable practical application.

Green et al. (1989b) quantitatively assessed the precision associated with extrapolation of the Laslett et al. (1987) model from laboratory to geological timescales, suggesting typical uncertainties of ~0.5 μm for mean lengths around 10 μm or less, and ~0.3 μm for mean lengths longer than 10 μm. These figures are equivalent to an overall uncertainty in estimates of maximum paleotemperature derived using this approach of around ±5° C (95% confidence limit [c.l.]).

Accuracy in this context means the degree to which model predictions reproduce the behavior of the natural system. This was also assessed in detail by Green et al. (1989b), who showed that predictions based on the Laslett et al. (1987) model agree well with observed AFTA parameters in samples from temperatures less than about 70° C from a series of reference wells in the Otway Basin of SE

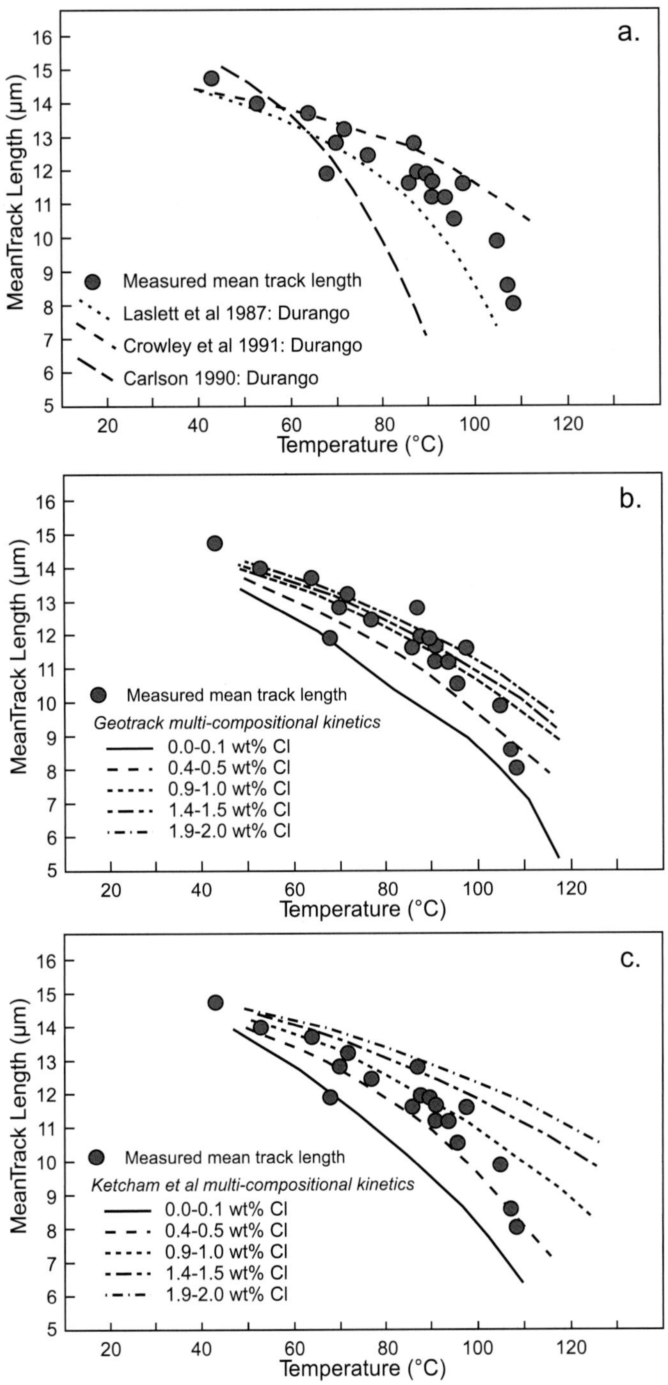

FIG. 10.—Measured mean track lengths in samples from a series of reference wells in the Otway Basin of SE Australia, compared with values predicted from various annealing models. In (a), three monocompositional annealing models, all based on laboratory annealing in Durango apatite, show various degrees of mismatch to the measured data, emphasizing the need for geological calibration in order to make reliable predictions. The Laslett et al. (1987) model provides a reasonable match to the data, while those of Crowley et al. (1991) and Carlson (1990) depart significantly from the measured values. In contrast, predictions from Geotrack's multicompositional model (b) show a good match to the measured

←

values (which should not be surprising, as data similar to these were employed in construction of the model). Predictions from the model of Ketcham et al. (1999), using chlorine as the kinetic parameter (c) also show a reasonable fit to the data, but they also show a greater dispersion over the range from 0.0 to 2.0 wt% Cl compared to the Geotrack model.

Australia (Fig. 10). A systematic mismatch above 70° C (Fig. 10) can be explained by the difference between the composition of the Durango apatite used in the Laslett et al. (1987) model and the apatites in samples from the Otway Basin, which contain a much broader range of Cl contents (see Fig. 6, which shows chlorine data in samples from the same data set). As also shown in Figure 10, other monocompositional models provide predictions that differ widely from those of the Laslett et al. (1987) model, and they are clearly not as accurate in geological conditions. Corrigan (1993) reached somewhat different conclusions from a similar analysis based on data from the US Gulf Coast, but the most important factor is that different laboratory models give different predictions in geological conditions. Any mismatch between predicted and observed geological behavior inevitably implies that use of such models to extract thermal history information will inevitably result in unreliable results.

In comparing model predictions with measured data from geological samples, it is vital to ensure that the models used are appropriate to the apatite compositions present in the sample. Figure 10 also shows predictions from our own multicompositional model, together with those from the model of Ketcham et al. (1999). Both these approaches take explicit account of the distribution of wt% Cl in the sample and the systematic change in annealing properties with increasing Cl content, and both provide a close match to the measured data, with the main difference being a greater dispersion of data in apatites from different Cl contents using the Ketcham et al. (1999) model. Both approaches are also calibrated by a combination of laboratory and geological annealing constraints, so the improved match for the multicompositional models in Figure 10 should perhaps not be too surprising. However, the improved accuracy of these models compared to the monocompositional models reflects the inherent limitations of models based on laboratory data alone and illustrates the benefits of incorporating geological constraints into model calibration.

Accuracy of Model Predictions at Low Temperatures

Application of the Laslett et al. (1987) kinetic model to extract thermal history information from AFT data in outcrop samples (using approaches discussed in a later section) has in many cases resulted in histories involving Late Cenozoic cooling from temperatures around 60° C. This is widely considered to represent an artifact, reflecting the inaccuracy of the model at low temperatures (e.g., Hendriks and Andriessen 2002, Stephenson et al. 2006).

Vrolijk et al. (1992) studied low-temperature annealing in detail, using data in apatites from samples of core from Ocean Drilling Program (ODP) sites in the western Pacific, which they suggested can confidently be shown to have never experienced temperatures in excess of 25° C at any time after deposition. In their key sample, MB-7, Vrolijk et al. (1992) reported a mean track length of 14.6 ± 0.1 μm, which is considerably shorter than the value of ~15.3 μm predicted from the Laslett et al. (1987) model for the reconstructed thermal history of these samples. Thus, they concluded that the Laslett et al. (1987) model overestimated the observed track length, suggesting that the model is over-retentive at low temperatures. Predicted values using a number of other models were also longer than the observed mean track length.

However, the comparison of predicted and measured values in the Vrolijk et al. (1992) study is compromised by several issues. For example, the Laslett et al. (1987) model strictly relates only to data generated in exactly the same manner as the data from which the model was originally derived (Green et al. 1986). Interlaboratory differences, and even within-laboratory differences between different analysts, can easily account for differences in mean track length of the order of 0.5 μm (Barbarand et al. 2003a, 2003b). In addition, the confined track length measurements reported by Vrolijk et al. (1992) were produced using collimated ^{252}Cf fission fragment tracks as hosts for "Tracks-IN-Track" (TINTS). Barbarand et al. (2003b) reported that measurements on TINTS tend to produce rather shorter mean lengths compared to measurements of "Track-IN-CLEavageS" (TINCLES), so some of the reported difference between measured and predicted mean length reported by Vrolijk et al. (1992) could have arisen from this source.

Perhaps the most important factor that could explain at least some of the reported difference is apatite composition. Vrolijk et al. (1992) reported a mean Cl content of 0.13 wt% for the MB-7 apatite, as compared with 0.43 wt% in the Durango apatite on which the Laslett et al. (1987) model is based. On the basis of evidence discussed earlier, the MB-7 apatite should thus be more easily annealed, and therefore should give a shorter mean track length than that predicted from the Laslett et al. (1987) model, exactly as observed. In fact, since most common apatites tend to contain less than 0.1 wt% Cl, it should be expected that the apatites analyzed in most published studies might be more sensitive than Durango apatite. Thus, it is possible that much of the so-called anomalous Late Cenozoic cooling reported in many studies could be due to such compositional effects, rather than any innate deficiencies in the model (Green 2004).

In addition to the factors discussed here, it seems far from clear that the thermal histories employed by Vrolijk et al. (1992) are as well controlled as they suggest, and changing the maximum temperature of a sample by as little as 10° C could have a major impact on the expected mean track length.

Spiegel et al. (2007) developed and extended the concept of analyzing samples from ODP cores, making explicit allowance for compositional influences on annealing. They compared measured mean track lengths in shallowly buried volcanic apatites with values predicted from reconstructed thermal histories based on the Laslett et al. (1987) model and the Ketcham et al. (1999) model, finding that each model successfully predicted some values and performed less well with others, with differences related to apatite composition. In Figure 11, we compare the measured mean track lengths with the predicted values

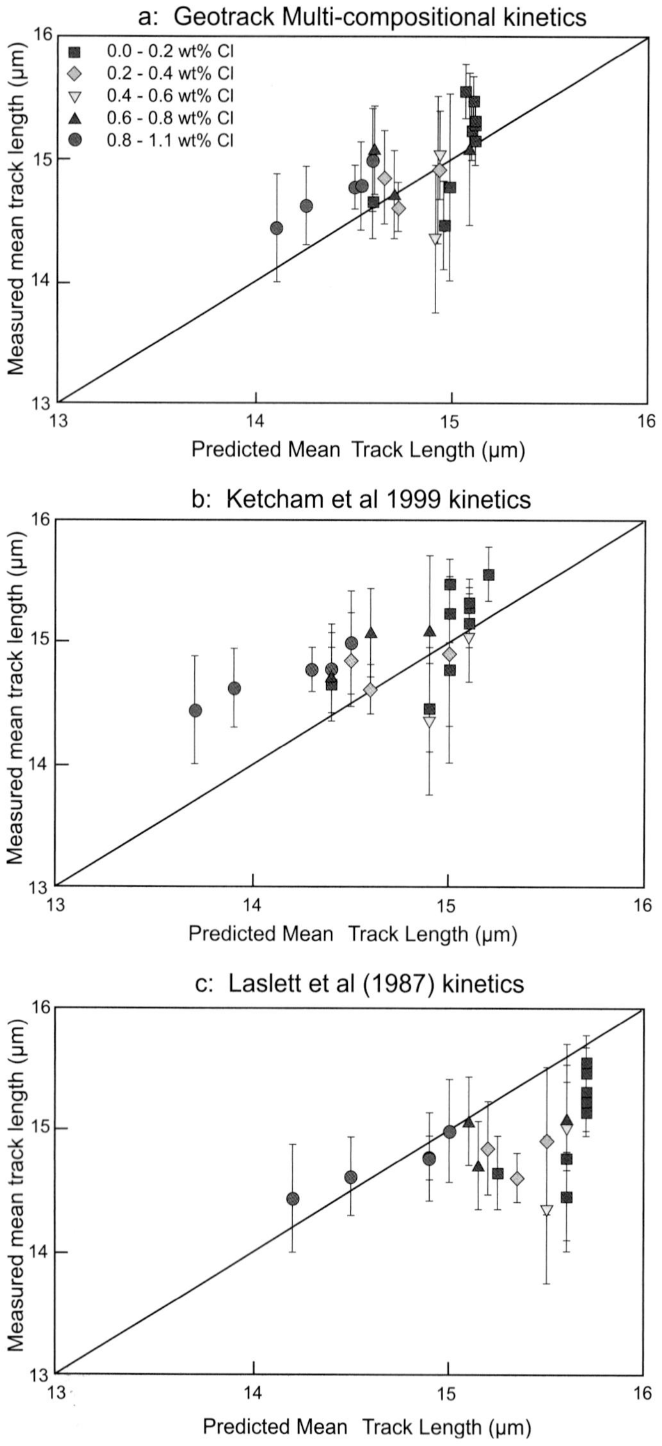

FIG. 11.—Measured mean track lengths in a series of samples designed to provide constraints on low-temperature annealing behavior in geological conditions (from Spiegel et al. 2007), plotted against the values predicted from various annealing models. Apatite compositions are shown, and kinetics appropriate to each composition have been used for the two multicompositional models in (a) and (b). Values predicted using the Geotrack model (a) show generally excellent agreement overall, while being up to ~0.25 μm less than the measured values at the highest and lowest mean Cl contents.

←

Values predicted using the Ketcham et al. (1999) model (b) are more consistently around 0.5 μm less than measured values over the entire compositional range, while values predicted using the Laslett et al. (1987) model (c) are around 0.5 μm longer than the measured values for apatites containing 0.0–0.2 wt% Cl, with the disparity increasing to slightly over 0.5 μm for compositions similar to Durango apatite (~0.4 wt% Cl) and then decreasing at higher Cl contents, such that around 0.8 wt% Cl, the agreement is excellent. While it is clear that multicompositional models provide better matches to the low-temperature constraints, the general level of agreement of values predicted from the Laslett et al. (1987) model should be regarded as impressive, since this model is based purely on extrapolation of laboratory data, with no geological control, yet the agreement is almost as good as for models that directly incorporate geological constraints. This disproves the common misconception that the Laslett et al. (1987) model is too sensitive at low temperatures, as discussed in the text.

from these two models and also with values predicted using the Geotrack multicompositional model. For the Geotrack and Ketcham et al. (1999) models, kinetics appropriate to the Cl content of each sample have been used. We note that in three of the samples reported by Spiegel et al. (2007) (samples 43-10, 47-3, and 47-7), the measured mean track lengths appear to be anomalously low compared to those in adjacent samples with similar compositions. This is most likely due to the presence of shorter track lengths from grains derived from older sources, which is supported by the larger than usual standard deviations reported for these three samples compared to the rest of the data set (Spiegel et al. 2007, Table 4). Results from these three samples have therefore been omitted from Figure 11.

Values predicted using the three models give rather different results in Figure 11. Values predicted using the Geotrack model are up to ~0.25 μm less than the measured values at the highest and lowest mean Cl contents, although the agreement is excellent for compositions between 0.2 and 0.8 wt% Cl. Values predicted using the Ketcham et al. (1999) model are more consistently around 0.5 μm less than measured values over the entire compositional range. In contrast, values predicted using the Laslett et al. (1987) model for apatites containing 0.0 to 0.2 wt% Cl are around 0.5 μm longer than the measured values (as expected from the previous discussion), and this difference first increases and then decreases with increasing wt% Cl, such that for Cl contents around 0.8 wt% Cl, the agreement is excellent. While mean track lengths for apatites of similar composition to Durango apatite (~0.4 wt% Cl) are shorter than predicted, the general level of agreement of values predicted from the Laslett et al. model in Figure 11 should be regarded as impressive, since this model is based purely on extrapolation of laboratory data, with no geological control, yet the agreement is almost as good as for models that directly incorporate geological constraints. It should be clear from Figure 11 that any failure of this model at low temperatures is relatively minor, and, in particular, the prediction from this model of significant length reduction even at temperatures as low as 10° C over geological timescales closely reproduces the natural system behavior, as discussed further later herein. We should also note that the comparisons by Spiegel et al. (2007) are subject to the same issues with respect to possible analytical differences as previously discussed in relation to the Vrolijk et al. (1992) study.

Misguided Revision of the Laslett et al. (1987) Model

In an attempt to make allowance for the perceived inaccuracy in the low-temperature behavior of the Laslett et al. (1987) model, Allen et al. (2002), Gunnell et al. (2003), and Kohn et al. (2002), amongst others, have used a revised version of this model, in which the original model has been renormalized to a lower value of L_0 (the unannealed mean track length) of 14.5 μm, which they considered to provide a more reliable representation of unannealed tracks in geological conditions. (Note that the results reported by Spiegel et al. [2007], as discussed already, show that this is inappropriate.) However, a revision of the model in this way also introduces a systematic error at higher temperatures, where the original model has been validated (Green et al. 1989b) and is generally considered to be accurate. For this reason, the denudation histories extracted by Allen et al. (2002), Gunnell et al. (2003), and Kohn et al. (2002) and others, using the revised model, can have little geological meaning.

Use of this approach has been further considered recently by Stephenson et al. (2006), who integrated such a model more directly with geological constraints in an attempt to produce a more accurate model at low temperatures. However, on the basis of the previous discussion, it should be clear that correcting the Laslett et al. (1987) model in this way results in a less accurate model, and the most effective route toward obtaining reliable results lies in making full allowance for the influence of apatite composition on annealing rates, and using kinetic equations that relate directly to the composition of the apatite being analyzed, as advocated by Ketcham et al. (1999).

The errors introduced by this misguided revision of the Laslett et al. (1987) model were illustrated in a recent study by Campanile et al. (2008), who compared the denudational history of the Western Indian passive margin derived from AFT data by Gunnell et al. (2003) with the pattern of offshore sediment accumulation through time. Whereas the denudational history onshore was reported as showing a strong peak in the Late Cretaceous and Paleocene, the offshore depositional record reported by Campanile et al. (2008) shows peaks in both the Early and Late Cenozoic. While the earlier pulse of sedimentation could be explained by the Paleocene acceleration in denudation onshore revealed by the AFT data, Campanile et al. (2008) could offer no explanation for the Pliocene episode of sedimentation offshore, which was regarded as enigmatic. In this regard, it is surprising that Campanile et al. (2008) did not mention that the denudational history reported by Gunnell et al. (2003) was derived using the revised version of the Laslett et al. (1987) annealing model discussed here, the specific purpose of which was to remove the supposedly anomalous late cooling indicated by AFT data. The version of the denudational history of the margin derived by Gunnell et al. (2003) using the original version of the Laslett et al. (1987) model shows an accelerated phase of Neogene denudation (in addition to a Late Cretaceous pulse), which can easily explain the renewed pulse of offshore sediment accumulation in the Late Cenozoic. Thus, the record of sedimentation offshore of the Western Indian margin provides good support for the accuracy of the original Laslett et al. (1987) model, and this example illustrates the folly of revision of this model adopted in recent studies (cited previously). We strongly recommend that use of this revised model should be abandoned, and results derived previously using the revised model should be disregarded.

Alternative Viewpoints on Fission-Track Annealing in Apatite

While opinions differ on the most appropriate form of kinetic model to use, and the best indicator of differential annealing between various apatite species, there is general agreement that the only major environmental control on fission tracks in apatite is the variation of temperature with time. However, two alternative factors have recently attracted renewed attention, and we review each in turn.

Wendt et al. (2002) suggested that contrary to previous ideas (e.g., Fleischer et al. 1975), pressure could play an important role in affecting the kinetics of the annealing process. Kohn et al. (2003) criticized many aspects of the ideas presented by Wendt et al. (2002), and despite contrary arguments by Vidal et al. (2003), the idea that pressure can affect fission-track annealing kinetics is not a widely held view. We note that if pressure does have a significant effect on annealing rates, it might be expected to be most obvious in results from deep drill holes in cratonic regions characterized by low thermal gradients, but results to date from such regions (e.g., Rohrman et al. 1995, Lorencak et al. 2004) show no such evidence.

More recently, Hendriks and Redfield (2005) suggested that AFT ages across Scandinavia were anomalously young because of nonthermal annealing resulting from the effects of long-term radiation dose from alpha decay of uranium and thorium at low temperatures over hundreds of millions of years. This suggestion, if true, would have severe implications for the interpretation of AFT data. One of the main lines of evidence used to justify this claim was the observation that AFT ages were younger than (U-Th)/He ages in the same apatites, contrary to expectation. However, Green et al. (2006) showed that this is due to anomalous behavior in the (U-Th)/He system, involving enhanced He retentivity due to the accumulation of radiation damage within the apatite lattice (subsequently demonstrated also by Shuster et al. 2006, Shuster and Farley 2009) and not anomalous fission-track annealing behavior. The remaining evidence used by Hendriks and Redfield (2005) to support the notion of "radiation enhanced" fission-track annealing can be interpreted in conventional fashion in terms of standard processes of thermal annealing (Green and Duddy 2006, Larson et al. 2006). The concept of "radiation-enhanced annealing" currently finds little support outside its direct proponents.

Despite these recent suggestions of unrecognized system behavior, the AFT system is one of the best-understood (at least in an empirical sense) methods of thermochronology currently available, and it provides results that match constraints from other techniques when applied in tandem (see later discussion). While approaches used by different groups may differ in detail, the response of the system is well defined, and thermal history information provided by the technique can be considered reliable within the constraints and limitations of the approaches used. We suggest that perceived problems with the technique, as discussed for example by Gunnell (2000), result from a failure to appreciate these limitations, rather than any fundamental problem with the technique itself.

EXTRACTING THERMAL HISTORY INFORMATION FROM AFTA DATA

Basic System Response

An understanding of the thermal history response of fission tracks in detrital apatites to heating and cooling is fundamental to appreciating how the technique can be applied in practice. The basic system response within a typical sedimentary basin framework, based on principles outlined in Figure 5, is illustrated in Figure 12. The nature of this response is common to all forms of kinetic models discussed earlier and is a fundamental property of the AFT system.

For sample A in Figure 12a, as temperature increases (representing increased burial depth in this example), the length of all tracks is progressively reduced, and because temperature dominates over time in the kinetics of annealing, at any time during this phase of the history, all but the most recently formed tracks at any given time have the same mean length (Fig. 12b) (but remember that each population of tracks has a finite spread in length). At the point when the maximum temperature is reached and the history changes from heating to cooling, all tracks formed up to that point in time are effectively "frozen" at the length to which they have been reduced. They do not undergo further shortening because annealing rates are much slower at the reduced temperatures now prevailing, and they do not get longer because the annealing process is irreversible. Those tracks formed after the onset of cooling remain longer because of the much lower annealing rates at the prevailing lower temperatures.

At the end of the history (i.e., the present day), two populations of tracks are present in sample A: a shorter population representing tracks formed up until the onset of cooling from the paleo–thermal maximum, and a longer population formed after the onset of cooling, resulting in a bimodal track length distribution (Fig. 12c). The shorter population of tracks will contribute a reduced component to the fission-track age, compared to the time interval over which tracks have been retained, while the contribution to the fission-track age of the longer population will be much closer to the time elapsed since the onset of cooling. The final measured fission-track age will represent the summed contributions of both components (Fig. 12d).

Sample B in Figure 12a reached a maximum temperature sufficient to reduce the length of all tracks produced up to that time to zero (Fig. 12b) (i.e., all the radiation damage has been repaired, and no etchable tracks remain). At the present day, this sample contains only one track population (Fig. 12c), formed after the sample cooled to temperatures at which tracks could be retained (~110° C for typical apatite compositions). The track length distribution in this sample will reflect the thermal history in the postcooling period, while the fission-track age will be determined by the time when the sample began to retain tracks, but moderated by the degree of length reduction of tracks formed during the cooling history (Fig. 12d).

Extending these basic principles through the vertical section results in the variation of fission-track age and mean track length with depth shown in Figure 12e, which is characteristic of a section that has cooled from higher temperatures. Fission-track age declines rapidly with increasing depth through the shallower section, as the mean length of the shorter population of tracks (formed up to the onset of cooling) is progressively shortened and the proportion of these tracks that can reach the polished grain surface to be revealed decreases. Similarly, the mean track length reduces due to the decreasing mean length of the shorter population. However, as the depth (and temperature) corresponding to total annealing of all tracks formed prior to the onset of cooling is approached, the mean track length begins to increase again, as the shorter population of tracks becomes increasingly difficult to reveal and therefore contributes less to the overall mean for the sample, which is increasingly dominated by the longer population of tracks formed after cooling. As the transition from partial to total annealing of tracks formed prior to cooling is crossed, the mean track length increases abruptly because the sample is now dominated only by longer tracks formed after the onset of cooling, and the fission-track age reduction shows a characteristic "break in slope," below which only a single component of tracks is present, with parameters controlled by the history after the onset of cooling. With further increase in depth, both the fission-track age and mean track length show progressive reduction to zero at a temperature controlled by apatite composition and the timescale of heating/burial, similar to the simple situations illustrated in Figure 3.

What Information Is Contained in Fission-Track Age and Length Data?

Figure 12 illustrates how samples at different depths through a vertical section contain different types of information regarding the underlying thermal history. In samples shallower than the transition from partial to total annealing, which we will refer to as the "paleo–110° C isotherm" (although we stress that, in detail, this temperature will depend on apatite composition and timescale of heating), the length of the shorter component of the track length distribution is determined by the maximum temperature to which each sample was heated, and the proportion of short to long tracks at any given horizon is determined by the ratio of the time prior to cooling to the time after cooling (moderated by the geometric biases in revealing tracks of different length as described by Laslett et al. 1982). The fission-track age is determined by a combination of these factors and has no fundamental significance in its own right (i.e., the fission-track age does not denote a time at which some specific event occurred). In contrast, samples from below the paleo–110° C isotherm contain only tracks formed after cooling and, therefore, only provide a minimum estimate of the maximum temperature. However, the fission-track age in such samples is controlled by the time at which the samples cooled through the paleo–110° C isotherm, again moderated by the track length reduction as a result of the history after the onset of cooling.

To summarize, for the style of history illustrated in Figure 12, which is typical of exhumed sedimentary basins, AFTA can provide the following information in samples shallower than the paleo–110° C isotherm:

(1) maximum paleotemperature from the mean length of the shorter component;

(2) onset of cooling from the proportion of short to long tracks; and

(3) additional refinement of these first two factors from the degree of fission-track age reduction; and

in samples deeper than the paleo–110° C isotherm, the following information can be obtained:

(1) minimum estimate of the maximum paleotemperature; and

(2) onset of cooling from the combination of fission-track age and track lengths.

Thus, we obtain different types of thermal history information from different facets of the data at different positions within the section in Figure 12. The practical consequence of this is that in order to obtain the maximum amount of information from AFTA, it is necessary to analyze a sequence of samples spanning a range of vertical horizons. As a corollary of this, the amount of information that can be obtained from a single sample may be limited.

What Information Is NOT Contained in AFTA Data?

Also implicit in Figure 12 is the fact that there is a definite limit to the amount of information that can be obtained from AFTA data in such situations. In any sample from a depth shallower than the paleo–110° C isotherm, when the paleo–thermal maximum is reached, track populations formed at different times through the history are reduced to more or less the same mean track length (but note that a finite spread in length will be present because of the inherent distribution of track lengths). For this reason, all information on the prior history of such samples has been lost (except for the total duration over which tracks have been retained, and this can be very difficult to reconstruct, as discussed later herein). Figure 13 illustrates the ways in which successive heating episodes overprint the effects of earlier episodes, leaving only evidence of the maximum temperature episode and the subsequent history after cooling from the paleo–thermal maximum. Thus, AFTA data are sensitive only to the magnitude of the maximum temperature and the timing of the onset of cooling (in relation to the overall time over which tracks have been retained), and they preserve no information on the prior history (except that temperatures must have been lower than at the paleo–thermal maximum).

Similar considerations apply to scenarios involving continuous cooling histories, as shown in Figure 14, which illustrates the insensitivity of such histories to the detailed variation of temperature with time. Perhaps most importantly, the final AFTA parameters are particularly insensitive to the earliest history, which provides such a minor degree of information to the fission-track age and track length distribution that huge differences in time produce insignificant changes in the expected parameters, which are beyond practical resolution. Thus, in monotonic cooling histories, the data are insensitive to the early history and specifically to the time at which the sample began to retain tracks.

Note that Figures 12, 13, and 14 are constructed for monocompositional apatites, but the principles involved are equally applicable to apatites containing a range of annealing properties (due to differing Cl contents). As illustrated in Figure 9, AFTA data from the Fresne-1 well in the Taranaki Basin of New Zealand (Crowhurst et al. 2002) provide a multicompositional counterpart involving real data for comparison with the notional monocompositional data in Figure 10.

Practical Implementation Strategies

Because of the high degree of redundancy in the data (i.e., many histories result in the same measured age and length parameters; Figs. 13, 14), extraction of explicit thermal history solutions directly from apatite fission-track data is not possible. Instead, the problem is normally approached by forward modeling of the parameters expected from a range of specified thermal histories and defining the range of conditions that provide predictions that are consistent with the measured data. Within this common philosophy, a range of approaches has been developed by different groups, each of which has its advantages and disadvantages. This disparity can often confuse the nonspecialist, so next we review the most commonly used approaches. Most published studies represent a variation on one of these basic themes.

Geotrack Approach (AFTA®): The approach adopted in our own studies, designed primarily for application to sedimentary basins, takes account of the fact that sedimentary horizons are deposited at the surface and then buried/heated to some maximum depth/temperature, after which they may be exhumed and cooled. By modeling expected AFTA parameters resulting from a range of possible thermal histories, we can define the range of values of maximum paleotemperature and the onset of cooling giving predictions that match the measured data within 95% confidence limits, using likelihood theory similar to that described by Gallagher (1995). The basic principles involved are illustrated in Figure 15 for a monocompositional apatite, while Figure 16 illustrates the extension of these principles to multicompositional data.

Note that no attempt is made to define the whole thermal history, because the history prior to the onset of cooling is overprinted by the thermal maximum (Fig. 12). Instead, we focus on determining those aspects of the thermal history that directly control the measured AFTA parameters, viz. the maximum paleotemperature and the time at which cooling from the paleo–thermal maximum began. Additional episodes of heating and cooling following the onset of cooling from the paleo–thermal maximum can often be resolved, as discussed in more detail in a later section.

The episodic heating and cooling approach is designed specifically for application to sedimentary basins, but we also believe that it is relevant to many (if not all) basement terrains, and such an approach is essential in basement regions where sedimentary outliers occur, revealing earlier cycles of exhumation, burial, and re-exhumation (e.g., Green and Duddy 2006, 2007).

In routine application of AFTA, a multicompositional annealing model is used that takes specific account of the influence of wt% Cl on annealing rates. This model consists of a series of parallel equations, each taking the form of a linear fanning Arrhenius plot with nonzero intercept (Laslett et al. 1987, Crowley et al. 1991, Ketcham et al. 1999), with coefficients that vary systematically with wt% Cl.

AFTA data are routinely integrated with data from other paleo–thermal indicators such as vitrinite reflectance, and/or indicators of burial such as sonic velocity (discussed later). Such data provide an independent check on the interpretation of the AFTA data and ensure that the resulting thermal histories and the information derived from them are not affected by artifacts of any individual technique.

One of the advantages of this approach is that determination of maximum paleotemperatures from AFTA data in a series of samples over a range of depths or elevations allows definition of the variation of paleotemperatures with depth, which in turn provides unique insights into the underlying mechanisms of heating and cooling, as described in a later section.

Monte-Trax Approach: Gallagher (1995) showed how forward modeling could be combined with a "genetic algorithm" to constrain the range of thermal histories that is consistent with the measured data and to converge on the most likely history. This approach, embodied in the "MonteTrax" software package, which has been employed in a large number of published studies, is commonly used to define the complete thermal history below ~110° C. Application is commonly within a framework of monotonic cooling, although this is not essential. The kinetic model of Laslett et al. (1987) is usually

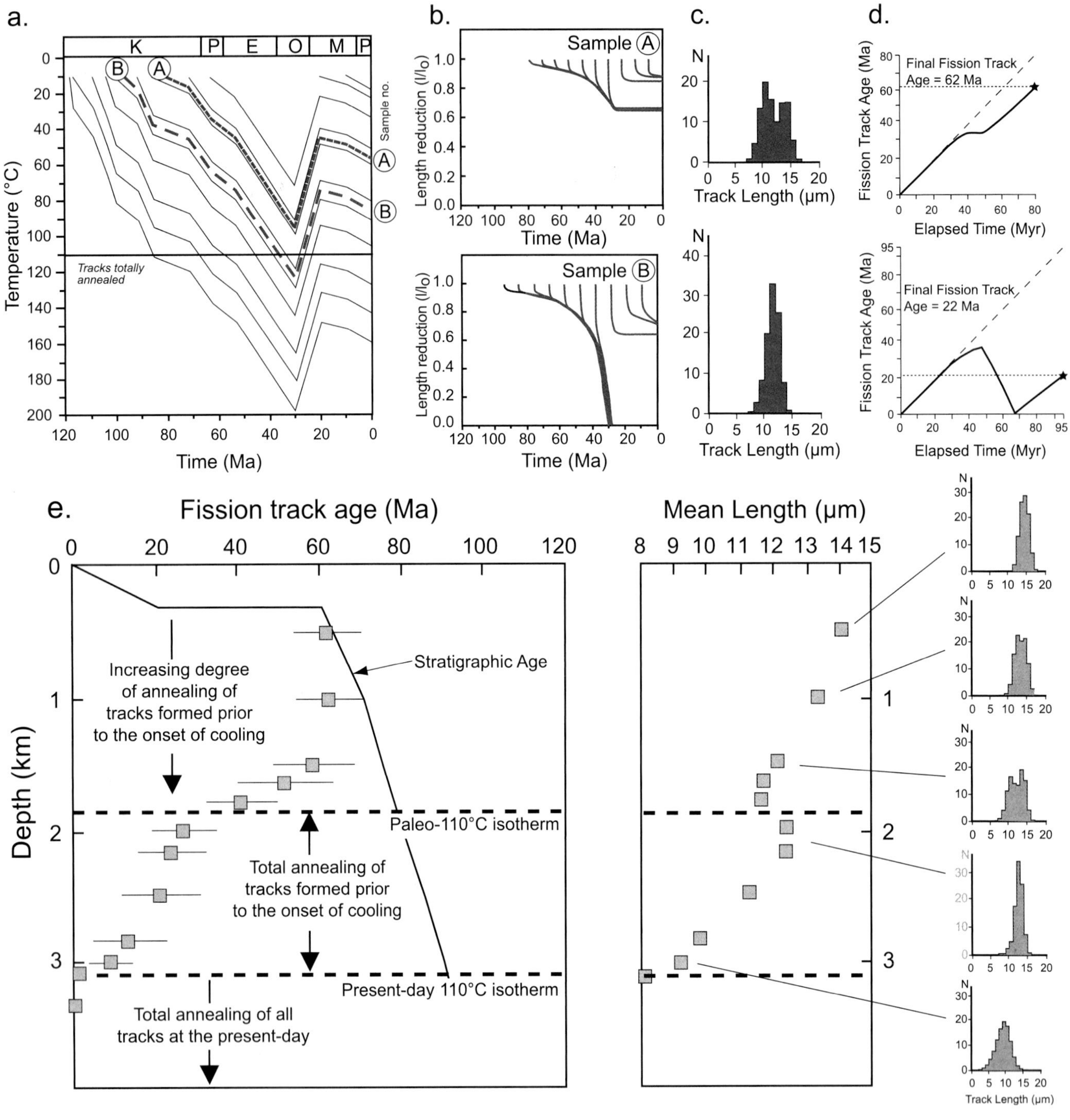

FIG. 12.—Thermal history response of fission tracks in apatite under geological conditions. While this is based on a monocompositional apatite of Durango composition using the Laslett et al. (1987) model, the nature of this response is common to all forms of kinetic models, and it is a fundamental property of the AFT system governed by a fanning Arrhenius plot. (a) Notional thermal history for a sedimentary sequence that underwent progressive burial through the Cretaceous to Middle Cenozoic, followed by cooling due to uplift and erosion commencing at 30 Ma and completed by 20 Ma, with minor reburial from 20 Ma to the present day. The thermal histories of two samples A and B are shown. (b) Track length shortening trajectories for tracks produced at different times in samples A and B. As temperature increases, the length of all tracks is progressively reduced, and because temperature dominates over time in the kinetics of annealing, at any time during this phase of the history, all but the most recently formed tracks at any given time have the same mean length (although each population of tracks has a finite spread in length). At the point when the maximum temperature is reached and the history changes from heating to cooling, all tracks formed up to that point in time are effectively "frozen" at the length to which they have been reduced. They do not undergo further shortening because annealing rates are much slower at the reduced temperatures now prevailing, and they do not get longer because the annealing process is irreversible. Those tracks formed after the onset of cooling remain longer because of the lower annealing rates at the prevailing lower temperatures.

employed, on the assumption that the Durango apatite, for which this model was developed, is representative of common apatites. In many studies using this approach, AFT data are used in isolation as the only constraint on thermal and denudation histories.

Adoption of the Laslett et al. (1987) model as the keystone of this approach introduces some problems, particularly with the low-temperature (<60° C) range of the thermal history solutions, which often contain apparently spurious cooling within the last 10 to 20 million years. As noted already, this is commonly attributed to inaccuracy in predictions of the model at low temperatures (e.g., Hendriks and Andriessen 2002). However, a large contribution to this effect probably arises from the fact that Durango apatite is not representative of apatite grains commonly analyzed from basement rocks, which typically contain <0.1 wt% Cl. In contrast, the Durango apatite contains ~0.4 wt% Cl, and it is more resistant to annealing than typical low-Cl basement-derived apatites (Ketcham et al. 1999, Barbarand et al. 2003a), which will therefore show a greater degree of annealing compared to the predictions of a model based on Durango apatite. This can readily explain the "anomalous Miocene cooling" commonly seen in Monte-Trax solutions (Green 2004). As discussed earlier, Gunnell et al. (2003) renormalized the Laslett et al. (1987) model in an attempt to account for these problems, but in addition to modifying the low-temperature behavior, this also made the model more sensitive at high temperatures. For this reason, coupled with the false premise on which revision of the model was based (as discussed earlier), any results obtained using this modified model therefore have little meaning.

Despite abundant evidence for compositional influences on annealing rates, and the advent of kinetic models that explicitly include such variation (Ketcham et al. 1999), compositional effects have yet to be incorporated into the Monte-Trax software. Thus, the resulting thermal histories can only be considered an approximation to the real history, at best.

Several variations of the "Monte-Trax" approach have been developed (e.g., Corrigan 1991, Lutz and Omar 1991, Willett 1997), some of which implement the Laslett et al. (1987) annealing model, while others have used alternative models (e.g., Crowley et al. 1991). However, extrapolation of these alternative models to geological timescales is not as accurate as that of the Laslett et al. (1987) model, as discussed earlier, and these models have not been widely adopted.

AFTSolve/HeFTy Approach: This approach typically allows for within-sample variation in annealing kinetics using etch pit size as the kinetic parameter, although wt% Cl may also be used (Ketcham et al. 1999). Thermal history information is extracted from AFT data using the "AFTSolve" software package (Ketcham et al. 2000) or more recently HeFTy (Ehlers et al. 2005, Ketcham 2005), in which statistical search procedures are employed to define a range of viable histories for which predictions provide "acceptable" fits to the measured data, including a restricted range of solutions providing "good" fits, which provide an assessment of the range of uncertainty around the best-fit history. Either monotonic cooling or episodic heating and cooling can be employed. Episodic heating and cooling are clearly more relevant to sedimentary basins, but the need to specify points or "temperature–time boxes" in the thermal history at which the history changes from heating to cooling (or vice versa) has the unfortunate result that the timing of the onset of cooling must be defined at the outset. Since this timing is one of the key pieces of information that is contained within the AFT data, this step effectively eliminates one of the major advantages of the apatite fission-track system—the ability to provide an independent estimate of the onset of cooling. This approach results in solutions that are defined up to 200° C (complete with defined uncertainty limits), but this is purely an artifact resulting from the assumed style of history, and it has no geological meaning, since apatite does not retain tracks above ~130° C for typical timescales and compositions.

Fission-Track Stratigraphy Approach: The qualitative use of AFT age vs. elevation patterns or "fission-track stratigraphy" to identify offsets between crustal blocks was pioneered in the Trans-antarctic Mountains (TAM) by Fitzgerald and Gleadow (1988), prior to

←

Sample B reached a maximum temperature sufficient to reduce the length of all tracks produced up to that time to zero. At the present day, this sample contains only one track population, formed after the sample cooled to temperatures at which tracks could be retained (~110° C for typical apatite compositions). (c) Track length distributions for samples A and B resulting from the thermal histories shown in (a). For sample A, two populations of tracks are present in this sample at the present day: a shorter population representing tracks formed up until the onset of cooling from the paleo–thermal maximum, and a longer population formed after the onset of cooling. For sample B, the measured track length distribution will reflect the thermal history in the postcooling period only. (d) Evolution of fission-track age with time resulting from the thermal histories shown in (a). For sample A, the final measured fission-track age will represent the summed contributions from the shorter of the two populations of tracks; the shorter tracks will contribute a reduced component to the fission-track age compared to the time interval over which tracks have been retained, while the contribution to the fission-track age of the longer population will be much closer to the time elapsed since the onset of cooling. For sample B, the final fission-track age will be determined by the time when the sample began to retain tracks, but moderated by the degree of length reduction of tracks formed during the cooling history. (e) Predicted variation of fission-track parameters with depth for a well section that has undergone a thermal history of the style shown in (a). Fission-track age declines rapidly with increasing depth through the shallower section, as the mean length of the shorter population of tracks (formed up to the onset of cooling) is progressively shortened, and the proportion of these tracks that can reach the polished grain surface and be revealed decreases. Similarly, the mean track length reduces due to the decreasing mean length of the shorter population. However, as the depth (and temperature of ~110° C) corresponding to total annealing of all tracks formed prior to the onset of cooling is approached, the mean track length begins to increase again, as the shorter population of tracks becomes increasingly difficult to reveal and therefore contributes less to the overall mean for the sample, which is increasingly dominated by the longer population of tracks formed after cooling. As the transition from partial to total annealing of tracks formed prior to cooling is crossed at the paleo-isotherm of ~110° C, the mean track length increases abruptly because the sample is now dominated only by longer tracks formed after the onset of cooling, and the fission-track age reduction shows a characteristic "break in slope," below which only a single component of tracks is present, with parameters controlled by the history after the onset of cooling. With further increase in depth, both the fission-track age and mean track length show progressive reduction to zero at the present-day ~110° C, although, in detail, this temperature is controlled by apatite composition and the timescale of heating/burial.

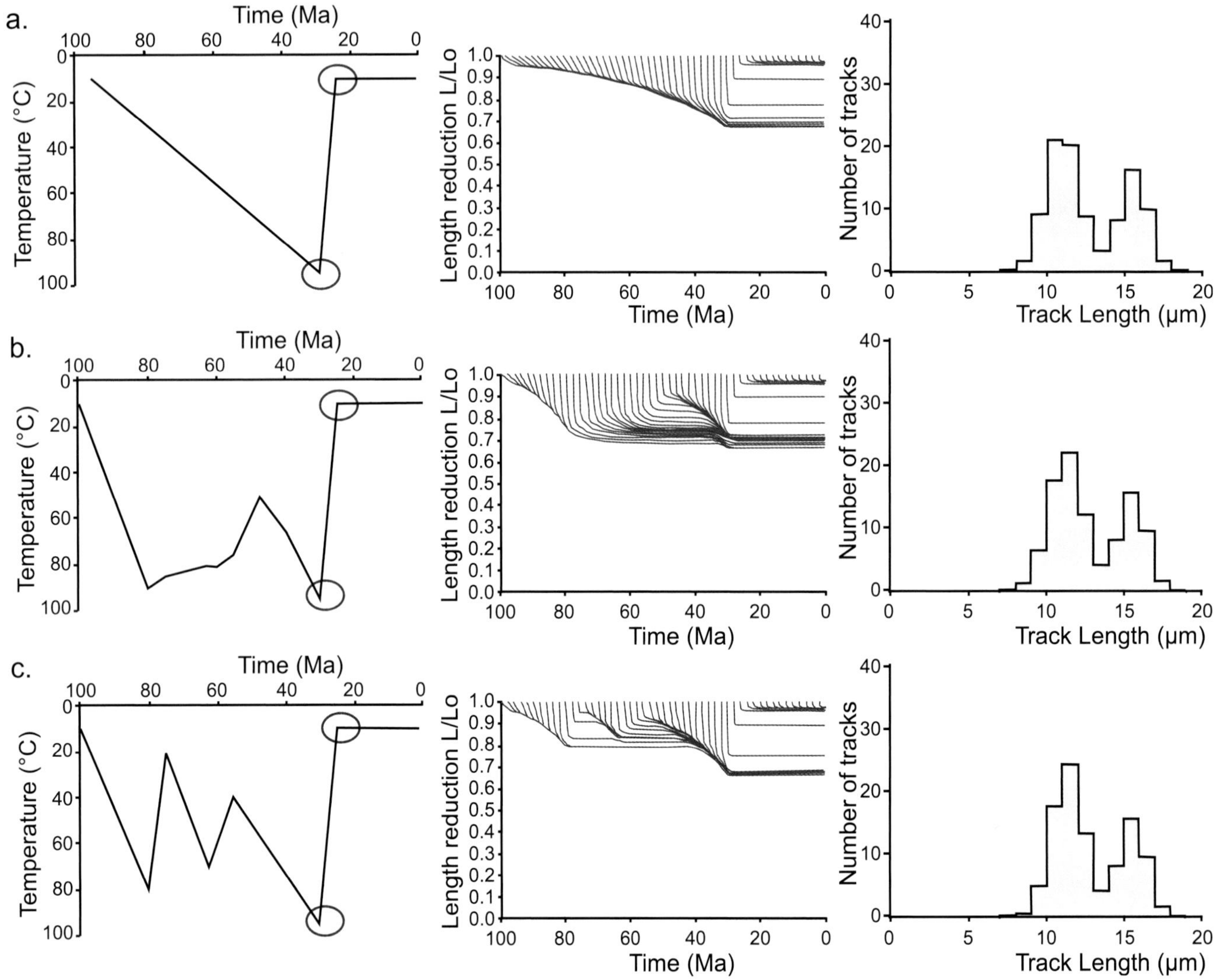

FIG. 13.—Shortening trajectories (center) for tracks produced at different times through three thermal history scenarios (left) representing increasing levels of complexity from (a) through (c). The resulting track length distributions are also shown (right). Despite the obvious differences in the thermal histories, the resulting track length distributions are effectively identical because the time and magnitude of maximum temperatures and the rate of cooling from maximum paleotemperatures are the same for each history (circled points on the thermal histories). This outcome reflects the fundamental kinetics of the AFT system such that the data are sensitive only to the magnitude of the maximum temperature and the timing of the onset of cooling (in relation to the overall time over which tracks have been retained), and they preserve no information on the prior history (except that temperatures must have been lower than at the paleo–thermal maximum). Thus, successive heating episodes overprint the effects of earlier episodes, leaving only evidence of the maximum temperature episode and the subsequent history after cooling from the paleo–thermal maximum. It is therefore not possible to discriminate between these three scenarios from apatite fission-track data.

the advent of quantitative methods. In early studies (e.g., Wagner and Reimer 1972), it was commonly assumed that the change in apatite fission-track age with elevation represented slow cooling associated with progressive denudation of the mountain range, and segments of the age–elevation trend with different slopes were interpreted as representing a change in the rate of denudation. Subsequent integration of confined track length information with the age data (Fitzgerald and Gleadow 1988) showed that the upper segment of such trends above the "break in slope" actually represented an uplifted (exhumed) partial annealing zone (cf. Fig. 12), while samples below the "break in slope" were totally annealed prior to the onset of exhumation and effectively dated the onset of cooling. This approach has been adopted in other regions; e.g., the Pyrenees (Fitzgerald et al. 1999) and Scandinavia (Redfield et al. 2005). While this approach may at first seem to be free of problems associated with the precise kinetic behavior, in fact it demands that all apatites have a similar annealing response (i.e., Cl content). Lorencak et al. (2004) provided a useful illustration of the way in which varying Cl contents can influence the variation of fission-track age with depth in a borehole from Canada.

Summary Comments: In all approaches that involve extracting quantitative thermal history constraints from apatite fission-track data, it should be appreciated that the resulting thermal history solution will only ever be an approximation to the true underlying history. Whether monotonic cooling histories or episodic heating and cooling is employed, for any but the simplest of histories, factors such as the natural spread in the track length distribution, which increases as tracks are progressively shortened (Green et al. 1986), plus the high level of redundancy in the data resulting from the basic nature of the system response, as discussed earlier, impose a fundamental limit to the detail that can be resolved. This factor must be borne in mind when considering information derived from AFT data in individual samples. We suggest that this factor lies behind many perceived problems with published AFT studies (e.g., Gunnell 2000), and, in many cases, too much is expected from the technique. We suggest that it is when attention is focused on the unique information that can be obtained that the power of the technique becomes apparent.

PRACTICAL ISSUES IN INTERPRETING AND UNDERSTANDING AFTA DATA

The Meaning of a Fission-Track Age

Fission-track ages from crustal sections are often discussed in terms of a zonation of ages (e.g., Naeser et al. 1989), with ages being unaffected at shallow depths ("zone of no annealing," <70° C), while at depths greater than ~3 to 4 km (temperatures in excess of ~125° C), no tracks are retained ("total annealing zone"). Between these two extremes, fission-track ages are progressively reduced to zero through a "partial annealing zone" or PAZ. While this zonation provided a simple conceptual basis for early studies based on ages alone, the combination of borehole data (Gleadow and Duddy 1981) and laboratory experiments (Green 1988) showed that fission-track age and length reduction proceed even at temperatures below 70° C, albeit more slowly than at higher temperatures. This can be seen most easily in confined track length data due to the higher precision of these measurements compared to fission-track ages (Fig. 3).

For this reason, a measured fission-track age only rarely indicates the time over which tracks have been retained in an apatite grain. In the same way, a fission-track age rarely reflects a "cooling age," and we recommend that this term should not be used in any circumstances. In slowly cooled terrains, radiometric ages are often referenced to a "closure temperature," below which the daughter product (in our case, a fission track) is effectively retained. However, given the lack of stability of fission tracks in apatite even at low temperatures, this concept is of dubious validity for fission-track ages in apatite, and it is not possible to literally relate a fission-track age to a specific point on a cooling path. Instead, an apatite fission-track age should be regarded as an integrated measure of the balance between the production of tracks by spontaneous fission and the reduction in track density due to the reduction in track length that results from the thermal history.

Only in samples that have cooled extremely rapidly to temperatures less than 50° C and subsequently remained at such temperatures will

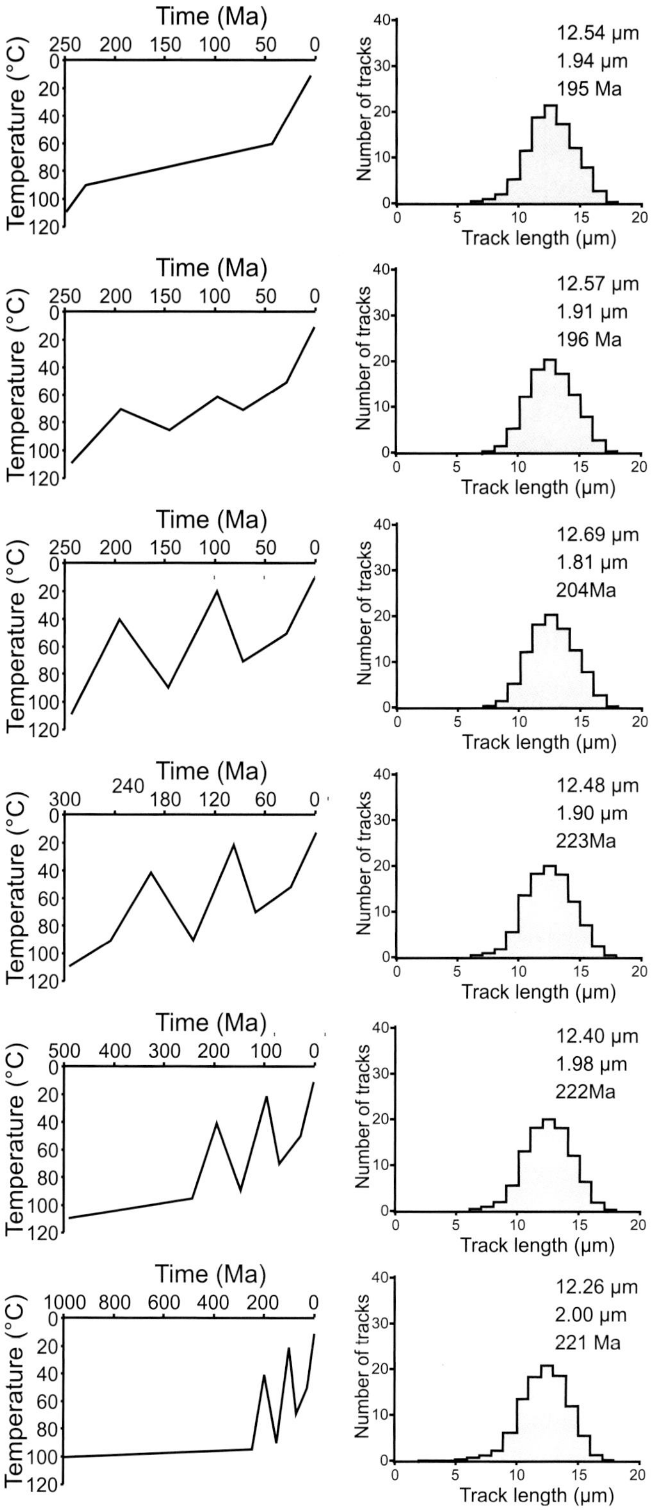

FIG. 14.—Predicted track length distributions (also listing the predicted mean track length, standard deviation of the length distribution, and apatite fission-track age) for a series of cooling histories of increasing complexity. The fission-track parameters resulting from these histories are indistinguishable, despite the time of cooling below 110° C varying from 250 Ma to 1000 Ma, and for histories ranging from progressive cooling to various more complex ← heating and cooling histories. The final AFTA parameters are particularly insensitive to the earliest history, which provides such a minor degree of information to the fission-track age and track length distribution that huge differences in time produce insignificant changes in the expected parameters, which are beyond practical resolution. Predictions are based on a monocompositional apatite of Durango composition using the Laslett et al. (1987) model.

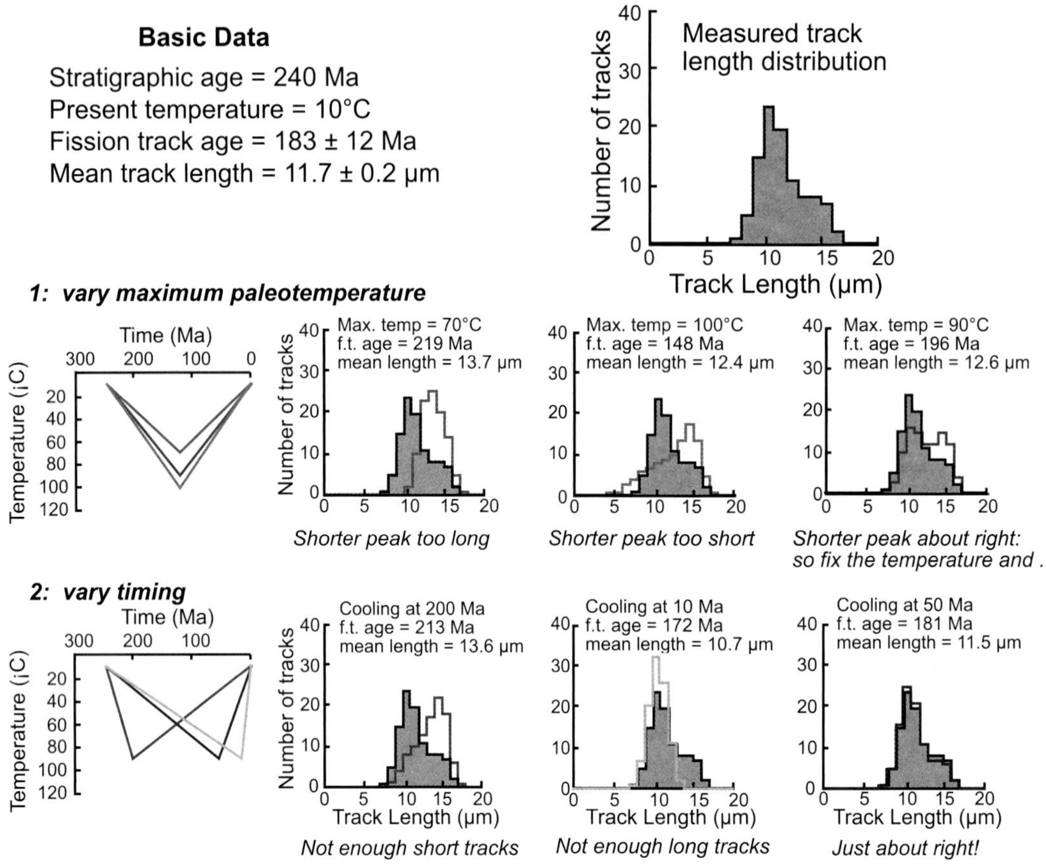

FIG. 15.—Principles of AFT interpretation illustrated for a monocompositional apatite, showing how a thermal history solution can be extracted from measured AFTA parameters (fission-track age, mean track length, and track length distribution). For samples of sedimentary rock, it is necessary to know the stratigraphic age and present temperature of the sample. In principle, the surface temperature at the time of deposition is also required, but because the maximum temperature is the major control on track length (e.g., Figs. 12–14), the influence of the depositional temperature is minimized once the sediment is heated above ~50° C. By predicting the AFTA parameters for various thermal history scenarios, we can define the best-fit thermal history. As a first step, we assume that cooling from the maximum paleotemperature occurred at the midpoint of the history (120 Ma in this case). By varying the maximum temperature and comparing measured and predicted parameters, we find a good match with the shorter population of tracks in the measured track length distribution at a maximum paleotemperature of 90° C. However, the predicted track length distribution contains too many long tracks. A good match between the predicted and measured track length distributions, as well as the fission-track age, is achieved with cooling commencing at 50 Ma, while keeping the maximum temperature constant at 90° C. Note that no attempt is made to define the whole thermal history, because the history prior to the onset of cooling is overprinted by the thermal maximum. Note also that by itself, the measured fission-track age of 183 ± 12 Ma provides no information on the time of cooling, which only comes from kinetic modeling of the details of the track length distribution together with the fission-track age. Predictions are based on a monocompositional apatite of Durango composition using the Laslett et al. (1987) model.

the fission-track age be close to the time over which the sample has been cool enough to retain tracks. Experience has shown that such situations are rare. More commonly, the fission-track age and length data must be assessed in tandem, using the quantitative modeling approach described earlier in order to extract information on the timing (and magnitude) of cooling events from the data.

Long-Term Residence in the Partial Annealing Zone vs. Heating and Cooling

It is a common fallacy that the presence of an exhumed partial annealing zone, as identified, for example, by a break in slope in the variation of fission-track age with depth (cf. Fig. 9), represents a prolonged period of residence in the partial annealing zone prior to exhumation. Consideration of Figure 12 shows that this is not necessary, as this type of data is easily produced by heating (e.g., by burial) of the sequence to temperatures characterizing the PAZ followed immediately by rapid subsequent cooling/exhumation. This misunderstanding arises, in part, because of the common adoption of monotonic cooling histories for explaining apatite fission-track data, which is discussed in more detail later.

"Boomerang Plots"

Green (1986) showed that, in a sequence of rocks from the NW of England that have undergone a single dominant episode of heating and subsequent cooling, with individual samples reaching different maximum temperatures prior to the onset of cooling, the relationship

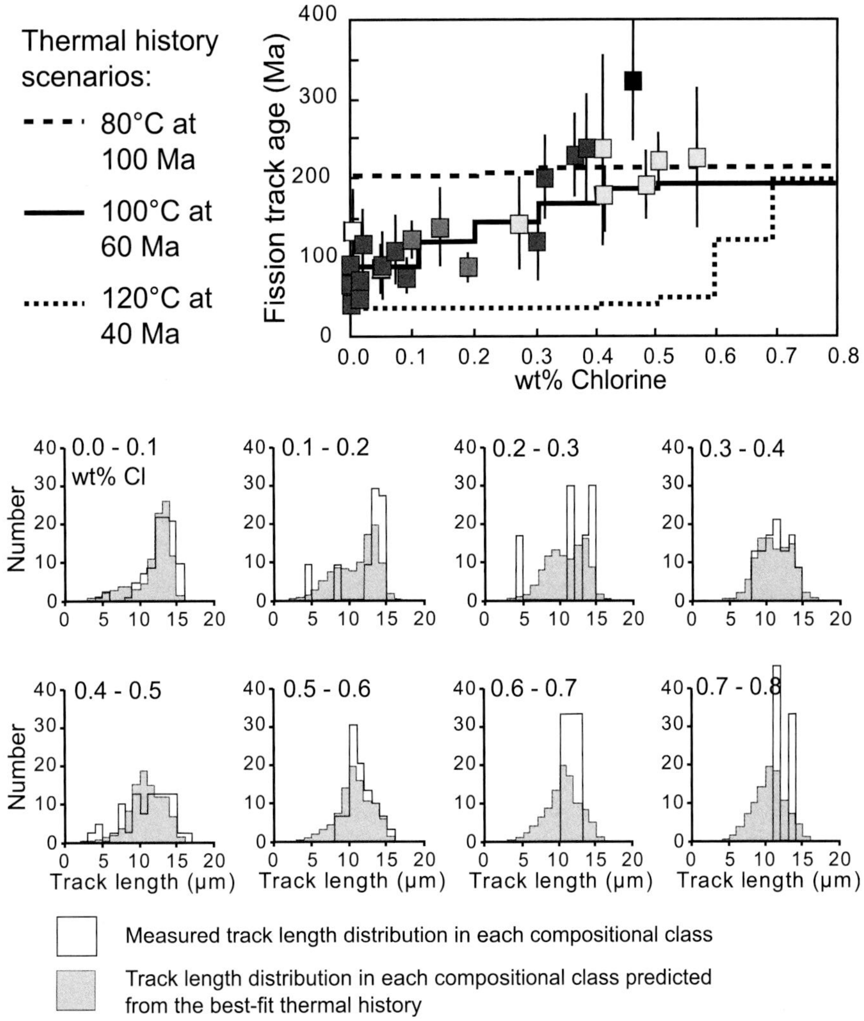

FIG. 16.—AFT interpretation methodology for a multicompositional apatite. The same basic information and interpretation strategy as described in Figure 15 is used for samples that contain apatite of different compositions, but supplemented by the wt% chlorine of each apatite grain (measured by electron microprobe) in which a fission-track age or track length is measured. Fission-track ages and track lengths are grouped according to the chlorine content into 0.1 wt% Cl intervals, and a multicompositional annealing model is used that takes specific account of the influence of wt% Cl on annealing rates. The matching procedure is the same as for a single composition, but now it involves the simultaneous matching of fission-track age and the details of the track length distribution in all compositional groups present in the sample. In the example, eight groups are present containing between 0 and 0.8 wt% Cl, and the best-fit match to the data is achieved for cooling from a maximum temperature of 100° C beginning at 60 Ma (real data in a Permian sandstone outcrop sample from NE England).

between mean confined track length and fission-track age showed a systematic variation defining a "boomerang-shaped" trend (Fig. 17). Samples that have undergone only minor thermal disturbance have old ages with relatively long mean track lengths, while samples in which all fission tracks were totally annealed prior to the onset of cooling give much younger fission-track ages ("reset ages"), also characterized by long (~14 μm) mean lengths. Between these two extremes, as the fission-track age decreases (representing increasing maximum paleotemperatures prior to the onset of cooling), the mean track length decreases as the partially annealed tracks are progressively shortened. This continues until the final stages of age reduction, when the partially annealed tracks become so short that their contribution to the mean length is diminished and the mean length increases with further reduction in fission-track age, trending upward towards the long mean length characterizing the reset ages. The variation in Figure 17 is analogous to that illustrated in Figure 10, except that samples from different depths/temperatures return to the surface at low temperature after the onset of cooling, and thus each sample contains a similar population of long tracks, in addition to the shorter population, which has been annealed to differing degrees in different samples, reflecting different maximum paleotemperatures prior to the onset of cooling.

Returning to the question of the meaning of a fission-track age, Figure 17 illustrates the way in which partially reset fission-track ages reflect the degree of shortening of those tracks formed prior to the onset of cooling. Only those samples for which all tracks were totally annealed give fission-track ages around 60 Ma, but ALL samples underwent cooling at this time. The measured fission-track age in any other sample provides no direct indication of the timing of any event in its own right.

Gallagher and Brown (1997) emphasized the usefulness of this type of plot in considering the implications of regional AFT data sets, albeit in the context of monotonic cooling. However, in many regions, age vs. length data show a very different type of trend than the simple pattern reported by Green (1986). For example, results from Norway (Rohrman et al. 1995) show an almost opposite relationship to the classic "boomerang" trend, while data from Africa and Brazil (Gallagher and Brown 1999a, 1999b) show wide dispersion with only a slight tendency toward longest lengths associated with the youngest ages. Compared to the simple situation in NW England where a single dominant heating/cooling episode has produced a well-defined trend, non-boomerang-style relationships imply a much more complex history, most likely involving a series of paleo–thermal episodes, each of which may vary in magnitude across the region. Therefore, such plots should be interpreted with care, and should always be considered together with the systematic change in the form of the track length distribution through the plot, which was central to the original description by Green (1986).

Monotonic Cooling vs. Episodic Heating and Cooling

While clearly not applicable to sedimentary basins, many thermochronological studies are carried out within a framework involving monotonic cooling from above ~110° C to surface temperatures. Such approaches are routinely applied to basement terrains, but even in such regions, the presence of the merest veneer of sedimentary cover means that the underlying basement was at the surface when that cover was deposited. Failure to understand this issue can result in quite erroneous interpretations (e.g., Persano et al. 2006; see discussion by Brown 2007, Gibson 2007, Green and Duddy 2007). While episodes of heating and cooling are clearly more realistic in sedimentary basins, as witnessed by the common occurrence of unconformities in sedimentary sequences, we suggest that such histories may also be quite reasonable for many basement terrains. This is clearly appropriate where sedimentary outliers are present, but we see no reason why it should not also be true in areas devoid of present-day cover, which could simply indicate that the former cover has been totally stripped.

Impact of Composition on Thermal History Interpretation of AFTA Data

Natural apatites essentially have the composition $Ca_5(PO_4)_3$ (F,OH,Cl). Most common detrital and accessory apatites are predominantly fluorapatites, but they can contain appreciable amounts of chlorine (Fig. 18). In most quartzofeldspathic sandstones, the majority of grains contain between 0 and 0.1 wt% Cl, while smaller numbers of grains give values up to ~0.5 wt% Cl, and occasionally grains contain up to 1 wt% Cl and above (Fig. 18a). In contrast, volcanogenic sandstones typically contain apatites showing a much broader spread of Cl contents up to 2 or even 3 wt% Cl and beyond (Fig. 18b). Cl contents in granitic basement samples and silicic high-level intrusives are typically much more dominated by compositions close to end-member fluorapatite (Fig. 18c), while apatites from gabbroic and other basic intrusive rock types may contain appreciable amounts of Cl, although many exceptions occur to these general rules.

As discussed earlier, the amount of chlorine in the apatite lattice exerts a subtle control on fission-track annealing rates, which is typically most pronounced in the temperature range 90 to 120° C (Figs. 9, 10). In samples heated to such temperatures, individual apatite grains may show a significant spread in the degree of annealing (i.e., length reduction and fission-track age reduction). Such within-sample variation can be very useful in identifying samples exposed to paleotemperatures in this range.

The effects of compositional variation are, however, ignored in many published studies, and this can lead to major errors in interpretation, particularly if analytical procedures are not designed to take such effects into account. To illustrate the potential problems that can arise, Figure 19a shows a data set in which grains were selected in a continuous transect across the grain mount, with a thermal history solution extracted by making due allowance for the variation of fission-track age and length with wt% Cl. Figure 19b illustrates the result of a biased analysis in which fission-track ages were measured only in high wt% Cl grains, which contain high track densities and which could represent a favorable "target" for counting, while track lengths have been measured in the more common (and more sensitive) apatites containing lower amounts of chlorine. In this case, the thermal history solution extracted from the biased data results in a wildly inaccurate interpretation (Fig. 19b).

Some workers prefer to use the size of etch pits as a kinetic parameter for resolving differences in annealing within samples (Burtner et al. 1994). However, as discussed earlier, etch pit dimensions provide only a very poor indicator of annealing sensitivity (Fig. 8), and due to the relatively small range of etch pit sizes for most common apatites, this approach lacks the resolution that is possible using chlorine content. In addition, use of etch pit sizes is extremely demanding of etching procedures, which must be reproducible to the standards employed in the original calibration experiments (Carslon et al. 1999). And in borehole studies involving conditions where fission tracks in some apatites are totally annealed, measurement of etch pit sizes in these grains is clearly not possible.

Therefore, to ensure accurate thermal history interpretations from AFTA, the following are essential steps in data collection: grains need to be analyzed without reference to track density, Cl content should be measured in every grain for which either fission track age or track length measurements are made, and the data should be assessed in terms of their variation with wt% Cl, as illustrated in Figure 16.

Influence of Tracks Inherited from Sediment Source Terrains

An apatite in which the fission-track age is older than the depositional age of the host sedimentary rock clearly retains tracks that were formed in the sediment provenance terrain, prior to deposition. The effect of such tracks is often posed as a potential problem for AFTA. However, this rarely poses practical problems, and it can sometimes prove advantageous.

The contribution to the measured fission-track age and the length distribution made by these "inherited tracks" will be characteristic of the thermal history of the sediment provenance terrains. The influence of such tracks on the ability to extract information on the postdepositional thermal history of a sedimentary rock from AFTA data will depend essentially on the relative duration and severity of heating during the predepositional and postdepositional history. For example, in apatites derived from an ancient basement terrain and deposited in Neogene sedimentary rocks, in which only a small proportion of tracks may have formed after deposition of the host sediment, AFTA data will be dominated by tracks formed prior to deposition. In samples of this nature, which have been heated to only moderate temperatures after deposition (say <60° C), it may not be possible to resolve the effects of this heating from the influence of the predepositional history. However, as the severity of postdepositional annealing increases (i.e., with increasing maximum postdepositional temperature), the effects of the predepositional history are progressively "overprinted," and the AFTA parameters become dominated by the postdepositional history.

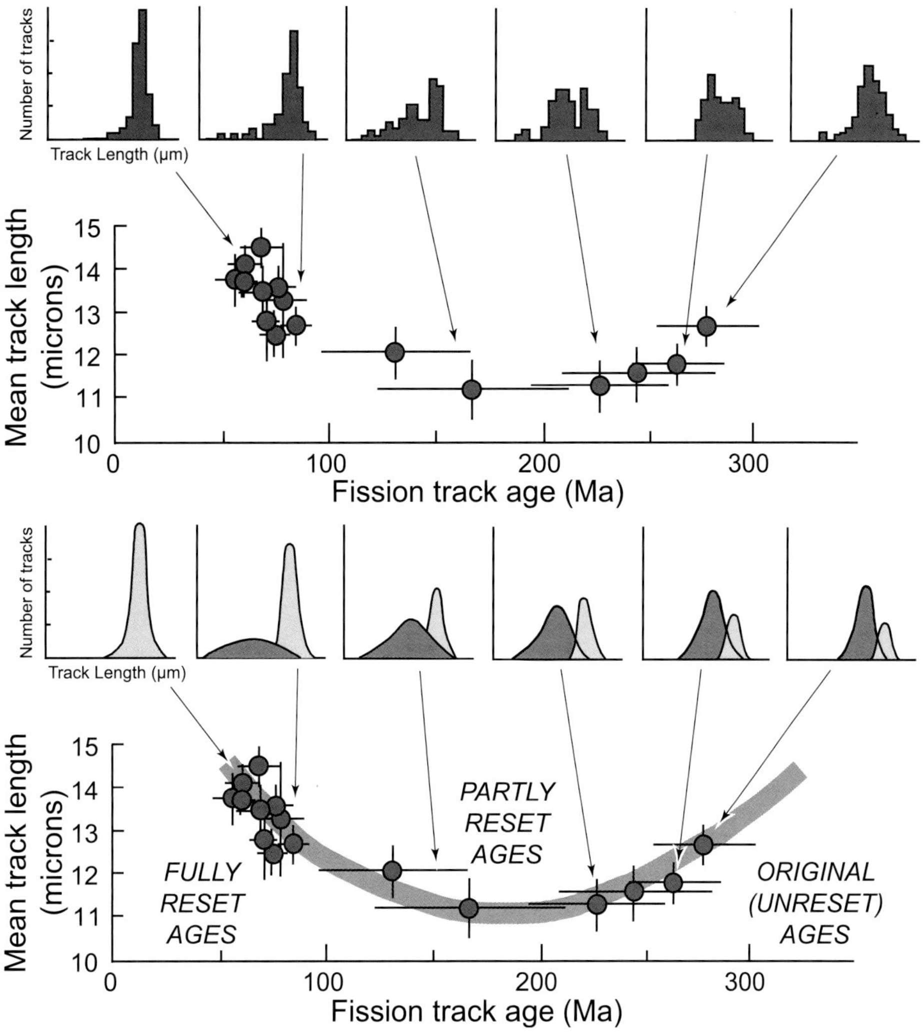

FIG. 17.—Boomerang plot: relationship between mean track length and fission-track age for a suite of outcrop samples that have undergone cooling from different maximum paleotemperatures at the same time (Green 1986). Samples that have experienced low maximum temperatures have old ages with relatively long mean track lengths, while samples in which all fission tracks were totally annealed prior to the onset of cooling give much younger fission-track ages ("reset ages"), also characterized by long (~14 μm) mean lengths. Between these two extremes, as the fission-track age decreases (representing increasing maximum paleotemperatures prior to the onset of cooling), the mean track length decreases as the partially annealed tracks are progressively shortened. This continues until the final stages of age reduction, when the partially annealed tracks become so short that their contribution to the mean length is diminished, and the mean length increases with further reduction in fission-track age, trending upward towards the long mean length characterizing the reset ages. This data set can be considered analogous to the situation illustrated in Figure 12, but with all samples cooling to low (near-surface) temperatures, such that each contains a population of long tracks formed after cooling.

For samples containing a mixture of tracks formed pre- and postdeposition, Figure 20 illustrates the way in which only the shortest tracks retain a provenance signature. The predepositional track length distribution is progressively overprinted from longer lengths to shorter lengths, as the latter have experienced higher predepositional temperatures, and require even higher postdepositional temperatures to be further shortened. This is a direct consequence of the principle of "equivalent time," discussed earlier. Once the effects of postdepositional annealing begin to dominate over the predepositional history, the resulting AFTA data can provide reliable constraints on the postdepositional history. As heating becomes sufficiently severe, the fission-track age will be reduced to a value less than the depositional age, and the age data will also become dominated by the effects of the postdepositional history.

In general, inherited tracks pose practical problems in extracting information on the postdepositional thermal history only for samples in which the majority of tracks were formed prior to deposition, perhaps in a Neogene sediment in which apatites were derived from a stable Paleozoic shield with fission-track ages of ca. 400 Ma or more, or in samples that have experienced only very minor postdepositional heating (say <50° C). In such cases, often the only information on the postdepositional thermal history that can be obtained from AFTA might be that the sample has not been heated above say 90° C at any time after deposition.

Resolving Multiple Paleo–thermal Episodes in Individual Samples

In samples that have undergone two episodes of heating and cooling, AFTA data can often provide constraints on both episodes, provided that the magnitude and timing of the paleo–thermal maximum and subsequent peak are sufficiently separated (Fig. 21). In rare cases (e.g., Green et al. 2001a, Turner et al. 2008), three discrete episodes can be resolved in data from a single sample. This is most likely when the earliest event involves a maximum paleotemperature sufficient to totally anneal all tracks (typically >110° C), followed by a subsequent peak around 90 to 100° C, which reduces tracks to a mean length around 10 μm, and then cooling to low temperature is followed by re-heating to around 70° C sufficient to reduce lengths of tracks formed after the second event to around 12 to 13 μm. In such circumstances, integration of AFTA data with results from other techniques (particularly vitrinite reflectance) can provide corroborative evidence of the earliest, maximum temperature episode, as discussed in the next section.

INTEGRATION WITH INFORMATION FROM OTHER SOURCES

Introduction

While many workers often use apatite fission-track data in isolation, significant advantages can be obtained by combining AFTA data with results from other methods. One reason for this is the inherent redundancy in the method, in that a large number of thermal histories can result in very similar AFTA parameters, due to the dominance of maximum paleotemperatures in the kinetics of annealing (Figs. 13, 14). In addition, the natural spread in the length distribution imposes limits on the recognition of low-temperature events, and can cause problems in resolving complex histories involving multiple episodes. Integration with independent techniques not only provides corroboration of conclusions derived from AFTA, but it can also refine the range of thermal history solutions defined from AFTA alone and provide a coherent thermal history framework. Here, we discuss some of the techniques that have been used in tandem with AFTA and discuss the benefits that can be obtained.

Geological Data

One form of independent information that is essential in ensuring that the information obtained from AFTA is meaningful, but is often underestimated, is basic geological information. AFTA data in isolation can often be explained by such a wide range of histories that unless the depositional age of a sample is specified (in as much detail as possible), together with the present-day temperature, no meaningful thermal history information can be obtained. By specifying the time at which the apatite was at the surface, and the present temperature, these fundamental constraints on two points of the temperature–time history provide a framework within which the postdepositional history of the host sedimentary rock can be defined.

This emphasizes that it is essential to interpret AFTA data within a context, rather than in isolation. As discussed in more detail later, to provide the required context, we define *a default thermal history*, which is that part of the history that can be defined on the basis of the available geological evidence (or alternatively—the history that would be appropriate if no paleo–thermal events have affected the sample). This can be particularly important for samples from sedimentary sections containing a number of unconformities separating thin units of different ages, as the presence of the overlying section shows that the deeper units were close to the surface when the shallower units were deposited.

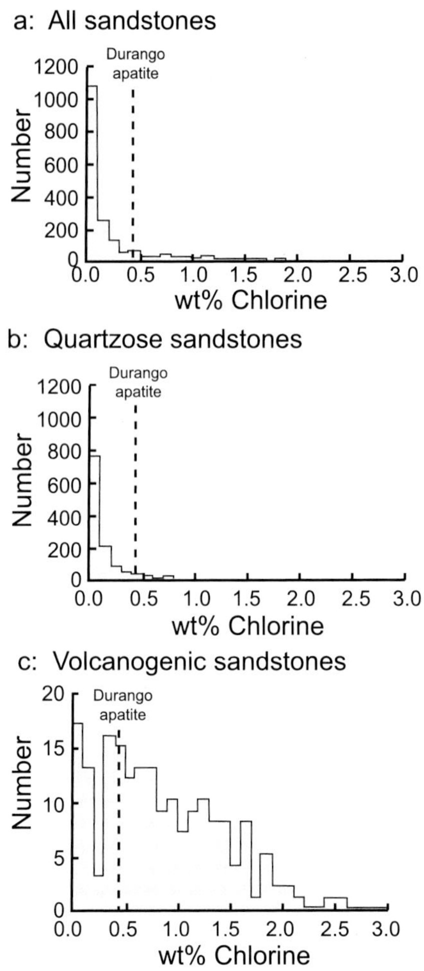

FIG. 18.—Typical distributions of wt% Cl in apatites from different sources. (a) Histogram of Cl contents (wt%) in over 1750 apatite grains from over 100 samples of various sedimentary and igneous rocks. Most samples give Cl contents below ~0.5 wt %, while those apatites giving higher Cl contents are characteristic of volcanogenic sandstones and basic igneous sources. (b) Histogram of Cl contents (wt%) in 1168 apatite grains from 61 samples that can loosely be characterized as "normal quartzofeldspathic sandstone." The distribution is similar to that in the upper figure, except for a lower number of grains with Cl contents greater than ~1%. (c) Histogram of Cl contents (wt%) in 188 apatite grains from 15 samples of volcanogenic sandstone. The distribution is much flatter than the other two, with a much higher proportion of Cl-rich grains.

Vitrinite Reflectance

Vitrinite reflectance (VR), based on the increase in reflectivity of the organic maceral vitrinite (a key constituent of coal) with temperature, is the standard measure of organic maturity for hydrocarbon exploration (e.g., Tissot and Welte 1984). The kinetics of this process are well understood (Burnham and Sweeney 1989), and are very similar to

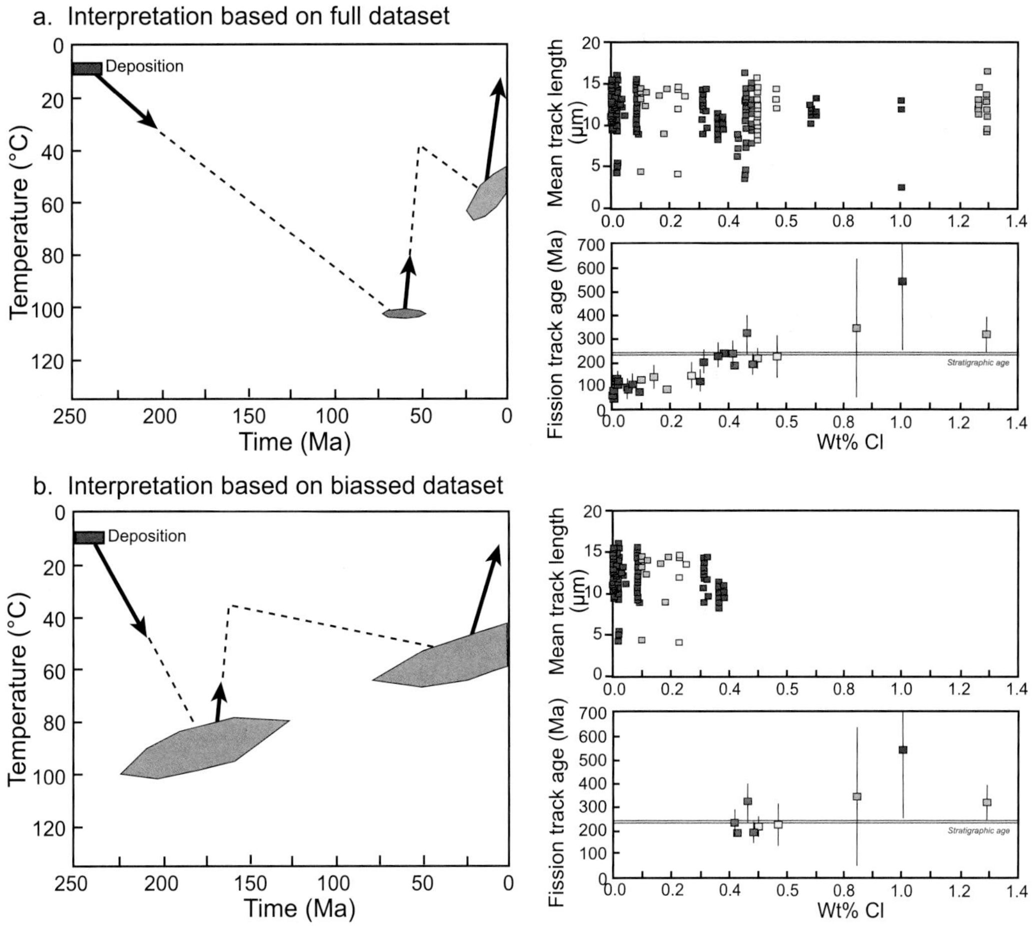

FIG. 19.—Thermal history solutions extracted from AFTA data from the same Triassic outcrop sample. In (a), complete allowance is made for the variation of fission-track age and track length with apatite chlorine content. In (b), the solution is based on track length data from only the most abundant (low wt% Cl) grains, while fission-track ages are measured only in the higher wt% Cl grains. Such a situation could easily arise if ages are only measured in grains containing high track densities, without making due allowance for the influence of wt% Cl. The solution in this case is very different from the correct solution (a), illustrating the problems that can arise if compositional effects are not correctly taken into account.

those of fission-track annealing in apatite (Duddy et al. 1994, 1998), with VR values of 0.65% to 0.7% corresponding to total annealing of fission tracks in typical apatites (Duddy et al. 1994). These factors make VR an ideal complement to AFTA data applied to sedimentary sequences, as demonstrated in a wide range of studies (e.g., Duddy 1997; Green et al. 2004; Japsen et al. 2005, 2007a; Turner et al. 2008).

In particular, VR provides independent determination of maximum postdepositional paleotemperatures, which can provide support for those indicated by AFTA. This is important, because data from either technique alone might be viewed with suspicion, but when two independent techniques provide consistent paleotemperatures, the conclusions can be regarded as reliable. A classic example is provided by studies of wells on the East Midlands Shelf in the southern North Sea of the UK, where early VR data were disregarded as indicating amounts of eroded section that were "not consistent with the known geological evolution" (Cope 1986). Subsequent application of AFTA to wells from this region (Green 1989) confirmed the heating suggested by the VR data, showing that the section in these wells had indeed been more deeply buried and then exhumed.

Integration of VR data from fine-grained units with AFTA data from sandstones also allows determination of paleotemperatures over a wider range of depths than possible from AFTA alone. The combination of both techniques can provide much tighter control on paleogeothermal gradients and amounts of removed section than would be possible from either technique on its own. In addition, integration of VR with AFTA data can be of great assistance in confirming earlier events soon after deposition, which may not be confidently defined from AFTA alone (e.g., Green et al. 2004).

Despite the importance of VR data to the oil exploration industry and its widespread routine application, in recent years it has become apparent that VR data generated by different analysts are not equivalent (see discussion in Green et al. 2002). We find a high degree of consistency between thermal history interpretations from AFTA and VR data generated using an approach involving measurement of maximum reflectance under oil ($R_o$max) in polished thick sections (Cook 1982). In this approach, which is recommended by the International Commission on Organic and Coal Petrography (www.iccop.org), identification of the indigenous vitrinite population is made on textural grounds. This allows independent assessment of the possible presence of reworked vitrinite populations from petrographic evidence, as well as allowing identification of caved material in subsurface samples. Alternation between reflectance and fluorescence modes allows checking for associated fluorescing liptinite, bitumen impregnation, or the presence, intensity, and source of oil-cut, which may affect the reading. An alternative approach, often encountered in hydrocarbon industry reports, involves measurements of random reflectance ($R_o$rand) in strewn slides of organic concentrates, with the indigenous vitrinite population often identified only after the analysis is complete by inspection of histograms of measurements and separation into perceived subpopulations. In our experience, this approach can lead to serious errors in determining maturity levels (usually resulting in underestimation). Integration of AFTA with VR data allows such cases to be readily identified.

A key factor in interpreting VR data is the possible suppression of reflectance levels, as often observed, for example, in H-rich source-rock horizons (e.g., Wilkins et al. 1992). Suppression can often be recognized by local departures from an overall trend, but where few data are available, this is less straightforward. This can be particularly problematical where VR data (and other organic maturity indicators) are only measured in source rock horizons. Again, integration with AFTA allows detection of anomalous vitrinite reflectance values.

Data from a range of additional organic-based methods for assessing maturity levels, such as biomarker reactions (e.g., MacKenzie and McKenzie 1983), Spore Colour Index (SCI; Fisher et al. 1980), or Rock Eval Tmax (Tissot and Welte 1984), can be converted to equivalent VR values, which can also be integrated with AFTA data. While biomarker reactions show promise as quantitative thermal indicators, the expense and difficulties with elucidating the true kinetic descriptions have so far limited their usefulness, while SCI values are less commonly available and Rock Eval Tmax values are subject to variability as a result of variation in organic facies. While these methods and others have some practical limitations, all may be used, where available, to complement VR and AFTA data.

Fluid Inclusions

Fluid inclusions are micron-sized vacuoles of liquid or gas that have been trapped within crystal imperfections during mineral growth. Univariant phase transitions observed during heating and freezing of fluid inclusions in the laboratory can yield valuable information concerning the composition and density of the fluids trapped within the inclusions and the temperature and pressure conditions at which the mineral precipitated. Taking advantage of the independent constraint on maximum paleotemperature provided from fluid inclusions, AFTA has been used to estimate the duration of heating related to diagenesis (O'Brien et al. 1996, Duddy et al. 1998, Parnell et al. 2005) and impact events (Parnell et al. 2007). Integration of AFTA data with information from fluid inclusions has also defined the timing of episodes of hot fluid circulation (Parnell et al. 1999) and has clarified issues related to the timing of hydrocarbon generation (Mark et al. 2008) in the UK West of Shetland region.

Sonic Velocities

The progressive compaction of sediments with increasing burial has been widely used as the basis for estimating former burial depths in exhumed basins (e.g., Marie 1975, Bulat and Stoker 1987, Hillis 1995). Comparison of compaction proxies such as sonic velocity in an exhumed formation with a reference curve defining the expected variation with depth in sequences at maximum burial depth provides an indication of the amount of net exhumation. Selection of appropriate reference curves has been problematical in some areas, leading to erroneous conclusions regarding the extent and magnitude of

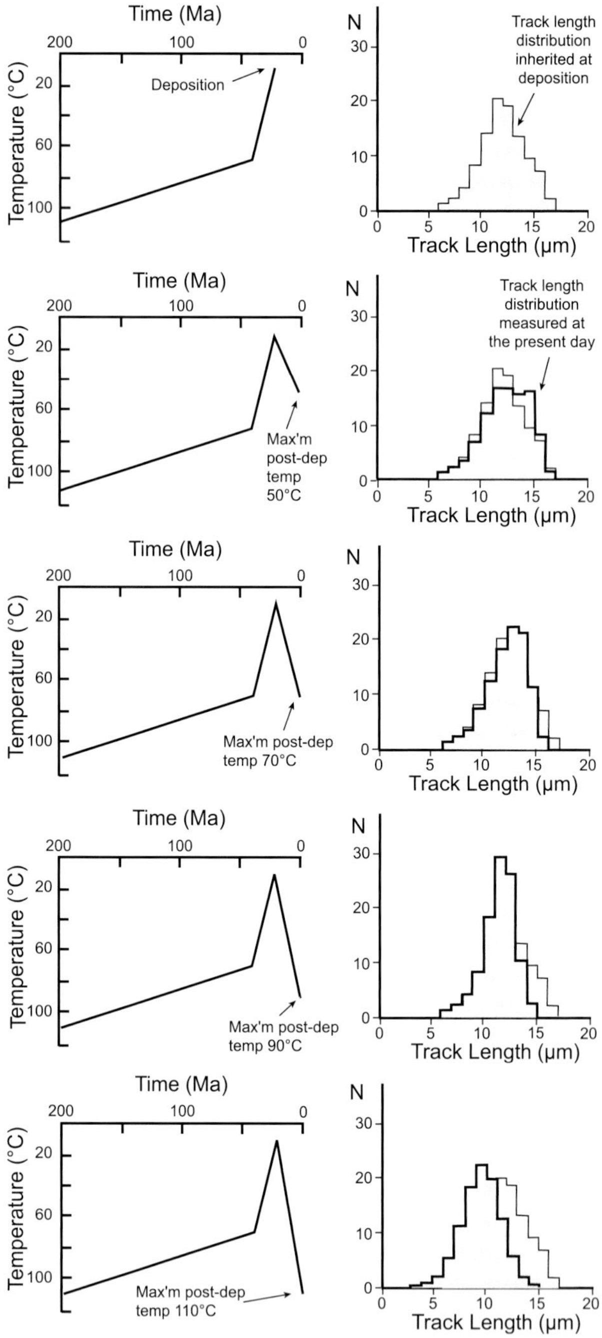

FIG. 20.—Progressive overprinting of inherited tracks by postdepositional heating. For samples containing a mixture of tracks formed pre- and postdeposition, only the shortest tracks retain a provenance signature. In the figure, as the postdepositional temperature increases, the predepositional track length distribution (in outline) is progressively overprinted from longer lengths to shorter lengths, as the latter have experienced higher predepositional temperatures, and require even higher postdepositional temperatures to be further shortened. Once the effects of postdepositional annealing begin to dominate over the predepositional history at temperatures above about 90° C in the example, the resulting AFTA data can provide reliable constraints on the postdepositional history.

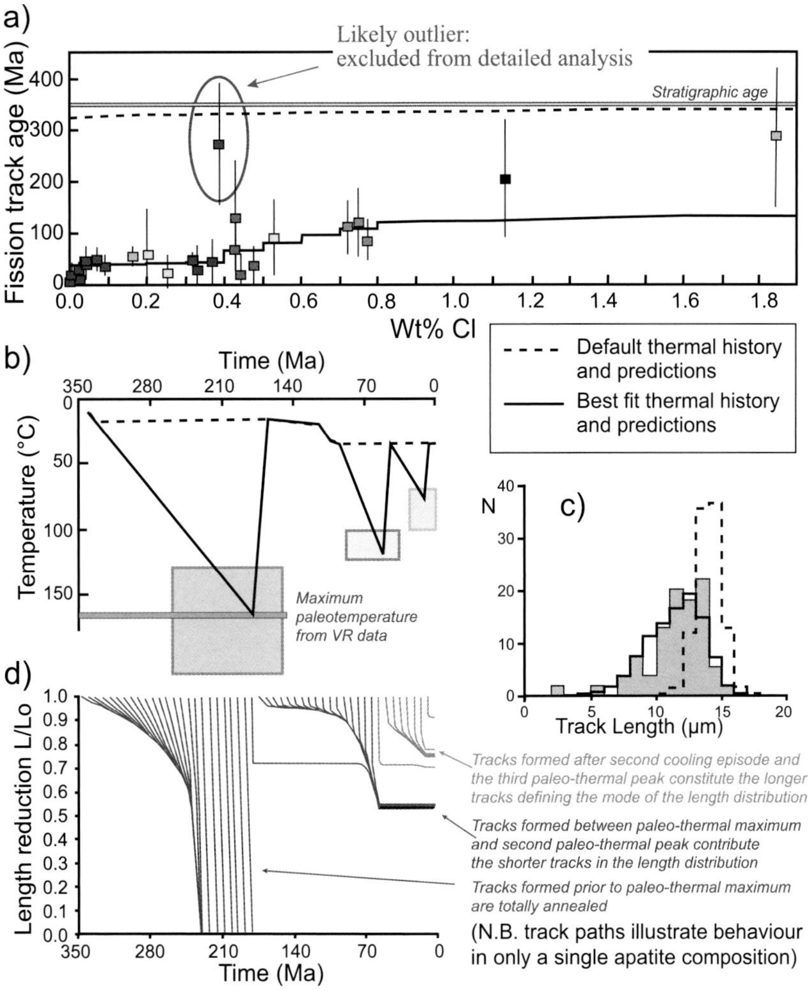

FIG. 21.—Resolution of multiple episodes from AFTA data in a single sample is possible where the events are sufficiently separated in temperature and time. In this example, based on AFTA data (a) in combination with vitrinite reflectance results in a sample from the Dodo Canyon K-03 well in the Mackenzie Valley, Northwest Territories, Canada, three thermal episodes are resolved (b). Predicted track shortening trajectories for the three-episode history (d) show how all tracks are erased in the earliest episode (prior to 180 Ma), a population of shorter tracks is produced at the paleo–thermal peak in the second episode, and a longer population of tracks (which constitutes the main mode of the distribution) is produced in the final episode. These tracks are still shorter than expected at surface temperatures, which allows resolution of this final episode (as illustrated by comparing the measured track length distribution in [c] with the distribution predicted by the default thermal history).

exhumation (Cope 1986, Green et al. 2001b). More rigorous definition of the necessary reference curves in recent years (Japsen et al. 2007b) allows more reliable estimation of former burial depths, and integration of such data with constraints from AFTA and VR data has provided highly consistent reconstructions of eroded section in Denmark (Japsen et al. 2007a) and Cardigan Bay, western UK (Holford et al. 2005).

Since these compaction-based methods are controlled primarily by maximum burial depths, in situations where they provide consistent indications of former burial depths with those derived from paleo–thermal methods such as AFTA and VR, the results can be regarded with confidence. In addition, while results from AFTA and VR can often be explained by a range of paleogeothermal gradients and amounts of removed section, additional constraints from compaction-based methods can significantly reduce the range of viable solutions, as illustrated, for example, by results from the Hans-1 well, offshore Denmark (Japsen et al. 2007b).

Zircon Fission-Track Analysis (ZFTA)

Zircon is another common uranium-bearing detrital mineral amenable to fission-track analysis. From both laboratory annealing studies and geological evidence (Hurford 1986, Tagami et al. 1996), fission tracks in zircon are known to be more resistant to annealing than fission tracks in apatite. Investigation of zircon data in samples with different levels of vitrinite reflectance suggests that no significant fission-track age reduction occurs in zircon at VR levels below ~4% (Geotrack, unpublished results). VR values in excess of 5% (equivalent to maximum paleotemperatures in excess of 300° C) are required in order to produce any significant age reduction. Some evidence suggests that the rock needs to reach the stage of greenschist facies metamorphism in order to produce observable effects in zircon. Therefore, in most sedimentary basin settings, ZFTA can provide little or no information on postdepositional heating, except where intrusions are present or in cases of extremely high heat flow (e.g., Logan and Duddy 1998). However, the relative stability of tracks in zircon makes ZFTA a useful tool for investigating sediment provenance. ZFTA data can also provide useful constraints on depositional ages (e.g., Morais Neto et al. 2008), since in the absence of postdepositional resetting, a zircon fission-track age must represent cooling of sediment source regions, thereby providing an upper limit on the depositional age.

Apatite (U-Th)/He Dating

The recent advent of apatite (U-Th)/He dating (Zeitler et al. 1987, Wolf et al. 1996) potentially opened up a lower temperature window for low-temperature thermochronology. Initial studies (e.g., House et al. 1997, Warnock et al. 1997, Wolf et al. 1997) illustrated the potential of the technique to provide useful information at temperatures in the 40 to 80° C range, giving this method possibly even greater potential than AFT methods for providing information relevant to the interpretation of landscape development, and suggesting that the combination of the two techniques should be particularly powerful.

Whereas fission-track dating is based on spontaneous fission of ^{238}U atoms, it is the alpha decay of uranium and thorium (and to some extent samarium) isotopes, leading to the accumulation of helium within the apatite lattice, that forms the basis of (U-Th)/He dating. Application of the method to elucidate thermal history information in geological conditions depends on the interplay between production of helium through alpha decay and the loss of He due to diffusion, which occurs at a rate that depends on temperature (as well as grain size). In similar fashion to AFT methods, comparison of measured ages with values predicted from various thermal history scenarios allows definition of the range of thermal histories giving predictions that are consistent with the results. This approach is best applied to vertical sequences of samples, where the variation of age with depth helps to further restrict the range of viable solutions.

Prediction of expected ages is based on the results of laboratory diffusion studies (Farley 2000) based on Durango apatite. Initially, it was widely assumed that these results could be applied to all common apatites, and the method was applied with apparent success in a wide range of environments from mountain belts to sedimentary basins. Integration of (U-Th)/He ages with AFT data from boreholes in the Otway Basin, SE Australia (House et al. 1999, 2002), illustrated the unique sensitivity of (U-Th)/He ages in apatite to temperatures in the range 20 to 80° C, and appeared to confirm the extrapolation of laboratory studies. Integration of apatite (U-Th)/He ages with AFTA data from the Fresne-1 well in the Taranaki Basin (Crowhurst et al. 2002) provided further demonstration of the consistency between the two techniques.

However, results from more recent studies in which (U-Th)/He data have been integrated with AFT data have revealed a systematic discrepancy between the two techniques. This evidence was reviewed by Green and Duddy (2006) and Green et al. (2006), who showed that the discrepancy arises from an increase in the retentivity of helium in apatite as either the (U-Th)/He or the fission-track age increases. This results in anomalously old (U-Th)/He ages compared to the expected response based on the systematics presented by Farley (2000) in samples with ages older than ca. 50 Ma. Shuster et al. (2006) confirmed, based on laboratory diffusion studies, that this effect is due to an increase in helium retention as the amount of radiation damage in the apatite lattice increases, and they provided a new quantitative model of the process. However, this model cannot explain the anomalous retention of He in granite pebbles within the Early Cretaceous section of the Otway Basin (Green et al. 2006).

Subsequent developments reported by Shuster and Farley (2009) and Flowers et al. (2009) have led to a more advanced treatment, in which modeling of He diffusion is integrated with modeling of fission-track retention as a surrogate for the degree of radiation damage. This approach offers the promise of providing more accurate models, but the application of these to elucidate accurate thermal histories, particularly in complex situations, has yet to be proven. Until this is achieved, we suggest that apatite (U-Th)/He results should be treated with caution, particularly when no AFT data are available from the same samples.

The results presented by Green et al. (2006) clearly show that (U-Th)/He ages in samples where the product of the age (in Ma) and the equivalent uranium content in ppm is greater than ~2500 cannot be explained in terms of the Farley (2000) systematics, which should be applied only to low-uranium apatites with relatively young (<50 Ma) fission-track ages. Note that this applies even though these anomalously old He ages often show a high degree of reproducibility.

In the absence of quantitative modeling, (U-Th)/He ages can, in principle, be used in a qualitative sense, similar to the "fission-track stratigraphy approach" described earlier, by comparing age vs. elevation trends. However, in addition to the systematic effects just described, Green et al. (2006) also reported excessive scatter in He ages in many samples, beyond that expected from purely analytical uncertainties, which can cause problems in comparing He ages from different samples. Even in the absence of such effects, because of the systematic effect of increasing levels of radiation damage, it is essential for only similar apatites (in terms of U content and grain size) to be compared.

In some studies, (U-Th)/He ages have been interpreted as representing the time at which samples cool through a "closure temperature," typically around 50° C, although, in detail, this will vary with radiation damage (Shuster et al. 2006). However, as highlighted by Green and Duddy (2006), "open system" behavior is to be expected in the upper 1 to 2 km of the crust, and long-term residence at low temperatures, or even histories involving reburial (i.e., reheating) followed by exhumation (cooling), can produce results that can be easily mistaken for slow cooling. In practice, interpretations of apatite (U-Th)/He ages are most reliable when accompanied by AFT data providing consistent interpretations.

APPLICATION OF AFTA TO THERMAL HISTORY RECONSTRUCTION IN SEDIMENTARY BASINS

AFTA data in individual samples can provide constraints on the paleo–thermal maximum as well as one or possibly two subsequent paleo–thermal peaks. However, the amount of information that can be obtained from a single sample is limited, and the real strength of the technique emerges from application to sequences of samples over a range of depths in boreholes, or over a range of elevations in an outcrop section. In such cases, integrated AFTA and VR analyses can provide a well-defined thermal history framework, involving quantitative definition of the timing and magnitude of major paleo–thermal events as well as determination of paleogeothermal gradients, allowing unique insights into mechanisms of heating and cooling. This information allows reconstruction of the thermal and burial/uplift histories of the sedimentary section, as explained in the following sections.

Default Thermal Histories

As discussed earlier, AFTA and VR data must be investigated within a context in order to provide meaningful information on the magnitude of possible paleo–thermal effects (i.e., events in which a rock sample was hotter in the past than it is today, due to either deeper burial, elevated basal heat flow, local igneous intrusion, or hot fluid movements). In sedimentary sections, this context is provided by a default thermal history, which represents the history that can be constructed in the absence of any paleo–thermal effects. For subsurface samples, this is calculated by combining the burial history derived from the preserved sedimentary section with the present-day geothermal gradient and surface temperature. Because both AFTA and VR data are dominated by the maximum temperatures experienced, this provides a basic point of reference for the expected degree of fission-track annealing and the organic maturity. If measured AFTA and/or VR data are consistent with the values predicted from the default thermal history, then the sample is presently at or close to its maximum postdepositional temperature, and the data retain little or no information on any paleo–thermal effects (because the data are dominated by maximum temperature, as explained earlier). However, if AFTA data show a greater degree of fission-track annealing or VR data show higher maturity than expected on the basis of the default thermal history, then the sampled horizon must have been hotter in the past. In this case, AFTA allows determination of the time at which cooling began, and both AFTA and VR can define the magnitude of the maximum paleotemperature reached by individual samples.

A key factor in defining default thermal histories for subsurface samples is definition of the present-day thermal regime. Information on present-day temperatures in hydrocarbon exploration wells usually comes in the form of bottom-hole temperatures (BHTs) from logging runs, or in less common circumstances from drill stem tests (DSTs), which sample formation fluids directly. A key problem in using such information to reconstruct present-day temperatures is the disturbance to the thermal field imposed by drilling of a well due to the introduction of drilling fluids, which produce significant cooling. BHT values, which are normally recorded within a short time after drilling must be "corrected" to estimate the true formation temperature. Conversely, DST temperatures are generally thought to more closely reflect true ambient temperatures and can be used directly.

The most common BHT correction method is the "Horner correction," based on an exponential increase toward the true temperature with time, but this requires repeated temperature measurements from multiple logging runs, as well as the time between runs. This information is not always available, so simpler methods have been suggested. In our own work, we use a simple method based on observations of borehole temperatures over timescales of years (Andrews-Speed et al. 1984). In this approach, quoted BHT values are corrected by increasing the difference between the surface or sea-bed temperature and the uncorrected BHT by 20% for uncorrected temperatures below 150° F (66° C), and by 25% above 150° F. In wells where multiple temperature measurements are available at a given depth, the earliest recorded BHT value is used. If no circulation times are available, the lowest temperature value at each depth is used. Whilst simplistic, this procedure has the advantage of allowing a common approach in all cases, and results in present-day temperatures that are consistent with the kinetic descriptions of AFTA and VR employed to extract information on paleo–thermal histories. Thus, a self-consistent framework is achieved, which is essential in any approach to this subject.

Paleotemperature Profiles

The variation of paleotemperatures with depth, or the "paleotemperature profile," provides key information on likely mechanisms of heating and cooling (Bray et al. 1992, Duddy et al. 1994, Green et al. 2002) (Fig. 22). Provided that heterogeneities in lithology through the section are sufficient to smooth out any potential large-scale variations in thermal conductivity, heating due solely to deeper burial should produce a more-or-less linear paleotemperature profile with a similar gradient to the present temperature profile. In contrast, heating due primarily to increased basal heat flow (perhaps also with some component of deeper burial) should produce a more-or-less linear paleotemperature profile with a higher gradient than the present temperature profile. Nonlinear profiles may be produced by contact heating around intrusive bodies, or in the absence of intrusions, they may be diagnostic of lateral introduction of heat, most likely due to passage of hot fluids within confined aquifer horizons.

Estimating Additional Burial (Removed Section)

Where heating can be attributed to some degree of deeper burial, possibly combined with elevated heat flow, fitting a linear profile to paleotemperatures as a function of depth allows the paleogeothermal gradient at the paleo–thermal maximum to be determined, and extrapolation of the paleogeothermal gradient from the depth of the appropriate unconformity to an assumed paleo-surface temperature (Fig. 23) provides an estimate of the amount of additional section that was once present and was removed during cooling (see Bray et al. 1992; Duddy et al. 1994; Green et al. 2002, 2004). The inverse correlation between these two parameters results in a hyperbolic ellipsoid region of allowed values as shown in Figure 23. If a preferred value of paleogeothermal gradient is available (e.g., if the present-day gradient of heat flow has not changed through time) or if amounts of additional burial can be independently constrained (e.g., from sonic velocity data), then the complementary parameter can be defined with greater confidence from such plots.

It should be stressed that estimating amounts of removed section by extrapolating a linear paleotemperature profile assumes that the additional section had the same average thermal conductivity as the preserved section. If independent evidence suggests that this assumption is not appropriate, then a more detailed analysis using suitable thermal conductivities is required in order to provide a more accurate solution. However, we submit that this approach is inherently more reliable than extrapolation of VR profiles to values of 0.2% to estimate amounts of removed section, since the underlying principles are more explicit, and problems such as those highlighted by Dow (1977) and Katz et al. (1988) are more easily avoided.

This method also requires use of a preferred value for the paleo-surface temperature, which can be obtained from paleo-climate studies (e.g., Zachos et al. 2001). The influence of this factor can be assessed by dividing the change in temperature by the appropriate paleogeothermal gradient. For instance, for a paleogeothermal gradient of 30° C/km, a 10° C increase in paleo-surface temperature is equivalent to a reduction of 333 m of removed section. Thus, subtle changes in surface temperature may be equivalent to major changes in amounts of removed section. This, combined with discussion of possible nonlinear paleotemperature profiles presented earlier herein, emphasizes that estimation of removed section in this way is not a precise method, and the results can only ever be regarded as providing a general indication of the true amount.

Thermal History Reconstruction

By integrating the information on the timing and magnitude of the main paleo–thermal events that have affected the sedimentary section with information provided by the preserved section (i.e., the default thermal history), a more complete thermal and burial/uplift history can be reconstructed (e.g., Green et al. 2004).

It should be evident from the previous discussion that rigorous control on paleogeothermal gradients and therefore on amounts of additional burial is usually possible only at limited times in the history (i.e., the main paleo–thermal maxima and possibly one or two subsequent paleo–thermal peaks). It is not possible to define the complete burial or thermal history, and instead we focus on defining the key paleo–thermal events in as much detail as possible. In doing so, it is important to realize that processes other than burial may play key roles in controlling the thermal history. Elevated basal heat flow may produce a paleo–thermal maximum that does not correlate with the time of maximum burial and may in fact occur within a period of continuing burial, while heating as a result of hot fluid movement can occur at any time (Duddy et al. 1994, 1998). In addition, variation in surface temperature may cause significant cooling of a sedimentary

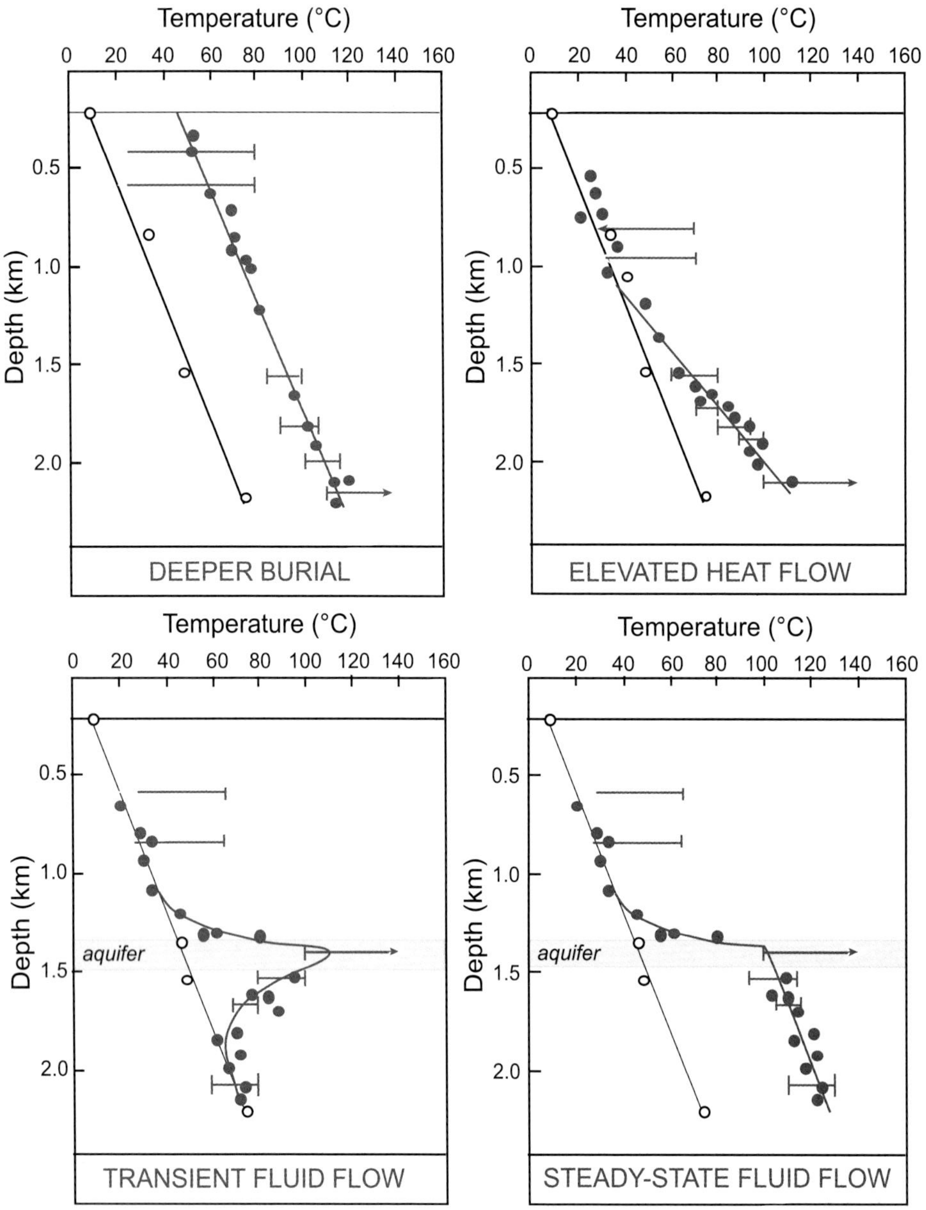

FIG. 22.—The way in which paleotemperatures characterizing a particular paleo–thermal episode vary through a vertical sequence of rocks (the "paleotemperature profile") provides key information on the mechanisms of heating and cooling. Deeper burial followed by exhumation, with no change in basal heat flow, will result in paleotemperatures defining a linear profile parallel to the present-day thermal profile but offset to higher temperatures. Elevated heat flow results in a linear paleotemperature profile with a higher slope compared to the present-day profile. In contrast, transient hot fluid flow through a localized aquifer results in a markedly nonlinear profile with a maximum centered on the aquifer, while prolonged fluid flow can result in a linear profile below the aquifer as the deeper section reaches a "steady-state" situation. Combinations of these four simple cases are possible.

section during progressive burial, as illustrated by Japsen et al. (2007a) in the eastern North Sea during the Cenozoic.

Precision and Accuracy in Thermal History Reconstruction

In discussing precision and accuracy in thermal history reconstruction, it is important to distinguish between the degree to which thermal history constraints can be defined, on the one hand, and the way in which this translates to uncertainty in the corresponding burial history on the other hand. Green et al. (2002) discussed these issues at some length, and only the basic points are reproduced here. Precision in this context describes the narrowness of the 95% confidence intervals on maximum paleotemperatures, corresponding amounts of additional burial, and the onset of cooling, while accuracy deals with how close the estimates might be to the true values.

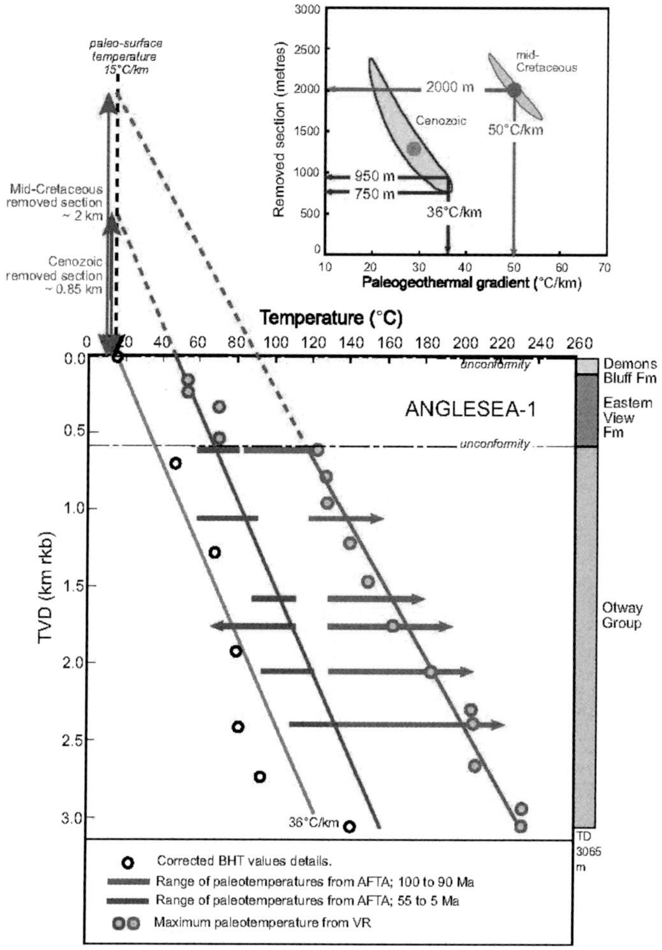

FIG. 23.—Extrapolation of paleotemperature profiles to estimate eroded section in the Anglesea-1 well, Otway Basin (after Green et al. 2004). AFTA and VR results define linear paleotemperature profiles for two thermal episodes: Mid-Cretaceous (cooling beginning between 100 and 90 Ma) associated with the unconformity between the Otway Group and the Eastern View Formation, and Tertiary (cooling beginning between 55 and 5 Ma) associated with the unconformity at the top of the Demons Bluff Formation at the ground surface. Fitting of linear profiles using maximum likelihood methods defines the range of paleogeothermal gradients consistent with the observed paleotemperatures. Linear extrapolation of these paleo-gradients to an appropriate paleo-surface temperature (in this case 15° C) defines the magnitude of additional section required to explain the paleotemperatures in each episode (subsequently removed by erosion). A cross-plot of allowed values of paleogeothermal gradient and removed section shows the allowed ranges (±95% confidence limits) for the Mid-Cretaceous and Tertiary thermal episodes. The paleogeothermal gradient for the mid-Cretaceous episode was between 45 and 55° C/km, significantly higher than the present-day gradient of 36° C/km, and was associated with 2000 ± 300 m of additional burial. The results allow a wider range of paleogeothermal gradient for the Tertiary episode (~20 to 37° C/km), encompassing the present-day gradient of 36° C/km, for which 750–950 m of additional burial are required to explain the Tertiary paleotemperature constraints. Note that this ← analysis assumes that the thermal conductivity of the eroded sequences was the same as that of the preserved sequences.

Accuracy of the thermal history constraints derived from AFTA is determined by how well the kinetic model used to define those constraints describes the natural system. As discussed earlier and as illustrated in Figures 7 and 11, multicompositional models provide reasonably accurate predictions when compared to data from geologically well-controlled situations. Consistent indications of maximum paleotemperatures from multiple techniques provide an additional check on accuracy. The precision of paleotemperatures determined from AFTA depends critically on the paleotemperature itself. Below around 60° C, mean track length varies only slowly with temperature, and it is often only possible to place an upper limit on the maximum paleotemperature. At higher temperatures, track length changes more rapidly with increasing temperature and the fission-track age reduction becomes increasingly pronounced (Fig. 2), and maximum paleotemperatures in this range become more precise, such that maximum paleotemperatures around 100° C can be defined within an overall uncertainty (95% confidence limits) of ~5° C. In other words, it is possible to define that a sample reached a paleotemperature between 100 and 105° C. However, as the paleotemperature increases further, and tracks become totally annealed, the only constraint that is possible is a minimum estimate of the maximum paleotemperature.

Translation of paleotemperatures defined from AFTA to estimates of former burial depth is much less precise, and also potentially a lot less accurate. Precise estimation requires control on paleogeothermal gradient, which can only be obtained by paleotemperature constraints over a range of depths, ideally from both AFTA and VR data. Even with high-quality data sets with consistent constraints from AFTA and VR over a depth interval of ~3 km, a 95% confidence interval of around 10° C/km around a best-fit paleogeothermal gradient of ~50° C/km is the best that can be achieved (i.e., 50 ± 5° C/km). This translates to a confidence interval of around 600 m or more for best-fit values of removed section around 2 km (i.e., 2.0 ± 0.3 km). So even for the most tightly constrained data sets, estimation of amounts of removed section in this way is not a precise method, and for less well-defined data sets, uncertainties can be much higher. This is exacerbated by considering the accuracy of predicting amounts of removed section by extrapolation of paleotemperature profiles, since assumptions regarding such factors as the paleo-surface temperature and the linearity of the profile through the removed section play a key role in the process (see Green et al. [2002] for a thorough review of these factors).

In terms of the timing information that can be obtained from AFTA, the numbers of fission tracks that can be counted for the fission-track age determination and the numbers of track lengths that can be measured play a major role in determining precision. In any given sample, these are determined by the apatite abundance, the quality of the grains, and the thermal history of the sample. In a best-case scenario, the onset of cooling from paleotemperatures sufficiently high to totally anneal all tracks may be defined with an uncertainty of around ±5% (e.g., 200 ± 10 Ma). So this technique does not provide the sort of high-precision analyses that can be obtained from mass spectrometer–based techniques. AFTA does provide information that cannot be obtained from any other approach, and since the systematics of the method are extremely well understood, the information is highly reliable, provided that appropriate procedures are employed at every step of the process. However, as we have tried to highlight in the foregoing discussion, failure to correctly implement any step of the process can result in highly misleading information.

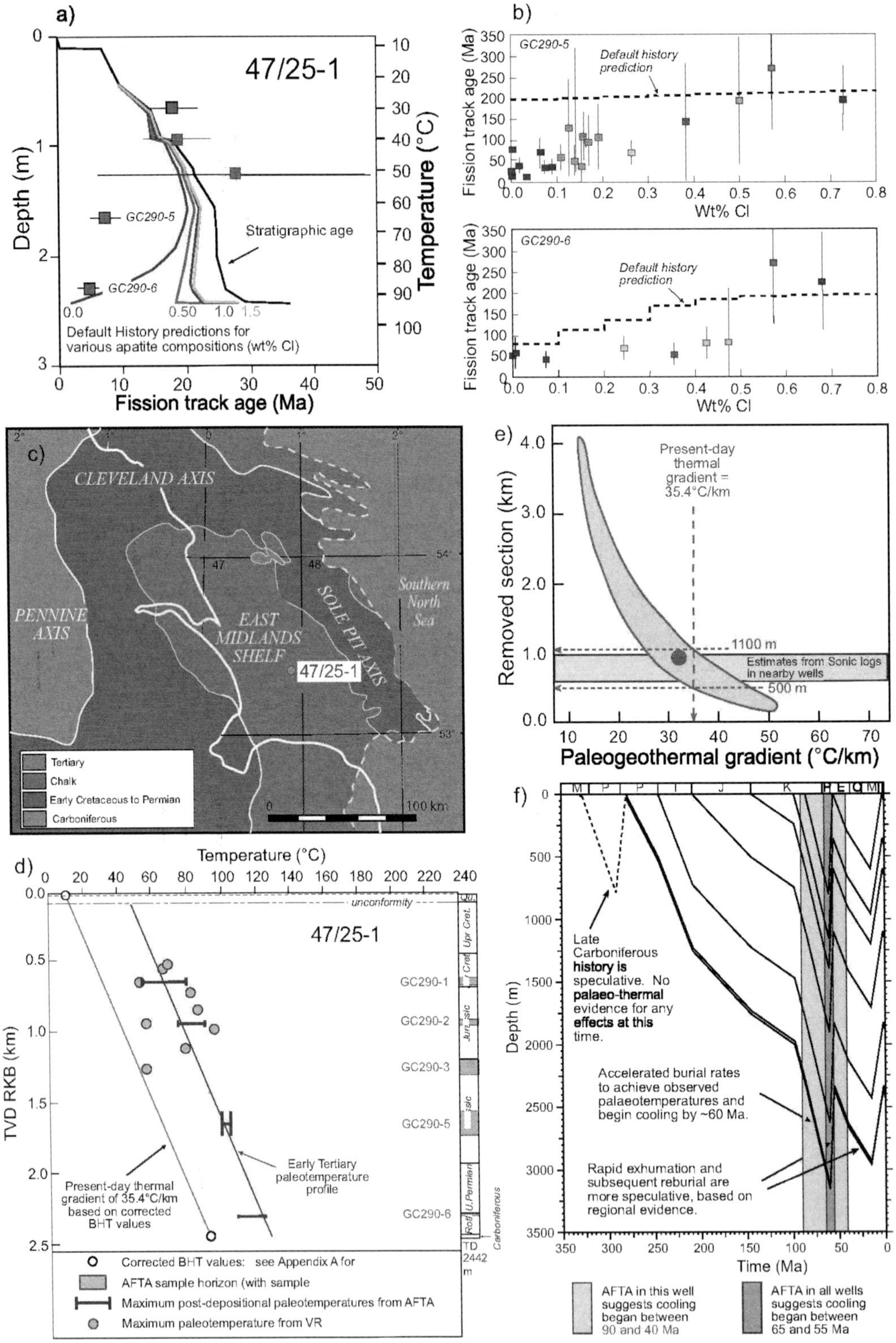

FIG. 24.—AFTA data from UK southern North Sea well 47/25–1 and resulting interpretations and reconstruction. (a) Measured fission-track ages versus depth, also showing the variation of stratigraphic age with depth together with values predicted from the default thermal history. The variation of fission-track age with wt% Cl in individual apatite grains within each sample is also shown for the two deepest samples (b). These data reveal cooling from maximum paleotemperatures around 40° C higher than present-day temperatures, which began between 90 and 40 Ma. Other wells in the region refine this timing for the onset of cooling to between 65 and 55 Ma. (c) Location map. (d) Paleotemperatures derived from AFTA and VR plotted against depth below kelly bushing (kb). VR data from this well show considerable scatter, which is common in this area due to varying degrees of suppression of reflectance levels in H-rich organic facies within the Jurassic section. (e) Allowed

Integration of AFTA Data with Basin Modeling Software

Whereas vitrinite reflectance (VR) data are used routinely in several commercially available basin modeling packages to calibrate models, AFTA has yet to be adopted routinely for this purpose. One reason for this is the complexity involved in modeling the AFTA system, particularly in terms of the possible presence of tracks in samples prior to deposition (which may differ between grains within a single sample), and variation in apatite composition and its effect on annealing kinetics. For this reason, we advocate the following: Rather than implementing quantitative modeling of the AFTA system within basin modeling software, a more appropriate approach is to use the temperature–time constraints derived from AFTA to calibrate viable models. In addition to being simpler and more straightforward, this has the advantage of maintaining a focus on the nature of the thermal history, and the factors that control it.

Sample Requirements for AFTA and Sampling Strategies

Apatite is a common accessory phase in many crystalline rocks, and, being among the most resistant of detrital minerals (Morton 1984), it forms a common detrital constituent of most sandstones and other coarse sedimentary rocks, which can readily be extracted using standard heavy mineral separation methods. AFTA can be applied either to outcrop samples or to ditch cuttings or core for subsurface samples, and in both situations, similar considerations apply regarding sampling. In sedimentary sequences, medium-grained sandstones are the most suitable lithologies for analysis, although coarse grits or coarse silts can be used. It is advisable to collect around 1 kg of material from the most mineralogically immature units available. In general, experience shows that 80 to 90% of sandstone samples collected on this basis contain sufficient apatite for analysis.

Note that since a high-quality analysis can be derived from as few as 20 grains of apatite, the overall abundance of apatite required is very low. Inspection of thin sections or even heavy mineral separates may not reveal apatite in samples that contain sufficient material for analysis, so the ultimate test is to refine the apatite fraction as far as possible and then to prepare, polish, and etch a grain mount. Since apatite is the only mineral that can be etched by dilute nitric acid, the etched apatite grains can easily be identified within a mass of other grains (multimineral composites, carbonates, sulfides) with similar density. The presence of these additional grains on the grain mount has no effect on the resulting analysis.

In downhole studies, samples of cuttings can be composited over a depth interval representing a range in downhole temperature of up to ~5° C, corresponding to a depth range of about 150 m for a typical geothermal gradient of 30° C/km. Samples should not be composited across an unconformity or major stratigraphic boundary. Where core samples are available, an integral solid piece of core is not necessary, since the sample will be crushed anyway. Offcuts, rubble, or scraps remaining from previous sampling can all be used, and again compositing over a range of depths is possible.

The most appropriate sampling strategy depends to some extent on the problems to be addressed, but in general, in order to obtain as much information as possible on key aspects of the paleo–thermal history, such as paleogeothermal gradients, removed section, etc., a vertical sequence of samples is required, spanning as wide a range of depths as possible from near surface to a depth where present downhole temperatures exceed about 110° C (corresponding to around 3 to 3.5 km for a typical geothermal gradient of 30° C/km), where tracks are totally annealed in the present-day thermal regime. Integration with VR data is recommended as routine practice, for reasons discussed earlier. VR samples should be distributed more or less evenly through a well section, including the temperature realm above 110° C; integration of information from this depth range with AFTA data provides a more complete reconstruction that would otherwise be available.

CASE STUDIES

Introduction

AFTA has been applied in a wide range of settings, sometimes in isolation, but more often in tandem with a variety of other techniques. Such studies are of direct importance to hydrocarbon exploration; e.g., in providing firm constraints on the timing of hydrocarbon generation in relation to formation of structures and thereby defining areas that are most prospective for hydrocarbons. These studies have highlighted a number of key issues with important implications for sedimentary basin evolution and basin modeling, as well as regional tectonism. In the following, we outline a small selection of these studies, highlighting the importance of the outcomes and stressing the general conclusions that emerge, under a series of headings related to the nature of the underlying processes.

Regional Exhumation in Intraplate Settings and Not-So-Passive Margins

Many AFTA studies in areas that can be broadly categorized as intraplate settings have revealed major paleo–thermal episodes that are best explained dominantly in terms of heating due to deeper burial and cooling due to subsequent exhumation. Well-documented examples include the UK East Midlands Shelf (Green et al. 2001b, Green 2005), UK East Irish Sea Basin (Green et al. 1997, Holford et al. 2005), and eastern North Sea Basin (Denmark) (Japsen et al. 2007a). Figure 24 illustrates results from UK southern North Sea well 47/25–1, drilled where the Late Cretaceous Chalk occurs at the seabed, showing that the Chalk and underlying units were buried by around 1 km of additional Late Cretaceous to Paleogene section, which was subsequently removed during Cenozoic exhumation.

Similarly, results from a number of passive margins have revealed that, far from the simple history of progressive deposition following rifting, subsidence has been interrupted by episodes of exhumation, with the resulting breaks in the section often erroneously interpreted as representing periods of stability (Fig. 25). Such results are often regarded with skepticism, as they conflict with established interpretations, but consistent indications of the degree of burial/exhumation from multiple techniques (including paleo-thermal and other methods) mean that the evidence cannot be dismissed as an artifact of any one method and must be considered reliable.

With increasing amounts of data becoming available, a pattern has

←

ranges of paleogeothermal gradient and removed section derived from the paleotemperature constraints in (d) using methods described by Bray et al. (1992), compared with independent estimates of removed section from sonic velocity analysis (Japsen 2000). Both approaches give highly consistent results. (f) Resulting burial history reconstruction for the well section, showing onset of exhumation at 60 Ma. An additional Miocene phase is included by analogy with nearby wells, although data from this well show no direct evidence for this episode.

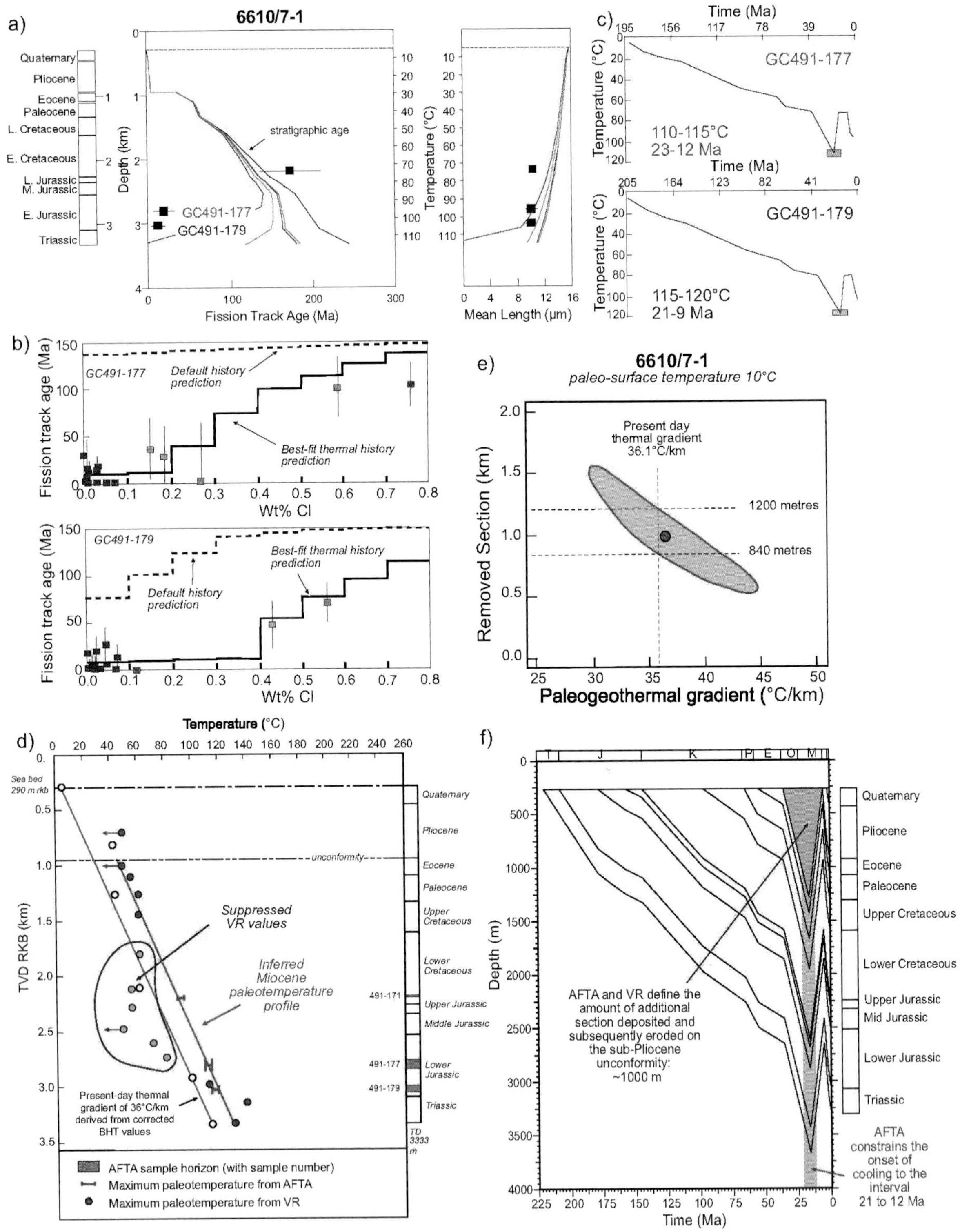

FIG. 25.—AFTA and VR data in mid-Norway well 6610/7–1 and resulting interpretations and reconstruction. (a) Measured fission-track ages versus depth, also showing the variation of stratigraphic age with depth together with values predicted from the default thermal history. Also shown for the two deepest samples are the variation of fission-track age with wt% Cl in individual apatite grains within each sample (b) and the best-fit thermal histories derived from the AFTA data (c). These data reveal cooling from maximum paleotemperatures around 20° C higher than present-day temperatures, which began between 21 and 12 Ma. This timing correlates closely with the base-Pliocene unconformity in this well, which spans the Late Eocene to Early Pliocene interval (ca. 40 to 4 Ma). (d) Paleotemperatures derived from AFTA and VR plotted against depth below kb. VR data from the Jurassic section in this well are anomalously low, due to of suppression of reflectance levels in H-rich

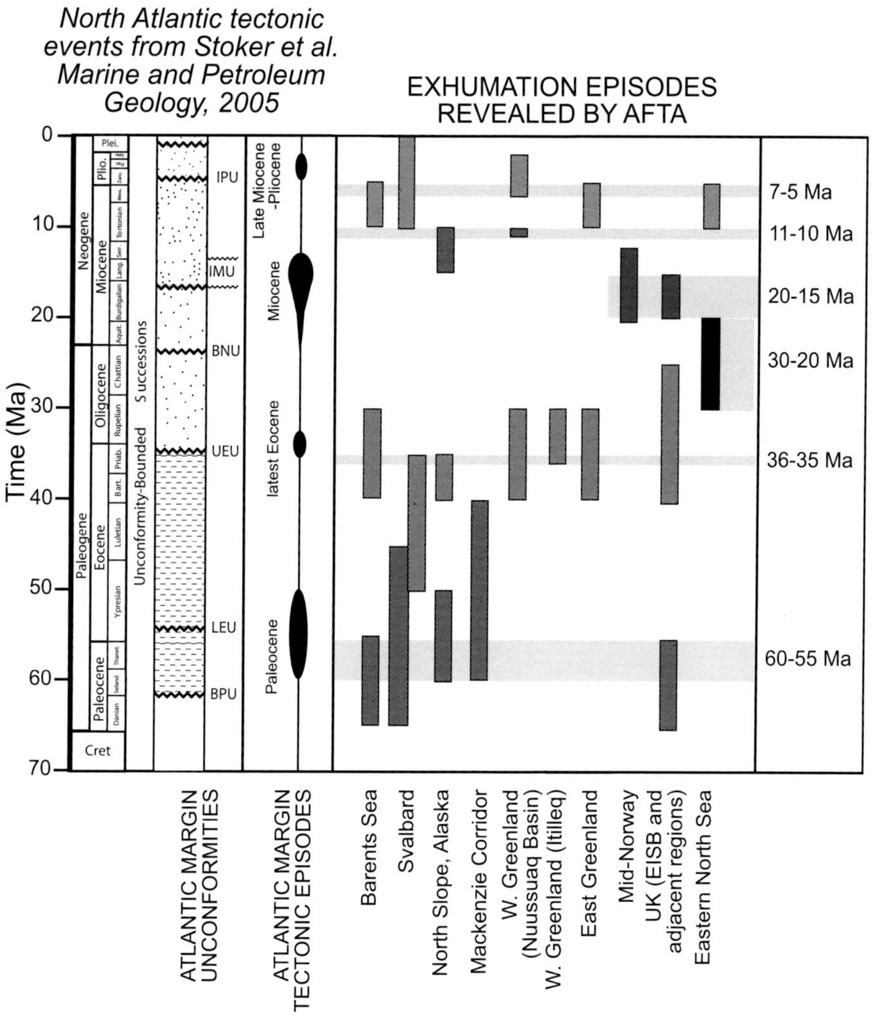

FIG. 26.—Synthesis of Cenozoic stratigraphy along the NW European Atlantic margin has revealed a series of regional unconformities representing regional tectonic events (Praeg et al. 2005, Stoker et al. 2005). Cenozoic exhumation episodes revealed by AFTA in samples from the UK and NW Europe correlate closely with these unconformities and the underlying tectonic movements, which are interpreted as reflecting major periods of deformation at proximal plate boundaries. See Holford et al. (2009a) for further details.

emerged suggesting that the onset of exhumation in these episodes is broadly synchronous across wide regions, with Paleogene, end-Eocene, and Miocene events recognized across wide areas of the North Atlantic and Arctic regions (Holford et al. 2009a). These exhumation episodes also appear to correlate with periods of regional tectonic activity related to events at plate boundaries (Fig. 26), expressed as unconformities that can be recognized along the Atlantic margin from Rockall to Lofoten (Praeg et al. 2005, Stoker et al. 2005). Thus, these events, which are of relatively minor magnitude along the continental margin, such that they can be recognized within the preserved stratigraphic section, have had much more pronounced effect inboard of the margins in intraplate settings, where all stratigraphic evidence has been removed by erosion.

Similar synchronicity of events has also been identified in the South

←

organic facies within this section. However, paleotemperatures derived from shallower and deeper VR data fall on a common profile with those from AFTA, defining a linear profile subparallel to the present-day temperature profile. (e) Allowed ranges of paleogeothermal gradient and removed section derived from the paleotemperature constraints in (d) using methods described by Bray et al. (1992). The best-fit paleogeothermal gradient is almost identical to the present-day value, confirming that heating was due to deeper burial, with around 1 km of section deposited and then removed in the interval 40 to 4 Ma. The onset of exhumation, in the interval 21 and 12 Ma, correlates with a pronounced regional mid-Miocene unconformity recognized along the Atlantic margin from Lofoten to Porcupine (Stoker et al. 2005; see Fig. 26). (f) Resulting burial history reconstruction for the well section, showing onset of exhumation at 15 Ma. Note the significant thickness of Pliocene section deposited after the episode of exhumation, which diminishes the overall difference between present-day and paleotemperatures, making definition of this event less straightforward than it otherwise might be.

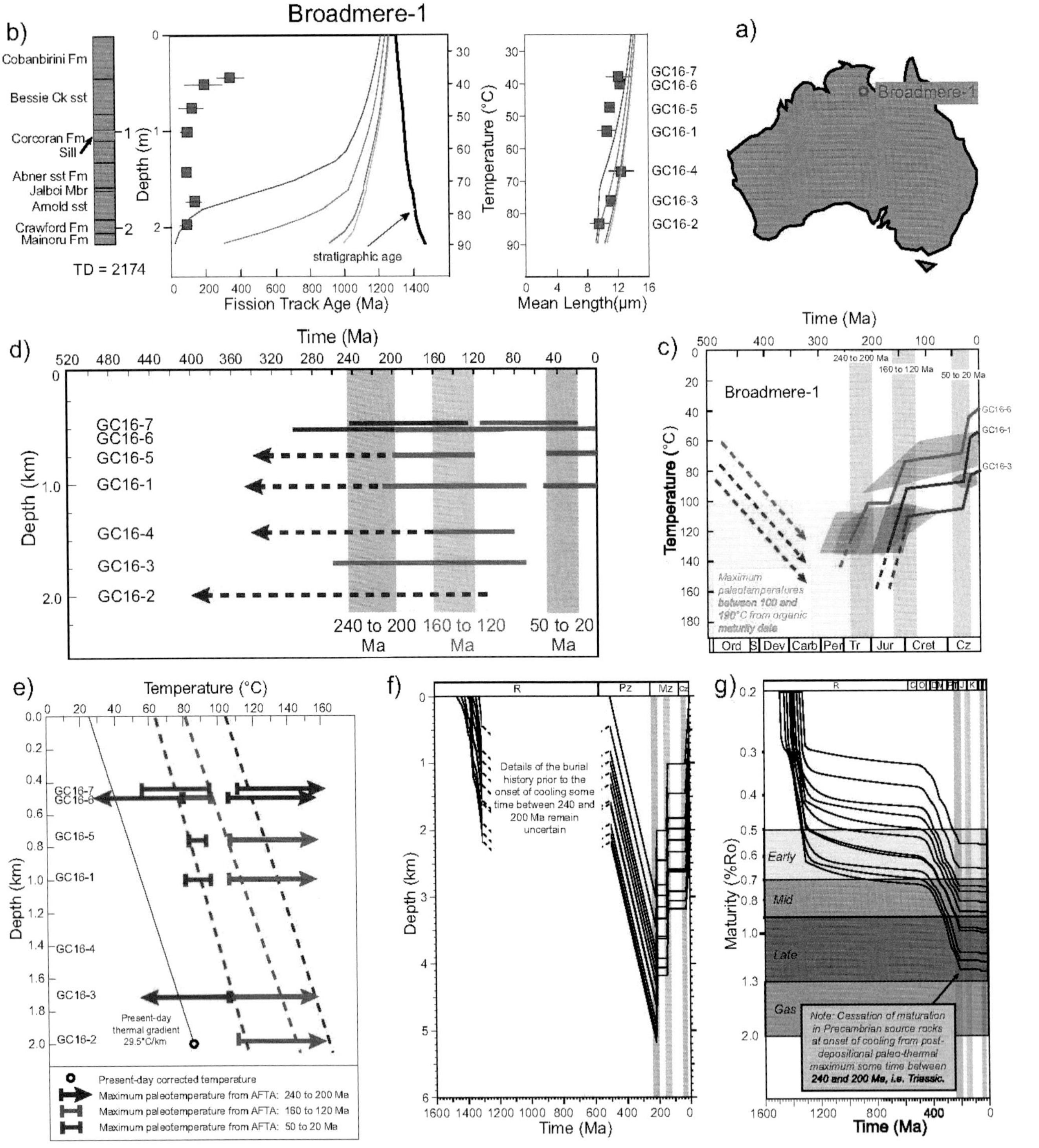

FIG. 27.—AFTA data in McArthur Basin (Australia) well Broadmere-1 and resulting interpretations and reconstruction (after Duddy et al. 2004). (a) Location map. (b) Basic AFTA parameters plotted against sample depth and present temperature. Fission-track ages in all samples are much younger than the stratigraphic age at low present-day temperatures, showing that they have been subjected to maximum paleotemperatures higher than present temperatures at some time after deposition. (c) AFTA thermal history solutions for individual samples showing that all samples cooled through ~100° C at various times since the Triassic, followed by subsequent cooling from lower temperatures. (d) Timing constraints on three thermal episodes (Triassic, Jurassic–Cretaceous, and Tertiary) derived from the overlap of AFTA timing constraints in individual samples. (e) AFTA paleotemperature estimates (±95% confidence limits) and possible paleotemperature depth profiles for three thermal episodes. Gradients for the Tertiary and Jurassic–Cretaceous episodes are generally similar to the present-day

Atlantic (Turner et al. 2008). Further, Late Miocene exhumation revealed by AFTA in southeastern Australia (Green et al. 2004) appears to have been synchronous with the onset of compression of the Australian plate transmitted from plate boundaries in Papua New Guinea and New Zealand (Dickinson et al. 2001). Thus, all these studies point to the conclusion that intraplate stresses transmitted from plate boundaries across often huge distances can cause significant episodes involving vertical motion in which significant amounts of section, often 1 to 2 km in thickness, can be removed. The underlying mechanisms of such episodes are as yet only poorly understood. This is due in part to the fact that the evidence that defines such episodes has often been disregarded as not consistent with the accepted geological evolution, and often reduced to the simplistic argument that no sedimentary cover is present now, so none was ever deposited. Sufficient information is now available to confirm that such episodes have indeed affected wide regions of the world, and further work is required in order to fully understand the nature of the processes responsible.

Basin Inversion

Many applications of AFTA relate to inverted sedimentary basins in which compressional reactivation of former extensional faults has led to expulsion and erosion of former rift sections. The study by Kamp and Green (1990) of the Taranaki Basin provides an excellent example of use of AFTA in this type of setting, subsequently followed up by Crowhurst et al. (2002), who integrated apatite (U-Th)/He dating with AFTA and VR data to refine the detail of the timing of inversion.

Japsen et al. (2007a) showed that inversion of the Sorgenfrei–Tornquist Zone offshore southern Norway was initiated in the Mid-Cretaceous, beginning between 110 and 95 Ma, and a similar timing was identified in the Sole Pit inversion axis in the UK southern North Sea by Green et al. (2005b). In both locations, various conflicting estimates have been suggested for the onset of inversion based on investigations of the preserved stratigraphic section. Consistent results from both paleo-thermal (AFTA and VR data) and paleo-burial (sonic velocity) techniques resolves this former conflict and provides firm constraints on the timing of inversion, which allows more accurate basin models for hydrocarbon exploration, and a more rigorous understanding of the relationship between inversion and regional stress patterns.

Another major outcome of the application of AFTA has been the recognition of inversion in basins where regional exhumation has removed any direct evidence of reverse movement on former extensional faults. Holford et al. (2009b) coined the term "cryptic inversion" to describe this situation in the East Irish Sea Basin, UK, and pointed out that this process may explain why the magnitude of shortening is frequently underestimated in such basins.

Phanerozoic Events in Proterozoic Basins

One of the most important outcomes from application of AFTA to Proterozoic sedimentary basins in Australia and other parts of the world is recognition that in many regions, the sedimentary section has been exhumed more recently than previously thought (Tingate and Duddy 2002, Duddy et al. 2004, Hegarty et al. 2007). This has the important implication that in Proterozoic basins, the main phase of hydrocarbon generation may have been much more recent than previously considered. Figure 27 typifies results in this sort of setting, based on the Broadmere-1 well from the Macarthur Basin of northern Australia. AFTA data from this well show that Proterozoic units cooled from their paleo–thermal maximum in Mesozoic times, in contrast to previous ideas involving progressive exhumation from Cambrian times onward (e.g., Silverman et al. 2007). Similarly, Hegarty et al. (2007) showed that the Proterozoic section of the US Midcontinent Rift System underwent a series of Phanerozoic exhumation episodes, in contrast to previous ideas involving exhumation that began in the Late Proterozoic. In such settings, results from AFTA imply that sequences of younger sedimentary units have been deposited and subsequently eroded away over vast areas. Such conclusions can attract skepticism, normally for no better reason than the cover units are not present now, so they never were deposited. Such circular conservative reasoning is now giving way to the realization of the true nature of the history of such regions, but as noted earlier, the underlying mechanisms by which these episodes of regional burial and exhumation occur are still only poorly understood.

Fluid Effects

The extent to which heat can be transported around sedimentary basins has been controversial in the past. However, studies based on the combination of AFTA, VR, and fluid inclusion measurements in the UK West of Shetland region (Parnell et al. 1999, 2005) and the North West Shelf of Australia (O'Brien et al. 1996) have quite clearly shown the effects of basin-scale heating due to hot fluids. Duddy et al. (1994, 1998) provided further documentation of this process, including results from the Canning Basin of NW Australia, where emplacement of intrusions into unconsolidated sediments produced regional heating due to hot fluid circulation. Such effects are commonly most strongly pronounced where thin quartzose sandstones forming regional aquifer horizons occur within thick muds, which serve to seal and constrain fluid movement.

Landscape Studies

Integration of AFTA and VR data with independent information provided by landscape studies has been particularly useful in defining the development of the West Greenland continental margin (Japsen et al. 2005). Among the most important outcomes of these studies is the conclusion that the present-day morphology is unrelated to processes of rifting and continental separation, postdating the cessation of seafloor spreading in the adjacent ocean basin by ~30 million years. This has important implications for the tectonic development of similar margins around the world (Japsen et al. 2009), as well as for hydrocarbon exploration. In West Greenland, AFTA data show that the sedimentary units of the Nuussuaq Basin began to cool from their maximum postdepositional paleotemperatures at around 35 Ma, much later than Late Cretaceous–Paleocene rifting. This implies that structures related to rifting in this region are likely to be prospective for hydrocarbons, since they were available for charging during the main phase of generation, in contrast to the situation in many rift basins, where the main phase of hydrocarbon generation occurred

←

geothermal gradient, although the best-fit values are lower and higher, respectively. The gradient for the Triassic episode is not constrained by the results, but a gradient similar to the present-day gradient is allowed. (f) Reconstructed burial history based on the AFTA data with 3 km of total of additional burial prior to uplift and erosion in a step-wise manner beginning in the Triassic. Reconstruction of the magnitude of eroded section assumes that the eroded section had the same average thermal conductivity as the preserved sequence in the well. (g) reconstructed pattern of maturation through time, based on the thermal history reconstruction in this well.

early, prior to formation of structures related to rifting, as seen for example in the Otway Basin of SE Australia (Green et al. 2004). The possibility of seal breach and loss of charge as a result of exhumation onshore West Greenland, which most likely accounts for the common oil seeps in the area (Pedersen et al. 2006), must be taken into account, but where seals have retained their integrity, the timing relationships established from the combination of AFTA, VR, and landscape studies in this region suggest that the region remains an attractive exploration target.

General Implications for Hydrocarbon Exploration

One of the major outcomes from application of AFTA to hydrocarbon exploration is the observation that in many sedimentary basins, the process that dominates hydrocarbon prospectivity is deeper burial and subsequent uplift and erosion (exhumation). Implications for exploration include enhanced maturity levels (compared to assessments that underestimate former burial depths), possible loss of reservoired hydrocarbons as a result of seal breach, and phase changes due to pressure reduction (see Doré et al. [2002] for an extensive review of the influence of exhumation on hydrocarbon systems). In addition, the timing of generation becomes a key issue in such areas, particularly for plays involving structures formed in the initial stages of exhumation, as these structures were not available for charging during the main phase of generation, which occurred during burial prior to the onset of exhumation.

Although exhumation is often considered as a negative factor for prospectivity, exhumed basins in many parts of the world contain significant quantities of hydrocarbons, including the East Irish Sea Basin and southern North Sea of the UK, the Reconcavo Basin of Brazil, and much of North Africa. In these basins, discovering hydrocarbons requires detailed knowledge of magnitude and timing of exhumation and associated effects, and AFTA can provide unique constraints on these crucial issues.

It is a common mantra in basin modeling studies that once the depositional history has been constrained, on the basis of preserved geological evidence, then varying the heat flow through time to obtain a match with measured maturity data results in a viable description of the basin history. While we have encountered a number of situations where elevated heat flow is required to explain at least some of the heating indicated by AFTA and other paleo–thermal constraints (e.g., Duddy 1997, Green et al. 2004, Japsen et al. 2005, Turner et al. 2008), it is far more common for a constant heat-flow scenario involving heating solely due to deeper burial to provide the best match to the available data (e.g., Green et al. 2001a, 2001b; Holford et al. 2005; Japsen et al. 2007a). Even in those situations where elevated heat flow is required, at least some degree of deeper burial and subsequent exhumation is also usually required. The deeper burial required to explain the AFTA data is often regarded as controversial, because little or no trace of the required sedimentary sequences now remains. However, as stressed here, consistent evidence from a range of different techniques all pointing to a common conclusion leaves little room for doubt that the required burial must have occurred. As the oft-quoted epithet has it—absence of evidence is not evidence of absence.

CONCLUSIONS

Of the common methods of low-temperature thermochronology, only apatite fission-track analysis has reached the stage of routine application as a research technique and as a commercial tool for hydrocarbon exploration. AFTA allows determination of the maximum paleotemperature of a sedimentary unit and the time at which the unit cooled from that paleo–thermal maximum, as well as possibly defining one or even maybe two subsequent episodes. AFTA cannot be used in isolation and requires a certain minimum amount of geological background information to provide a reliable thermal history solution. When used in tandem with other methods such as vitrinite reflectance, determination of paleogeothermal gradients allows identification of mechanisms of heating and cooling, and determination of amounts of removed section. Combined with the timing constraints uniquely available from AFTA, this comprehensive suite of quantitative information allows more accurate reconstruction of thermal and burial/uplift histories than would otherwise be possible. Such information can provide unique information on hydrocarbon prospectivity, not only in frontier regions, but also in well-studied areas, where new insights can often be obtained.

REFERENCES

Allen P, Carter A, Cunningham M, Densmore A, Galewsky J, Gallagher K, Lazzaretti E, Phillips, A. 2002. The post-Variscan thermal and denudational history of Ireland. *In* Doré AG, Cartwright J, Stoker MS, Turner JP, White N (Editors). *Exhumation of the North Atlantic Margin: Timing, Mechanisms and Implications for Petroleum Exploration*: Geological Society, London. Special Publication 196, p. 371–399.

Andrews-Speed CP, Oxburgh ER, Cooper BA. 1984. Temperatures and depth-dependent heat flow in western North Sea. *American Association of Petroleum Geology Bulletin* 11:1764–1784.

Argent JD, Stewart SA, Green PF, Underhill JR. 2002. Heterogeneous exhumation in the Inner Moray Firth, UK North Sea: constraints from new AFTA® and seismic data. *Journal of the Geological Society* 159:715–729.

Barbarand J, Carter A, Wood I, Hurford AJ. 2003a. Compositional and structural control of fission track annealing in apatite. *Chemical Geology* 198:107–137.

Barbarand J, Hurford AJ, Carter A. 2003b. Variation in apatite fission-track length measurement: implications for thermal history modelling. *Chemical Geology* 198:77–106.

Bhandari N, Bhat SG, Lal D, Rajagopalan G, Tamhane AS, Venkatavaradan VS. 1971. Fission fragment tracks in apatite: recordable track lengths. *Earth and Planetary Science Letters* 13:191–199.

Bray RJ, Green PF, Duddy IR. 1992. Thermal history reconstruction using apatite fission track analysis and vitrinite reflectance: a case study from the UK East Midlands and the southern North Sea. *In* Hardman RFP (Editor). *Exploration Britain: Into the Next Decade*: Geological Society, London. Special Publication 67, p. 3–25.

Brown MC. 2007. Discussion on 'Apatite (U- Th)/He age constraints on the Mesozoic and Cenozoic evolution of the Bathurst region, New South Wales: evidence for antiquity of the continental drainage divide along a passive margin'. *Australian Journal of Earth Science* 54:1011–1012.

Bulat J, Stoker SJ. 1987. Uplift determination from interval velocity studies, UK, southern North Sea. *In* Brooks J, Glennie KW (Editors). *Petroleum Geology of North West Europe*: Graham & Trotman, London. p. 293–305.

Burnham AK, Sweeney JJ. 1989. A chemical kinetic model of vitrinite reflectance maturation. *Geochimica et Cosmochimica Acta* 53:2649–2657.

Burtner RL, Nigrini A, Donelick RA. 1994. Thermochronology of lower Cretaceous source rocks in the Idaho–Wyoming thrust belt. *American Association of Petroleum Geologists Bulletin* 78:1613–1636.

Campanile D, Nambiar CG, Bishop P, Widdowson M, Brown R. 2008. Sedimentation record in the Konkan–Kerala Basin: implications for the evolution of the Western Ghats and the Western Indian passive margin. *Basin Research* 20:3–22.

Carlson WD. 1990. Mechanisms and kinetics of apatite fission-track annealing. *American Mineralogist* 75:1120–1139.

Carlson WD, Donelick RA, Ketcham RA. 1999. Variability of apatite fission-track annealing kinetics: I. Experimental results. *American Mineralogist* 84:1213–1223.

Cook A (Editor). 1982. *The Origin and Petrology of Organic Matter in Coals, Oil Shales and Petroleum Source Rocks*: The University of Wollongong, Wollongong, NSW, Australia. 106 p.

Cope MJ. 1986. An interpretation of vitrinite reflectance data from the southern North Sea Basin. *In* Brooks J, Goff JC, Van Hoorn B (Editors). *Habitat of Palaeozoic Gas in Northwest Europe*: Geological Society, London. Special Publication 23, p. 85–98.

Corrigan JD. 1991. Inversion of apatite fission track data for thermal history information. *Journal of Geophysical Research* 96:10347–10360.

Corrigan JD. 1993. Apatite fission-track analysis of Oligocene strata in south Texas, U.S.A.: testing annealing models. *Chemical Geology* 10:227–249.

Crowhurst PV, Green PF, Kamp PJJ. 2002. Appraisal of (U-Th)/He apatite thermochronology as a thermal history tool for hydrocarbon exploration: an example from the Taranaki Basin, New Zealand. *American Association of Petroleum Geologists Bulletin* 86:1801–1819.

Crowley KD. 1993. Mechanisms and kinetics of apatite fission-track annealing—discussion. *American Mineralogist* 78:210–212.

Crowley KD, Cameron M, Schaeffer RL. 1991. Experimental studies of annealing of etched fission tracks in fluorapatite. *Geochimica et Cosmochimica Acta* 55:1449–1465.

Dickinson JA, Wallace MW, Holdgate GR, Daniels J, Gallagher SJ, Thomas L. 2001. Neogene Tectonics in SE Australia: implications for petroleum systems. *Australian Petroleum Production and Exploration Association Journal* 41:37–52.

Donelick RA, Roden MK, Moores JD, Carpenter BS, Miller DS. 1990. Etchable length reduction of induced fission tracks in apatite at room temperature (~23° C): crystallographic orientation effects and "initial" mean lengths. *Nuclear Tracks and Radiation Measurements* 17:261–265.

Doré AG, Corcoran DV, Scotchman IC. 2002. Prediction of the hydrocarbon system in exhumed basins, and application to the NW European margin. *In* Doré AG, Cartwright J, Stoker MS, Turner JP, White N (Editors). *Exhumation of the North Atlantic Margin: Timing, Mechanisms and Implications for Petroleum Exploration*: Geological Society, London. Special Publication 196, p. 331–354.

Dow W. 1977. Kerogen studies and geological interpretations. *Journal of Geochemical Exploration* 7:79–99.

Duddy IR. 1997. Focussing exploration in the Otway Basin: understanding timing of source rock maturation. *Australian Petroleum Production and Exploration Association Journal* 37:178–191.

Duddy IR, Green PF, Bray RJ, Hegarty KA. 1994. Recognition of the thermal effects of fluid flow in sedimentary basins. *In* Parnell J (Editor). *Geofluids: Origin, Migration and Evolution of Fluids in Sedimentary Basins*: Geological Society, London. Special Publication 78, p. 325–345.

Duddy IR, Green PF, Gibson HJ, Hegarty KA. 2004. Regional palaeo-thermal episodes in northern Australia. *In* Ellis GK, Baillie PW, Munson TJ (Editors). *Timor Sea Petroleum Geoscience. Proceedings of the Timor Sea Symposium, Darwin, Northern Territory, 19–20 June 2003*: Northern Territory Geological Survey, Darwin, Australia. Special Publication 1, p. 567–591.

Duddy IR, Green PF, Hegarty KA, Bray RJ, O'Brien GW. 1998. Dating and duration of hot fluid flow events determined using AFTA® and vitrinite reflectance-based thermal history reconstruction. *In* Parnell J (Editor). *Dating and Duration of Hot Fluid Flow and Fluid-Rock Interaction*: Geological Society, London. Special Publication 144, p. 41–51.

Duddy IR, Green PF, Laslett GM. 1988. Thermal annealing of fission tracks in apatite 3. Variable temperature behavior. *Chemical Geology (Isotope Geoscience Section)* 73:25–38.

Ehlers TA, Chaudri T, Kumar S, Fuller C, Willett SD, Ketcham RA, Brandon MT, Belton DX, Kohn BP, Gleadow AJW. 2005. Computational tools for low-temperature thermochronometer interpretation. *Reviews in Mineralogy and Geochemistry* 58:589–622.

Farley KA. 2000. Helium diffusion from apatite: general behaviour as illustrated by Durango fluorapatite. Journal of Geophysical Research 105(B2):2903-2914.

Fisher MJ, Barnard PC, Cooper BS. 1980. Organic maturation and hydrocarbon generation in the Mesozoic sediments of the Sverdrup Basin, Arctic Canada, Volume 2. In Proceedings of the 4th International Palynology Conference; 1976-1977; p. 581-588.

Fitzgerald PG, Gleadow AJW. 1988. Fission track geochronology, tectonics and structure of the Transantarctic Mountains in Northern Victoria Land, Antarctica. *Chemical Geology (Isotope Geoscience Section)* 73:169–198.

Fitzgerald PG, Muñoz JA, Coney PJ, Baldwin SL. 1999. Asymmetric exhumation across the Pyrenean orogen: implications for the tectonic evolution of collisional orogens. *Earth and Planetary Science Letters* 173:157–170.

Fleischer RL, Price PB, Walker RM. 1975. *Nuclear Tracks in Solids*. University of California Press, Berkeley, California. 595 p.

Flowers RM, Ketcham RA, Shuster DL, Farley KA. 2009. Apatite (U-Th)/He thermochronometry using a radiation damage accumulation and annealing model. *Geochimica et Cosmochimica Acta* 73:2347–2365.

Galbraith RF. 2006. *Statistics for Fission Track Analysis*: Chapman and Hall/CRC, Boca Raton, Florida. 200 p.

Galbraith RF, Laslett GM. 1988. Some calculations relevant to the thermal annealing of fission tracks in apatite. *Proceedings of the Royal Society of London* A419:305–321.

Galbraith RF, Laslett GM. 1990. Apatite fission track analysis: geological thermal history analysis based on a three-dimensional random process of linear radiation damage. *Philosophical Transcripts of the Royal Society of London* 332:419–438.

Gallagher K. 1995. Evolving temperature histories from apatite fission-track data. *Earth and Planetary Science Letters* 136:421–435.

Gallagher K, Brown R. 1997. The onshore record of passive margin evolution. *Journal of the Geological Society* 154:451–457.

Gallagher K, Brown R. 1999a. Denudation and uplift at passive margins: the record on the Atlantic margin of southern Africa. *Philosophical Transactions of the Royal Society of London* 357:835–859.

Gallagher K, Brown R. 1999b. The Mesozoic denudation history of the Atlantic margins of southern Africa and southeast Brazil and the relationship to offshore sedimentation. *In* Cameron NR, Bate RH, Clure VS (Editors). *The Oil and Gas Habitats of the South Atlantic*: Geological Society, London. Special Publication 153, p. 41–53.

Gibson DL. 2007. Discussion on 'Apatite (U-Th)/He age constraints on the Mesozoic and Cenozoic evolution of the Bathurst region, New South Wales: evidence for antiquity of the continental drainage divide along a passive margin'. *Australian Journal of Earth Science* 54:1012–1015.

Gleadow AJW, Duddy IR. 1981. Fission track analysis: a new tool for the evaluation of thermal histories and hydrocarbon potential. *Australian Petroleum Exploration Association Journal* 23:93–102.

Gleadow AJW, Duddy IR, Green PF, Lovering JF. 1986. Confined fission track lengths in apatite—a diagnostic tool for thermal history analysis. *Contributions to Mineralogy and Petrology* 94:405–415.

Goswami JN, Jha R, Lal D. 1984. Quantitative treatment of annealing of charged particle tracks in common rock minerals. *Earth and Planetary Science Letters* 71:120–128.

Green P. 2004. The importance of validating annealing models. *In* 10th International Conference on Fission Track Dating and Thermochronology Abstracts, August 9–13, 2004, Amsterdam, The Netherlands (abstract), p. 53.

Green PF. 1980. On the cause of the shortening of spontaneous fission tracks in certain minerals. *Nuclear Tracks* 4:91–100.

Green PF. 1986. On the thermo-tectonic evolution of northern England: evidence from fission track analysis. *Geological Magazine* 123:493–506.

Green PF. 1988. The relationship between track shortening and fission track age reduction in apatite: combined influences of inherent instability, annealing anisotropy, length bias and system calibration. *Earth and Planetary Science Letters* 89:335–352.

Green PF. 1989. Thermal and tectonic history of the East Midlands Shelf (onshore UK) and surrounding regions assessed by apatite fission track analysis. *Journal of the Geological Society* 146:755–773.

Green PF. 2005. Post-Carboniferous burial and exhumation histories of Carboniferous rocks of the southern North Sea and adjacent onshore UK. *In* Collinson JD, Evans DJ, Holliday DW, Jones NS (Editors). *Carboniferous Hydrocarbon Geology: The Southern North Sea and Surrounding Onshore Areas*: Yorkshire Geological Society, Leicester, UK. Occasional Publication 7, p. 25–34.

Green PF, Crowhurst PV, Duddy IR. 2004. Integration of AFTA and (U-Th)/He thermochronology to enhance the resolution and precision of thermal history reconstruction in the Anglesea-1 well, Otway Basin, SE Australia. *In* Eastern Australian Basins Symposium II; September 19–22, 2004; Adelaide, South Australia: Petroleum Exploration Society of Australia, Perth, Australia. p. 117–131.

Green PF, Crowhurst PV, Duddy IR, Japsen P, Holford SP. 2006. Conflicting (U-Th)/He and fission track ages in apatite: enhanced He retention, not anomalous annealing behavior. *Earth and Planetary Science Letters* 250:407–427.

Green PF, Duddy IR. 2006. Interpretation of apatite (U-Th)/He ages and fission track ages from cratons. *Earth and Planetary Science Letters* 244:541–547.

Green PF, Duddy IR. 2007. Discussion on 'Apatite (U-Th)/He age constraints on the Mesozoic and Cenozoic evolution of the Bathurst region, New South Wales: evidence for antiquity of the continental drainage divide along a passive margin'. *Australian Journal of Earth Science* 54:1009–1011.

Green PF, Duddy IR, Bray RJ. 1997. Variation in thermal history styles around the Irish Sea and adjacent areas: implications for hydrocarbon occurrence and tectonic evolution. *In* Meadows NS, Trueblood SP, Hardman M, Cowan G (Editors). *Petroleum Geology of the Irish Sea and Adjacent Areas*: Geological Society, London. Special Publication 124, p. 73–93.

Green PF, Duddy IR, Bray JR. 2005b. Resolution of Early Cenozoic regional exhumation and Mid-Cretaceous basin inversion in the southern North Sea, using AFTA. *In* American Association of Petroleum Geologists (AAPG) Meeting Abstract Volume, September 11–14, 2005, Paris, (abstract).

Green PF, Duddy IR, Bray RJ, Duncan WI, Corcoran DV. 2001a. The influence of thermal history on hydrocarbon prospectivity in the central Irish Sea Basin. *In* Shannon PM, Haughton PDW, Corcoran DV (Editors). *The Petroleum Exploration of Ireland's Offshore Basins*: Geological Society, London. Special Publication 188, p. 171–188.

Green PF, Duddy IR, Gleadow AJW, Lovering JF. 1989a. Apatite fission track analysis as a palaeotemperature indicator for hydrocarbon exploration. *In* Naeser ND, McCulloh T (Editors). *Thermal History of Sedimentary Basins—Methods and Case Histories*: Springer-Verlag, New York. p. 181–195.

Green PF, Duddy IR, Gleadow AJW, Tingate PR, Laslett GM. 1985. Fission-track annealing in apatite: track length measurements and the form of the Arrhenius plot. *Nuclear Tracks* 10:323–328.

Green PF, Duddy IR, Gleadow AJW, Tingate PR, Laslett GM. 1986. Thermal annealing of fission tracks in apatite 1. A qualitative description. *Chemical Geology (Isotope Geoscience Section)* 59:237–253.

Green PF, Duddy IR, Hegarty KA. 2002. Quantifying exhumation from apatite fission-track analysis and vitrinite reflectance data: precision, accuracy and latest results from the Atlantic margin of NW Europe. *In* Doré AG, Cartwright J, Stoker MS, Turner JP, White N (Editors). *Exhumation of the North Atlantic Margin: Timing, Mechanisms and Implications for Petroleum Exploration*: Geological Society, London. Special Publication 196, p. 331–354.

Green PF, Duddy IR, Hegarty KA. 2005a. Comment on "Compositional and structural control of fission track annealing in apatite" by J. Barbarand, A. Carter, I. Wood, and A.J. Hurford (Chemical Geology, 198, 2003, 107–137). *Chemical Geology* 214:351–358.

Green PF, Duddy IR, Laslett GM. 1988. Can fission track annealing be described by first-order kinetics? *Earth and Planetary Science Letters* 87:216–228.

Green PF, Duddy IR, Laslett GM, Hegarty KA, Gleadow AJW, Lovering JF. 1989b. Thermal annealing of fission tracks in apatite 4. Quantitative modelling techniques and extension to geological timescales. *Chemical Geology (Isotope Geoscience Section)* 79:155–182.

Green PF, Laslett GM, Duddy IR. 1993. Mechanisms and kinetics of fission-track annealing. *American Mineralogist* 78:441–445.

Green PF, Thompson K, Hudson JD. 2001b. Recognising tectonic events in undeformed regions: contrasting results from the Midland Platform and East Midlands Shelf, central England. *Journal of the Geological Society* 158:59–73.

Gunnell Y. 2000. Apatite fission-track thermochronology: an overview of its potential and limitations in geomorphology. *Basin Research* 12:115–132.

Gunnell Y, Gallagher K, Carter A, Widdowson M, Hurford AJ. 2003. Denudation history of the continental margin of Peninsular India since the early Mesozoic—reconciling apatite fission-track data with geomorphology. *Earth and Planetary Science Letters* 215:187–201.

Harrison TM, Burke K. 1989. $^{40}Ar/^{39}Ar$ thermochronology of sedimentary basins using detrital feldspars: examples from the San Joaquin Valley, California, Rio Grande Rift, New Mexico, and North Sea. *In* Naeser ND, McCulloh T (Editors). *Thermal History of Sedimentary Basins—Methods and Case Histories*: Springer-Verlag, New York. p. 141–155.

Hegarty KA, Foland SA, Cook AC, Green PF, Duddy IR. 2007. Direct measurement of timing: underpinning a reliable petroleum system model for the Mid-Continent Rift system. *American Association of Petroleum Geologists Bulletin* 91:959–979.

Hendriks BWH, Andriessen P. 2002. Pattern and timing of the post-Caledonian denudation of northern Scandinavia constrained by apatite fission-track thermochronology. *In* Doré AG, Cartwright J, Stoker MS, Turner JP, White N (Editors). *Exhumation of the North Atlantic Margin: Timing, Mechanisms and Implications for Petroleum Exploration*: Geological Society, London. Special Publication 196, p. 117–137.

Hendriks BWH, Redfield TF. 2005. Apatite fission track and (U-Th)/He data from Fennoscandinavia: an example of underestimation of fission track annealing in apatite. *Earth and Planetary Science Letters* 236:443–458.

Hillis RR. 1995. Quantification of Tertiary exhumation in the United Kingdom southern North Sea using sonic velocity data. *American Association of Petroleum Geologists Bulletin* 79:130–152.

Holford SP, Green PF, Duddy IR, Turner JP, Hillis RR, Stoker MS. 2009a. Regional intraplate exhumation episodes related to plate boundary deformation. *Geological Society of America Bulletin* 121:1611–1628.

Holford SP, Green PF, Turner JP. 2005. Palaeothermal and compaction studies in the Mochras borehole (NW Wales) reveal Early Cretaceous and Neogene exhumation and argue against regional Paleogene uplift in the southern Irish Sea. *Journal of the Geological Society* 162:829–840.

Holford SP, Turner JP, Green PF, Hillis RR. 2009b. The signature of cryptic sedimentary basin inversion revealed by shale compaction data in the Irish Sea, western British Isles. *Tectonics* 28. DOI:10.1029/2008TC002359

House MA, Farley KA, Kohn BP. 1999. An empirical test of helium diffusion in apatite: borehole data from the Otway Basin, Australia. *Earth and Planetary Science Letters* 170:463–474.

House MA, Farley KA, Kohn BP. 2002. Evaluating thermal history models for the Otway Basin, southeastern Australia, using (U-Th)/He and fission track data from borehole apatites. *Tectonophysics* 349:277–295.

House MA, Wernicke BP, Farley KA, Dumitru, TA. 1997. Cenozoic thermal evolution of the central Sierra Nevada, California, from (U-Th)/He thermochronometry. *Earth and Planetary Science Letters* 151:167–179.

Hurford AJ. 1986. Cooling and uplift patterns in the Lepontine Alps, south central Switzerland and an age of vertical movement on the Insubric fault line. *Contributions to Mineralogy and Petrology* 92:413–427.

Japsen P. 2000. Investigation of multi-phase erosion using reconstructed shale trends based on sonic data. Sole Pit axis, North Sea. *Global and Planetary Change* 24:189–210.

Japsen P, Bonow JM, Green PF, Chalmers JA, Lidmar-Bergström K. 2009. Formation, uplift and dissection of planation surfaces at passive continental margins. *Earth Surface Processes and Landforms* 34:683–699.

Japsen P, Green PF, Chalmers JA. 2005. Separation of Palaeogene and Neogene uplift on Nuussuaq, West Greenland. *Journal of the Geological Society* 162:299–314.

Japsen P, Green PF, Nielsen LH, Rasmussen ES, Bidstrup T. 2007a. Mesozoic–Cenozoic exhumation in the eastern North Sea Basin: a multi-disciplinary study based on palaeo-thermal, palaeo-burial, stratigraphic and seismic data. *Basin Research* 19:451–490.

Japsen P, Mukerji T, Mavko G. 2007b. Constraints on velocity–depth trends from rock physics models. *Geophysical Prospecting* 55:135–154.

Kamp PJJ, Green PF. 1990. Thermal and tectonic history of selected Taranaki Basin (New Zealand) wells assessed by apatite fission track analysis. *American Association of Petroleum Geologists Bulletin* 74:1401–1419.

Katz BJ, Pheifer RN, Schunk DJ. 1988. Interpretation of discontinuous vitrinite reflectance profiles. *American Association of Petroleum Geologists Bulletin* 73:926–931.

Ketcham RA. 2005. Forward and inverse modelling of low-temperature thermochronometry data. *Reviews in Mineralogy and Geochemistry* 58:275–314.

Ketcham RA, Donelick RA, Carlson WD. 1999. Variability of apatite fission-track annealing kinetics: III. Extrapolation to geological timescales. *American Mineralogist* 84:1235–1255.

Ketcham RA, Donelick RA, Donelick MB. 2000. AFTSolve: a program for multi-kinetic modeling of apatite fission-track data. *Geological Materials Research* 2:1–18.

Kohn BP, Gleadow AJW, Brown RW, Gallagher K, O'Sullivan PB, Foster DA.

2002. Shaping the Australian crust over the last 300 million years: insights from fission track thermotectonic imaging and denudation studies of key terranes. *Australian Journal of Earth Science* 49:697–717.

Kohn BP, Belton DX, Brown RW, Gleadow AJW, Green PF, Lovering JF. 2003. Critical comment on: A.S. Wendt et al. 'Experimental evidence for the pressure dependence of fission track annealing in apatite' (Earth and Planetary Science Letters, 201, 2002, 593–607). *Earth and Planetary Science Letters* 215:299–306.

Larson SA, Cederbom CE, Tulborg E-L, Stiberg J-P. 2006. Comment on 'Apatite fission track and (U –Th)/He data from Fennoscandia: An example of underestimation of fission track annealing in apatite' by Hendriks and Redfield (Earth and Planetary Science Letters, 236, 443–458). *Earth and Planetary Science Letters* 248:561–568.

Laslett GM, Galbraith RF. 1996. Statistical modelling of thermal annealing of fission tracks in apatite. *Geochimica et Cosmochimica Acta* 24:5117–5131.

Laslett GM, Green PF, Duddy IR, Gleadow AJW. 1987. Thermal annealing of fission tracks in apatite 2. A quantitative analysis. *Chemical Geology (Isotope Geoscience Section)* 65:1–13.

Laslett GM, Kendall WS, Gleadow AJW, Duddy IR. 1982. Bias in measurement of fission track length distributions. *Nuclear Tracks* 6:79–85.

Logan P, Duddy IR. 1998. An investigation of the thermal history of the Ahnet and Reggane Basins, central Algeria, and the consequences for hydrocarbon generation and accumulation. *In* MacGregor DS, Moody RT, Clark-Lowes DD (Editors). *Petroleum Geology of North Africa*: Geological Society, London. Special Publication 132, p. 131–155.

Lorencak M, Kohn BP, Osadetz KG, Gleadow AJW. 2004. Combined apatite fission track and (U-Th)/He thermochronometry in a slowly cooled terrane: results from a 3440-m-deep drill hole in the southern Canadian Shield. *Earth and Planetary Science Letters* 227:87–104.

Lutz TM, Omar G. 1991. An inverse method of modelling thermal histories from apatite fission-track data. *Earth and Planetary Science Letters* 104:181–195.

MacKenzie AS, McKenzie D. 1983. Isomerization and aromatization of hydrocarbons in sedimentary basins formed by extension. *Geological Magazine* 120:417–528.

Marie JPP. 1975. Rotliegendes stratigraphy and diagenesis. *In* Woodland AW (Editor). *Petroleum and the Continental Shelf of North-West Europe:* Applied Science, London. p. 205–211.

Mark D, Green PF, Parnell J, Kelly SP, Lee MR, Sherlock SC. 2008. Late Palaeozoic hydrocarbon migration through the Clair field, West of Shetland, UK Atlantic margin. *Geochimica et Cosmochimica Acta* 72:2510–2533.

Morais Neto JM, Green PF, Karner GD, Alkmim FF. 2008. Age of the Serra do Martins Formation, Borborema Plateau, northeastern Brazil: Constraints from apatite and zircon fission track analysis. *Boletim Geociencias Petrobras* 16:23–52.

Morton AC. 1984. Stability of detrital heavy minerals in Tertiary sandstones from the North Sea Basin. *Clay Minerals* 19:287–308.

Naeser CW, Faul H. 1969. Fission track annealing in apatite and sphene. *Journal of Geophysical Research* 74:705–710.

Naeser CW, Forbes RB. 1976. Variation of fission track ages with depth in two deep drill holes (Abstract). *EOS* 57: 363.

Naeser ND, McCulloh TH (Editors). 1989. *Thermal History of Sedimentary Basins—Methods and Case Histories*: Springer-Verlag, New York. 319 p.

Naeser CW, Naeser ND, McCulloh TH. 1989. The application of fission track dating to the depositional and thermal history of rocks in sedimentary basins. *In* Naeser ND, McCulloh TH (Editors). *Thermal History of Sedimentary Basins—Methods and Case Histories*: Springer-Verlag, New York. p. 157–180.

O'Brien GW, Lisk M, Duddy I, Eadington PJ, Cadman S, Fellows M. 1996. Late Tertiary fluid migration in the Timor Sea: a key control on thermal and diagenetic histories? *Australian Petroleum Exploration Association Journal* 36:399–426.

Parnell J, Bowden SA, Osinski GR, Lee P, Green PF, Taylor C, Baron M. 2007. Organic geochemistry of impactites from the Haughton impact structure, Devon Island, Nunavut, Canada. *Geochimica et Cosmochimica Acta* 71:1800–1819.

Parnell J, Carey PF, Green P, Duncan W. 1999. Hydrocarbon migration history, West of Shetland: integrated fluid inclusion and fission track studies. *In* Fleet AJ, Boldy SAR (Editors). *Petroleum Geology of Northwest Europe: Proceedings of the 5th Conference: Geological Society*: Geological Society, London. p. 613–625.

Parnell J, Green PF, Watt G, Middleton D. 2005. Thermal history and oil charge on the UK Atlantic margin. *Petroleum Geoscience* 11:99–112.

Pedersen GK, Andersen LA, Lundsteen EB, Petersen HI, Bojesen-Koefoed JA, Nytoft HP. 2006. Depositional environments, organic maturity and petroleum potential of the Cretaceous coal-bearing Atane Formation at Qullissat, Nuussuaq Basin, West Greenland. *Journal of Petroleum Geology* 29:3–26.

Persano C, Bishop P, Stuart FM. 2006. Apatite (U-Th)/He age constraints on the Mesozoic and Cenozoic evolution of the Bathurst region, New South Wales: evidence for antiquity of the continental drainage divide along a passive margin. *Australian Journal of Earth Science* 53:1041–1050.

Praeg D, Stoker MS, Shannon PM, Ceramicola S, Hjelstuen B, Laberg JS, Mathiesen A. 2005. Episodic Cenozoic tectonism and the development of the NW European 'passive' continental margin. *Marine and Petroleum Geology* 22:1007–1030.

Redfield TF, Braathen A, Gabrielsen RH, Osmundsen PT, Torsvik TH, Andriessen PAM. 2005. Late Mesozoic to Early Cenozoic components of vertical separation across the More-Trondelag Fault Complex, Norway. *Tectonophysics* 395:233–249.

Rohrman M, van der Beek P, Andriessen P, Cloetingh S. 1995. Meso-Cenozoic morphotectonic evolution of southern Norway: Neogene domal uplift inferred from apatite fission track thermochronology. *Tectonics* 14:704–718.

Shuster DL, Farley KA. 2009. The influence of artificial radiation damage and thermal annealing on helium diffusion kinetics in apatite. *Geochimica et Cosmochimica Acta* 73:183–196.

Shuster DL, Flowers RM, Farley KA. 2006. The influence of natural radiation damage on helium diffusion kinetics in apatite. *Earth and Planetary Science Letters* 249:148–161.

Silverman MR, Landon SM, Leaver JS, Mather TJ, Berg E. 2007. No fuel like an old fuel: Proterozoic oil and gas potential in the Beetaloo Basin, Northern Territory, Australia. *In* Munson TJ, Ambrose GJ (Editors). *Proceedings of the Central Australian Basins Symposium (CABS)*; Alice Springs, Northern Territory; 2005: Northern Territory Geological Survey, Perth, Australia. Special Publication 2, p. 205–215.

Spiegel C, Kohn BP, Raza A, Raiuner T, Gleadow AJW. 2007. The effect of low-temperature exposure on apatite fission track stability: a natural annealing experiment in the deep ocean. *Geochimica et Cosmochimica Acta* 71:4512–4537.

Stephenson J, Gallagher K, Holmes C. 2006. A Bayesian approach to calibrating apatite fission track annealing models for laboratory and geological timescales. *Geochimica et Cosmochimica Acta* 70:5183–5200.

Stoker MS, Praeg D, Hjelstuen BO, Laberg JS, Nielsen T, Shannon PM. 2005. Neogene stratigraphy and the sedimentary and oceanographic development of the NW European Atlantic margin. *Marine and Petroleum Geology* 22:977-1005.

Tagami T, Carter A, Hurford AJ. 1996. Natural long term annealing of the zircon fission-track system in Vienna Basin deep borehole samples: constraints upon the partial annealing zone and closure temperature. *Chemical Geology* 130:147–157.

Tingate PR, Duddy IR. 2002. The thermal history of the eastern Officer Basin (South Australia): evidence from apatite fission track analysis and organic maturity data. *Tectonophysics* 349:251–275.

Tissot BP, Welte DH. 1984. *Petroleum Formation and Occurrence*: Springer-Verlag, New York. 699 p.

Turner JP, Green PF, Holford SP, Lawrence SR. 2008. Thermal history of the Rio-Muni (West Africa)–NE Brazil margins during continental breakup. *Earth and Planetary Science Letters* 270:354–367.

Vidal O, Wendte AS, Chadderton LT. 2003. Further discussion on the pressure dependence of fission track annealing in apatite: reply to the critical comment of Kohn et al. *Earth and Planetary Science Letters* 215:307–316.

Vrolijk P, Donelick R, Queng J, Cloos M. 1992. Testing models of fission track annealing in apatite in a simple thermal setting; Site 800, Leg 129. *In* Larson R, Lancelot Y (Editors). *Proceedings of the Ocean Drilling Program, Science Results*, Vol. 129: Ocean Drilling Program, College Station, Texas. p. 169–178.

Wagner GA. 1968. Fission track dating of apatites. *Earth and Planetary Science Letters* 4:411–415.

Wagner GA, Reimer GM. 1972. Fission track tectonics: the tectonic interpretation of fission track ages. *Earth and Planetary Science Letters* 14:263–268.

Wagner GA, Van den Haute P. 1992. *Fission-Track Dating*: Kluwer Academic Publishers, Dordrecht. 285 p.

Warnock AC, Zeitler PK, Wolf RA, Bergman SC. 1997. An evaluation of low-temperature apatite U-Th/He thermochronometry. *Geochimica et Cosmochimica Acta* 61:5371–5377.

Wendt AS, Vidal O, Chadderton LT. 2002. Experimental evidence for the pressure dependence of fission track annealing in apatite. *Earth and Planetary Science Letters* 201:593–607.

Wilkins RWT, Wilmshurst JR, Russell NJ, Hladky G, Ellacott MV, Buckingham C. 1992. Fluorescence alteration and the suppression of vitrinite reflectance. *Organic Geochemistry* 18:629–640.

Willett SD. 1997. Inverse modeling of annealing of fission tracks in apatite 1: a controlled random search method. *American Journal of Science* 297:939–969.

Wolf RA, Farley KA, Silver LT. 1996. Helium diffusion and low-temperature thermochronometry of apatite. *Geochimica et Cosmochimica Acta* 60:4231–4240.

Wolf RA, Farley KA, Silver LT. 1997. Assessment of (U-Th)/He thermochronometry: the low temperature history of the San Jacinto Mountains, California. *Geology* 25:65–68.

Young D, Myers AT, Munson EL, Conklin NM. 1969. Mineralogy and Geochemistry of Fluorapatite from Cerro de Mercado, Durango, Mexico: US Geological Survey. Professional Paper 650-D, Washington DC. p. D84–D93.

Zachos JC, Pagani M, Sloan LC, Thomas E, Billups K. 2001. Trends, rhythms, and aberrations in global climate 65 Ma to present. *Science* 292:686–693.

Zeitler PK, Herczig AL, McDougall I, Honda M. 1987. U-Th-He dating of apatite: a potential thermochronometer. *Geochimica et Cosmochimica Acta* 51:2865–2868.

BASIN THERMAL HISTORY ANALYSIS USING (U-TH)/HE THERMOCHRONOMETRY

RICHARD A. KETCHAM

Department of Geological Sciences, Jackson School of Geosciences, University of Texas, Austin, Texas 78712, USA
e-mail: ketcham@jsg.utexas.edu

Abstract: The (U-Th)/He technique is based on the accumulation and diffusive loss of helium generated by alpha decay. The temperatures required to drive diffusion vary with mineral type and characteristics and can range from <50° C to >200° C, an ideal range for basin thermal history analysis. Although promising due to its conceptual simplicity and capacity for quick analysis, (U-Th)/He has proven to be quite challenging in practice due to a variety of complications, such as alpha particle ejection and implantation, grain-size effects, and the influence of radiation damage on helium diffusivity. Recent research has begun to quantitatively characterize many of these processes and their effects, allowing the technique to be applied with increasing confidence and robustness. This paper reviews the current state of knowledge of the physics and chemistry underlying the (U-Th)/He system, and the modeling approaches that allow thermal history information to be extracted from various kinds of data. A new approach for simultaneously accounting for grain-shape effects on both diffusion and alpha particle ejection is proposed. Various examples of basin and other field studies employing (U-Th)/He are reviewed and in some cases reinterpreted using up-to-date tools and approaches. As the (U-Th)/He system is further explored and tested in the field and laboratory, it will become an increasingly robust and versatile tool for basin analysis.

Key words: Helium, (U-Th)/He, basin analysis, thermochronology, radiation damage

INTRODUCTION

Thermal history analysis using thermochronometry is a powerful tool for petroleum exploration because it provides a time element that is absent when purely thermal indicators are used. It can thus be used to address important questions such as the timing of thermal maturation for particular sedimentary packages versus the development of structural traps. It can also provide information about the thermal evolution of a basin, such as changes in sedimentation rates and heat flow over time.

Over the past decade, the (U-Th)/He technique has attracted a great deal of interest for basin analysis, due to the low closure temperatures for several common accessory minerals, the conceptual simplicity of the physical system, and analytical capacity for high sample throughput. However, interest in this technique dates to the earliest years of radiometric dating. It was among the first radiometric geochronometers proposed and attempted (Rutherford 1906, Strutt 1908), and even in this early work, the poor correlation of He/U ratio in phosphate minerals with stratigraphic position was attributed to imperfect retention of helium. A subsequent study of zircon (Strutt 1910) resulted in much better correspondence between stratigraphic age and helium content, indicating that zircon was more resistant to He leakage. However, the inability to anticipate and account for the geological circumstances surrounding He leakage eventually led to abandonment of the technique in favor of other isotopic methods. Various workers reinvestigated the system roughly every decade during the twentieth century, with similar results (Reiners [2005] provides a good review of this period), though some of them recognized that the young ages may have resulted from He loss due to metamorphism, or more generally, a thermal event.

With the creation of the theoretical framework for thermochronometry (Fechtig and Kalbitzer 1966; Wagner and Reimer 1972; Dodson 1973, 1979), the utility of isotopic systems that lose their daughter products in a systematic way as a function of temperature was recognized, and the means to exploit it were developed. Early work concentrated on then-current techniques, such as K-Ar and fission-track. However, these developments paved the way for the ground-breaking and, in many ways, prescient study by Zeitler et al. (1987), which proposed reconsideration of the (U-Th)/He system in apatite as a potential low-temperature thermochronometer and provided the first measurements allowing estimates of closure temperatures below 100° C.

Part of the early appeal of (U-Th)/He thermochronometry was that it is based on bulk diffusion, a well-understood physical process for which the relevant material parameters can be obtained via standard, straightforward experimental procedures. It thus compared favorably with the other principal low-temperature thermochronometer, fission-track analysis; as the mechanism behind fission-track annealing is not yet well understood, all fission-track calibrations are empirical. However, as the (U-Th)/He method has become more widely applied and as the diffusion behavior of certain minerals has been studied in increasing detail, its complexity has become more apparent, owing to a variety of natural factors.

This paper reviews the current state of knowledge of (U-Th)/He thermochronometry in general and as it applies in specific mineral systems, with particular attention paid to the latest developments and the ways in which they may affect data interpretation. The technique is then reviewed in relation to the way in which it has been employed for thermal history reconstruction in a variety of scenarios, and how new understanding of the physics and chemistry underlying helium diffusion can impact the data analysis in a number of recent studies.

THE PHYSICS AND CHEMISTRY OF THE (U-Th)/He SYSTEM

This section outlines the principles and experimental observations that underlie (U-Th)/He thermochronology. Several recent, good geochemically based reviews are available (Farley 2002, Farley and Stockli 2002, Reiners 2005), and readers seeking more depth than is presented here are referred to these. The objective of the present work is to summarize briefly the relevant concepts for the scientist wishing to engage in or evaluate (U-Th)/He–based thermal history modeling and also to provide an update on the most recent results and the ways in which they impact thermal history analysis.

Analyzing the Thermal History of Sedimentary Basins: Methods and Case Studies
SEPM Special Publication No. 103, Copyright © 2012
SEPM (Society for Sedimentary Geology), Print ISBN 978-1-56576-315-9, CD/DVD ISBN 978-1-56576-317-3, p. 105–123.

Physical Processes

Helium Production: The (U-Th)/He chronometer is based on the accumulation of ^{4}He in a mineral due to alpha decay. The principal alpha emitters are ^{238}U, ^{235}U, and ^{232}Th, which respectively produce eight, seven, and six alpha particles over the course of their decay chains. An additional potential alpha source is ^{147}Sm, which can be very abundant in some minerals but only produces a single alpha particle and has a much longer half-life. Production of ^{4}He from these isotopes (typically specified in nmol/g) is described by the equation:

$$^4\mathrm{He} = 8^{238}\mathrm{U}(e^{\lambda_{238}t} - 1) + 7^{235}\mathrm{U}(e^{\lambda_{235}t} - 1) + 6^{232}\mathrm{Th}(e^{\lambda_{232}t} - 1) + ^{147}\mathrm{Sm}(e^{\lambda_{147}t} - 1) \quad (1)$$

where t is time since ^{4}He accumulation began, and the λ terms are the decay constants for the isotopes indicated by their subscripts. The Sm term is frequently omitted in cases where it is considered likely to be unimportant. Given measurements of helium and the parent isotopes, the age is determined by iteratively solving Eq. 1 for t.

Diffusion: Because helium is a noble gas, it does not chemically bond to other atoms in the mineral structure. It is thus free to diffuse through the mineral via a simple stochastic mechanism driven by concentration gradients and system energy; i.e., thermally activated diffusion, which is typically characterized as:

$$\frac{\partial v}{\partial t} = \nabla(D\nabla v) + A \quad (2)$$

where v is the diffusing quantity (He), D is the diffusivity, and A is production, which can be calculated by an extension of Eq. 1. Because all that is required to fully describe this system is the diffusivity of the mineral with respect to helium, there was early optimism that this method would be simple, easily calibrated, and robust.

In most cases the dominant control on diffusivity is temperature, characterized by the Arrhenius relationship:

$$\frac{D}{a^2} = \frac{D_0}{a^2}e^{-E/RT} \quad (3)$$

where D_0 is the diffusivity at infinite temperature (also called the frequency factor), E is the activation energy, T is absolute temperature, and R is the universal gas constant. The variable a is the characteristic dimension of the diffusion domain, the meaning of which varies with the presumed domain geometry; typical theoretical diffusion domains are a sphere, an infinite cylinder, and a slab, in which case a represents a radius, radius, and half-thickness, respectively. For almost all He diffusivity work, a sphere model has been found to be adequate, although in minerals with significant anisotropy in diffusivity, a cylinder is sometimes appropriate (e.g., Cherniak et al. 2009). Values for D_0 and E can be derived in a number of ways, the most common being step-heating experiments; Harrison and Zeitler (2005) provide a succinct review.

Another geometric issue is how to identify the scale of the "diffusion domain," or the region of a mineral grain that Eq. 3 characterizes. A diffusion domain can either be an entire mineral grain, or some subregion within the grain bounded by internal higher-diffusivity features, such as cleavage planes, twin planes, or internal defects. In theory, a mineral can have multiple diffusion domains at varying scales, with the most notorious example being the multi-diffusion-domain (MDD) model for argon behavior in potassium feldspar (Lovera et al. 1997, 2002). Most minerals used for (U-Th)/He dating have been found to be best characterized by a diffusion domain that corresponds to the entire mineral grain; specific cases will be discussed later herein.

An important assumption implicit in the diffusion calculation is that the diffusion pathways at the high-temperature conditions used in laboratory step-heating experiments are the same as those present during low-temperature conditions over geological timescales. This assumption continues to inspire energetic debate concerning the ^{40}Ar/^{39}Ar MDD technique (Parsons et al. 1999, Heizler et al. 2008). It has also been brought up for helium diffusion in zircon. Reich et al. (2007) used molecular dynamics simulations to model the diffusion of helium in the zircon crystal structure and found that it was very anisotropic at low temperature, with preferential flow along open channels along the crystallographic **c**-axis (or normal to the [100] plane), and more isotropic at high temperature. However, such simulations presume and characterize a pristine crystal structure, and several lines of evidence indicate that radiation damage plays a central role in determining net diffusivity (Reiners et al. 2004, Shuster et al. 2006, Guenthner and Reiners 2009). An undamaged structure in any natural zircon is unlikely in virtually any situation due to its characteristically high abundance of radioactive elements, and even in apatite, radiation damage effects appear to dominate diffusivity in most natural cases. The practical applicability of direct insights obtained from molecular dynamic simulations may thus be limited.

Effects of Long Alpha Stopping Distances: Alpha decay imparts considerable kinetic energy to the emitted particle and recoiling nucleus, leading to substantial complications in (U-Th)/He thermochronometry. Most produced alpha particles have energies in the 4 to 5 MeV range, and as a result travel on the order of ~20 μm through a mineral grain before coming to a rest. Because this distance is, in most cases, of the same order as the typical grain size of the trace phases used in (U-Th)/He, particle projection can lead to a substantial proportion of the helium produced within a grain being deposited outside of it, lowering its apparent age. Correction for this effect in spheres with uniformly distributed U and Th is straightforward (Farley et al. 1996). More complex shapes and zoning patterns can be simulated using analytical and Monte Carlo approaches (Farley 2002, Hourigan et al. 2005, Gautheron and Tassan-Got 2010), although these require obtaining data to characterize shape and/or zoning. Shape effects on modeling will be examined more closely later in this paper.

The variable describing the total fraction of alpha particles from all isotopes that are not ejected (i.e., retained within the grain) is called F_T (Farley et al. 1996). Originally, this was used as a direct, linear correction to the age (i.e., corrected age = measured age/F_T). Min et al. (2003) improved this treatment to correct instead the amount of helium used to calculate the age, which accounts for the nonlinearity of age with helium accumulation. However, even this amendment does not provide a complete correction due to the variation in stopping distances among different isotopes. The mean alpha stopping distance for ^{238}U decay products is ~15% smaller than for ^{235}U, and so if the helium accumulation time is long enough to significantly change the proportion of ^{235}U to ^{238}U (hundreds of millions of years or more), then an independent F_T value for each isotope is necessary. Both this effect and zoning in U and Th (if known) can be accounted for by calculating over a series of annular shells i centered at the grain core. The relevant calculation for ^{238}U is:

$$F_{T,238} = \frac{\sum_{i=0}^{rad} v_i[^{238}\mathrm{U}_i]F_{i,238}}{\sum_{i=0}^{rad} v_i[^{238}\mathrm{U}_i]} \quad (4)$$

where v_i is the volume of shell i, $[^{238}\mathrm{U}_i]$ is the concentration of ^{238}U in shell i, and $F_{i,238}$ is the fraction of alpha particles generated in shell i by ^{238}U and its daughters that are ejected from the mineral grain. The corrected age is then calculated by solving

$$^4He = 8F_{T,238}\,{}^{238}U(e^{\lambda_{238}t} - 1) + 7F_{T,235}\,{}^{235}U(e^{\lambda_{235}t} - 1) + 6F_{T,232}\,{}^{232}Th(e^{\lambda_{232}t} - 1) + F_{T,147}\,{}^{147}Sm(e^{\lambda_{147}t} - 1) \quad (5)$$

for t (Ketcham et al. 2011).

Detailed attention is paid to F_T and its calculation because of its large, first-order effect on age determinations. Typical F_T values for trace phases range from 0.65 to 0.90 (Farley 2002), indicating corrections of 35% to 10% to measured ages. Grains with lower F_T values are typically discarded owing to increasing uncertainty in the F_T determination, though the particular cutoff point varies from laboratory to laboratory.

Another consequence of long alpha stopping distances is that alpha particles can be implanted into a mineral grain from the outside, artificially increasing the apparent age. Although previously considered a very unlikely scenario, some recent observations suggest that this phenomenon may occur frequently enough to warrant concern. Kohn et al. (2008) documented a number of instances in which excess scatter and unreasonably high values among apatite (U-Th)/He ages were markedly reduced by abrading off the outer 30 μm of material. Closer study of these cases revealed that apatite grains in situ were located preferentially in proximity to higher-actinide minerals, both primary (igneous) and secondary (metamorphic and diagenetic). A related observation by Phillips et al. (2007) is that kimberlitic diamonds in contact with groundwater often have green surface tinting caused by alpha implantation (Shelkov et al. 1998). Measurements reveal extremely high ^4He concentrations in these rims of up to 625 nmol/g; this is two orders of magnitude more than is observed, for example, in natural Durango apatite. Alpha implantation can not only add helium to the margin of a crystal but can also diffuse inward or prevent diffusive loss of internally generated helium, rendering abrasion a partial fix.

Examples: The combined effects of generation, diffusion, and long alpha stopping distances can be seen by looking at calculated helium concentration profiles for various thermal histories (Fig. 1). Figure 1a shows three example thermal histories: fast cooling from 120° C to 0° C followed by 80 million years of isothermal holding; linear cooling from 120° C at 80 Ma to 0° C at present; and linear cooling from 120° C until 40 Ma followed by 35 million years of residence at 60° C, in the midst of the apatite (U-Th)/He partial retention zone (PRZ), where helium accumulation is roughly matched by diffusive loss, followed by linear cooling to 0° C. Figure 1b shows the corresponding helium zoning profiles for Durango apatite according to the kinetics reported by Farley (2000). The fast-cooling profile reflects a system with virtually no diffusive loss of helium. The flat diagonal cutoff toward the grain rim corresponds to helium that has been lost from the edge of the grain due to alpha ejection.[1] The "extended PRZ residence" profile is much more rounded, reflecting rapid He loss at high temperature and only limited accumulation time after cooling. The linear cooling profile can be seen as a mixture of these two profiles: The flattened alpha-ejection region is apparent, but it has been rounded by diffusion.

Figure 1c shows an example of the effect that zoning can have on diffusional profiles. The linear cooling path is modeled for an apatite grain in which the outer 25% contains 50% more U and Th than the core (dashed line). The effect on the helium profile is clear and makes it evident that even fairly subtle zoning will confound any attempt to utilize diffusion profiles unless it is measured and accounted for. A secondary feature of the profile is that the peak He concentration is at a radial coordinate of about 0.65, even though the U-Th enrichment spans radial coordinates from 0.75 through 1.0. This is another manifestation of long alpha stopping distances: The He enrichment represents alpha particles that were "shot in" from the U-Th–rich rim region of the grain when they were generated (Ketcham 2005). In general, long alpha stopping distances serve to smooth out all zoning effects in this way. Helium concentrations in the rim region itself are partially depleted by alpha particle ejection.

Figure 1d shows the linear cooling He profile for the set of Durango kinetic parameters proposed by Wolf et al. (1996), which differ from Farley (2000) in that they assume subgrain rather than whole-grain diffusion domains. The principal difference in this case is that there is no alpha ejection effect in the modeled profile. This is not because ejection is no longer active, but rather because it cannot be incorporated into the diffusion calculation. The concept of subgrain domains implies that there are many such domains within a crystal, and the radial coordinate of the profile graph refers to the radius of these domains, rather than the radius of the crystal itself. Domains near the rim of a grain may be heavily impacted by alpha ejection, while domains near the core may suffer no alpha ejection effects at all—net transfer out will be matched by net transfer in. As a result, alpha effects can only be approximated by a postmodel calculation. Similarly, radial zoning in U and Th cannot be incorporated directly into these models (unless it is the subgrain domains that are zoned).

Measurement

Measurements for (U-Th)/He thermochronometry typically consist of separate steps of bulk He extraction and U+Th (+Sm) determination, along with measurements of size and shape to determine the F_T parameter. Recent developments center on measuring zoning in either or both of these components: U+Th to improve alpha ejection correction, and He to measure the shape of the zoning profile to extract further thermal history information.

Grain Selection and Measurement: A critical first step in all (U-Th)/He thermochronometry is grain selection. For minerals with relatively low U-Th like apatite, it is very important to find grains that are free of both mineral and fluid inclusions. Inclusions of higher–U-Th phases may result in spuriously old ages if their He is measured but their U and Th are not, as may occur if the inclusion phase is resistant to the dissolution protocol used for apatite. A "point source" of He also contaminates diffusion profiles and generally violates the assumptions underlying diffusion calculations, rendering them less reliable. Finally, a high–U-Th inclusion may create a concentrated zone of radiation damage, which, as will be discussed in the next section, can have a large effect on diffusivity. Fluid inclusions can potentially host excess helium (e.g., Stockli et al. 2000), either because some amount was present in fluids while the mineral was forming, or because they are traps for internally diffusing helium; due to partitioning effects, helium within a fluid inclusion may need to achieve concentrations five to six orders of magnitude higher than in the apatite crystal before outflow equals inflow (Baxter et al. 2007).

Euhedral mineral grains are also greatly preferable to avoid ambiguity in F_T corrections. Grain dimensions are typically measured under a microscope to determine length, width, and breadth. If the grains are broken, lengths may be multiplied by a small factor (1.5) to compensate for missing material; the effect of this uncertainty on F_T corrections is minor (Farley et al. 1996). Empirical relations for determining F_T for hexagonal and tetragonal prisms without pyramidal terminations are provided by Farley (2002); Hourigan et al. (2005) provide additional solutions for tetragonal prisms with pyramidal terminations; this additional detail alters F_T values, and thus ages, by 1 to 3%. Resources

[1] Careful inspection reveals a slight change in slope of the cutoff line at a fractional radius of about 0.7. This corresponds to the location where alpha loss from the ^{238}U decay chain ends while loss from ^{232}Th daughter products continues, due to the difference in their respective mean alpha stopping distances (19.68 μm and 22.46 μm; Farley et al. 1996). This is in fact a computational simplification, as there are 15 alpha particles with distinct energies in these chains with stopping distances ranging from 12.60 to 34.14 μm, and so the true transition will be more gradual.

diffusivity to be accurately measured for virtually any apatite (Shuster et al. 2003). New analyses of a large number of apatites corroborated the range and diversity of T_c values indicated by the earlier studies, with the crucial additional observation that T_c was positively correlated with ^{4}He concentration (Shuster et al. 2006). Because classical diffusivity is independent of concentration, the natural conclusion was that the actual mechanism for increasing diffusivity was radiation damage. The principal types of damage are from fission tracks and from nucleus recoil after alpha particle emission. Fission tracks are much larger (~18 μm vs. ~30–40 nm) and reflect much more energy deposition into the mineral structure (>100 MeV vs. <1 MeV), but they are much less plentiful; significant numbers are only formed by ^{238}U and not ^{232}Th, and for every ^{238}U nucleus that undergoes spontaneous fission decay, two million undergo alpha decay. In terms of the total ionizing energy deposited into the mineral volume, or *kerma* (Shuster and Farley 2009), the ratio of spontaneous fission kerma to alpha kerma is $<10^{-5}$. As a result, the damage currently considered to be most responsible for increasing diffusivity is alpha recoil damage, although further research is required on this front.

This work led to the formulation of a "trapping" model for helium diffusion in apatite (Shuster et al. 2006), in which zones of radiation damage were postulated to constitute traps into which diffusing helium atoms could fall easily and exit only with difficulty. The system can be considered as being characterized by two activation energies, one for standard diffusion through the crystal and one for escaping from traps. The variable used to estimate total trap volume was [^{4}He] concentration, due to its observed correlation with T_c and its ready availability as part of the standard (U-Th)/He analysis. However, even at that time, it was apparent that radiation damage could be retained even if helium was lost by diffusion, imparting a significant one-sided bias to using [^{4}He] to estimate damage.

Recent work by Shuster and Farley (2009) has shown that direct introduction of radiation damage via neutron bombardment in a nuclear reactor can indeed increase diffusivity, and that diffusivity in Durango apatite can be lowered by heating. T_c changes above 55° C were found to correspond to conditions for thermal annealing of apatite fission tracks (Carlson et al. 1999), leading to the possibility that existing fission-track calibrations and modeling procedures may be appropriate for characterizing helium diffusivity development. This is somewhat surprising, as the atomic-scale interactions behind fission damage are different than those for alpha recoil damage (Fleischer et al. 1975), and in some minerals, alpha damage has been demonstrated to anneal at significantly lower temperatures. For example, the retention temperature for etchable alpha-recoil tracks in biotite is about 100° C lower than that of fission tracks (Saini and Nagpaul 1979, Hashemi-Nezhad and Durrani 1983, Yuan et al. 2008). Zircon etchability is believed to be a function of alpha damage (Krishnaswami et al. 1974, Gleadow et al. 1976), and it has been shown that damage affecting etchability can be removed by laboratory heating without affecting fission tracks (Braddy et al. 1975). On the other hand, the congruence between the diffusivity results for Durango apatite and other apatites is markedly improved by accounting for the unusually high ^{232}Th content of Durango apatite (~180 ppm), which would not be the case if diffusivity changes were predominantly caused by fission damage.

A new diffusivity model based on damage annealing via a fission-track analogue has been proposed by Flowers et al. (2009). The Radiation Damage Accumulation and Annealing Model (RDAAM) accounts for radiation damage using effective spontaneous fission-track density ($e\rho_s$), the spontaneous density measured during fission-track analysis modified to include the damage contribution of ^{232}Th.[2] Annealing of radiation damage is characterized using the apatite fission-track annealing model of Ketcham et al. (2007). The RDAAM provides an explanation for a number of formerly troubling observations, such as the inversion of AFT and AHe dates in many old terrains. It also provides a potential new approach to thermal history analysis by intentionally utilizing apatites of varying U-Th content. For example, Flowers (2009) used high–U-Th apatites from felsic dikes and low–U-Th apatites from mafic ones to constrain the Paleozoic burial and unroofing history in the East Lake Athabasca region of northern Saskatchewan.

A similar damage-enhanced diffusivity model has been proposed by Gautheron et al. (2009), which is based on less experimental data than the RDAAM but utilizes a different parameterization of damage effects. Both models seem to make similar predictions, and they have not yet been thoroughly compared.

Despite the strides that have been made in the understanding and calibration of the AHe system, it is clear that important challenges remain (Farley and Clark 2006). The principal indication of our incomplete understanding is the poor reproducibility that is frequently encountered in AHe determinations in multiple aliquots from single samples, which is commonly much poorer than expected based on analytical uncertainty. In most but not all cases, the contaminating bias seems to be toward inappropriately old ages, based on comparison with other thermochronometers (e.g., AFT), higher-temperature geochronometers and stratigraphic relationships, and AHe determinations among samples with straightforward structural relations (Green et al. 2006). Known mechanisms for such excess variation have been discussed already, including: alpha implantation, whether by "bad neighbors" or fluids; unaccounted-for zoning in U and Th; partial resetting by wildfires (Reiners et al. 2007); fluid inclusions as He traps; and mineral inclusions, as sources of excess helium and possibly inducers of radiation damage in their vicinities, changing local diffusivity. Another crucial variable is the extent of diffusivity-altering radiation damage, the retention of which precedes retention of helium. This last consideration renders AHe, in essence, a non-self-contained system: AHe ages can be heavily influenced by a variable quantity, radiation damage, which is not constrained by AHe data.

Recognizing these problems, Farley and Clark (2006) proposed a demanding set of criteria for generating reliable AHe ages, including dating 6 to 10 >80-μm-grain aliquots per sample and only trusting the result if scatter is less than 8%. If excess scatter is encountered and the cause can be confidently identified, the data may remain interpretable, but only with great care. These criteria have not yet been widely adopted, however.

Another widely used remedy to some of the issues confronting AHe thermochronology is to simultaneously obtain AFT data; this provides both direct and indirect benefits. Most importantly, it supplies the currently most reliable estimate of radiation damage from which helium diffusivity can be estimated. Because fission-track analysis produces in essence a direct radiation damage map of apatite grains, it can serve as a test for U zoning and the prevalence of harmful inclusions and "bad neighbors." This test can either be applied directly by plucking grains from AFT mounts for AHe analysis (e.g., Reiners et al. 2007), or indirectly by analyzing different aliquots from the same sample by both methods and using the fission-track analysis to flag sample likely to have problems.

Zircon: Zircon is another very widely used mineral for (U-Th)/He analysis (ZHe), due to its ubiquity, toughness, and typically high U+Th content in the hundreds to thousands of parts per million. The currently accepted T_c range of 175 to 193° C is based on experimental results by Reiners et al. (2004). As with apatite, the grain is the principal diffusional domain, although there is some evidence for smaller diffusional domains housing some of the helium released early in step-heating experiments, probably in rim regions. The high U+Th content of zircon allows it to be used in very young terrains, but it also makes radiation damage an important issue. The principal recognized

[2]The combined effects of U and Th are typically characterized as effective U (eU = U + 0.235Th).

effect is to render a grain exceptionally leaky with respect to helium once enough alpha damage has accumulated to essentially make the grain metamict (2–4 × 10^{18} α/g). Based on studies of synthetic zircon-structure minerals, Farley (2007) estimated undamaged zircon to have a T_c of ~120° C; the higher values observed in natural specimens are most likely caused by radiation damage, which accumulates rapidly because of the high U-Th content. Efforts are under way to develop a quantitative characterization encompassing these observations, similar to and extending from the apatite RDAAM schema (Guenthner and Reiners 2009). A good review of the technique is provided by Reiners (2005).

Titanite: Titanite is another commonly occurring mineral used for (U-Th)/He analysis. It has a closure temperature slightly above that for zircon, with values ranging from 191 to 218° C (Reiners and Farley 1999). As with zircon, the grain is the principal diffusional domain, though again there is some evidence for a small mass fraction of helium being retained in smaller domains. Typical U and Th contents are in the hundreds of parts per million. No first-order radiation damage effects on He diffusivity have been recognized, but this may be due to dearth of data.

Xenotime: Xenotime is another good candidate for (U-Th)/He dating, because it typically contains very high U and Th contents in the hundreds to tens of thousands of parts per million (e.g., England et al. 2001). Only limited data are currently available for characterizing its diffusive behavior, however. Farley and Stockli (2002) cite preliminary data indicating a T_c of about 115° C, roughly equivalent to the AFT method, but subsequent work has evidently pointed to higher T_c values: Farley (2007) cites revised kinetic parameters pointing to a T_c closer to 200° C for small grains. Synthetic xenotime has a low T_c of 60° C (Farley 2007), and thus it seems likely that again radiation damage effects exert a dominant influence on its helium retentivity. It is not yet clear whether the full grain is the diffusional domain. Although xenotime can be a primary igneous mineral, it can also be deposited during diagenesis or as a result of metamorphic or hydrothermal processes, and thus its origin needs to be established to attach proper meaning to age determinations.

Monazite: Monazite is a very interesting mineral for (U-Th)/He dating due to its typically high Th content (often at the wt% level) and its unusual feature of annealing radiation damage at very low temperature, allowing it to recover from a level of persistent bombardment that renders other phases metamict (Gleadow et al. 2002). Due to the interplay between helium diffusion and damage annealing observed in other minerals, full understanding of diffusive behavior in this phase may be very challenging. However, Farley and Stockli (2002) report a T_c value of ~220° C, and Farley (2007) obtained very similar diffusion parameters for synthetic monazite, indicating a T_c of ~205° C,[3] suggesting relatively little variation. However, the frequent occurrence of severe sector zoning and overgrowth in monazite indicates that caution should be exercised when analyzing this mineral.

Magnetite: Magnetite is a very common mineral that has started to attract attention for (U-Th)/He analysis. Thus far, the only application has been to dating primary grains in intermediate to mafic volcanic rocks (Blackburn et al. 2007). It tends to have low U and Th (0.1–2 ppm), and so relatively large grains are required to provide sufficient helium for analysis. Grain abrasion is recommended to avoid implantation effects. The closure temperature, based on a single step-heating experiment, is 250 ± 40° C; there are not yet data bearing on whether the grain is the diffusion domain. Magnetite can also occur diagenetically, but such occurrences have not yet been investigated for (U-Th)/He.

Hematite: Hematite was the first mineral found to have good retention of helium over geological time scales (Strutt 1908). Its potential as a geochronometer also was reassessed in the 1980s, before that of apatite. It can have a variety of origins, and it has received attention as a potential chronometer for mineralization events (Wernicke and Lippolt 1993, Lippolt et al. 1995). In many cases, hematite concentrates considerable U in the tens to hundreds of parts per million; however, zoning in U by up to a factor of five has been observed. Lippolt et al. (1993) used step heating on specular hematite to derive T_c values that ranged from 200 to 300° C, depending on grain dimension and cooling rate.

Rutile: Limited work has been reported on rutile thus far, pointing to a closure temperature in the 220 to 235° C range (Stockli et al. 2005). As with other cases discussed here, rutile can be both a primary and secondary phase, and so interpretation of such data will require establishing petrologic context.

Goethite: Goethite is a very interesting mineral for (U-Th)/He analysis because it typically occurs as a supergene product caused by weathering, giving rise to the possibility of dating the time since formation in near-surface conditions. In diffusion experiments, Shuster et al. (2005) found that goethite grains typically have two diffusional domains, a retentive one (T_c ~ 50–70° C) and a very leaky one that retains virtually no helium. The leaky domain comprised 4 to 8% of the mass of grains they analyzed, and determination of this percentage for any material dated is a required to compute a proper correction.

Calcite: Another potentially important mineral for (U-Th)/He analysis in basins is calcite, particularly for the frequently encountered case in which no clastic components are present in the rock to provide dateable accessory phases. Calcite U-Th content tends to be low (tenths to single digits in parts per million), but this obstacle can be overcome by using large grains. Attempts to use step-heating experiments suggest an approximate T_c of ~60° C, but diffusive behavior is often quite erratic, implying that multiple diffusion domains may be present (Copeland et al. 2007).

Because calcite can be primary or secondary, and it is subject to recrystallization, petrographic analysis to determine the nature of occurrence is crucial for meaningful interpretation. One approach is to try to use fossils, which can have diagnostic original structures (Copeland et al. 2008). Studies to date have shown mixed results, with some ages being in reasonable agreement with stratigraphy and others reflecting ages that are either too old, possibly due to common He, or too young, with possible explanations including U uptake and partial resetting due to a thermal event.

THERMAL HISTORY RECONSTRUCTION FROM (U-TH)/HE DATA

This section reviews some of the primary methodologies for quantitatively applying (U-Th)/He thermochronometry to basin analysis, including some illustrative applications. The discussion starts at the single-sample level, discussing both forward modeling issues and attempts to derive additional thermal history information through inverse modeling. Following this, multiple-sample approaches are considered.

Computational Methods for Single-Sample Modeling

The crux of the computational problem for (U-Th)/He analysis is solving the system of equations describing the production and diffusion of helium atoms as a function of time and temperature while

[3]Farley (2007) reports much higher closure temperatures in the 300 to 350° C range, but these are based on 1-cm crystals; the T_c values listed here reflect a 100-μm radius, as is more typical for basin studies.

also accounting for alpha stopping effects. Solving the production and diffusion equations is relatively straightforward, and several solutions have been published (Wolf et al. 1998, Ehlers 2001, Meesters and Dunai 2002, Ketcham 2005, Herman et al. 2007). Because in most cases a crystal shape can be sufficiently closely approximated as being spherical (Dunai 2005), the system can be solved efficiently in one dimension due to symmetry, but solutions for three-dimensional cases are also available if special situations require them. Meesters and Dunai (2002) give solutions for a variety of simple shapes using eigenmode decomposition. Herman et al. (2007) utilize a three-dimensional finite difference grid to solve the diffusion equation for the true grain shape as derived from μCT analysis. Another promising method is a Monte Carlo solution to the production–diffusion equations that provide a relatively fast and robust solution for an arbitrary geometry (Gautheron and Tassan-Got 2010).

For a spherical geometry, the HeFTy software (Ketcham 2005) provides a fast, flexible, and robust implementation using the finite difference method. HeFTy computes ages and zoning profiles for arbitrarily complex time–temperature histories accounting for alpha ejection and zoning. It can simultaneously model multiple (U-Th)/He systems, as well as the AFT and vitrinite systems.

Two computational issues will be discussed in the context of the way they are implemented in HeFTy. First, there is the extent to which a sphere provides an adequate approximation to more complex geometries, with respect to both alpha ejection and diffusion, to facilitate efficient forward modeling. Second, the inverse problem of finding the set of thermal histories that match measured data is considered.

Geometric Effects: As discussed already, a spherical geometry is the most efficient for computation. However, insofar as most of the minerals of interest for (U-Th)/He thermochronology have somewhat elongated habits, some accounting for the suitability of using a spherical geometry is in order. For both diffusion and alpha ejection, the first-order control is the surface area to volume ratio (S/V), as it is the surface through which helium is lost via both processes.

In the case of diffusion, early attempts to use a spherical model were flawed in that they assumed that the grain minimum dimension would be the proper sphere radius to use (e.g., Wolf et al. 1998, House et al. 2000, Reiners and Farley 2001). However, it has since been shown that cubes, blocks, and cylinders are much more accurately approximated by a sphere with the same S/V as the more complex shape (Meesters and Dunai 2002, Dunai 2005). For a range of thermal histories, Dunai (2005) demonstrated that age calculation errors are generally less than 2% when this measure is taken for cylinders; the worst case is for a square cube (due to the relatively high proportion of volume being close to a corner), for which equivalent-sphere ages were 6% too low for thermal histories geared to maximize possible misfit. Blocks with a higher aspect ratio were between these two cases. In all cases, the bias leads to slightly too-low ages for the S/V-equivalent spheres. It should be noted, however, that Dunai's analysis utilized the kinetic parameters from Wolf et al. (1996), and so did not incorporate grain-size effects, which will affect these comparisons. Furthermore, alpha ejection was evidently not included in these calculations.

Because alpha ejection correction is a necessary step for virtually all (U-Th)/He analyses, it has received more detailed attention than the diffusion calculation. To account for typical crystal forms, Farley (2002) provides polynomial functions to calculate F_T values for hexagonal prisms (i.e., apatite) and tetragonal prisms (i.e., zircon) for ^{238}U and ^{232}Th mean stopping distances, based on an empirical fit to results from Monte Carlo modeling of these shapes. In both cases, these shapes have pinacoidal (i.e., flat) terminations, lacking the pyramidal terminations that are more typical of apatite and zircon habits. Hourigan et al. (2005) expanded on this analysis for zircon by providing a slightly better fit for pinacoidally terminated tetragonal prisms, and also providing parameters for the more general case of a prism with terminations varying from pinacoidal to pyramidal over a limited range of crystal aspect ratios. In all cases the F_T value is posed as a function of S/V for each shape, based on an equation of the form:

$$F_T = 1 + a_1(S/V) + a_2(S/V)^2 \tag{5}$$

Values for a_1 and a_2, along with equations to calculate S/V, are provided in Table 1.

Actual, more-complex crystal shapes can be measured using μCT analysis, from which F_T parameters can be calculated (Evans et al. 2008), but there has not yet been an attempt to derive a generalized conversion. The fact that a single conversion can be used for a tetragonal prism with any type of termination suggests that a reasonable optimization exists, however. Furthermore, the inefficiency of the three-dimensional calculations required to solve the diffusion equation for arbitrary shapes makes them less favorable for routine use.

As matters currently stand, researchers commonly use a customized shape-based function for F_T calculation but an equivalent-S/V-sphere conversion for diffusion modeling. This may not be an optimal state of affairs, especially insofar as diffusion modeling must also account for alpha ejection. Furthermore, standard data tables for (U-Th)/He commonly include F_T values but not the dimensional information from which they were calculated, making it unclear how to correctly model such data. In light of the foregoing observations, this paper now explores an alternative conversion from a complex shape to a sphere by maintaining a constant F_T correction rather than a constant S/V ratio.

The misfits between the F_T values for hexagonal and tetragonal prisms vs. the F_T values for the corresponding equivalent-S/V spheres

TABLE 1.—Conversion factors to derive F_T corrections and equivalent sphere radii from shape and size measurements.

Mineral, geometry	Series	a_1 [1]	a_2 [1]	c_1 [2]	c_2 [2]
Hexagonal prism (apatite), pinacoidal[3] (Farley 2002)	^{238}U	−5.13	6.78	0.964	3.25
	^{232}Th	−5.90	8.99	0.957	3.67
Tetragonal prism (zircon), pinacoidal[4] (Hourigan et al. 2005)	^{238}U	−4.354	5.474	1.16	2.68
	^{232}Th	−4.938	6.881	1.15	3.59
Tetragonal prism (zircon), pyramidal[5] (Hourigan et al. 2005)	^{238}U	−4.281	4.372	1.16	2.68
	^{232}Th	−4.869	5.605	1.15	3.59

[1] From Eq. 5.
[2] From Eq. 6.
[3] S/V calculated from physical measurements as $(2.31L + 2R)/(RL)$, where L is grain length, and R is half-distance between opposed apices.
[4] S/V calculated as $(4L + 2W)/(LW)$, where L is grain length, and W is prism width.
[5] S/V calculated as $4([L - 2h] + [h^2 + W^2/4]^{0.5})/(W[L - 2h + 2h/3])$, where h is pyramid height.

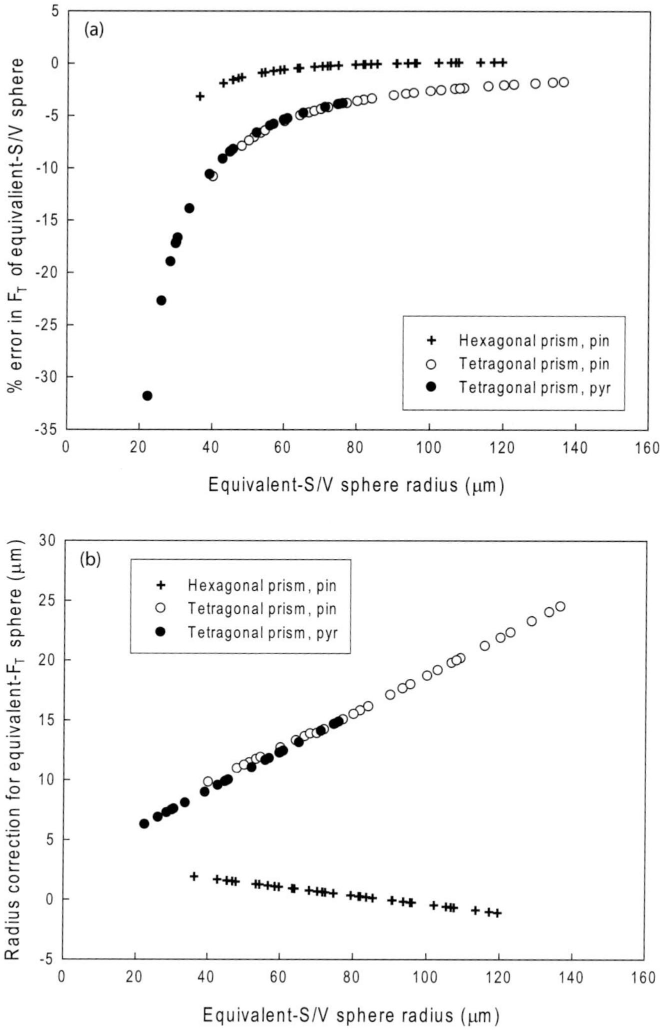

FIG. 2.—(a) Percent error in F_T based on equivalent-S/V sphere vs. customized shape-based model calculations. Hexagonal pyramid with pinacoidal terminations based on parameters from Farley (2002), using aspect ratios from 1:1 (cube) to 5:1 and short dimension half-widths from 40 to 100 μm. Tetragonal prisms are based on parameters from Hourigan et al. (2005), with pinacoidally terminated prisms having same range of dimensions as used for hexagonal prisms, and pyramidally terminated prisms having an aspect ratio of 2:1, lengths from 100 to 250 μm, and termination height ranging from zero (pincoidal) to half of the crystal length (dipyramids). (b) Sphere-radius corrections required to convert equivalent-S/V sphere to equivalent-F_T sphere. All results shown are for ^{238}U; relations for ^{232}Th are similar, but in all cases offset lower due to the longer mean alpha stopping distance.

based on the polynomials of Farley (2002) and Hourigan et al. (2005) are shown in Figure 2a. For all tetragonal prisms, the F_T of the S/V-equivalent sphere is 2 to 32% too small; for hexagonal prisms, the divergence is smaller, ranging from −3 to <1%. The systematic relationship of these misfits vs. equivalent-S/V-sphere radius (r_{esv}) points to a simple conversion. For each of these shapes, an equivalent-F_T sphere radius r_{eft} can be obtained with:

$$r_{eft} = c_1 r_{esv} + c_2 \tag{6}$$

Values for empirical parameters c_1 and c_2 for each shape and isotope are provided in Table 1, and the final r_{eft} value can be calculated as the eU-contribution-weighted average of the isotopes. The optimized radii lead to spheres having F_T values matching the shape-specific value to within 0.1% for all F_T values above 0.6. In the case of hexagonal prisms, the radius corrections are relatively small, on the order of 1 to 2 μm, which is within the error of the measurements used to determine the S/V ratio. Thus, using r_{eft} is essentially equivalent to using r_{esv} for apatite.

The corrections for tetragonal prisms are larger (additions of ~5–25 μm) owing to the 90° corners that constitute a sharper departure from sphericity. Corrections of this scale will have a somewhat more significant effect on diffusion calculations. The observed divergence between tetragonal prisms and spheres reported by Dunai (2005) indicates that a radius slightly larger than r_{esv} is necessary to have a sphere precisely replicate a prism. A series of test cases examined with Monte Carlo modeling indicates that although r_{eft} sometimes overshoots the mark, it almost always results in a more accurate correction than r_{esv} (Ketcham et al. 2011). However, any adverse effect is limited; the change in T_c for zircon caused by using r_{eft} instead of r_{esv} is about 3° C, which is well within the uncertainty of the relevant kinetic parameters.

Another advantage of the equivalent-F_T method is that it allows for straightforward modeling of (U-Th)/He ages from the literature in which F_T values are provided but other dimensional data are not. This approach is used for a number of the examples later herein where grain dimensions are only reported as cross-section radii of apatite hexagonal prisms, making F_T the only source of sufficient dimensional information. All that is required is to calculate the grain radius that corresponds to the reported F_T value. Because the conversion depends on a number of factors (stopping distances and proportion of U to Th), there is no straightforward conversion, but it is fairly easy to type candidate radii into HeFTy once these parameters have been entered, and quickly converge to a value that replicates F_T by observing its effect on the measured age.

Finally, all of these calculations are predicated on having accurate alpha particle stopping distances. Generally, these values are not measured directly for minerals of interest and are instead obtained via theory by extrapolating from observations on other, generally simpler materials. All previous work, including this paper, has been based on stopping distances reported by Farley et al. (1996) and Farley and Stockli (2002). Stopping distances have recently been revised, with new figures up to 1.5 μm shorter for some minerals (Ketcham et al. 2011). The impact on F_T corrections, and by extension ages, is on the order of 1 to 1.5%. Incorporating these revisions into the everyday practice will require reformulating the polynomial corrections, an effort that has recently been completed (Ketcham et al. 2011).

Inverse Modeling of Age Data

The inverse problem for thermochronology consists of using measured data to derive information about some aspect of the thermal history of a sample or region, such as cooling rates, time and/or temperature of peak heating, or timing and degree of erosional episodes. Computationally, this is done by linking the forward model solution to a time–temperature (t-T) path generator and finding the set of paths that is statistically consistent with measured data.

The statistical problem is straightforward when only age data are being fitted. The standard approach is to use a least-squares difference between model outcome and measured age, weighted by the measurement uncertainty (Ketcham 2005). However, in the practical case of (U-Th)/He data, a complicating factor is determining the uncertainty of the age measurement, given the frequent occurrence of

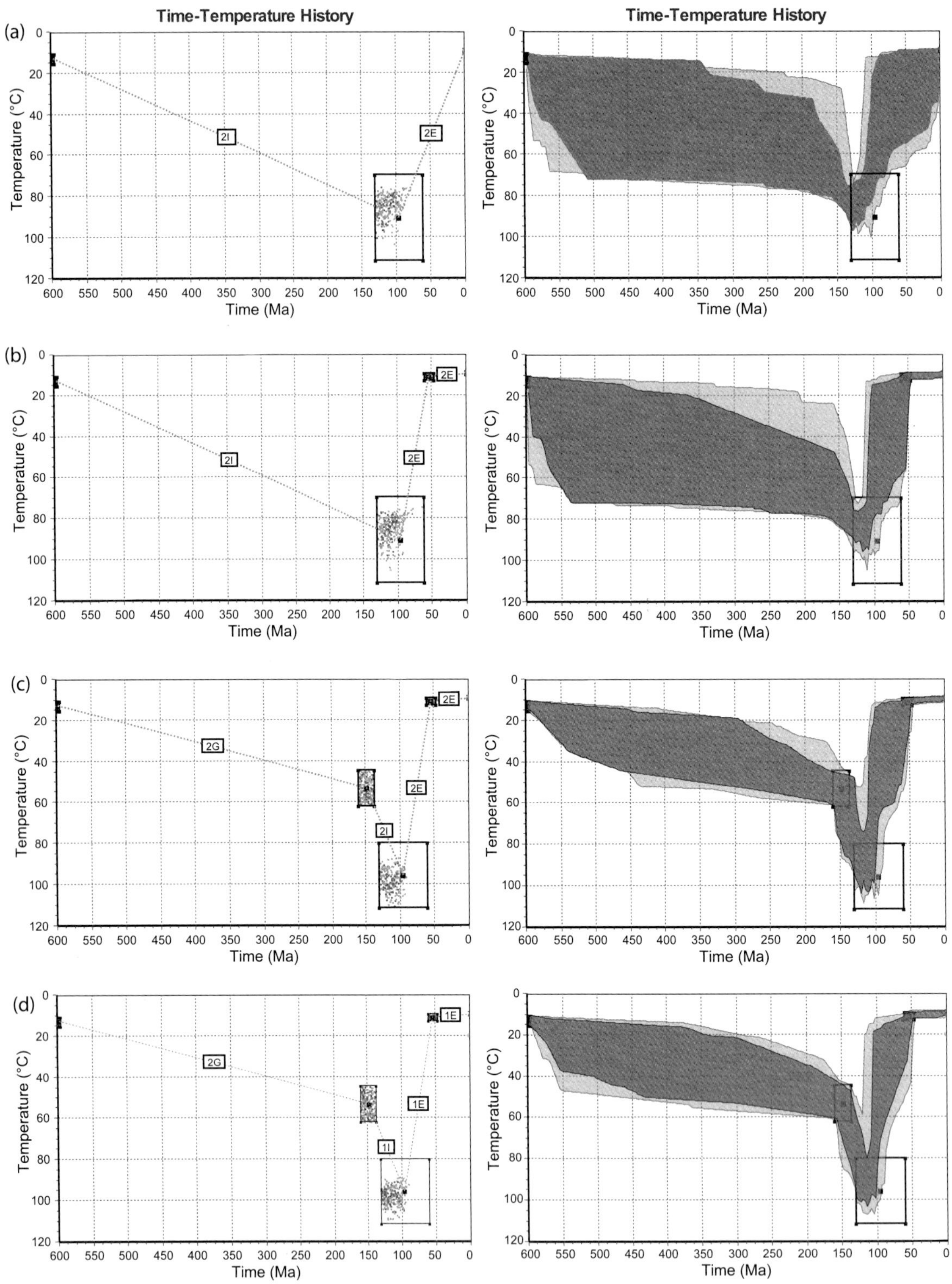

FIG. 3.—Thermal history inversions of AHe data from the Bighorn Mountains (Reiners and Farley 2001) using HeFTy software (Ketcham 2005) and the RDAAM diffusivity model. Graphs in the left column show boxes through which time–temperature paths are forced to pass, and codes on the lines connecting boxes indicate how those path segments are allowed to vary. Paths between boxes are monotonic cooling or heating, depending on relative position of boxes. The number indicates how many additional nodes can be interpolated between the boxes, as $2^N - 1$; i.e., a value of 2 means the $2^2 - 1 = 3$ nodes are interpolated, allowing heating/cooling rates to vary. The letter code "E" means heating/cooling rates are subject to episodic (i.e., sudden) change, "G" means gradual, and "I" is intermediate between the two. Points in boxes are inversion results showing where paths with statistically good (dark gray) and acceptable (light gray) fits to data passed. Graphs on right show envelopes

spread in ages that substantially exceeds analytical uncertainties. When single or small numbers of analyses are being modeled, the effect of increasing or decreasing the assigned uncertainty is mainly to expand or contract the area of the confidence region in t-T space. In the context of multisample and/or multichronometer modeling, the relative weighting given to different analyses becomes a more important question, as least-squares approaches are susceptible to undue influence from outliers, which, given current knowledge of the AHe system (among others), can be considered the rule rather than the exception. One common practice is to add an increment of uncertainty based on repeated analysis of age standards, but even this is likely to be an underestimate in many cases because standards are by definition analytically favorable specimens.

The time–temperature path set generated by the modeling software constitutes the family of histories that are consistent with both the data and the limits imposed by the software user and the software itself. These latter points can be as important as the assignment of data uncertainty, or even the data themselves, in influencing the modeling result.

In HeFTy, the family of t-T paths to consider is defined by entering (1) boxes in time–temperature space through which the paths must pass and (2) parameters controlling the complexity along a monotonically cooling or heating path between these boxes (e.g., Fig. 3). These parameters include the number of nodes to interpolate (i.e., the number of times the heating or cooling rate can change), maximum heating or cooling rate along a segment, and whether the paths generated tend toward gradual changes, episodic ones, or an intermediate degree between the two. Paths are generated randomly within these bounds (i.e., the Monte Carlo method), and each is evaluated against the data. Those with a goodness-of-fit (GOF) value of 0.5 are termed as "good," as 0.5 is the expected value for a correct model, and constitutes a reasonable measure of the statistical limit of precision. Those with a GOF value of 0.05 are termed "acceptable," as these are the models that do not fail the 95% confidence test. Those with a GOF value below 0.05 are rejected (Ketcham 2005).

A brief reexamination of an AHe study of burial and unroofing in the Bighorn Mountains, Wyoming (Reiners and Farley 2001, Crowley et al. 2002), provides an illustrative example of inverse modeling of age data and the influence of model parameters. The Bighorn Mountains are a Laramide range with exposures of Precambrian basement in their core and Phanerozoic strata with a thickness of ~2.8 to 3.9 km in the adjacent basins, suggesting only a modest amount of burial since the Precambrian. Roughly half or more of the sedimentary cover was deposited in the Cretaceous. Deformation and exhumation are thought to have commenced in the latest Cretaceous or the early Paleocene, based on sedimentary indicators (paleocurrent directions, strandlines, isopach maps) that show no evidence of uplift in the Late Cretaceous. Exhumation continued through the early Tertiary, with a late pulse of deformation indicated by thick Eocene orogenic clastic rocks (Crowley et al. 2002).

The geological history of the Bighorns make them an excellent candidate for testing and taking advantage of the variation in AHe T_c with grain size, as the long-term residence of apatite grains at low temperatures would be expected to amplify the effect of otherwise subtle differences in diffusive systems. Accordingly, Reiners and Farley (2001) analyzed two granitic basement samples, separating grains into size aliquots based on their approximate cylindrical radius. As expected, they found a correlation between grain size and measured age, with ages ranging from ca. 100 Ma for 30- to 50-μm grains to 250 Ma for 100-μm grains of one sample and 350 Ma for the other. However, in their attempt to model their data, they arrived at two unexpected results. First, the indicated maximum temperatures were very low (67–86° C), requiring a combination of low geothermal gradient (20° C/km) and burial at the minimum depth indicated by adjacent Phanerozoic basin sediments, or even less burial with a higher geothermal gradient. Second, the trend of age vs. size was better fit by a scenario in which exhumation began at 120 Ma and proceeded linearly until 55 Ma, rather than a sudden event at 65 Ma. The former observation was corroborated by examining multiple-sample transects, leading Crowley et al. (2002) to propose, perceptively, that accumulated radiation damage may have raised the closure temperature of these apatites.

The modeling employed by Reiners and Farley (2001) had two shortcomings that were not fully understood at the time. First, they used the hexagonal half-width rather than the equivalent-S/V sphere radius for their calculations, resulting in too-small spheres that artificially lowered the closure temperature. Second, the radiation damage control of helium diffusivity in apatite had not yet been experimentally established or calibrated.

The new RDAAM AHe diffusivity calibration (Flowers et al. 2009) provides the means to reevaluate the Bighorn data. Figure 3 shows a series of inversions using the RDAAM implementation in HeFTy that simultaneously fit the large, medium, and small aliquots from one of their samples (PRBH17), which have respective ages of 232.0±14.0, 156.0±9.4, and 107.0±6.4 Ma, and mass-weighted mean radii of 130, 88, and 56.2 μm (these latter values have been recomputed to result in an equivalent-F_T sphere based on the F_T values reported by Reiners and Farley 2001). All results are based on Monte Carlo simulations that continue until 50 statistically good time–temperature paths are found. The graphs on the left show the peak time and temperature for all good and acceptable paths, and the graphs on the right shows envelopes surrounding the good and acceptable paths for four different modeling iterations.

The first inversion (Fig. 3a) is the simplest, only positing formation in the Late Precambrian and maximum burial in the Cretaceous, ostensibly to find the maximum extent of burial and timing of the onset of unroofing and form of the cooling path. Broadly, the inversion indicates a maximum burial temperature of 75–100° C, and a pre–90 Ma exhumation onset followed by a wide range of potential cooling paths. These results are more consistent with the expected burial history (for example, they are consistent with 2.6–3.6 km of burial with a geothermal gradient of 25° C/km). However, it is also clear that the data do not do much to narrow the range of possible cooling histories, and some of the cooling paths used to map out the peak temperature conditions feature very late exhumation, which is clearly at odds with known geological constraints.

The inversions in Figure 3b and Figure 3c refine the picture by adding constraints based on geological data. In Figure 3b, cooling is forced to end between 60 and 50 Ma, as indicated by observations in adjoining basins. This modification has no recognizable effect on peak conditions predicted by the modeling. In Figure 3c, another constraint has been added to enforce that a significant portion of burial should occur in the Cretaceous, as required by stratigraphic thicknesses. This adjustment has a more significant effect on predicted peak conditions,

←

surrounding all good and acceptable paths, in dark and light gray, respectively. (a) Only Precambrian deposition, Cretaceous peak heating, and present-day temperature constrained. (b) Additional constraint enforces cooling by 60–50 Ma. (c) Additional constraint separates pre-Cretaceous and Cretaceous sedimentation, shifting peak inversion temperatures higher. (d) Same constraints as (c) but with reduced path complexity allowed between boxes, restricting range of thermal histories found by the inversion algorithm.

pushing them somewhat earlier and to a higher temperature range of 90–110° C, or 3.2 to 4.0 km of burial at 25° C/km. It is noteworthy that the higher peak temperature paths are consistent with the constraints used earlier for Figure 3a and Figure 3b, but they had the appearance of being excluded because they were less likely to be generated by the inversion algorithm due to the wide swaths of time–temperature space it was tasked with exploring. As additional geological information is brought in, the search is fine-tuned to better analyze a more restricted range of possibilities; this bears witness to a truism of modeling, which is that it is never done and can always be improved with new data.

The inversion result in Figure 3c is consistent with the expected burial history of the area, owing to the higher T_c predicted by the RDAAM due to the long near-surface residence of the basement from the Precambrian through the Cretaceous. However, it repeats the modeling result of Reiners and Farley (2001) in that the (U-Th)/He data are better matched by cooling that commences in the mid-Cretaceous, rather than the end. This may be an artifact of some aspect of the AHe system that is still not understood, or it may reflect a pulse of unroofing that has not been recognized in the surrounding strata. However, the inversion does permit ~40 to 50° C of rapid cooling starting as late as ca. 65 Ma and extending through the Eocene, consistent with geological observations.

Another aspect of inversion conditions is explored in Figure 3d. The constraint boxes are the same, but the complexity of the paths between them has been reduced to only allow one additional degree of freedom along the final three path segments rather than three. The range of peak conditions is essentially unchanged, but the cooling paths are smoother, supporting only modest amounts of unroofing after 65 Ma. This is a direct result of the limitations imposed by the modeler: The component of fast Tertiary cooling permitted by the Figure 3c inversion has been excluded, but only because of the restrictions on complexity embedded in the algorithm and its input parameters, not the data. Reducing degrees of freedom can be a tempting means of "simplifying" inversions and obtaining seemingly tighter bounds on thermal histories, but unless such simplifications are supported by the geology, they can be artificial and dangerous. This observation applies to modeling any type of thermochronometric data, including fission track (Ketcham 2005).

Inverse Modeling of Diffusion Profile Data

Obtaining diffusion profile data, whether by $^4He/^3He$ analysis or excimer laser ablation, opens the potential to derive more detailed thermal history information than is available from ages alone. As shown in Figure 1, the form of the profile contains information about the relative amounts of time spent in the helium partial retention zone (PRZ) vs. lower temperatures where helium is retained.

Figure 4 shows an example implementation of profile-based inversion using data obtained from $^4He/^3He$ analysis. The data (provided by K. Farley, personal communication, November 2008) are for apatite from a granodiorite in the King's River Gorge on the western side of the Sierra Nevada in California, probably primarily reflecting incision of the canyon. Figure 4a shows confidence regions defined by a set of randomly generated helium diffusion profiles that were consistent with the step-heating analysis. The profile outlined by the region is typical of a low-temperature history, with its principal feature being the alpha-ejection zone at the rim. Figure 4b shows the inversion based on these data, and a non-F_T-corrected AHe age of 50±4 Ma. The inversion appears to show that the profile data exert very tight control over the low-temperature part of the thermal history. This tightness is to some extent an artifact of the modeling however; Figure 4c shows an inversion in which some reheating is permitted in the low-temperature part of the thermal history, revealing that considerably more leeway is possible if nonmonotonic paths are allowed. Nevertheless, the revelatory power of profile data is evident.

An interesting consequence of the dominance of the alpha ejection rim is that the relative dimensions of the grain radius vs. the alpha ejection radius become critical; solutions are impossible to find unless these dimensions, or their ratios, are reasonably close to the proper value. In the case of Figure 4, for example, statistically good solutions are only possible if the grain radius is between 49 and 67 μm, and a smaller range (52–66 μm) is permitted if only monotonic cooling paths are considered. This brings up the question: How might the simplifications that allow us to go from real data to model input, such as averaging multiple grains and equivalent-S/V or equivalent-F_T sphere conversion, impact the form and scaling of the zoning profile? The fact that adequate fits can be found is encouraging, but this is an area requiring further study.

Another alternative for this type of modeling is to directly fit the step-heating data, rather than profiles obtained from it, as is typically done for inverse modeling of $^{40}Ar/^{39}Ar$ spectra. Schildgen et al. (2010) adapted this approach for $^4He/^3He$ spectra, and a modified version of their formulation is also available in HeFTy.

Multiple-Sample Approaches

The single-sample methods discussed in the previous section can be further leveraged by analyzing multiple samples with known geographic or stratigraphic relation to each other to obtain regional and/or spatial patterns. Natural applications for multiple-sample studies include constraining paleogeothermal gradients and the position and development of helium PRZs, and studying patterns of erosion and unroofing.

The helium PRZ is based on a hypothetical stable crustal section that is instantaneously emplaced and remains at constant temperature for some time interval. For a given mineral and a given isothermal holding time, the He PRZ is defined as the temperature interval over which He ages range from 90 to 10% of the holding time. Figure 5 shows an example AHe PRZ for the Durango apatite kinetic parameters of Farley (2000) for a 60-μm grain and an isothermal holding time of 50 million years; in this case, the PRZ is from about 31 to 62° C. This should be taken as a range for a "typical" apatite, but not an expected value for a random one unless U-Th content is known and taken into account; the RDAAM diffusivity calibration (Flowers 2009) predicts a PRZ of 22 to 54° C for a low–U-Th apatite (2 ppm U, 5 ppm Th) and 36 to 68° C for an apatite on the higher end of the range of observed values (20 ppm U, 100 ppm Th). With a longer holding time allowing more damage accumulation, both ranges rise, but their disparity increases.

From a more practical standpoint, one would expect to see some sort of PRZ if AHe ages were obtained for a series of down-well samples in a stable region. If a stable region is uplifted and exhumed, one would expect to see a "relict" PRZ that could provide data on the timing of exhumation and the geothermal gradient during holding. For example, Stockli et al. (2000) documented an AHe fossil PRZ on a tilted block in the White Mountains, California. The structurally deeper age of 12 Ma at the bottom of the PRZ established the time of rapid exhumation, and the shallower ages of ca. 57 Ma at the top of the PRZ implied about 45 million years of residence time. About 2200 m of stratigraphic section separated top and bottom of the PRZ. A presumed PRZ span of ~40° C (derived in part from the kinetics of Wolf et al. [1996] and comparison with simultaneously collected AFT data) implied a predeformation geothermal gradient of ~18° C/km.

Also shown in Figure 5 for comparison is the AFT partial annealing zone (PAZ), which shows a less symmetric trend, reflecting various aspects of the fission-track annealing process. The upper part of the apatite PAZ is generally defined less quantitatively by the break in slope between the relatively slowly changing upper-section ages and the faster-changing midsection ages.

In general, the detailed computational approaches used for single samples have not been applied much to multiple-sample studies, due in

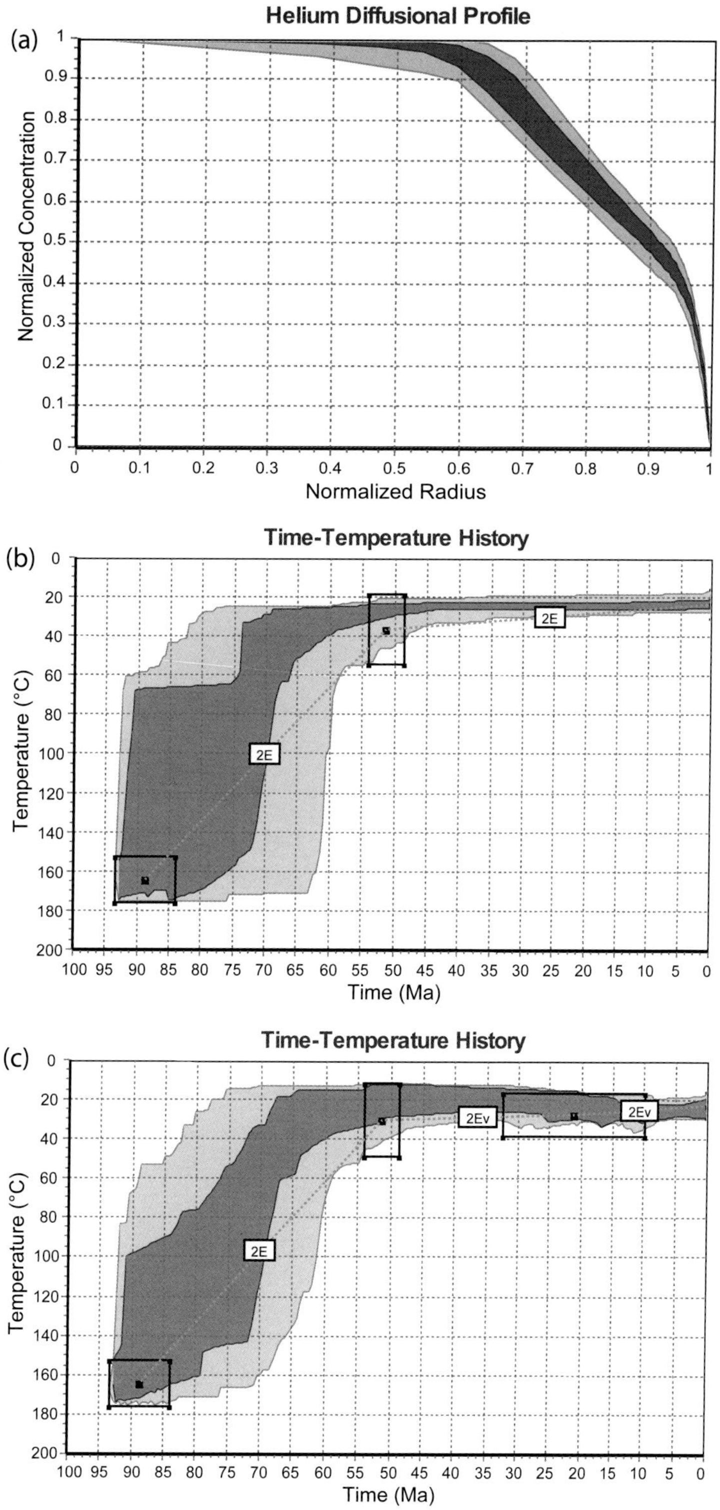

FIG. 4.—Example of inverse modeling based on He diffusion profile data. (a) Envelopes of diffusion profile forms that fit a data set from a $^4He/^3He$ step-heating experiment. Envelopes were created by comparing 50,000 randomly generated diffusion profiles to measured data, and accepting those with an adequately close fit; 50% of all accepted solutions are within darker inner envelope, and 95% are within lighter outer envelope. (b) Thermal history inversion using RDAAM diffusivity model in HeFTy, assuming a monotonic, cooling-only history. (c) Thermal history inversion that allows for reheating, showing wider solution envelope. Letter codes are as in Figure 3 caption, with additional code "v" indicating that the path between boxes is variable, and may be either monotonic cooling or monotonic heating in a time–temperature path.

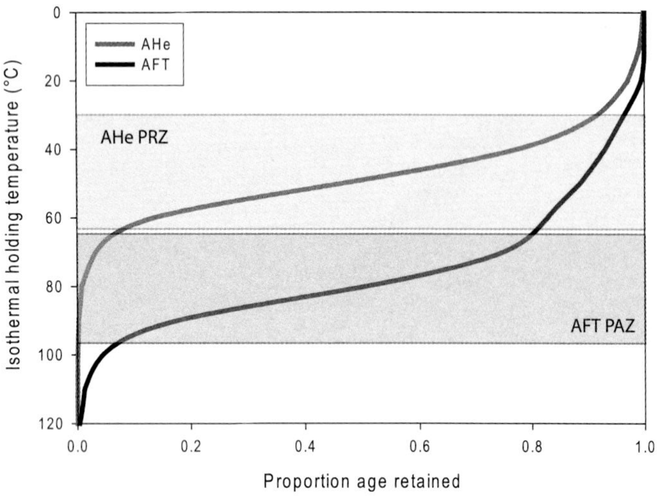

FIG. 5.—Trends illustrating the AHe PRZ and AFT PAZ, for an isothermal holding time of 50 million years.

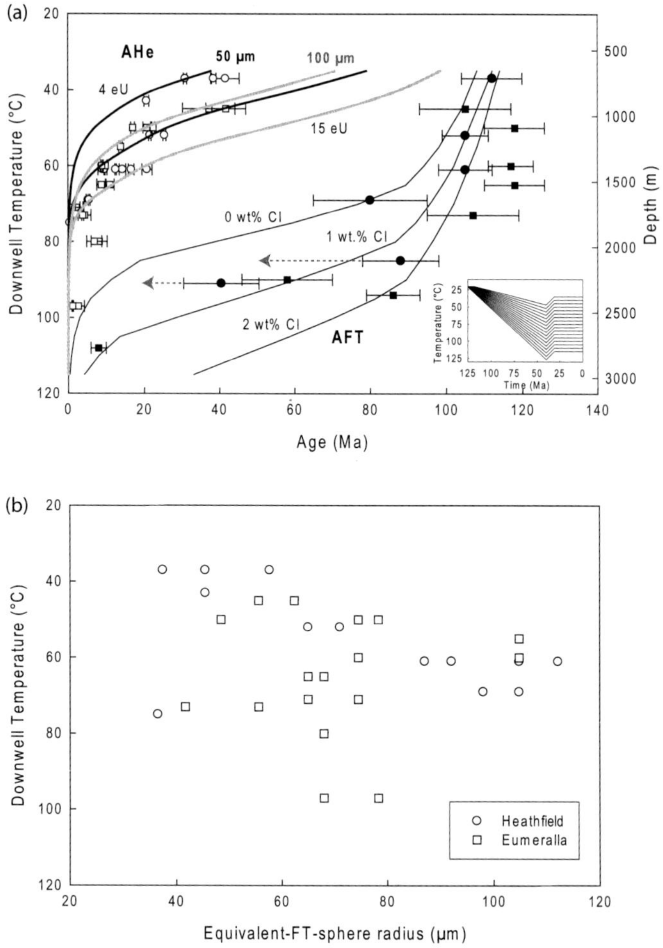

FIG. 6.—Data and modeling for the western Otway Basin. (a) Age vs. down-well temperatures from AHe (open symbols) and AFT (filled symbols) data in the Heathfield (circles) and Eumeralla (squares) wells in the western Otway Basin. Error bars are 2σ. Inset shows a hypothetical down-section thermal history that leads to the trends superimposed on the data. Two pairs of trends are used to illustrate effects of various parameters on AHe dates. Each pair bounds low-eU and high-eU apatites. The black trend pair bounds small grains (50-μm radius), and the gray pair bounds large grains (100-μm radius). Three lines show predicted AFT ages for kinetics corresponding to 0, 1, and 2 wt% Cl, to represent the variability present in annealing kinetics. For two AFT samples, arrows indicate more likely end-member F-apatite ages based on data shown in House et al. (2002) (Fig. 5). (b) Size of analyzed grains vs. down-well temperature in Heathfield and Eumeralla wells. Data are from House et al. (2000, 2002).

part to the additional layer of complexity required to jointly constrain the calculations among different spatial–temporal positions. Instead, the usual method is to calculate a set of ages for a presumed thermal history based on geological hypothesis testing. This may be done in the context of simple time–temperature scenarios, or integrated with more sophisticated two- or three-dimensional thermal models simulating basin evolution (e.g., Duddy 1997), landform development (Braun 2005, Valla et al. 2010), or other crustal processes (Ehlers 2005). These approaches basically calculate multiple solutions by allowing certain geological parameters to vary, such as geothermal gradient, exhumation rates, topography development, and even material parameters and elastic thickness, finding sets of parameters that yield solutions that are compatible with the measured ages.

With respect to basin modeling, multiple-sample methods have experienced longer development for AFT than (U-Th)/He, and most basin studies with AHe data have also included AFT analyses (e.g., Crowhurst et al. 2002, Reed et al. 2005). There is in fact a dearth of basin studies that utilize (U-Th)/He data exclusively, and thus all of the examples listed in the next sections include some reference to AFT data. The main focus of this discussion is on the interpretation of (U-Th)/He data, and the ways in which recent discoveries impact previous modeling and interpretation, but the following examples also show how these different thermochronometers can be complementary or, occasionally, in conflict.

The Western Otway Basin: A noteworthy example of the use of the AHe PRZ (and AFT PAZ) to illuminate basin thermal history is the studies by House et al. (2000, 2002) of the Otway Basin in southeastern Australia. The Otway Basin was initiated by the continental breakup of Australia and Antarctica in the Late Jurassic through Early Cretaceous. Rapid subsidence during rifting led to deposition of 3 to 5 km of sediments of the Otway Supergroup between ca. 123 and 106 Ma, mostly from contemporaneous volcanic sources. Basin modeling based on AFT and vitrinite data suggests that the eastern Otway Basin underwent rapid heating followed by about 3.5 km of Cretaceous unroofing, followed by lower-magnitude burial and unroofing events in the Tertiary (Cooper and Hill 1997). Much of the western Otway Basin, by contrast, is thought to have undergone near-monotonic burial since deposition, with only minor Mesozoic and Cenozoic inflections, although there may have been a thermal high during the Cretaceous owing to temporary effects associated with rifting (Duddy 1997).

AHe data acquired in the eastern Otway generally accord with basin evolution models based on AFT and vitrinite data. In the western Otway Basin, however, AHe ages appear to reflect the lower part of a PRZ, and they are generally young (<42 Ma), at modest down-well temperatures of 37 to 45° C (Fig. 6a). These data suggest that, rather than gradual, monotonic burial and heating through the Tertiary, parts of the western basin may have experienced a shallow thermal event that was not sufficient to strongly affect the apparent PAZ in the AFT data

(House et al. 2000, 2002). The particular form and magnitude of such an event were not postulated, however.

One area for concern in the AHe data was significant extra-analytical scatter among grains from the same well intervals, and House et al. (2002) found that they could attribute some of this variation to grain-size effects, with larger grains having, effectively, higher closure temperatures. Although size does explain a significant component of the scatter, it appears that radiation damage is also an important effect, in more ways than one. Western Otway Basin apatites are quite low in U (mostly 2–8 ppm) and Th (8–30 ppm), limiting the amount of radiation damage they have accrued. Thus, although they are old, they have less damage than the Durango apatite underlying the model that was used to interpret their data, and thus they require lower temperatures to reset.

By taking radiation damage effects into account using RDAAM, both AHe and AFT data can be honored in a number of scenarios. The inset to Figure 6a shows one very rudimentary model, which is intended principally for illustrative purposes. It posits gradual heating until 40 Ma, followed by 12° C of cooling over the next 9 million years, and constant temperature until present day. This simplified scenario is broadly consistent with the accepted history of the basin, though it notably omits inheritance of predepositional age information. The principal necessary component is a small thermal high at the correct time to reset low-damage AHe samples.

AHe model ages were calculated for the end-member cases of low-eU (2 ppm U, 8 ppm Th) vs. high-eU (8 ppm U, 30 ppm Th), and small (50-μm radius) vs. large (100-μm radius) apatites, bounding the ranges observed in the Heathfield and Eumeralla wells studied by House et al. (2000, 2002). The resulting model age profiles are superimposed on the data in Figure 6a, and they show reasonable general accordance with both the AFT and AHe data. The model trends show an enormous range of possible AHe ages due to variation in size and eU. However, within these uncertainties, the patterns match well with the systematics of the way in which apatite size varies down-well (Fig. 6b): At low temperatures from 37 to 50° C, the ages lie within the small-grain bounds, and at 60° C, they are within the large-grain bounds.

Interpretation of the Otway fission-track data is also complicated by substantial variability in kinetics, in this case, those of annealing. The apatites measured by House et al. (2002) have a wide range of chlorine contents, from 0 to 2.4 wt%, and in many samples, there is a good to fair correlation between Cl and age; in fact, it was the Otway Basin that provided some of the first convincing evidence that Cl can increase the T_c of the apatite fission-track system (Green et al. 1985). House et al. (2002) attempted to control for kinetic variation by only pooling data from apatites with Cl $<$ 0.5 wt%, but this measure has significant shortcomings. First, even within this range, there is substantial variation in ages, with some samples failing the χ^2 test, and the resulting mixed ages have little meaning. To predict F-apatite behavior for modeling purposes, it would have been more accurate to use only the very lowest-Cl apatites; in the Heathfield data shown in their Figure 5, this would result in ages of ca. 20 Ma and ca. 50 Ma at 91° C and 85° C, respectively, rather than 40 $\pm$ 10 Ma and 88 $\pm$ 10 Ma, as shown with gray arrows in Figure 6a. Second, the noisiness of the Cl–age correlations suggests that other compositional factors, such as cation substitutions (Fe, Mn; Ketcham et al. 1999), are also likely to be contributing to variation in annealing kinetics, and so modeling based on the measured Cl content alone is not sufficient. To account for variation in kinetics that probably exceeds that inferred from Cl alone, AFT ages were calculated using the Ketcham et al. (2007) annealing model assuming Cl contents of 0, 1, and 2 wt% Cl. The resulting trends successfully encompass the major features of the AFT data.

Although a gross simplification, the thermal scenario and calculations demonstrate that the AHe and AFT data need not be incompatible in this area, and they highlight the utility of AHe data to detect and constrain low-temperature thermal events that may escape detection when AFT is used alone. They also illustrate the high degree of variability in ages expected even with fairly subtle changes in diffusion (and annealing) kinetics, and thus the importance of thoroughness when attempting to deduce these kinetics.

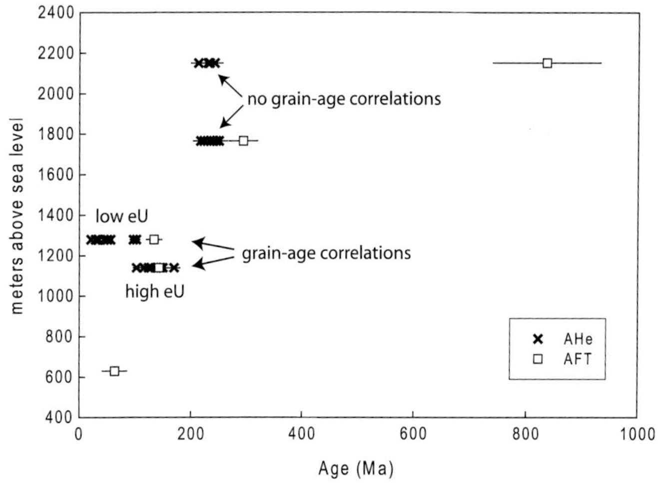

FIG. 7.—Apatite (U-Th)/He and fission-track age–elevation profiles from the Kangertittivatsiaq region, eastern Greenland. Data are from Hansen and Reiners (2006).

Eastern Greenland: A study by Hansen and Reiners (2006) of the exhumation history of the southeast Greenland continental margin, while not itself concerned with basin evolution, illustrates a number of points applicable to basin analysis. The margin between 67°N and 62°N is characterized by high relief due to glacial processes, with elevations ranging from 800 m at the coast to 2 km inland at the ice-sheet margin, providing a natural means to sample different crustal levels. Lithology in the study area consists exclusively of ancient crystalline basement. No remnants remain of any sedimentary overburden.

Figure 7 shows combined AHe and AFT data along an elevation transect in the Kangertittivatsiaq area, which is likely to have experienced more than 3 km of unroofing since the Caledonian Orogeny. At higher elevations, nearer the ancient peneplain, AFT ages are much higher than AHe ages, and the AHe ages are fairly consistent and show no grain size–age correlations. The latter was taken as evidence for cooling through the top of the AHe PRZ at about 250–220 Ma, allowing all apatites to retain their helium regardless of size (or damage). Lower elevations, sampling rocks that remained warmer longer, do have AHe size–age correlations, suggesting more prolonged residence within the PRZ.

Hansen and Reiners (2006) modeled AHe data for the final 250 million years of history assuming either isothermal holding at depth and exhumation starting between 20 and 0.5 Ma or near-surface residence followed by shallow burial and re-exhumation around 20 Ma. These end-member scenarios served to constrain the maximum holding temperature and the maximum burial temperature, respectively, at ~35 to 53° C and ~53 to 70° C for the lower-elevation samples. However, these models were not consistent with the AFT data: These models predicted too-old AFT ages, and models based on AFT predicted too-young AHe ages. The possibility of a radiation damage effect was cited as a potential explanation for this discrepancy.

The RDAAM diffusivity calibration helps explain some features of these data, but it appears to bring up new questions as well. The two

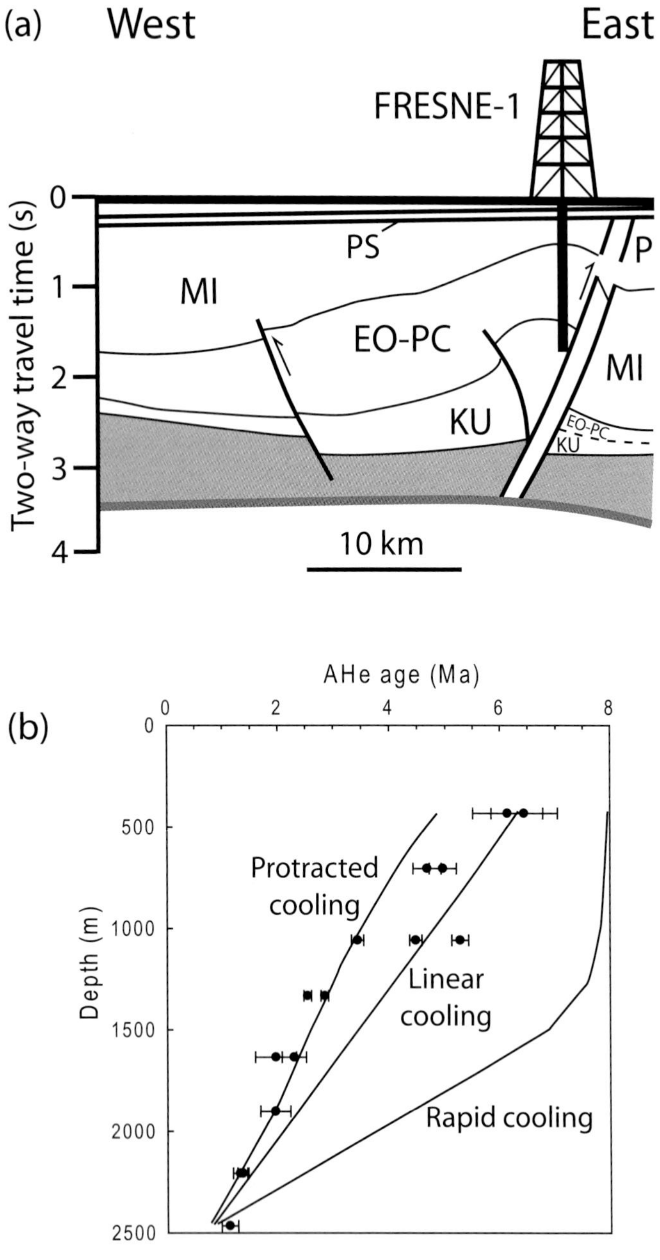

FIG. 8.—(a) Schematic cross section through Taranaki Basin at Fresne-1 well, showing inversion structures. PS = Pleistocene; P = Pliocene; MI = Miocene; EO–PC = Eocene–Paleocene; KU = Upper Cretaceous. Gray is basement (top seismic horizon). (b) AHe dates from Fresne-1 well, with model trends for linear cooling from a Miocene maximum (ca. 8 Ma to 2 Ma), rapid cooling, and protracted cooling that begins slowly and ends quickly. Adapted from Crowhurst et al. (2002).

sets of AHe data showing size–age correlations have markedly different U-Th contents; the samples that are concordant with AFT at 1140 m have 30 to 65 eU, and those with younger ages at 1280 m mostly have eU in the 4 to 20 range, although two have 35 and 65 eU, and the latter is not among the older of this cluster. Forward modeling with HeFTy using the eU values for the 1140-m samples indeed predicts rough concordance between AHe and AFT ages for the scenarios posed, and younger AHe ages for the lower-eU samples at 1280 m, thus appearing to resolve the worst of the discrepancy observed by Hansen and Reiners (2006). The more difficult-to-explain sample is the highest-elevation one, for which the ancient AFT age of 836±48 Ma indicates very long-term retention of radiation damage; calculations indicate that even low-eU apatites should have similar or even older AHe ages for any scenario that results in this AFT age.

Ancient terranes such as Greenland test all aspects of our understanding of radiation damage annealing, diffusivity development, and thermal history modeling more severely than younger areas (e.g., Hendricks and Redfield 2005). However, this study serves as an instructive example of detection of extended time spent in the He PRZ using size–age and eU–age correlations.

Taranaki Basin, New Zealand: The Taranaki Basin along the western margin of North Island, New Zealand, contains most of the country's known hydrocarbon reserves (Crowhurst et al. 2002). Early basin history was dominated by Cretaceous to Paleocene extension and locally rapid, fault-controlled subsidence, followed by regional thermal subsidence in the Eocene–Oligocene. Most structures in and around the basin formed during the Neogene, but several hydrocarbon-hosting inversion structures formed via Miocene reactivation of older normal faults.

Crowhurst et al. (2002) conducted a detailed study of the development of the Wakamarama anticline, using integrated AHe, AFT, and vitrinite data from the Fresne-1 well (Fig. 8a) in an attempt to constrain the timing and extent of burial, unroofing, and potential trap development. AFT and vitrinite data indicate rapid burial in the Miocene and subsequent cooling by ~70° C, corresponding to removal of ~2675 m of section at the Miocene–Pleistocene unconformity. However, those data do not provide very tight limits on the form of the final cooling path from ca. 8 to 2 Ma. Modeling of predicted down-well AHe ages for various scenarios (Fig. 8b) indicates that final unroofing was not rapid, but could have ranged from linear cooling to a more protracted history consisting of initial slow cooling by ~30° C from 8 to 4 Ma followed by the final 40° C of cooling from 4 to 2 Ma.

The apatites were of middling U-Th composition and Cretaceous to Eocene in age, and thus were reasonably well characterized with the Durango-based kinetic parameters available at the time of this study. Reconsideration of these data in the context of the RDAAM is thus not likely to lead to a significantly different result. The primary source of ambiguity in these data is the excess scatter in the ~1100-m-depth sample. Since the youngest age corroborates the protracted cooling scenario suggested by the overall data trend, this sample might be a good candidate for abrasion analysis to determine whether a "bad neighbor" effect could explain the older, off-trend ages.

CONCLUSION

The (U-Th)/He system has proven to be a powerful but challenging tool for basin thermal history analysis, as it has the often unique potential to uncover information about passage through an important temperature window. However, although initially thought to be a simple physical system, it has proven to be anything but that. This fact is most clearly demonstrated by the ironic development that the newest characterization of helium diffusivity in apatite has at its core an empirical model of fission-track annealing, many physical details of which remain inscrutable.

It is also clear that, despite recent advances, important challenges remain for the apatite (U-Th)/He system. Reproducibility of AHe ages is still frequently poor, and it is evident that even the latest models and methodologies cannot provide an explanation for many such cases. This may turn out to be simply an endemic trait that must be lived with—some known mechanisms, such as the "bad neighbor" effect, occur with unknown frequency and are difficult to detect when they do. For this reason, plus the importance of radiation damage for proper diffusivity characterization, AHe analysis should be combined with other techniques, notably AFT, when possible.

An important silver lining to the difficulties vexing AHe thermochronometry is that the continuing and intensive scrutiny given to the system continues to lead to new discoveries and new insights into its workings. As a result, the toolbox for thermal history analysis is expanding in parallel with these discoveries. As each factor that impacts helium behavior is understood and quantified (e.g., grain size, radiation damage), it passes from being a source of scatter to being a means of teasing out more subtle thermal history detail, as demonstrated in a number of cases in this chapter. Likewise, insights gained from the apatite system provide the groundwork for understanding other (U-Th)/He systems as they are further studied and applied. These salutary trends are certain to continue.

ACKNOWLEDGMENTS

The patience and forbearance of the editors in awaiting this manuscript is gratefully acknowledged. Thanks go to K. Farley, R. Flowers, C. Gautheron, P. Reiners, D. Shuster, and D. Stockli for variously provided discussion, analysis, modeling, and/or data that aided this manuscript. Constructive reviews by H. Dembicki and N. Harris improved the final product.

REFERENCES

Baxter EF, Asimow P, Farley KA. 2007. Grain boundary partitioning of Ar and He. *Geochimica et Cosmochimica Acta* 71:434–451.

Blackburn TJ, Stockli DF, Walker JD. 2007. Magnetite (U-Th)/He dating and its application to the geochronology of intermediate to mafic volcanic rocks. *Earth and Planetary Science Letters* 259:360–371.

Boyce JW, Hodges KV. 2005. U and Th zoning in Cerro de Mercado (Durango, Mexico) fluorapatite: Insights regarding the impact of recoil distribution of radiogenic ^{4}He on (U-Th)/He thermochronology. *Earth and Planetary Science Letters* 219:261–274.

Boyce JW, Hodges KV, Olszewski WJ, Jercinovic MJ, Carpenter BD, Reiners PW. 2006. Laser microprobe (U-Th)/He geochronology. *Geochimica et Cosmochimica Acta* 70:3031–3039.

Braddy D, Hutcheon ID, Price PB. 1975. Crystal chemistry of Pu and U and concordant fission track ages of lunar zircons and whitlockites. *In* Merrill RB, Hubbard NJ, Mendell WW, Williams RJ (Editors). *Proceedings of the 6th Lunar Science Conference: United States*: Pergamon Press, New York. p. 3587–3600.

Braun J. 2005. Quantitative constraints on the rate of landform evolution derived from low-temperature thermochronology. *Reviews in Mineralogy and Geochemistry* 58:351–374.

Brichau S, Ring U, Ketcham RA, Carter A, Stockli DF, Brunel M. 2006. Constraining the long-term evolution of the slip rate for a major extensional fault system in the central Aegean, Greece, using thermochronology. *Earth and Planetary Science Letters* 241:293–306.

Carlson WD, Donelick RA, Ketcham RA. 1999. Variability of apatite fission-track annealing kinetics I: Experimental results. *American Mineralogist* 84:1213–1223.

Cherniak DJ, Watson EB, Thomas JB. 2009. Diffusion of helium in zircon and apatite. *Chemical Geology* 268:155–166.

Cooper GT, Hill KC. 1997. Cross-section balancing and thermochronological analysis of the Mesozoic development of the eastern Otway Basin. *Australian Petroleum Production and Exploration Journal* 37:390–414.

Copeland P, Watson EB, Thomas JB. 2008. Fossils as alpha thermochronometers: *Geological Society of America Abstracts with Programs*, 2008 Joint Annual Meeting, Houston, abstract 222-13.

Copeland P, Watson EB, Urizar SC, Patterson D, Lapen TJ. 2007. Alpha thermochronology of carbonates. *Geochimica et Cosmochimica Acta* 71:4488–4511.

Crowhurst PV, Green PF, Kamp PJJ. 2002. Appraisal of (U-Th)/He apatite thermochronometry as a thermal history tool for hydrocarbon exploration: an example from the Taranaki Basin, New Zealand. *American Association of Petroleum Geologists Bulletin* 86:1801–1819.

Crowley PD, Reiners PW, Reuter JM, Kaye GD. 2002. Laramide exhumation of the Bighorn Mountains, Wyoming: an apatite (U-Th)/He thermochronology study. *Geology* 30:27–30.

Dodson MH. 1973. Closure temperature in cool geochronological and petrological systems. *Contributions to Mineralogy and Petrology* 40:259–274.

Dodson MH. 1979. Theory of cooling ages. *In* Jager E, Hunziker JC (Editors). *Lectures in Isotope Geology*: Springer, New York. p. 194–202.

Duddy IR. 1997. Focussing exploration in the Otway Basin: understanding timing of source rock maturation. *Australian Petroleum Production and Exploration Journal* 37:178–191.

Dunai TJ. 2005. Forward modeling and interpretation of (U-Th)/He ages. *Reviews in Mineralogy and Geochemistry* 58:259–274.

Ehlers TA. 2001. Geothermics of exhumation and erosion in the Wasatch Mountains, Utah [unpublished PhD dissertation]: University of Utah, Salt Lake City, Utah, 106 p.

Ehlers TA. 2005. Crustal thermal processes and the interpretation of thermochronometer data. *Reviews in Mineralogy and Geochemistry* 58:315–350.

England GL, Rasmussen B, McNaughton NJ, Fletcher IR, Groves DI, Krapez B. 2001. SHRIMP U-Pb ages of diagenetic and hydrothermal xenotime from the Archaean Witwatersrand Supergroup of South Africa. *Terra Nova* 13:360–367.

Evans NJ, McInnes BIA, Squelch AP, Austin PJ, McDonald BJ, Wu Q. 2008. Application of X-ray micro-computed tomography in (U-Th)/He thermochronology. *Chemical Geology* 257:101–113.

Farley KA. 2000. Helium diffusion from apatite: general behavior as illustrated by Durango fluorapatite. *Journal of Geophysical Research* 105:2903–2914.

Farley KA. 2002. (U-Th)/He dating: techniques, calibrations, and applications. *Reviews in Mineralogy and Geochemistry* 47:819–844.

Farley KA. 2007. He diffusion systematics in minerals: evidence from synthetic monazite and zircon structure phosphates. *Geochimica et Cosmochimica Acta* 71:4015–4024.

Farley KA, Clark MK. 2006. Progress and problems with apatite (U-Th)/He dating. *Geochimica et Cosmochimica Acta* 70:A166.

Farley KA, Stockli DF. 2002. (U-Th)/He dating of phosphates: apatite, monazite, and xenotime. *Reviews in Mineralogy and Geochemistry* 48:559–577.

Farley KA, Wolf RA, Silver LT. 1996. The effects of long alpha-stopping distances on (U-Th)/He ages. *Geochimica et Cosmochimica Acta* 60:4223–4229.

Fechtig H, Kalbitzer S. 1966. The diffusion of argon in potassium-bearing solids. *In* Schaeffer O, Zähringer J (Editors). *Potassium Argon Dating*: Springer-Verlag, New York. p. 68–107.

Fleischer RL, Price PB, Walker RM. 1975. *Nuclear Tracks in Solids; Principles and Applications*: University of California Press, Berkeley, California. 605 p.

Flowers RM. 2009. Exploiting radiation damage control on apatite (U-Th)/He dates in cratonic regions. *Earth and Planetary Science Letters* 277:148–155.

Flowers RM, Ketcham RA, Shuster DL, Farley KA. 2009. Apatite (U-Th)/He thermochronometry using a radiation damage accumulation and annealing model. *Geochimica et Cosmochimica Acta* 73:2347–2365.

Gautheron C, Tassan-Got L. 2010. A Monte Carlo approach to diffusion applied to noble gas/helium thermochronology. *Chemical Geology* 273:212–224.

Gautheron C, Tassan-Got L, Barbarand J, Pagel M. 2008. Alpha damage annealing effects on apatite (U-Th)/He thermochronology. *In* Garver J (Editor). Abstract volume, 11th International Conference on Thermochronometry, September 12–19, Anchorage, Alaska. p. 90–91.

Gautheron C, Tassan-Got L, Barbarand J, Pagel M. 2009. Effect of alpha-damage annealing on apatite (U-Th)/He thermochronology. *Chemical Geology* 266:157–170.

Gleadow AJW, Belton DX, Kohn BP, Brown RW. 2002. Fission track dating of phosphate minerals and the thermochronology of apatite. *Reviews in Mineralogy and Geochemistry* 48:579–630.

Gleadow AJW, Hurford AJ, Quaife RD. 1976. Fission track dating of zircon:

improved etching techniques. *Earth and Planetary Science Letters* 33:273–276.

Green PF, Crowhurst PV, Duddy IR, Japsen P, Holford SP. 2006. Conflicting (U-Th)/He and fission-track ages in apatite: enhanced He retention, not anomalous annealing behavior. *Earth and Planetary Science Letters* 250:407–427.

Green PF, Duddy IR, Gleadow AJW, Tingate PR, Laslett GM. 1985. Fission-track annealing in apatite: track length measurements and the form of the Arrhenius plot. *Nuclear Tracks and Radiation Measurements* 10:323–328.

Guenthner WR, Reiners PW. 2009. He diffusion in zircon: observations from (U-Th)/He suites and ^{4}He diffusion experiments and implications for radiation damage and anisotropic effects. *Eos, Transactions of the American Geophysical Union* 90:V52C-06.

Hansen K, Reiners PW. 2006. Low temperature thermochronology of the southern East Greenland continental margin: evidence from apatite (U-Th)/He and fission track analysis and implications for intermethod calibration. *Lithos* 92:117–136.

Harrison TM, Zeitler PK. 2005. Fundamentals of noble gas thermochronology. *Reviews in Mineralogy and Geochemistry* 58:123–149.

Hashemi-Nezhad SR, Durrani SA. 1983. Annealing behaviour of alpha-recoil tracks in biotite mica: implications for alpha-recoil dating method. *Nuclear Tracks* 7:141–146.

Heizler MT, Parsons I, Sanders RE, Heizler LL. 2008. K-feldspar microtexture and argon transport: *Geological Society of America Abstracts with Programs* 2008 Joint Annual Meeting, Houston, abstract 222-11.

Hendricks BWH, Redfield TF. 2005. Apatite fission track and (U-Th)/He data from Fennoscandia: an example of underestimation of fission track annealing in apatite. *Earth and Planetary Science Letters* 236:443–458.

Herman F, Braun J, Senden TJ, Dunlap WJ. 2007. (U-Th)/He thermochronometry: mapping 3D geometry using micro-X-ray tomography and solving the associated production–diffusion equation. *Chemical Geology* 242:126–136.

Hourigan JK, Reiners PW, Brandon MT. 2005. U-Th zonation-dependent alpha ejection in (U-Th)/He chronometry. *Geochimica et Cosmochimica Acta* 69:3349–3365.

House MA, Farley KA, Kohn BL. 2000. An empirical test of helium diffusion in apatite: borehole data from the Otway Basin, Australia. *Earth and Planetary Science Letters* 170:463–474.

House MA, Kohn BL, Farley KA, Raza A. 2002. Evaluating thermal history models for the Otway Basin, southeastern Australia, using (U-Th)/He and fission-track data from borehole apatites. *Tectonophysics* 349:277–295.

Ketcham RA. 2005. Forward and inverse modeling of low-temperature thermochronometry data. *Reviews in Mineralogy and Geochemistry* 58:275–314.

Ketcham RA, Carter AC, Donelick RA, Barbarand J, Hurford AJ. 2007. Improved modeling of fission-track annealing in apatite. *American Mineralogist* 92:799–810.

Ketcham RA, Donelick RA, Carlson WD. 1999. Variability of apatite fission-track annealing kinetics III: Extrapolation to geological time scales. *American Mineralogist* 84:1235–1255.

Ketcham RA, Gautheron C, Tassan-Got L. 2011. Accounting for long alpha-particle stopping distances in (U-Th-Sm)/He geochronology: Refinement of the baseline case. *Geochimica et Cosmochimica Acta* 75:7779–7791.

Kohn BL, Spiegel C, Phillips D, Gleadow AJW. 2008. Rubbing out apatite helium-age spread in fast-cooled rocks. *In* Garver J (Editor). Abstract volume, 11th International Conference on Thermochronometry, September 12–19, Anchorage, Alaska. p. 144.

Krishnaswami S, Lal D, Prabhu N, MacDougall D. 1974. Characteristics of fission tracks in zircon: applications to geochronology and cosmology. *Earth and Planetary Science Letters* 22:51–59.

Lippolt HJ, Wernicke RS, Bähr R. 1995. Paragenetic specularite and adularia (Elba, Italy): concordant (U+Th)-He and K-Ar ages. *Earth and Planetary Science Letters* 132:43–51.

Lippolt HJ, Wernicke RS, Boschmann W. 1993. ^{4}He diffusion in specular hematite. *Physics and Chemistry of Minerals* 20:415–418.

Lovera OM, Grove M, Harrison TM. 2002. Systematic analysis of K-feldspar ^{40}Ar/^{39}Ar step heating results II: relevance of laboratory argon diffusion properties to nature. *Geochimica et Cosmochimica Acta* 66:1237–1255.

Lovera OM, Grove M, Harrison TM, Mahon KI. 1997. Systematic analysis of K-feldspar ^{40}Ar/^{39}Ar step heating results I: relevance of laboratory argon diffusion properties to nature. *Geochimica et Cosmochimica Acta* 61:3171–3192.

Meesters AGCA, Dunai TJ. 2002. Solving the production–diffusion equation for finite diffusion domains of various shapes I: implications for low-temperature (U-Th)/He thermochronology. *Chemical Geology* 186:333–344.

Min K, Farley KA, Renne PR, Marti K. 2003. Single grain (U-Th)/He ages from phosphates in Acapulco meteorite and implications for thermal history. *Earth and Planetary Science Letters* 209:323–336.

Parsons I, Brown WL, Smith JV. 1999. ^{40}Ar/^{39}Ar thermochronology using alkali feldspars: real thermal history or mathematical mirage of microtexture? *Contributions to Mineralogy and Petrology* 136:92–110.

Phillips D, Kohn BL, Gleadow AJW, Harris JW. 2007. ^{4}He implantation in natural diamond: implications for apatite (U-Th)/He thermochronometry. *Eos, Transactions of the American Geophysical Union* 88:52.

Reed JS, Spotila JA, Eriksson KA, Bodnar RJ. 2005. Burial and exhumation history of Pennsylvanian strata, central Appalachian basin: an integrated study. *Basin Research* 17:259–268.

Reich M, Ewing RC, Ehlers TA, Becker U. 2007. Low-temperature anisotropic diffusion of helium in zircon: implications for zircon (U-Th)/He thermochronometry. *Geochimica et Cosmochimica Acta* 71:3119–3130.

Reiners PW. 2005. Zircon (U-Th)/He thermochronometry. *Reviews in Mineralogy and Geochemistry* 58:151–179.

Reiners PW, Farley KA. 1999. Helium diffusion and (U-Th)/He thermochronometry of titanite. *Geochimica et Cosmochimica Acta* 63:3845–3859.

Reiners PW, Farley KA. 2001. Influence of crystal size on (U-Th)/He thermochronology: an example from the Bighorn Mountains, Wyoming. *Earth and Planetary Science Letters* 188:413–420.

Reiners PW, Spell TL, Nicolescu S, Zanetti KA. 2004. Zircon (U-Th)/He thermochronometry: He diffusion and comparisons with ^{40}Ar/^{39}Ar dating. *Geochimica et Cosmochimica Acta* 68:1857–1887.

Reiners PW, Thompson SN, McPhillips D, Donelick RA, Roering JJ. 2007. Wildfire thermochronology and the fate and transport of apatite in hillslope and fluvial environments. *Journal of Geophysical Research* 112. doi: 10.1029/2007JF000759

Rutherford E. 1906. *Radioactive Transformations*: Archibald Constable & Co., London. 287 p.

Saini HS, Nagpaul KK. 1979. Annealing characteristics of fission tracks in minerals and their applications to earth sciences. *International Journal of Applied Radiation and Isotopes* 30:213–231.

Schildgen TF, Balco G, Shuster DL. 2010. Canyon incision and knickpoint propagation recorded by apatite ^{4}He/^{3}He thermochronometry. *Earth and Planetary Science Letters* 293:377–387.

Shelkov DA, Verchovsky AB, Milledge HJ, Pillinger CT. 1998. The radial distribution of implanted and trapped ^{4}He in single diamond crystals and implications for the origin of carbonado. *Chemical Geology* 149:109–116.

Shuster DL, Farley KA. 2003. ^{4}He/^{3}He thermochronometry. *Earth and Planetary Science Letters* 217:1–17.

Shuster DL, Farley KA. 2005. ^{4}He/^{3}He thermochronometry: theory, practice, and potential complications. *Reviews in Mineralogy and Geochemistry* 58:181–203.

Shuster DL, Farley KA. 2009. The influence of artificial radiation damage and thermal annealing on helium diffusion kinetics in apatite. *Geochimica et Cosmochimica Acta* 73:183–196.

Shuster DL, Farley KA, Sisterson JM, Burnett DS. 2003. Quantifying the diffusion kinetics and spatial distributions of radiogenic ^{4}He in minerals containing proton-induced ^{3}He. *Earth and Planetary Science Letters* 217:19–32.

Shuster DL, Flowers RM, Farley KA. 2006. The influence of natural radiation damage on helium diffusion kinetics in apatite. *Earth and Planetary Science Letters* 249:148–161.

Shuster DL, Vasconcelos PM, Heim JA, Farley KA. 2005. Weathering geochronology of (U-Th)/He dating of goethite. *Geochimica et Cosmochimica Acta* 69:659–673.

Stockli DF, Farley KA, Dumitru TA. 2000. Calibration of the apatite (U-Th)/He thermochronometer on an exhumed fault block, White Mountains, California. *Geology* 28:983–986.

Stockli DF, Farley KA, Walker JD, Blackburn TJ. 2005. He diffusion and (U-

Th)/He thermochronometry of monazite and rutile. *Geochimica et Cosmochimica Acta* 69:8.

Strutt RJ. 1908. On the accumulation of helium in geological time. *Proceedings of the Royal Society of London A* 81:272–277.

Strutt RJ. 1910. The accumulation of helium in geological time III. *Proceedings of the Royal Society of London A* 83:298–310.

Valla PG, Herman F, Van Der Beek PA, Braun J. 2010. Inversion of thermochronological age–elevation profiles to extract independent estimates of denudation and relief history I: theory and conceptual model. *Earth and Planetary Science Letters* 295:511–522.

Van Soest MC, Boyce JW, Monteleone B, Hodges KV. 2008. Laser microprobe depth profiling of ^{4}He diffusion in Durango apatite. *Geochimica et Cosmochimica Acta* 72:A976.

Wagner GA, Reimer GM. 1972. Fission track tectonics: the tectonic interpretation of fission track apatite ages. *Earth and Planetary Science Letters* 14:263–268.

Warnock AC, Zeitler PK, Wolf RA, Bergman SC. 1997. An evaluation of low-temperature apatite U-Th/He thermochronometry. *Geochimica et Cosmochimica Acta* 61:5371–5377.

Wernicke RS, Lippolt HJ. 1993. Botryoidal hematite from the Schwartzwald (Germany): heterogeneous uranium distributions and their bearing on the helium dating method. *Earth and Planetary Science Letters* 114:287–300.

Wolf RA, Farley KA, Kass DM. 1998. Modeling the temperature sensitivity of the apatite (U-Th)/He thermochronometer. *Chemical Geology* 148:105–114.

Wolf RA, Farley KA, Silver LT. 1996. Helium diffusion and low-temperature thermochronometry of apatite. *Geochimica et Cosmochimica Acta* 60:4231–4240.

Yuan W, Gao S, Dong J, Bao Z, Jia X. 2008. The etching behavior of alpha-recoil tracks in phlogopite. *Radiation Measurements* 43:S353–S356.

Zeitler PK, Herczeg AL, McDougall I, Honda M. 1987. U-Th-He dating of apatite: a potential thermochronometer. *Geochimica et Cosmochimica Acta* 51:2865–2868.

THERMAL HISTORY ANALYSIS OF SEDIMENTARY BASINS: AN ISOTOPIC APPROACH TO ILLITIZATION

NORBERT CLAUER

Laboratoire d'Hydrologie et de Géochimie de Strasbourg, (CNRS/UdS), 1 rue Blessig, 67084 Strasbourg, France
e-mail: nclauer@unistra.fr

AND

ABRAHAM LERMAN

Department of Earth and Planetary Sciences, Northwestern University, Evanston, Illinois, USA

ABSTRACT: The postdepositional history of clay minerals is critical to understanding the diagenetic evolution of sedimentary rocks. Diagenetic changes occurring in sedimentary K-bearing clay minerals, known as illitization, are routinely identified by a changing trend in the mineralogy from smectite to illite. This change is typically gradual, with the smectite-rich components changing into illite-rich components affected by the addition of K, generally occurring during progressive burial. Understanding of illitization is also germane to the use of authigenic illite as temperature and age indicators of the thermal history of sedimentary rock sequences. Oxygen and hydrogen isotope geothermometry and radiogenic isotope dating (by K-Ar or other methods) of authigenic illite represent tools that can potentially help decipher the thermal histories of sedimentary basins.

For these analytical techniques to be meaningful, a particularly adequate mineral separation and an accurate mineral identification are necessary, which may be difficult and not systematically successful because of the small sizes and varied origins of the studied minerals. Combined with detailed X-ray analyses and electron microscope observations, stable and radiogenic isotopic data have proven useful in distinguishing burial-induced from hydrothermally-induced temperature increases in sedimentary sequences. Some of the difficulties in the application of isotopic methods to separate clay-sized materials (which may also contain other silicates of identical size) can arise from uncertain values of the isotope fractionation factors of oxygen and hydrogen and from mixtures of detrital and authigenic components in the analyzed size fractions.

Conflicting isotopic data from illite-rich size fractions may also result from difficulties associated with demonstrating chemical and isotopic equilibrium between minerals and fluids, especially in the low-temperature domain. The intimate physical association of detrital and authigenic illitic particles of varied sizes, which may not be physically separated, can also contribute to the drawbacks. In order to provide a coherent overview of the input of isotope studies to the understanding of illitization—and therefore of the thermal history of sedimentary basins—burial-induced and more sporadic rock–hot fluid interactions, such as those of Northern Germany, Paris, East Slovak, and of the Mahakam Delta, are evaluated on the basis of combined mineralogical and isotopic databases of the illitic and associated materials.

KEY WORDS: illitization, clay mineralogy, oxygen and hydrogen isotope geochemistry, K-Ar isotope dating, geothermometry, thermal history of sedimentary basins

INTRODUCTION

It is generally agreed that many minerals initially deposited in sediments display a continuum of recrystallization during burial, from early diagenesis that starts soon after sediment deposition, to late diagenesis that occurs during burial, and finally to metamorphism. The rate and extent of mineralogical transformations depend on factors such as local or regional temperature and pressure gradients, chemistry of the interacting fluids present in the rocks, lithology of the host rocks, and duration of the diagenetic process. In this respect, it is important to keep in mind that a given sedimentary horizon may provide multiple clues to its diagenetic evolution depending on the material studied, as, for instance, clay material and organic matter respond differently to the same factors (Kübler 1997). The ability to identify the postdepositional alteration of mineral and/or organic assemblages also depends on the techniques used in the studies, and this identification is likely more accurate if a combination of different techniques is used.

One aim of this contribution is to illustrate why and how low-temperature isotope geochemistry, consisting primarily of the study of the oxygen and, to a lesser extent, of the hydrogen isotope fractionation, combined with K-Ar dating of illitic clay materials is able to provide reliable information for reconstructing the thermal histories of sedimentary basins. Illitization, which is a common crystallization and recrystallization process in sedimentary basins, can provide information about the maximum temperature and the timing of associated illite crystallization.

During the last five decades, significant progress has been made in the understanding of diagenetic to low-grade metamorphic alteration of minerals and organic matter in rocks of varied compositions. For example, the ordering of the illite crystalline structures was first evaluated as a simple index based either on the measurement of the width of the 10-Å illite peak at half height (Kübler 1966) or on the shape of the basis of the 10-Å peak as the sharpness ratio (Weaver 1961), both on X-ray diffraction (XRD) patterns. Reduction in the width of the peak and change in the basal shape of the 10-Å peak were attributed to the increasing quantity of 10-Å layers in a given structural domain of the mineral layering (e.g., Frey 1987). This qualitative approach was later refined to a more precise "illite crystallinity index," which likely does not represent the best nomenclature, as it referred mainly to an increase in the structural organization of the minerals and not to an increased crystallinity. Nonetheless, for many years it has been very useful to evaluate the degree of transformation of illite (Merriman et al. 1990); this activity has been supported more recently by the application of scanning electron microscopy (SEM), transmission electron microscopy (TEM), and high-resolution transmission electron microscopy. Additional studies of illite crystallinity established a relationship between this parameter and evolutionary stages in the transition from smectite-rich particles to either illite or chlorite (e.g., Velde 1995, Kübler and Goy-Eggenberger 2001) along the so-

Analyzing the Thermal History of Sedimentary Basins: Methods and Case Studies
SEPM Special Publication No. 103, Copyright © 2012
SEPM (Society for Sedimentary Geology), ISBN Print ISBN 978-1-56576-315-9, CD/DVD ISBN 978-1-56576-317-3, p. 125–146.

called "illite–smectite mixed-layer" trend (I/S), which applies to all illitic particles containing a more or less K-rich layering, according to the kinetic model of a theoretical reaction path (Essene and Peacor 1995).

The stable isotope methods ($\delta^{18}O$ and δD) applied to clay transformations aim primarily to establish the temperature and the physical and chemical conditions under which illite and other mineral phases, such as quartz, calcite, and oxides, formed in sedimentary rocks. The usefulness of these methods (e.g., reviews by Savin and Lee 1988, Sheppard and Gilg 1996, Sharp 1997) has been demonstrated, especially when oxygen and hydrogen isotope fractionation in illite are used as tracers of interactions with circulating fluids. Isotopic dating methods, particularly K-Ar because of its easy application, also provide an approach to obtain historic benchmarks for diagenetic reactions that could not be evaluated otherwise (e.g., reviews by Clauer and Chaudhuri 1995, 1997).

The primary goal of this contribution is therefore to review the published evidence of how illite can be best used for geothermometric estimations of sedimentary rocks by combining XRD analyses, SEM and TEM observations, stable isotope analyses for thermal evaluations, and isotopic dating of the authigenic illite. Although most of the case studies reviewed here do not include thermal history analyses on organic matter or accessory minerals, such analyses should, whenever possible, supplement the isotopic determinations on illite and strengthen the approach.

THE ILLITIZATION PROCESS

Illite has been reported to form in a variety of environments, from soils to deeply buried sediments. Evidence that high temperatures promote the authigenic precipitation of illite or the diagenetic conversion of I/S into illite, both termed "illitization," comes from experimental work as well as geologic studies in locations with well-documented geothermal regimes (e.g., Frey et al. 1980) or local temperature increases resulting from circulation of hot hydrothermal solutions at depth (e.g., Lampe et al. 2001, Meunier and Velde 2004, Timar-Geng et al. 2004). The formation of illite under different physical and chemical conditions has been variably attributed to such processes as direct mineral precipitation associated with local or remote dissolution of K-containing micas and feldspars and the uptake of K^+ ions from interstitial solutions by existing I/S, with structural rearrangement of the crystal lattices. In all cases, the process can be identified as illitization, even if this term appears more appropriate for a progressive than for an instantaneous crystallization. Also, whatever the process, and even if it can be attributed to local dissolution of detrital K-feldspars and micas in the coarser size fractions of sediments or to K import from outside the rock volume, the sources of K are still not well established (e.g., Freed and Peacor 1992, Harris 1992, Milliken 1992, Pollastro 1993). Theoretical models of illite formation were also developed for such mechanisms as mineral dissolution and precipitation in porous rocks, solid-state alteration in rocks with very low permeability, and generic transformation of I/S into illite (e.g., Velde and Vasseur 1992, Renac and Meunier 1995, Altaner and Ylagan 1997). Based on experimental studies, the upper temperature limit of illite stability was reported to be $\leq 250°$ C (relative to muscovite) or 360° C (Yates and Rosenberg 1996, Rosenberg 2002). Commonly reported ranges of temperatures and lithostatic pressures are shown in a theoretical sediment column 6 km thick at geothermal gradients of 25 to 45° C/km (Fig. 1).

The availability of K is clearly one determining control on the formation of illite. In this respect, the K^+/Na^+ concentration ratio in deep formation waters increases with increasing temperature in many locations (e.g., Fournier and Truesdell 1973). The K^+/Na^+ ratio in formation waters also increases as a result of the exchange reaction between K- and Na-feldspar—$KAlSi_3O_8 + Na^+ = NaAlSi_3O_8 + K^+$—

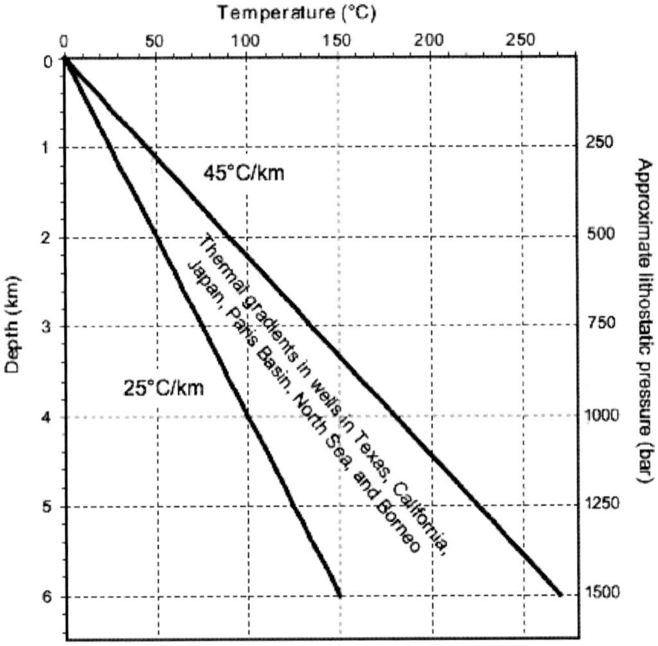

FIG. 1.—Plot of temperature as a function of depth for two geothermal gradients of 25 and 45° C/km. This range includes the linear gradients in wells of Borneo (Clauer et al. 1999), the Bergen High section in the North Sea (Glassmann et al. 1989), the Texas Gulf Coast (Hower et al. 1976), and those of other localities (Velde and Vasseur 1992) in which illitization was studied. Lithostatic pressure gradient shown for a rock of constant bulk density of 2.55 g/cm^3.

as a result of a change in the equilibrium constant, while the solubility of K-feldspar and albite both increase from 0 to 350° C with an increase of the atmospheric pressure below 100° C to water-vapor saturation at higher temperatures. K-feldspar (microcline) solubility increases more strongly than does that of albite, resulting in higher concentrations of the K^+ ions in solution (Arnórsson and Stefánsson 1999). Other experimental results and thermodynamic databases show a different behavior of temperature-related equilibria of K-containing minerals. A two-phase equilibrium in aqueous solutions of illite and kaolinite, I/S and kaolinite, and smectite and kaolinite pairs (Aja 1995) shows that between 25 and 250° C, the equilibrium constant of the reactions between the mineral pairs and their aqueous solutions is represented mainly by the ion activity ratio K^+/H^+. This ratio decreases strongly with increasing temperature for the above mineral pairs that include illite, but shows an increase for the smectite–kaolinite pair. A decrease in the K^+/H^+ ratio in solution at equilibrium favors the formation of illite at lower K^+ concentrations. Similar results were published by Yates and Rosenberg (1996) with regard to decreasing K^+/H^+ activity ratios in solutions at equilibrium between low-K kaolinite and high-K illite in a temperature range from 100 to 250° C. They suggest that higher temperatures may compensate for a lower K content in solutions during illitization.

The effect of pressure on illitization is not well known. A study of the effects of a progressive pressure increase from 1 to 1000 atm on a smectite (montmorillonite) and an illite at 22° C in seawater (Dayal 1975) shows that with increasing pressure, 10-Å illite layers develop in the mineral structure at the expense of the expandable smectite. However, in the pressure range from 335 to 1000 atm (equivalent to about 1–4-km depth in an average sediment column; Fig. 1), kaolinite, smectite, and illite release increasing amounts of SiO_2 to the solution. It appears that increasing pressure lowers the uptake of K^+ by these minerals but has no effect on that of Na^+ and variable effects on those of Mg^{2+} and Ca^{2+}. Bruce (1984) stated that the progressive change from smectite into illite is a major cause of overpressure in shales and in porous reservoirs interbedded with shales in the Gulf Coast. Overpressured zones (sometimes called "undercompacted zones") of sedimentary basins are characterized at their top and bottom by transition zones in which pressure increases with depth at a rate exceeding the hydrostatic pressure. Gautier et al. (1987) reported a temperature-dependent trend of decreasing smectite-layer content in I/S and a correlative ordering in the I/S from normally pressured to abnormally pressured shales in Northeastern Alaska. On the other hand, Osborne and Swarbrick (1999) provided theoretical calculations of porosity that indicate that diagenetic reactions, including illitization and quartz cementation occurring in the Upper Jurassic Fulmar Formation in the Central North Sea, did not generate significant overpressure. As both the pressure and temperature increases are often linked, one may infer from this limited set of data that illitization occurs also in high-pressure, high-temperature conditions, but pressure seems not to play the dominant role, although it may enhance the development of the 10-Å illite component.

MINERALOGICAL AND ISOTOPIC SIGNATURES OF ILLITIZATION

This section outlines the strengths and weaknesses of some of the mineralogical and isotopic approaches for analyzing illite for the purpose of reconstructing the thermal history of sedimentary basins. It also compares these approaches to other paleothermometric methods in some of the discussed study cases, when available.

Any study applied to clay material that includes isotope geochemistry has basic requirements such as the following: (1) separation of the clay population into various size fractions, which often helps to purify a mineral type, (2) proper identification of the clay-types in each size fraction, and (3) separate analysis of the size fractions. The accuracy of such applications also depends on an accurate thermal calibration of the stable isotopic fractionation factors that are not always as well constrained in the low-temperature domain as they are at higher temperatures.

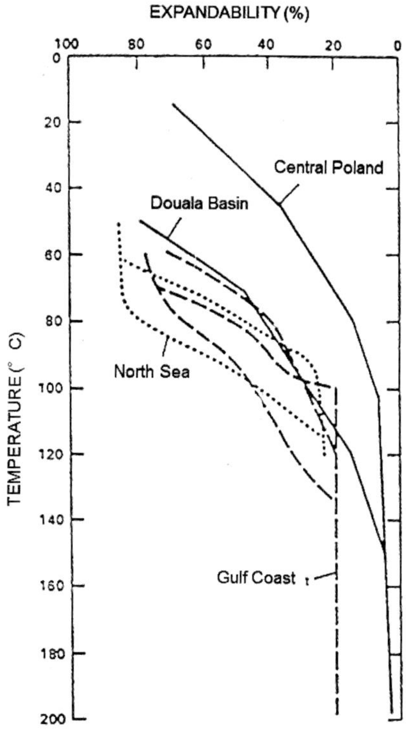

FIG. 2.—Expandability of smectite to illite-type mixed layers of different types of sediments from varied sedimentary basins relative to temperature estimated from burial depths. An expandability of 100 and 0 corresponds, respectively, to pure smectite and pure illite. The change in expandability relative to depth is dependent on the history of the basins and their thermal gradient. Illitization in the Central Poland and Douala basins ended at an expandability of about 5%; that of the Gulf Coast ended at an expandability of ~20% (after Środoń and Eberl 1987).

Mineralogical and Textural Characterization

Since the 1950s illitization has been largely characterized by a textural–mineralogical XRD approach (e.g., Weaver 1957, Burst 1959, Powers 1959, Dunoyer de Segonzac 1970, Hower et al. 1976, Velde 1985, Środoń and Eberl 1987). The early studies were based mainly on the measurement of the width of the 001 XRD peak, termed the "illite crystallinity index" (ICI), as suggested by Kübler (1966), with an increase of its value as the peak width decreases, that is to say when burial and temperature increase. Later, the same index was termed the "Full Width at Half Maximum" (FWHM), which is probably a more appropriate name, as it has no confusing connotation with particle crystallinity. In fact, the crystallinity criterion should only be used when the expandability of illitic material is less than 5%, meaning, in turn, that it is not strictly applicable to most I/S structures, making a precise determination of the smectite layers in the mineral structure very critical. The decrease in the FWHM numerical values actually relates to a decrease in expandability of the interlayer space of the smectite-type minerals, with the incorporation of K driving the composition toward an illite end-member. There is a consensus that this systematic and progressive conversion of smectite to illite (Fig. 2) is driven by temperature, even if the details of the process are still not well understood. The basic XRD analysis of clay minerals, including mathematical simulation and decomposition of the XRD peaks that are now part of the textural description, can be found in various books and articles (e.g., Reynolds 1985, Lanson and Besson 1992, Moore and Reynolds 1997). This mineralogical–textural approach is especially useful for sorting out the mineral assemblages in the smectite-to-illite trend. In the case of newly formed illite, its use is more conventional as a classical identification method.

Scanning electron microscope observations show the in situ shape of illite particles and their in situ textural relationship with the other minerals, detrital or authigenic, sometimes with organic matter. Transmission electron microscope observations are useful for verification of the presence or absence of detrital contaminants, even in minute amounts, and for a precise description of the shape of the particles after separation and purification: straight particle edges are typical for authigenic and irregular edges of detrital particles variously altered during transport and subsequent diagenesis (e.g., Peacor 1992); examples of authigenic and detrital particles are shown in Figure 3. High-resolution transmission electron microscopy observations may also show how illitization proceeds physically at the structural scale of the particles (e.g., Amouric and Olives 1991).

Of interest here is the fact that the commonly used ICI (FWHM) depends on the size of the analyzed particles and on the lattice strain (Klug and Alexander 1974, Árkai and Tóth 1983, Eberl and Velde 1989). Crystal size has been much less frequently studied than has ICI, although the former is probably more useful and meaningful, as it represents directly a crystallographic parameter reflecting the crystal-

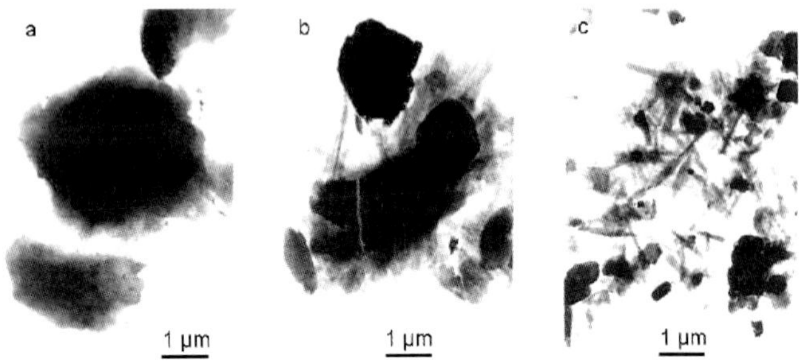

FIG. 3.—Transmission electron microscope pictures of clay particles from the Neogene Mahakam Delta Basin, Eastern Borneo. (a) Irregular shapes of detrital particles from the upper part of the sedimentary sequence (depth: 183 m); (b) Thin euhedral and transparent overgrowths of authigenic lath-shaped particles on larger, thicker detrital particles from shales in a deeper part of the basin (depth: 4232 m); (c) Euhedral, transparent laths and fibers of authigenic particles mixed with irregularly shaped bulky detrital particles from sandstones in a deeper part of the basin (depth: 4228 m) (from Furlan 1994).

lization or recrystallization episode of a clay population (Eberl et al. 1990). Several approaches for determining the crystal size of illitic material are available, but a widely used method combines the computational procedure of Eberl and Środoń (1988) and that of Eberl et al. (1990), both of which are based on the Warren and Averbach (1950) method. This procedure determines not only the mean crystal size of a particle population but also the crystal size distribution in the population. Minor recurrent difficulties in applying both the ICI and the crystal size methods arise because the results strongly depend on sample preparation, equipment conditions, and the analytical methods used, so that comparison of results from different investigators requires caution. Nonetheless, both ICI and crystal size distribution are important tools for a precise identification of the illitic particles and of the size distribution in any sample or in each of its size fractions.

However, for illitization to be used as a reliable indicator of geothermometry and geochronology in sedimentary sequences, it is also necessary to understand its progression during the course of sediment burial (e.g., Środoń 1999). Two basic models for illitization are presently available: the older one describes a solid-state transformation with a progressive fixation of K in smectite interlayers (Hower et al. 1976), and a more recent dissolution–precipitation model postulates dissolution of smectite layers, followed by nucleation and growth of illite crystals consisting of layers bound permanently by fixed K (Nadeau et al. 1984, Inoue et al. 1987). From a strict isotopic point of view, the radiogenic and the stable isotope systems of illite will behave differently depending on whether the process is a solid-state transformation or a dissolution–crystallization; it is therefore of prime importance to distinguish between the two potential processes.

Radiogenic Isotope Geochemistry

One fundamental aspect in understanding basin-scale fluid flows is the determination of the timing of such events. Fluid flows that, for example, relate to rifting activity often induce mass transfers that are recorded by minerals such as authigenic illite. These minerals can be used to reconstruct the timing of such activities through the application of radiogenic isotope methods (e.g., Awwiller 1994, Girard and Barnes 1995, Cathelineau et al. 2004). Isotopic dating of illite-rich size fractions has been practiced for many years using parent–daughter pairs such as $^{87}Rb/^{87}Sr$, $^{40}K/^{40}Ar$, $^{39}Ar/^{40}Ar$, and $^{147}Sm/^{144}Nd$. The principles and general applications of these methods and the techniques for sample preparation and mineral characterization are available in a variety of books (e.g., Faure and Mensing 2005). More specific aspects of isotope dating of illite and other clay minerals—in environments from continental deposits to metamorphosed sedimentary rocks, in which interactions occurred between migrating fluids and host rocks—can be found in Clauer and Chaudhuri's (1995 [and references therein]) review. The basic assumptions and the application of isotopic dating methods to illite authigenesis were also evaluated and discussed, for example, by Liewig et al. (1987), Lee et al. (1989), Hamilton et al. (1989, 1992), and Clauer et al. (1992).

Stable Isotope Geochemistry

Our understanding of large-scale fluid flows through sedimentary and metasedimentary sequences has often been hampered by the lack of understanding of the magnitude of the fluid fluxes and of the interactions with the host rocks, in addition to the poor time constraints (e.g., Halliday et al. 1991, Person et al. 1996). When applied to clay-type material, stable isotope geochemistry generally covers two broad fields: (1) geothermometry, in which the isotope fractionation of oxygen or hydrogen of the analyzed mineral is related to its crystallization temperature, and (2) tracing of fluid–rock interactions, in which the isotope composition of a rock or a fluid is used to evaluate the sources of the elements, the extent of the interactions, or the ratio of the interacting fluid and rock volumes. It is also appropriate to note here that the whole isotopic approach of illitization is based on the fact that cogenetic minerals and fluids were in isotopic equilibrium at the time of crystallization (e.g., Sheppard and Gilg 1996). This is probably the case for direct precipitation of illite particles, especially from migrating hydrothermal fluids, but it may not be true for a progressive burial-induced illitization of preexisting clay particles, especially when they occur in shales with low porosity and permeability. In this case, the situation can best be addressed by a consistent evaluation of the data, considering that the same process provides the same types of minerals with the same type of isotopic signatures.

It is useful to have some knowledge of the $\delta^{18}O$ and δD values of the major subsurface water reservoirs (Fig. 4) in order to estimate the

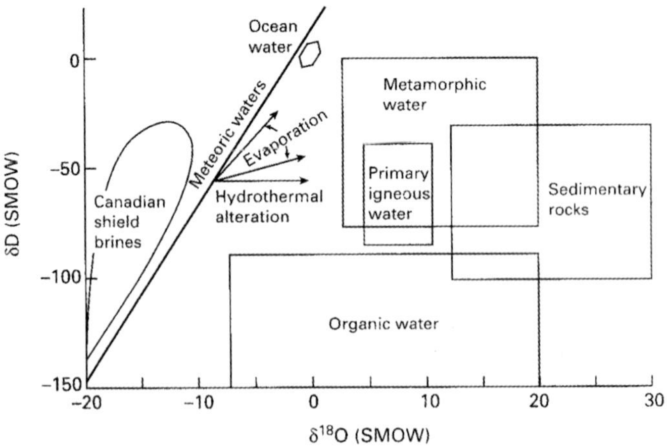

FIG. 4.—Plot of δD vs. $\delta^{18}O$ values for characteristic waters of different environments. The fluids from sedimentary rocks are significantly enriched in $\delta^{18}O$ relative to the meteoric waters. Hydrothermal fluids are characterized by constant δD when the $\delta^{18}O$ increases. Organic waters stand for interstitial fluids having interacted with organic matter (after Sheppard 1986).

extent of the interactions between fluids and minerals. For example, as can be seen in Figure 4, hydrothermal waters with compositions that differ from those of meteoric waters may have almost-constant δD values because the hydrogen fluid:rock ratio is often so high that the hydrogen isotopic exchange induced by the fluid–rock interaction does not modify the δD composition of the fluids.

Bottinga and Javoy (1973) proposed a general temperature-related equation for oxygen fractionation between two phases in liquids, solids, and gases:

$$1000 \ln \alpha_{x-y} = a \times 10^6/T^2 + b,$$

where α is the fractionation factor between $^{18}O/^{16}O$ in the mineral (phase x) and $^{18}O/^{16}O$ in the fluid (phase y), T the temperature (in Kelvin), and a and b are empirically determined constants. The fractionation factor α is defined as a ratio or a difference between the $\delta^{18}O$ values of the mineral (s) and the liquid (l):

$$\alpha = (^{18}O/^{16}O)s/(^{18}O/^{16}O)l = (1 + \delta^{18}O_s)/(1 + \delta^{18}O_l)$$
$$\ln \alpha \approx \alpha - 1 = \delta^{18}Os - \delta^{18}O_l,$$

where $\delta^{18}O$ values are fractional values or per-mil (‰) values.

The weakness of this isotopic approach to geothermometry is that experimental studies are lacking for most minerals in the temperature domain below 350° C (e.g., Bottinga and Javoy 1975, Chiba et al. 1989) because of very slow oxygen isotopic exchange rates. Consequently the results obtained at higher temperatures were often extrapolated lower, which does not work for clay minerals, as they are only stable at temperatures below about 250 to 350° C.

The application of oxygen isotope fractionation as a tracer for paleothermometric investigations in natural environments is generally based on the determination of the $\delta^{18}O$ value, either in two mineral phases that are believed to be in isotopic equilibrium, meaning that they formed at the same time from the same fluid, or in a mineral and the coexisting fluid. For the clay minerals, the basic approach is more complicated, even if coexisting mineral phases can be identified as authigenic. First, it is often difficult to separate the clay particles from each other and from the other detrital framework minerals because of their small size and because they are often intergrown with the detrital framework grains. Second, clay particles are systematically associated with three types of water: surface adsorbed water, water held in the interlayer positions, and water that is constitutive of the mineral framework. Oxygen and hydrogen of the adsorbed and interlayer water typically exchange isotopically very rapidly with ambient water, at rates of less than a few days between 100 and 500° C. This makes the $\delta^{18}O$ and δD of these waters useless as records of paleotemperatures. The rates of the isotopic exchange between oxygen in structural positions and ambient water are generally slow, depending on the temperature and the type of clay minerals. For example, the results of a 2-year laboratory experiment on illite and kaolinite indicate that oxygen isotopic exchange between ambient water and structural water, at 100° C or less, is nearly insignificant (James and Baker 1976, O'Neil and Kharaka 1976). In the laboratory, at temperatures of approximately 200° C, the extent of isotope exchange between illite (and kaolinite) structural oxygen and its fluid is small, in the range of a few percent, while it measures 19% for the structural oxygen of smectite after about 8 months, indicating that $\delta^{18}O$ determined on illite at temperatures as high as 200° C is preserved and may reflect the initial crystallization conditions.

In contrast to oxygen, hydrogen–isotope exchange between ambient water and structural hydrogen of the same clay minerals is relatively rapid, taking a few days to weeks, at temperatures as low as 60° C. There are a great deal of data concerning the exchange rate of oxygen and hydrogen from various clay minerals in natural environments, especially in marine settings (see Savin and Lee 1988), that lead to similar conclusions, even if in the detail it can be seen that study of natural samples for the purpose of defining fractionation rates is more difficult than laboratory experiments because of the mineral heterogeneity of the studied clay fractions.

Oxygen Isotope Fractionation of Illite–Fluid Pairs: When fluids are in isotopic equilibrium with minerals during crystallization, the minerals may preserve the records of the crystallization temperature, given that no further altering interactions occurred. However, unequivocal data on isotopic equilibrium and further preservation in the low-temperature domain are difficult to assess for several reasons. First, as already mentioned, it is difficult to obtain pure authigenic illite fractions and to separate pure contemporaneous authigenic silicate grains such as quartz overgrowths from detrital grain cores. Second, if there is a good agreement between the experimental curves relating temperature and oxygen isotopic fractionation of illite (and kaolinite) with water (Fig. 5), these experimental curves diverge significantly at crystallization temperatures below 150° C, which is precisely the domain of interest here, meaning, in turn, that the interpretations will be less accurate.

Because the experimental evidence indicates that isotope exchange between the structural oxygen of clay minerals and ambient water is extremely slow in buried sediments, the isotope composition of the structural oxygen may be considered to reflect the crystallization conditions of the minerals. In this case, the $\delta^{18}O$ values of authigenic illite can be used to calculate the composition of the ambient fluids and/or the crystallization temperature of the mineral fractions, depending on which of these two variables either is known or can be reasonably postulated.

Oxygen Isotope Fractionation from Illite-Bond Calculations: Because of the mentioned difficulties in applying oxygen isotope

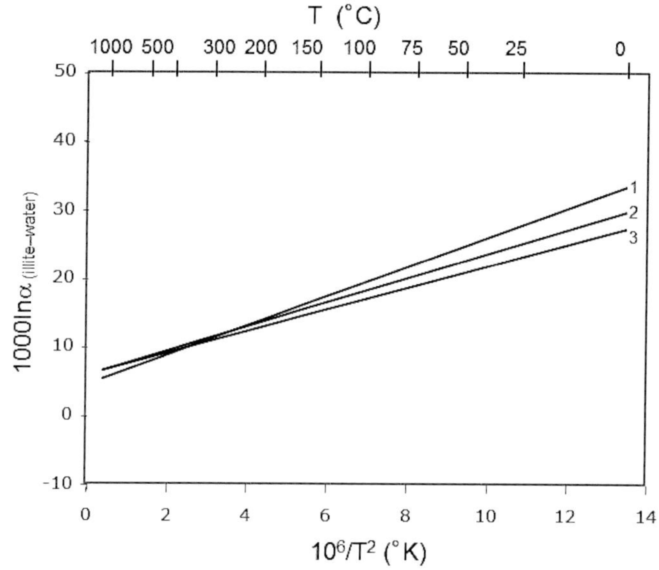

Fig. 5.—Estimates of oxygen isotope fractionation between illite and water. Curve 1 is from the bond-type approach (Savin and Lee 1988), curve 2 from a combined bond-type approach and analytical determinations (Lee 1984), and curve 3 from previously published analytical data (Eslinger and Savin 1973). α is the fractionation factor with the $\delta^{18}O$ of the water subtracted from that of illite. Additional explanations are in the text (figure after Savin and Lee 1988).

geothermometry to illite-mineral and illite-fluid pairs, another approach was explored for a more consistent evaluation of the oxygen isotope fractionation factor of illite and other clay minerals. This method utilizes empirical "bond-type" calculations (Savin and Lee 1988, Zheng 1993) and is based on the assumption that oxygen of any chemical bond in a mineral structure is characterized by a certain value of isotopic fractionation, regardless of the type of mineral. Taylor and Epstein (1962) assumed that the oxygen isotopic fractionation of a mineral is simply the weighted sum of the fractionation factors of the bonds they contain, and they reported good approximations to the measured oxygen isotope fractionation factors for high-temperature minerals. Refinements of this concept were published by Savin (1967) and Savin and Lee (1988) for low-temperature phyllosilicates of interest here. Although this approach to the isotope fractionation of minerals is not systematically very accurate, it provides values that agree reasonably with other calculations or extrapolations from experiments, as, for example, in the case of the kaolinite–water fractionation (Fig. 6). In other cases, as shown for the smectite–water pair, the bond-type approach gives fractionation curves that provide temperatures differing from each other by as much as 50° C (Fig. 7). For the purpose of evaluating crystallization temperatures between water and I/S, the bond-type oxygen fractionation curves are close to those reported for other methods, indicating that the approach is as reliable as the others (Savin and Lee 1988). Its use is reasonably constrained, provided the $\delta^{18}O$ of the fluids that interact with the nucleating and growing particles is known, although this is not always the case.

Thus, if the $\delta^{18}O$ differences between oxygen in the hydroxyl sites and nonhydroxyl sites of clay minerals can be measured experimentally, they have the potential to be used as single mineral thermometers (Hamza and Epstein 1980, Ustinov and Grinenko 1985, Girard and Savin 1996). Bechtel and Hoernes (1990) demonstrated such use of oxygen isotope fractionation for low-temperature geothermometry. Later, Delgado and Reyes (1996) suggested a combination of the δD and $\delta^{18}O$ of clay minerals for determination of the crystallization temperature based on the stable isotope values of the theoretically

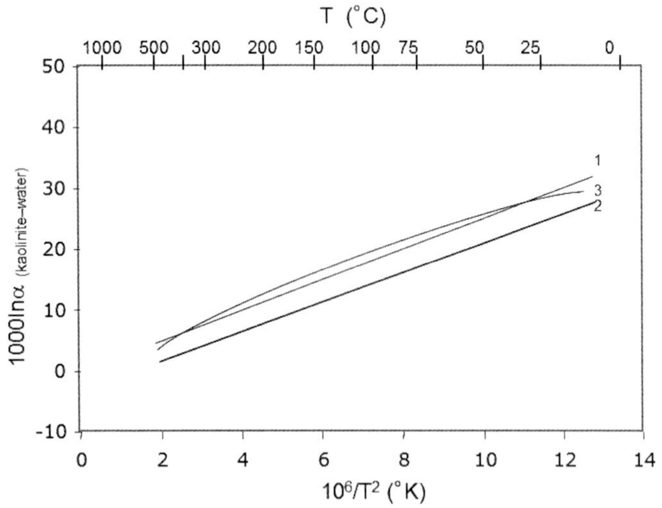

FIG. 6.—Oxygen isotope fractionation between kaolinite and water from bond-type calculations of interactions (curve 3; Savin and Lee 1988) and experimentally determined values (curve 1: Land and Dutton 1978; curve 2: Kulla and Anderson 1978) as a function of temperature. α is the fractionation factor with the $\delta^{18}O$ of the water subtracted from that of kaolinite.

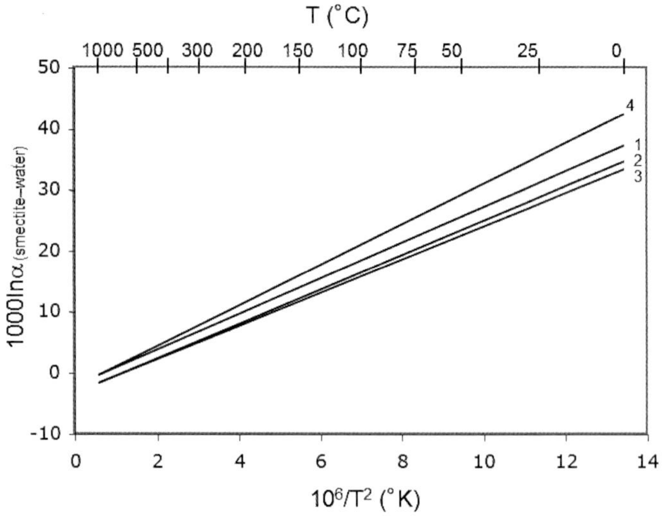

FIG. 7.—Oxygen isotope fractionations between smectite and water as a function of temperature. Curves 1, 2, and 3 were calculated using the bond-type approach (Savin and Lee 1988); curve 4, significantly above the three others, is from experimental determination. α is the fractionation factor with the $\delta^{18}O$ of the water subtracted from that of smectite.

contemporaneous meteoric water, as shown for smectite (Fig. 8). Again, application of the method depends on the isotopic equilibrium of both oxygen and hydrogen with the reacting fluid at the same temperature.

The Effects of Water–Mineral Volume Ratio on Isotopic Equilibrium: The oxygen and hydrogen isotope fractionation between a mineral and the fluid from which it precipitated indicates equilibrium only if the temperature at which the mineral crystallized has not subsequently changed. A further temperature increase induces a change in this equilibrium by additional authigenesis or recrystallization of already–newly formed clay minerals in the rock volume. Also, if the mass of water exchanging $^{18}O/^{16}O$ with the crystallizing mineral is large, as in the case of a tectonically induced continuous volume flow, the isotopic composition of the minerals at equilibrium will be in accordance with the isotopic composition of the solutions, reflecting a nearly open-system behavior. If the water–mineral mass ratio is low, the equilibrium values may differ considerably from the initial, as is discussed in this section. The oxygen and hydrogen isotopic trends of pore waters of sedimentary sequences may plot on lines with slopes somewhat less steep than that of the Meteoric Water Line, as reported for the Appalachian and Michigan Basins (Fig. 9). These lines were interpreted as resulting from some evaporation, as is usually the case for waters deviating similarly from the Meteoric Water Line in a δD vs. $\delta^{18}O$ graph. In the case of the Paleozoic sediments of the Ottawa Embayment of the Appalachian Basin (Ziegler and Longstaffe 2000), the isotopic composition of illite indicates an isotopic equilibrium at 40 to 60° C with waters that isotope composition falls around the Meteoric Water Line and the line of the Appalachian Basin. A higher equilibrium temperature would make the interstitial water heavier in ^{18}O. Because the K-Ar ages of the clay material are significantly younger than the stratigraphic age, indicating a late authigenic nature (Ziegler and Longstaffe 2000), it is conceivable that the oxygen and hydrogen isotope composition of the pore waters may have been modified by late interactions with the crystallizing clay

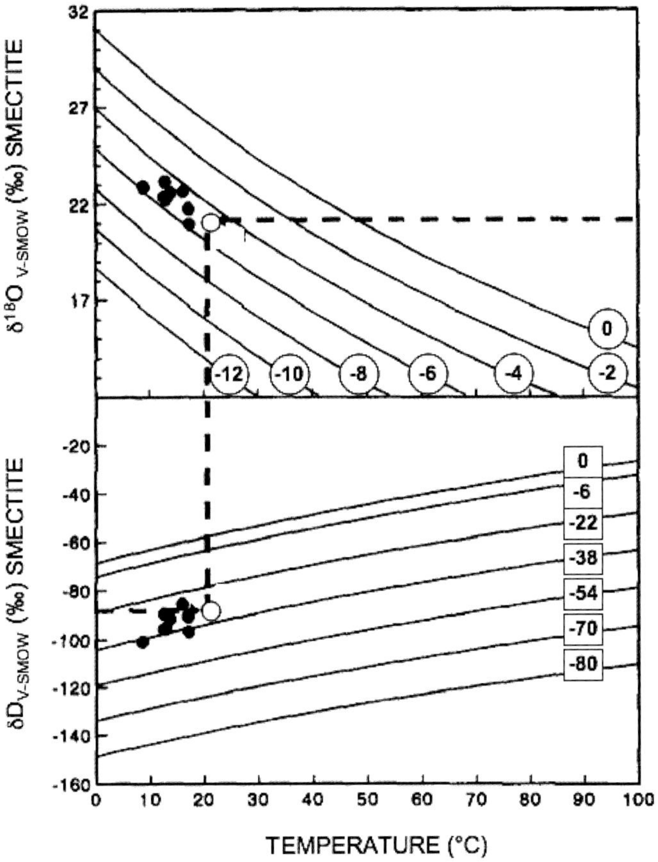

FIG. 8.—Plots of the $\delta^{18}O$ and δD values of smectite in equilibrium with present-day meteoric water as a function of temperature. The dashed lines show the Meteoric Water Line (MWL) for oxygen (a) and deuterium (b). The curves represent constant δ values for the water with the $\delta^{18}O$ given in the circles at the bottom of diagram (a), and the δD values in the boxes at the right of diagram (b) for each curve in both diagrams. The small circles correspond to data of Delgado and Reyes (1996) and the last circle on the right is an example: for a measured $\delta^{18}O$ at 21‰ and a measured δD of −85‰ for a smectite, there is only one corresponding water temperature of 21° C that falls on the MWL at −5‰ for $\delta^{18}O$ and −30‰ for δD (after Delgado and Reyes 1996).

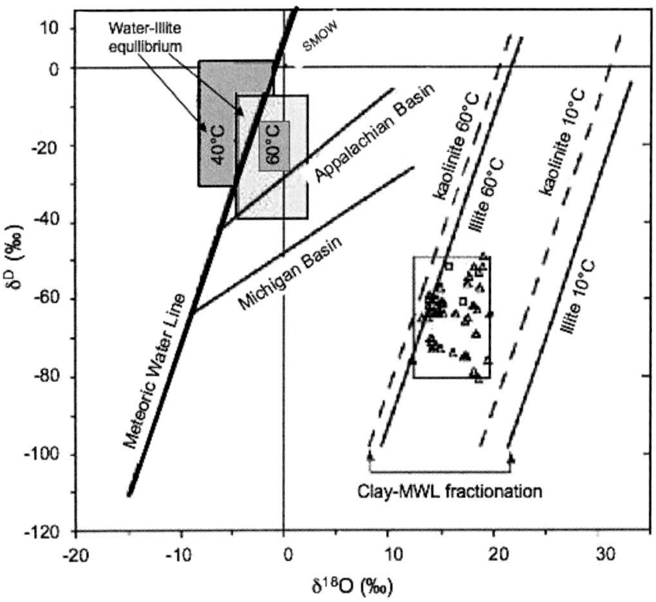

FIG. 9.—Plots of the Meteoric Water Line (MWL) and the general trends of the $\delta^{18}O$ and δD of subsurface waters in the Michigan and Appalachian basins. The calculated equilibrium δD and $\delta^{18}O$ of the illite-water and kaolinite-water pairs at 10 and 60° C are shown relative to the MWL. The isotopic compositions of Paleozoic illite (triangles in the rectangle) and kaolinite (squares in the rectangle) from the Ottawa Embayment of the Appalachian Basin plot in between the lines for the two pairs at 10 and 60° C. All δD and $\delta^{18}O$ of the clays would be in isotopic exchange equilibrium with waters of composition shown by the shaded rectangles at 40 and 60° C (modified from Ziegler and Longstaffe 2000).

material at a limited water–rock ratio, modifying the initial δD and $\delta^{18}O$ of the formation water.

The $^{18}O/^{16}O$ fractionation between pore-water and co-existing illite in equilibrium is determined by the temperature-dependent isotope fractionation factor between the two phases, the initial isotopic composition of each, and their masses. This is a straightforward problem of partitioning a component (^{18}O) between two phases, the steps for which are given in the Appendix. From consideration of the pore-water composition, its dissolved solid contents, and the mass proportions of the solution and newly forming illite, it is unlikely that an illite of a stoichiometric composition, such as, for example, $K_{0.6}Al_{2.2}Si_{3.7}O_{10}(OH)_2$, would precipitate as a congruent phase from single ions or complexes in solution without relicts of previous mineral fragments. Instead, it is more likely that its constituent species are derived, at least in part, from preexisting smectite or I/S and/or from dissolution of components such as micas, in which case the initial illite inherits some of the isotopic composition of the detrital silicate phase and continues to act toward an isotopic equilibrium with the solution. A simple interpretation of the water–rock volume ratio (V_l/V_s) is the ratio of the interconnected porosity (φ, fraction of the bulk sediment volume) to the solid fraction 1 − φ, also known as the void ratio φ/(1 − φ). However, if newly formed illite accounts for only a small fraction of the sediment solids, then the water–rock ratio of interest may be considerably larger than V_l/V_s for the bulk sediment. Another point here is that the amount of oxygen to consider and the direction of isotopic shifting for the oxygen-containing fluids and minerals, as they approach mutual equilibrium, are not related to the total mass of oxygen in the rock, but only to that of the oxygen in the reacting minerals.

The calculated trends of the $\delta^{18}O$ of illite ($\delta^{18}O_{illite}$) and water ($\delta^{18}O_{liq}$) for different water–mineral volume ratios, from 0.05 to 10 at the two temperatures of 60 and 150° C, are shown in Figure 10. This temperature range corresponds to the depths at which illite forms in some of the sedimentary basins discussed later. The $^{18}O/^{16}O$ exchange is shown between $H_2O_{(liq)}$ and two oxygen atoms in $(OH)_2$ in the stoichiometric formula of illite, as given above (Fig. 10A), and for all 12 O of the stoichiometric formula (Fig. 10B). Typical initial values of $\delta^{18}O_{liq}$ based on data from the Appalachian Basin (Fig. 9) are in the range of −6 to 0‰, and typical $\delta^{18}O$ values of micas in plutonic rocks range from +2 to +9‰. For the study case, the initial values taken are $\delta^{18}O_{liq} = -3‰$ and $\delta^{18}O_{illite} = +9‰$; thus, the liquid is isotopically

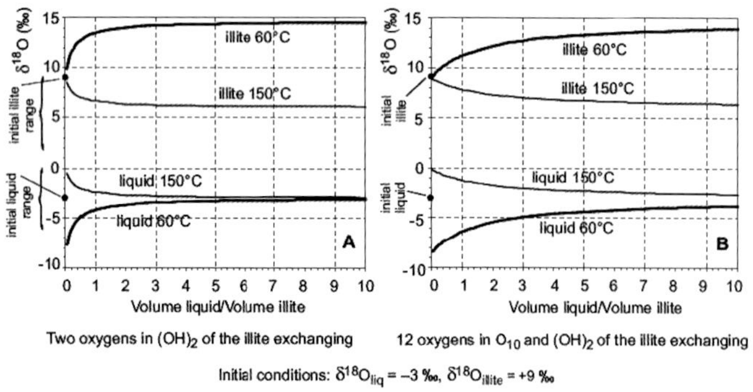

FIG. 10.—Calculated isotopic composition ($\delta^{18}O$) of water and illite, initially not at equilibrium, after attainment of isotope exchange equilibrium at 60 and 150° C, as a function of the water–mineral volume ratio in the sediment. (A) Isotope exchange between $^{18}O/^{16}O$ of H_2O in solution with 2 oxygens in $(OH)_2$ in illite. (B) Isotope exchange between water and 12 oxygens in the illite stoichiometric formula. The mathematical treatment is developed in the Appendix.

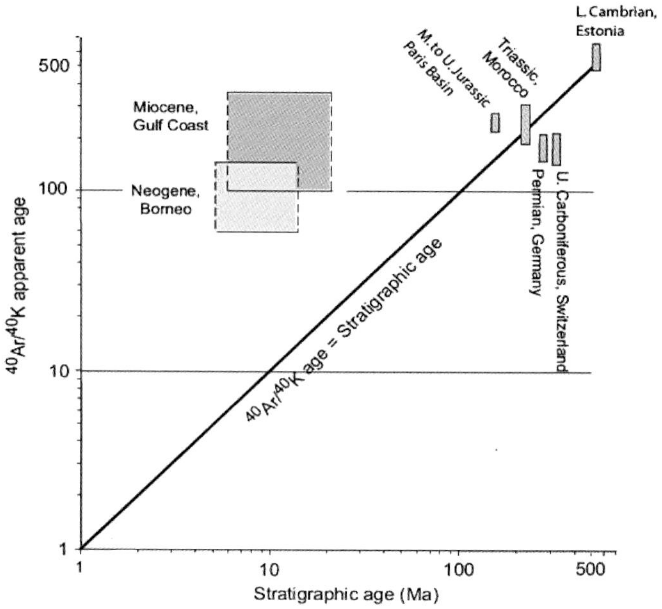

FIG. 11.—Stratigraphic and K-Ar age of illite and various illite–smectite mixed layers from identified sedimentary sequences. Note that the two age scales are logarithmic. Minerals to the left of the diagonal line of concordant stratigraphic and isotopic ages have a predominant predepositional history (the apparent ages are older than the depositional ages), and those on the right of the concordance line have a predominantly postdepositional history (the apparent ages are younger than the depositional ages) (from Lerman et al. 2007).

lighter than the illite, and the two phases are not at equilibrium. The direction of change, either positive or negative, in the isotopic composition of the liquid and illite at low water–mineral volume ratios, from an initial nonequilibrium ratio $(1+0.009)/(1-0.003)=1.012$ or $9-(-3)=12‰$, depends on the temperature and the value of the fractionation factor α. At a higher α (=1.0177) at 60° C, the solution loses ^{18}O to illite; at a lower α (=1.0091) at 150° C, the solution gains ^{18}O from illite (see Appendix). At high water–mineral volume ratios, the isotopic composition of the solution exerts the dominant effect by driving the illite to isotopic equilibrium. The higher water–mineral volume ratios represent circulating solutions in the sediment pore space, and the nearly constant values of the isotopic composition of the solution and illite can be interpreted as indicating that the exchange reaction is controlled by the fluid flow. One conclusion that can be drawn from the relationships shown in Figure 10 is that in the case of authigenic illite being in isotopic composition with its corresponding interstitial fluid when crystallizing, the respective values are significantly different depending on the water–mineral ratio. The hypothesis of a small water–mineral ratio does not apply when large volumes of hydrothermal solutions are circulating. As oxygen is held in two different sites of the illite mineral structure, the hydroxyl and nonhydroxyl sites, it is possible that the extent of oxygen exchange between the two sites, depends on either one not exchanging, or one exchanging more than the other. To the best of our knowledge, no studies are available yet with which to check this hypothesis.

ILLITIZATION: WHEN AND AT WHAT TEMPERATURE?

A major question motivating the isotopic dating of illitization has to do with determining whether illitization is related to a continuous change during burial and progressive heating or to episodic hydrothermal events associated with sudden temperature increases. Lerman et al. (2007) showed that K-Ar ages of illite particles are either similar or younger than the stratigraphic age of the host rocks in old sedimentary basins, being significantly older than the stratigraphic age of the rocks in younger basins (Fig. 11). This distinction indicates that the ICI of illite from sedimentary rocks of old basins typically reflects a greater degree of diagenetic alteration than does ICI for illite from sedimentary rocks in younger basins. In other words, old sedimentary sequences record more complex thermal histories than are typically found in younger basins, and the complex evolution of the former might be more difficult to decipher.

In the following section we describe results from some extensively studied sedimentary basins, in which illitization has been well described in deeply buried stratigraphic sections: at 4200 m for the Brent of the North Sea (Glasmann et al. 1989), at 4750 m for the Permian of Germany (Zwingmann et al. 1999), at 5500 m for the Miocene of the Gulf Coast (Hower et al. 1976), and at 4000 m for the Neogene of Borneo (Clauer et al. 1999). The Triassic of the Paris Basin was less buried (Clauer et al. 1996a), at 2700 m, and the Lower Cambrian of Estonia was even less buried, between 800 and 1200 m (Chaudhuri et al. 1999). It is clear from mineralogical results reported in these studies that illite is better crystallized in older sediments than in younger sediments. It is also significant that depth and the associated temperature increase do not necessarily determine the extent of illitization. For example, illitization is more pronounced in the Triassic sandstones of the Paris Basin than in the Neogene sandstones of the Mahakam Basin of Borneo or the Miocene shales of the Gulf Coast, although the latter two are buried deeper. Three case studies are discussed in greater detail in the following sections as examples of different illitization processes and, consequently, of the different thermal histories of the studied basins, on the basis of illite isotope data. These sedimentary sequences were selected because their previous studies combined varied approaches that are adequate to address and discuss the question of how to distinguish burial- from hydrothermal-induced illitization.

The Thermal History of the Permian Sandstones in Northern Germany

The southern Permian gas province that extends from the southern North Sea to Poland includes the northern German Basin. The lower Rotliegend (Permian) sedimentary sequences of this German Basin consist of continental clastics with important gas reservoirs. Early petrographic and mineralogical investigations indicate that diagenesis significantly affected their reservoir qualities (e.g., Glennie et al. 1978). Illitization was investigated in several studies (e.g., Drong 1979, Gaupp et al. 1993, Platt 1993) that showed that illite has a filamentous or platy shape and that these pore-bridging particles have a deleterious effect on the rock permeability.

Different size fractions of illite from Rotliegend sandstones were collected from a series of wells located in NW–SE-trending horst and graben structures to the East of Bremen, Germany. The authigenic origin of illite was confirmed by the fact that the illite particles have straight idiomorphic edges, the K-Ar ages of which range from 210 to 165 Ma, most being between 200 and 175 Ma (Zwingmann et al. 1999). These K-Ar ages are significantly older than the age of the maximum burial depth at approximately 5 Ma (Fig. 12). It is clear from this that illitization cannot result solely from burial, because if it had done so, at least some of the measured illite K-Ar ages should significantly shift toward the 5-Ma limit. From what is known about the behavior of the K-Ar system in clays minerals (e.g., see discussion in Clauer and Chaudhuri 1995), the illitization described here had to occur at a temperature above that reached at maximum burial, as any temperature above crystallization temperature affects the K-Ar system of authigenic illite. Therefore, the K-Ar system recorded an illite crystallization at 210 to 165 Ma that stayed unaffected only because the burial-induced temperature from 3000 to 3500 m in crystallization depth to the 4700-m maximum depth did not increase above that reached 210 to 165 Ma ago. In addition, if illitization were affected by burial, the K-Ar ages would not fit in a narrow range during Early Jurassic time, but instead would be scattered toward the age of maximum burial. In other words, the coherency in the K-Ar ages can only be explained by the fact that the K-Ar isotopic system became closed after illite crystallization, which implies in turn that crystallization necessarily occurred at a temperature higher than the present-day down-hole 165° C measured directly in the wells.

A postcrystallization increase in temperature of a mineral promotes either an alteration of the mineral, which can result in a preferential escape of such radiogenic isotopes as ^{40}Ar (e.g., Hunziker et al. 1986 for varied illite fractions), or sometimes even recrystallization of the previous generation of authigenic clay minerals. Backstripping of the depositional evolution of the North German basin (Fig. 12) shows that the burial history of the Rotliegend sandstones was complex, with varying sedimentation rates, some of which were very high (as was the case at the end of the Cretaceous time), and also a long sedimentation inversion during the second half of the Jurassic. In this respect, fluid-inclusion microthermometric determinations on quartz overgrowths gave paleotemperatures in the range of 175 to 235° C (Rieken 1988); these temperatures are higher than the maximum temperature reached at maximum burial. Illitization was, therefore, probably induced by migrating, tectonically driven fluids, when the sandstones were buried between 3000 and 3500 m, long before maximum burial, which also allows estimation of the geothermal gradient and therefore of the burial-related temperature. This is consistent with the subsidence history of the whole region, which is characterized by a long period of tectonic quiescence and by a steady subsidence punctuated by distinct periods of tectonic and volcanic activities (Berthelsen 1992).

Liewig and Clauer (2000) pursued the same issue, albeit on a smaller

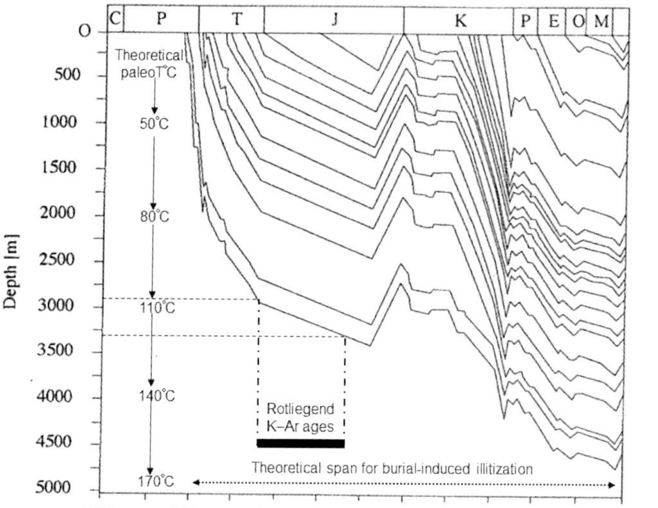

FIG. 12.—Decompacted burial of the whole sedimentary sequence of Northern Germany relative to time and depth. The Rotliegend sandstones are the first sediments on the basement with a thickness corresponding to the width of the first layer on the diagram. The burial history appears very complex as the result of an active tectonic activity in the region, with a maximum burial depth reached only very recently, to the lower right side of the diagram. The K-Ar ages of the illite-rich fractions are also plotted between 165 and 210 Ma, at a period during which burial was between 3000 and 3300 m (after Zwingmann et al. 1999).

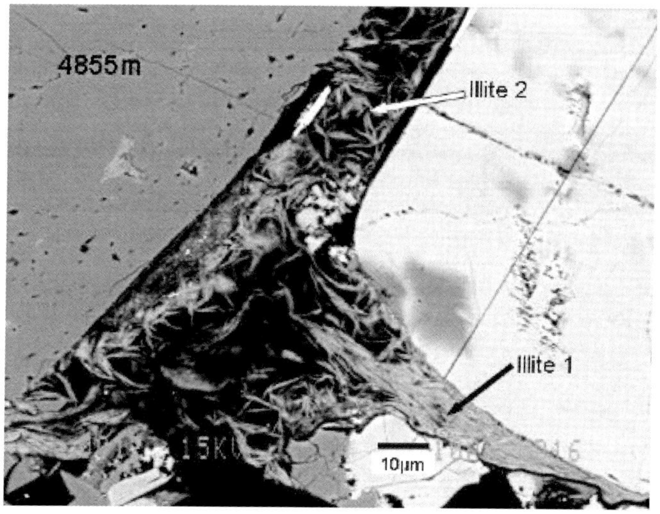

FIG. 13.—Scanning electron microscope picture of a thin section in a Rotliegend sandstone buried at 4855 m. The mineral grains are in white or pale gray, and the pore system is in black. Illite occurs either as particles that are tangential to the quartz overgrowths, labeled "*Illite 1*" to the right lower side of the photo, or as radial filaments growing into the open pore space and labeled "*Illite 2*." For more explanation, see the text. The black bar is the scale (after Liewig and Clauer 2000).

scale, by comparing isotopic results with detailed shapes of illite in the pore space of the same sandstones. The SEM observations on thin sections showed two forms of authigenic particles with straight edges, one occurring as laths tangential to the quartz overgrowths in the pore system and labeled "illite 1" in Figure 13 and the second occurring as radial filaments in the same pore system, labeled "illite 2." The former yield K-Ar ages of approximately 200 Ma and an $\delta^{18}O$ value of about +18‰, and the latter yield K-Ar ages of approximately 180 Ma and an $\delta^{18}O$ value ranging from +15 to +16‰. The two types of illite can be interpreted as forming either during one single event that extended over 20 Ma or during two distinct events, at ~200 and 180 Ma, at similar temperatures but different water–rock ratios. The difference in $\delta^{18}O$ for the two illite generations can be related to different crystallization temperatures or to an increase in the water–rock ratio during crystallization. The appearance of the illite particles in the pore system of the rocks is significant: a tangential growth along the pore walls at 200 Ma is indicative of a restricted pore system during a compressional tectonic activity, while radial growth of illite particles is consistent with higher water–rock ratios during the 180 Ma event in a more open pore system during an extensional tectonic activity. Although the two types of illite particles are purposely shown in the same MEB picture (Fig. 13), they often occur separately depending on the location of the samples in the regional setting. The tangential shape was mainly described and studied in sandstones taken close to the well-identified horst structures (Zwingmann et al. 1999), while the radial shape was mainly identified in rocks farther from the faults. This two-step interpretation of the data and observations is preferred because it is difficult to conceive a continuous migration of hot hydrothermal fluids into varied strata of a sedimentary sequence, driven by a continuous, 20-Ma lasting pulse of tectonic activity.

The Thermal History of Upper Triassic Sandstones in the Paris Basin

Mossmann et al. (1992) and Clauer et al. (1995) studied the Rhaetian clay minerals of the Lower Triassic sandstones from the deeper central section of the Paris Basin. The sediments consist of sandstones and sandy-to-silty beds deposited in tidal flats and buried at varied depths, now occurring from 2700 m deep in the southern central basin to outcrops in the eastern basin. Similar to the North German Basin, the burial history of the Lower Triassic sandstones of the Paris Basin was complex, with varying subsidence rates: an extremely high rate during the Upper Jurassic, another one during the Lower Jurassic, and an erosion period in the middle of the Cretaceous. The clay fraction in the sandstones consists mainly of kaolinite, chlorite, illite, I/S, and smectite in various proportions, with kaolinite dominant in the coarser size fractions and I/S in the finer fractions. Three types and generations of illite and I/S assemblages from the small size fractions of the sandstones were identified and dated. The two older generations of illite crystallized at about 190 and 150 Ma; they were considered to have been induced by fluids released by tectonic adjustments of the basement, an interpretation that agrees with the occurrence of structural features described in the basement of the basin (Brunet and Le Pichon 1982, Curnelle and Dubois 1986). These basement adjustments probably induced the dramatic changes in the sedimentation rate about 150 to 160 Ma ago. The youngest illite generation of about 80 Ma could have resulted from burial diagenesis at an expected lower temperature. Illitization described in the sandy units was extensively discussed in Clauer and Chaudhuri (1995), Spötl et al. (1996), and Clauer et al. (1996a), especially with respect to its timing and geothermometric application.

Clauer et al. (1995) reported distinctly different correlations between K-Ar and $\delta^{18}O$ data of the illite-rich material in two populations of samples, one of the samples collected from a fault zone and one from

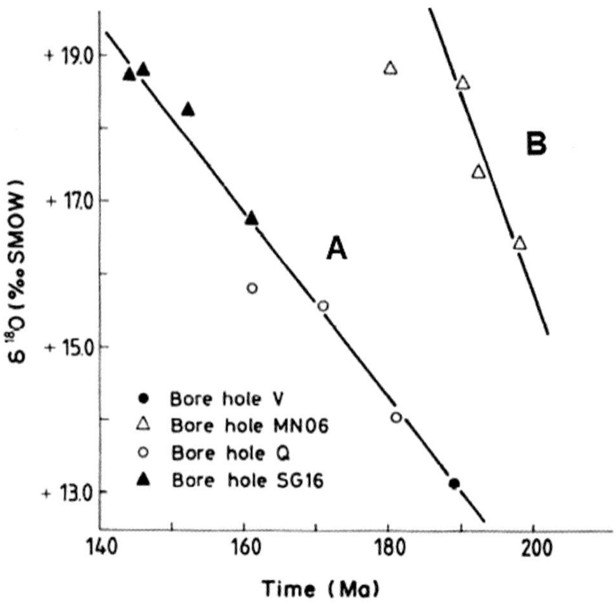

FIG. 14.—Relationship between K-Ar ages and $\delta^{18}O$ values of <0.4-μm illite–smectite mixed layers in Triassic sandstones from Paris Basin. The line A to the left of the diagram corresponds to illite-type minerals containing more and more illite–smectite mixed layers when moving to the left, with decreasing K-Ar ages and increasing $\delta^{18}O$. The line B to the right in the diagram that fits through the triangles corresponds to size fractions extracted from a sandstone sample visually cut by a fault. The K-Ar ages are closed to that of the black circle from the former line, but with significantly higher $\delta^{18}O$. For further explanations, see the text (from Clauer et al. 1995).

undisturbed rocks located tens of meters away from the fault. The size fractions from undisturbed rocks show an increase in $\delta^{18}O$ from 13.0 to 19.0‰ as the K-Ar ages decrease from 190 to 145 Ma (line A in Fig. 14). The size fractions of the samples from the fault (line B in Fig. 15) yield K-Ar ages that are similar to the oldest ages of illite from samples collected away from fault, but samples from the fault have significantly heavier $\delta^{18}O$, from +16 to +18.5‰, as compared to about +13 to 14‰ for the undisturbed samples. If the undisturbed rocks were characterized by a constant water–mineral ratio and a constant $\delta^{18}O$ for the interacting fluids, then the first correlation (line A) indicates that the younger illite formed at lower temperatures, as shown in Figure 7. In the case of the illite from the fault (line B), it is reasonable to consider that it crystallized in an environment characterized by a higher water–rock ratio than that of the undisturbed matrix. In addition, the formation temperature of the fault fractions could have been different or similar to that of the illite-rich fractions from nearby tectonically undisturbed matrix. In sum, the illite crystallization temperature, the $\delta^{18}O$ of the fluids, and the water–rock ratio were different in the fault than in the undisturbed matrix, but it is difficult to evaluate the impact of each on the illite $\delta^{18}O$.

The oldest illite ages were obtained from particles similar in shape to illite crystals growing from authigenic quartz. Clauer et al. (1995) suggested, based on the $\delta^{18}O$ of both the 190-Ma-old illite (line A in Fig. 14) and the quartz overgrowth, that these minerals crystallized at 220 to 250° C about 190 Ma ago. As the $\delta^{18}O$ of the quartz overgrowths was calculated from mixtures with varying proportions of detrital and authigenic quartz, the uncertainty for the temperature

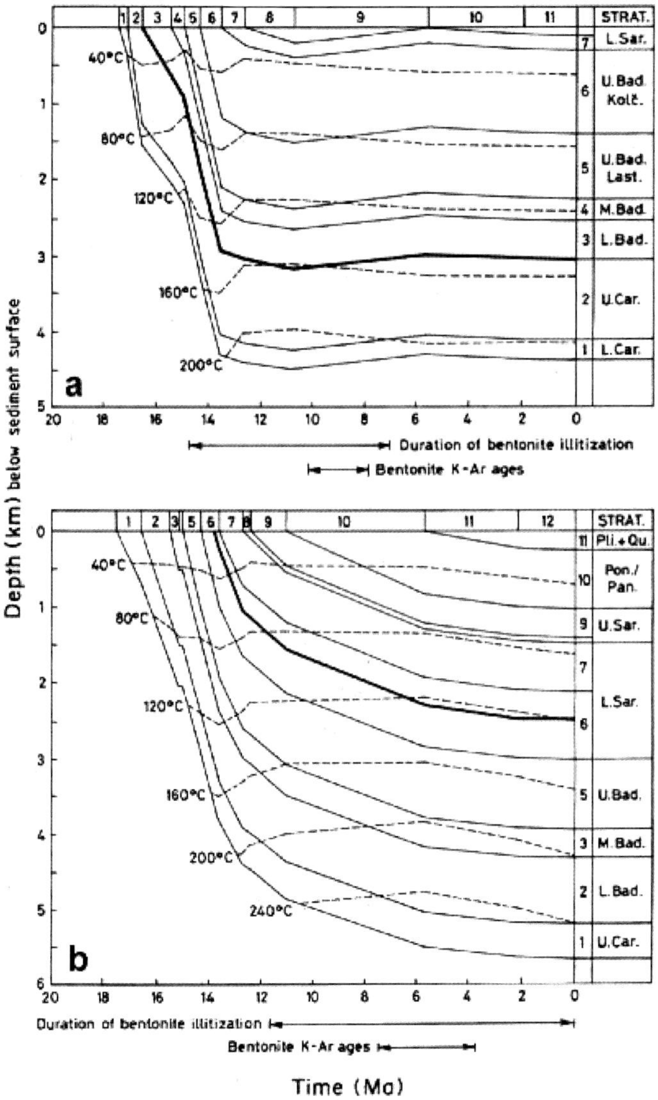

FIG. 15.—Burial history diagram for the East Slovak sedimentary sequence extrapolated from results of cores from a drill hole, with the studied bentonite beds in bold black. Time is plotted on the horizontal axis (in Ma) and depth on the vertical axis (in km). To the right of each diagram are the sample location and the stratigraphic sequence. The temperature isotherms are from organic maturation; they are discussed in the text. (A) The Trhovište sample; (B) the Čičarovče sample. Under each diagram are estimates of the duration of illitization in the bentonite unit and the hosting shales as well as the span of the K-Ar ages. For more explanations, see text (after Clauer et al. 1997).

estimate is higher than if the $\delta^{18}O$ were measured on pure quartz overgrowths. Regardless of possible uncertainties in the age dating and the $\delta^{18}O$ of the pure quartz overgrowths, the interpretation of an abnormal thermal event in the Paris Basin during the Liassic is consistent with the observations of many such events in other West European Triassic strata (Clauer et al. 1996b).

Determination of paleotemperatures not combined with independent geothermometers may provide an incomplete or erroneous reconstruction of the thermal history of a sedimentary basin, exemplified by controversial interpretations for the Triassic sandstone beds from Paris Basin that clearly depend on the used data set. Three publications on Triassic-to-Jurassic rocks of the Paris Basin (Guilhaumou and Gaulier 1991, Guilhaumou 1993, Demars and Pagel 1994) reported homogenization temperatures from 90 to 160° C for fluid inclusions in carbonate and quartz authigenic cements, after pressure correction, that appear, at first glance, to be inconsistent with the above calculated temperatures of 220 to 250° C by illite-quartz geothermometry. This discrepancy indicates that the higher temperatures calculated from the oxygen isotope fractionation factor between the authigenic pair illite–quartz could have resulted from local and/or short-term thermal activity. The temperature difference becomes much less inconsistent if one considers that three generations of illite nucleated and grew at about 190, 150, and 80 Ma at progressively decreasing temperatures from oldest to youngest, as discussed in the case of the Rotliegend sandstones in Northern Germany. As the authors of the quartz and calcite fluid inclusion studies never established any link between these minerals and the three generations of illite that formed nearby, the fluid inclusion paleotemperatures from 90 to 160° C could, in fact, correspond to any of the two younger illite crystallization episodes.

An annealing fission-track temperature of about 135° C in detrital apatite of the Upper Triassic from 90 to 50 Ma (Pagel and Meyer 1991) provides an additional temperature constraint. This age range is inconsistent with a model for a maximum temperature increase in the rocks resulting from burial only, because the maximum paleotemperatures obtained from fluid inclusion determinations do not match timely with those from the fission-track annealing of apatite. Backstripping of the sedimentary strata of the Paris Basin places the maximum burial about 30 Ma ago, not between 90 and 50 Ma ago. A simple calculation based on this backstripping also shows that the maximum temperature associated with burial would have been approximately 115° C at the most for a geothermal gradient of 35° C/km and 140° C for a slightly higher gradient of about 45° C/km. Backstripping also identifies dramatic increases in the sedimentation rates at about 190 and 150 Ma (Clauer et al. 1995) resulting from adjustments in the basement, as reported above, that released basement hot fluids into the overlying sedimentary sequence, with a significant impact on illitization.

Another indirect argument in favor of an episodic illitization in the Triassic sandstones of the Paris Basin is provided by Elliott and Matisoff (1996), who presented different models describing kinetic rates of a burial-induced transformation of smectite into illite. They found that individual models fit data from the Denver Basin, the Gulf Coast, and the Salton Sea system, but not those from the Paris Basin. The calculated age measurements that they found for the deepest buried I/S range from 99 to 78 Ma depending on the model used, which is precisely the K-Ar age range determined for the latest generation of I/S considered to be burial induced by Clauer et al. (1995).

In summary, isotopic dating of size fractions consisting of authigenic illite, particularly in conjunction with other paleothermometric techniques and burial history modeling, points to the fact that in ancient sedimentary basins, illitization can relate to tectonic activities that may induce migrations of hot fluids in porous sedimentary lithologies, rather than to burial alone. The recorded paleotemperatures calculated directly from illite by oxygen isotope determinations, or by any other method, need to be integrated and compared to results based on other independent methods, as they do not necessarily record a burial temperature. The extent and temperature of illitization within a rock volume need to be carefully evaluated, as local and episodic events could have occurred either before or after maximum burial, locally or regionally modifying significantly the history of any given sedimentary basin. Maximum paleotemperature is an important parameter in the reconstruction of the thermal history of a sedimentary basin, but it is not the only one: knowing when this maximum temperature occurred is also an essential part of the interpretation.

Another point to raise here is that the first step in applying a simple burial model is to postulate that illitization and concomitant resetting of the stable isotope and K-Ar signatures of I/S are due to continuous recrystallization of I/S. If that model fits a basin evolution subsiding at some rate, perhaps even at discontinuous rates, then the model is probably appropriate. But if the stable and radiogenic data of the authigenic illite do not fit such a burial diagenetic model, then it may be appropriate to determine whether "thermal" (or hydrothermal) overprint(s) can explain the data. Such an interpretation should be considered as possible and as ad hoc until there is independent evidence of such overprints. Even if such a relationship between a tectonic activity and an induced "release" of hydrothermal fluids consolidates the interpretation, it is not always easy to establish, depending on the origin of the fluids and their pathways. Convincing evidence of such relationships was described, for instance, in many studies on so-called "discordant" uranium ore deposits (e.g., Gustafson and Curtis 1983, Hoeve and Quirt 1987, Kotzer and Kyser 1995, Renac et al. 2002, Patrier et al. 2003).

The Thermal History of the Neogene Bentonites of the East Slovak Basin

Šucha et al. (1993) described the conversion of smectite to illite in progressively buried bentonite and shale beds of the Neogene East Slovak Basin using mainly XRD analysis. The temperature dependence of this smectite-to-illite conversion was later generalized by Środoń (1995). The interest in comparing I/S of bentonites and shales is that the former are generally considered to be free of detrital minerals, clays or others, while they are typically mixed with detrital phases in clay-rich sedimentary rocks. The authors reported a good correlation between decreasing expandability of the I/S with increasing depth; i.e., temperature, in both the bentonites and the shales. The onset of illitization was established at 60 to 80° C and its end at 150 to 160° C, which has since been shown to represent an adequate range (Honty et al. 2004). In addition, the diameter of the illite particles from both the bentonites and shales increases with decreasing expandability and increasing depth, K fixation during nucleation and crystal growth appearing to be the driving process of the smectite-to-illite conversion. The temperature limits of illitization at approximately 70 and 155° C are of special interest, as they were obtained from analysis of natural samples, and the whole process could be followed, from start to finish, because of an unusually high geothermal gradient of ~52° C/km.

Šucha et al. (1993) estimated the timing and duration of burial-induced illitization in the East Slovak Basin through an empirical approach that relates illite expandability of the bentonites by XRD to temperature isotherms calculated from organic matter maturation of the shales by Francú et al. (1990). In other cases, the duration of illitization was estimated on the basis of the K content, such as in the Denver Basin (Elliott et al. 1991). Clauer et al. (1997) dated nanometric illite-rich fractions (also called fundamental particles) of sizes as small as 0.02 µm using the K-Ar method and found, for example, that in a bentonite bed sampled at 3015 m in a drilling well near Trhovište, illitization lasted from 15 Ma when the progressively buried bentonite bed reached a temperature of 70 to 80° C, to 7 Ma when the same bed reached a temperature of 150 to 160° C (Fig. 15A). In a second drilling well located about 20 km away, illitization was estimated, on the same K-Ar basis, to have started at 11.5 Ma in the bentonite sample collected at 2495-m depth when the bed reached 70 to 80° C and to have lasted until present time at about 120° C (Fig. 15B). Illitization can therefore proceed differently in the same type of host rock located only about 20 km apart on a horizontal scale and 520 m apart on a vertical scale.

The shales that sandwich the bentonite beds were also studied for the nanometric illite particles of their I/S (Clauer et al. 1997, Honty et al. 2004). The obtained K-Ar ages are systematically older than the stratigraphic age of the units, showing that in such lithology authigenic illite nucleated either on detrital particles that were already present in the rocks when illitization began or incorporated chemical elements in their structure that were released from dissolution of the previous detrital micaceous components. In this case, newly formed illite nuclei could have also incorporated some radiogenic ^{40}Ar released from dissolving frame minerals, which did not escape but was scavenged by the new illite precipitates.

DURATION OF ILLITIZATION: NANOMETRIC ILLITE-TYPE PARTICLES AND ILLITIZATION RATES

In order to evaluate successfully the duration of illitization during burial diagenesis through isotopic approaches it is necessary to analyze pure authigenic particles separated, possibly free of detrital clay material, and therefore with initial mineral nuclei devoid in any radioactive and radiogenic isotopes. In practice, this means that the analyses should be made strictly on the illite interlayers that form next to authigenic smectite particles at the beginning of the process and become progressively dominant in the smectite fraction as nucleation and growth progress. The first studies were made on bentonite beds, because these are essentially uncontaminated by detrital minerals and because they contain varied authigenic I/S having a wide range of expandability depending on the samples (e.g., Elliott and Aronson 1987). Illite–smectite mixed-layer nuclei formed initially at low temperatures from devitrification of volcanic ash and are characterized by low percentages of illite layers (Altaner et al. 1984). Such an approach needs to take into account a progressive incorporation of K, a possible loss of SiO_2, and a continuous decay of the incorporated ^{40}K into ^{40}Ar for the mathematics involved in the modeling of K-Ar ages. The systematics of how this combination can be evaluated and even solved were described and evaluated by Lerman and Clauer (2005) and Lerman et al. (2007).

Isotopic Study of Illite-Type Nanometric Particles

The I/S consist of interlayers enriched in K that are dispersed in the smecture structure, therefore representing the initial nuclei of illite. These nucleating and growing K-rich illite interlayers were named "fundamental particles" by Nadeau et al. (1984). They can be separated from smectite-type interlayers by infinite osmotic swelling, dispersing small amounts of I/S powder in large amounts of distilled water. This dispersion allows removal of all cations from smectite-type interlayers of the I/S, consequently breaking the bonds of these interlayers, while those of the illite interlayers remain tightly collapsed. The dispersed elemental particles are then separated into various size fractions by continuous-flow ultracentrifugation. Depending on the mechanism of illitization, different oxygen and hydrogen isotope fractionation factors and different isotopic ages for such illite fundamental particles can be expected as they become thicker and larger while growing. An Ostwald ripening model (Eberl and Środoń 1988) predicts that illitization lasts longer for the separated, thicker illite particles than for the smaller ones that either stopped growing and were cannibalized during crystal growth or were nucleated later, meaning that their isotopic ages should be younger than those of the finer particles from which they grew. Recent studies on the thickness of such fundamental particles measured by different techniques report thickness values of between ~1 and 18 nm (Dudek et al. 2002, Shang et al. 2003).

Clauer et al. (1997) K-Ar dated illite-type nanoparticles separated from I/S collected in two bentonite beds from two bore-holes drilled in the East Slovak Basin, which were mentioned above. Based on the K-

Ar results, illitization of the nanoparticles from Trhovište drill hole lasted from 11 to 8 Ma (Fig. 15A), while the duration based on the empirical method of Šucha et al. (1993) was found to range from 15 to 7 Ma (Fig. 15A); the latter method relates the illite expandability of the I/S from bentonites to the temperature isotherms calculated from organic matter maturation in the shales. The K-Ar range from 7.5 to 4 Ma in the variable-sized nanoparticles from the second Čičarovče drill hole (Fig. 15B) relates to an illitization span from 11.5 Ma to present when based on the temperature isotherms (Fig. 15B). In the Trhovište bed, the coarsest fundamental particles are the oldest, while they are the youngest in the Čičarovče bed (Fig. 16), which is inconsistent with the Ostwald ripening mechanism suggested by Eberl and Środoń (1988) to explain the growing process of fundamental particles. The ripening process, whatever its type, is accompanied by an age decrease when the particles grow, which is not the case for the Trhovište size fractions, implying that another mechanism is at work here. Similar mineralogical, geochemical, and isotopic results on nanometric illite particles of I/S from other bentonite beds of the same East Slovak Basin are also available in Honty et al. (2004).

Clauer et al. (2003) also reported $\delta^{18}O$ values for the different size fractions of the illite-type nanoparticles from the same bentonite beds described above (Fig. 17). In the Trhovište-sized fundamental particles, the $\delta^{18}O$ increases with increasing particle size from 16.5 to 18‰, while the $\delta^{18}O$ decreases, still with increasing size, from 12 to 10‰ in the Čičarovče size fractions. The different K-Ar ages and $\delta^{18}O$ trends of the illite-type nanoparticles from I/S of the two bentonite beds indicate different nucleation and crystallization processes occurring only 20 km apart in the same basin. Because the porosity and permeability of bentonite beds are known to be very low (hydraulic conductivity $<10^{-10}$ m/s), the water–rock ratio was probably very small and nearly constant during the entire crystal growth process. The decrease in the $\delta^{18}O$ with increasing particle size at Čičarovče seems to relate to a slow temperature increase during the 3.5-Ma-long growing process, whereas the increase in $\delta^{18}O$ with particle size increase at

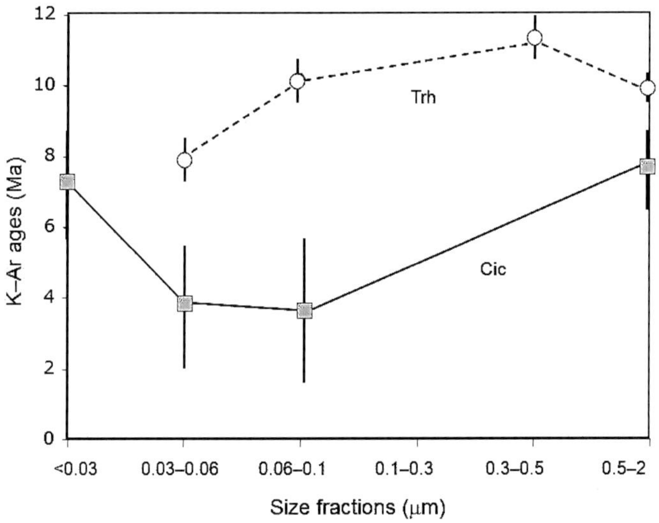

FIG. 16.—Diagram of K-Ar ages of the illite-type fundamental particles extracted at different sizes from illite–smectite mixed layers from bentonite units of the East Slovak Basin. The Trhovište sample is labeled "Trh" and the Čičarovče sample "Cic." Note that the smallest fraction of the Čičarovče sample is older than the coarser fractions, while the opposite is true for the Trhovište sample (after Clauer et al. 1997).

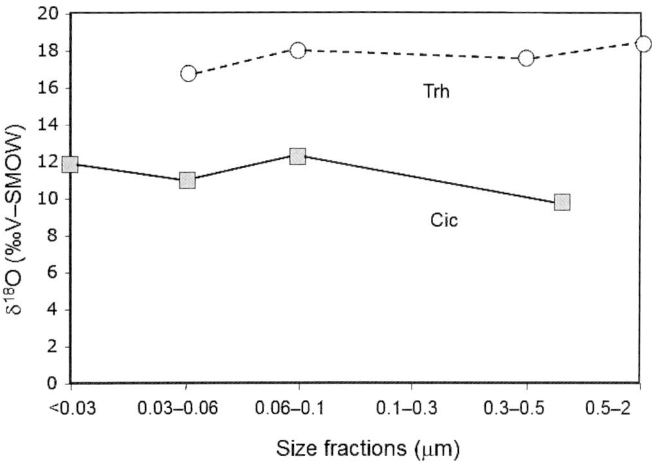

FIG. 17.—Diagram of the $\delta^{18}O$ of the illite-type fundamental particles extracted at different sizes from illite–smectite mixed layers from bentonite units of the East Slovak Basin. The Trhovište sample is labeled "Trh" and the Čičarovče sample "Cic." The analytical uncertainty ($\pm 2\ \sigma$) is about the size of the symbols. Note that the $\delta^{18}O$ is decreasing with size increase in the Čičarovče sample, while it increases in the Trhovište sample (after Clauer et al. 1997).

Trhovište may have been related to a temperature decrease between 11 and 8 Ma, which is consistent with paleotemperatures estimated by organic matter maturation shown as isotherms (Fig. 15A, B). In summary, the $\delta^{18}O$ of the fluids associated with the nucleation and crystal growth remained in a similar narrow range for the two sets of fundamental particles, with a calculated mean $\delta^{18}O$ of 0 ± 3‰ for the fluid at Trhovište and of 2 ± 1.5‰ for that at Čičarovče on the basis of the temperatures estimated from organic maturation data at about 120° C at Čičarovče and at about 160° C at Trhovište. The estimates based on stable isotopic tracing and radiogenic dating of the fundamental nanoparticles reasonably agree with the estimates made by Šucha et al. (1993) for the duration of illitization based on the crystallization temperatures that these authors derived from the maturation temperature of organic matter by Francú et al. (1990).

To pursue further the logic of tracing and dating illitization during burial by isotopic means, Clauer (2006) published a simplified diagram in which three theoretical pathways for the K-Ar data were qualitatively considered for the purpose of explaining the preliminary K-Ar ages obtained on the fundamental nanoparticles of the two above-described bentonite beds from East Slovak Basin (Fig. 18). The purpose was to gain some insight into the kinetics of illitization depending on where the K-Ar data plot on an arbitrary horizontal timescale and on a theoretical progression of crystal growth on the vertical axis, also called illite crystallite thickness by some (e.g., Eberl et al. 1996). Three different types of growth reactions were considered as (1) a constant linear growth shown by the straight line from the lower left side to the upper right side of the diagram, (2) an early rapid growth (at the upper left side of the graph) that slows down as the reaction progresses, and (3) an early slow growth (at the lower right side of the graph) that speeds up as the reaction progresses. The straight line describes a nearly constant growth with a nearly constant K supply. The nonlinear rates represent alternative possible scenarios in which K supply, and therefore particle growth, changes and even stops at any time. The model of early rapid crystal growth could apply, for example, if the pore throats became too small to allow the fluids to migrate, therefore reducing and stopping the flux of K that drives the reaction. In this

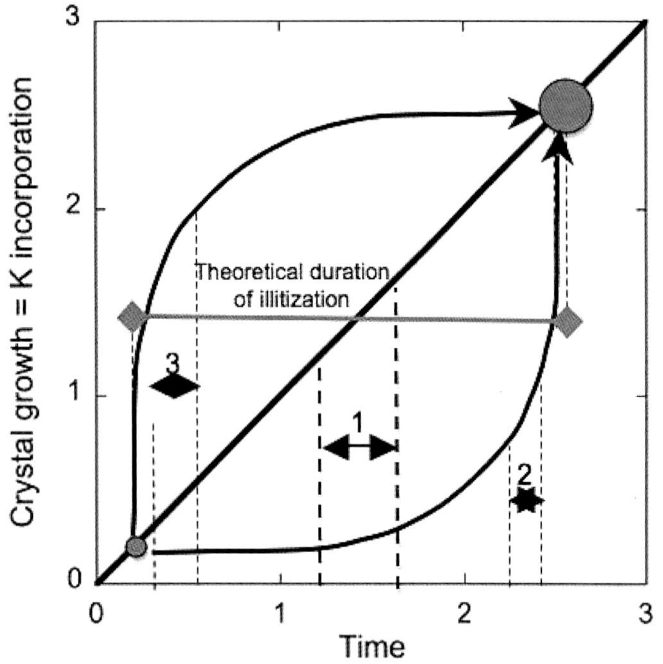

FIG. 18.—A qualitative diagram showing crystal growth (=particle size increasing upwards with K incorporation) on the vertical axis vs. time on the horizontal axis. Time starts on the left side with nucleation (black circle in the lower left side of the diagram) and increases to the right in the upper-right corner with the larger black circle. As shown, the theoretical duration of illitization is assumed to span from smallest to largest particle, from left to right. Three growth rates outlined by three arrows are assumed: one constant and two exponential. The K-Ar results obtained from the Čičarovče nanometric particles belong to span 1 and those from the Trhovište ones to span 2. For more explanation of the spans 1, 2, and 3, see text (after Clauer 2006).

case, the particles in such pore spaces remain small and yield the oldest K-Ar ages.

This simplified diagram (Fig. 18) assumes that illitization progresses only when available K drives the reaction, and it also assumes that temperature is not being considered. The reason to suggest that the prime parameter of illitization is the K availability in the pore solution, and not temperature, is based on the results of studies completed on progressively metamorphosed carbonates that show that smectite of these rocks remains unchanged when the interactive fluids are K-depleted (Dunoyer de Segonzac and Bernoulli 1976, Burkhard 1988), even during a temperature increase. If temperature rather than K availability were considered the main driver of illitization, the three curves can also be taken as respective records of (1) a constant deposition/burial rate, or a constant geothermal gradient in the case of the straight line; (2) an early high deposition/burial rate, which lowers with time, or an early relatively high geothermal gradient that decreases with time for the exponential curve at the top left side of the graph; or (3) an early low deposition/burial rate that increases with time or an early relatively low geothermal gradient that also increases with time for the exponential curve at the bottom right of the diagram.

In summary, the range of the K-Ar data of growing nanoparticles in rocks, like the bentonites here, will depend mainly on the K supply and will become younger as K availability increases. Since the range of K-Ar ages in the nanometric particles essentially depends on how fast the K enters the mineral structure, the data will plot in Figure 18 (1) either close to the middle of the theoretical period of illitization in the diagram sketch on the abscissa when the supply is relatively constant, (2) on the older side when the supply is large or the potassium flux is fast at the start of the process, or (3) on the younger side when the supply rate is slow at the beginning of the process and increases progressively. In the case study here, the K-Ar ages are located near the middle of the illitization duration estimated from organic maturation data (zone 1 in Fig. 18), indicating a constant growth rate after one nucleation episode. The range of the K-Ar ages can also plot either near the start of the estimated illitization period (zone 3 in the Fig. 18), indicating a fast initial growth rate that slowed down later and stopped near the end of the estimated illitization period (zone 2 in the Fig. 18), indicating an early slow growth rate that later increased. The different K-Ar ages in the estimated period of illitization appear to provide a new insight into illitization, when combined with the crystallization temperature of the fundamental particles obtained either by geothermometry or by maturation of the associated organic matter.

In combining such information, in the case of the nanometric-sized particles from bentonites of the East Slovak Basin, the contribution of the isotopic results to burial-driven illitization can be summarized as follows:

- At Čičarovče, the bentonite recovered at 2495 m records a burial diagenesis that reached a temperature of ~130° C during a single episode (Fig. 15B), which is supported by the K-Ar ages of the finest particles being older than those of the coarser particles. Nucleation was followed by nearly continuous crystal growth during the entire process, indicated by the K-Ar ages from 8 to 4 Ma (Fig. 16), which plot near the middle of the estimated duration of the process, interpreted to have lasted from 11.5 Ma to the present.
- At Trhovište, the bentonite bed at 3015 m records some differences: the burial diagenesis reached a higher temperature of approximately 165° C (Fig. 15A) that induced an episodic illitization, as indicated by the younger K-Ar ages of the finest particles compared to the coarser particles (Fig. 16). The younger K-Ar ages for the finest particles imply a renewed nucleation process. The growth of the initial nuclei seems to have been slow at the start of the process and faster toward the end, as the K-Ar ages range from 11 and 8 Ma, plotting toward the end of the independently estimated duration from 15 to 7 Ma. The increasing $\delta^{18}O$ values of the varied sizes of fundamental particles also indicate a decrease of either the temperature or the water–rock ratio, or perhaps both.

At this point, the preceding interpretation is based on a limited set of analytical data; it needs to be confirmed further, even if already supported by complementary studies of other bentonite beds in the East Slovak Basin (Honty et al. 2004) and of sandstones of the Neogene Mahakam Delta Basin (Clauer et al. 2004). If confirmed, this interpretation also raises another very important point, namely that results from studies of illite from a homogeneous rock lithology can differ in samples collected only a few tens of kilometers apart.

Illitization Reaction Rates

Numerous studies of the conversion rates of smectite to illite fall into two categories: studies done under laboratory experimental conditions and studies of the process recorded in natural sediment sequences. It is a common practice to report the fraction of expandable clays as a function of depth in sediment, where the expandable fraction usually decreases from some high value in shallow samples, such as 100 to 80%, to a smaller value as a result of the growth of crystalline illite in deeper samples. It may be mentioned that the measure of the smectite-to-illite conversion, taken as a percent or fraction of the expandable phase, is not identical to the increase in the illite fraction: for example, a decrease in the expandable fraction by a factor of 2, such as from 80 to 40%, represents an increase in the nonexpandable fraction by a factor of 3, from 20 to 60%. This relationship between the expandable and nonexpandable fractions may be borne in mind when a parameter such as reaction or conversion is discussed.

The rate of decrease in the expandable fraction with depth can vary greatly from one locality to another, depending on the local conditions, such as the geothermal gradient or temperature effects due to hydrothermal circulation, the presence of altering K-rich framework minerals such as K-feldspars and micas, the chemical composition of circulating fluids, and the occurrence of tectonic adjustments in the basement underneath the sedimentary basins that can propagate into the above deposited sedimentary rocks and represent potential pathways for fluid escape from the basement. The conversion of smectite to illite in different sections, as estimated from the reported concentration–depth profiles, ranges from a few hundred meters to perhaps as many as 2 to 3 km (e.g., Aronson and Hower 1976, Elliott et al. 1991, Velde and Vasseur 1992, Renac and Meunier 1995, Clauer et al. 1999). The conversion reaction is generally attributed to the effects of increasing temperature with depth and to the availability of the K^+ ion needed in the formation of illite. Precipitation of illite, as, for instance, related to and induced by migrating fluids in rock reservoirs, follows the same chemical reactions as during burial except for the duration of the reaction, which can be considered to be instantaneous. Many kinetic reactions have been proposed for the smectite-to-illite conversion, and their essential forms can be subdivided into two simple scenarios: a first-order conversion of smectite to illite, resulting in an exponential decrease of the smectite fraction with time, as proposed by several investigators, and a first-order sequential reaction of smectite reacting to a poorly crystallized illite (PCI), which in turn reacts to form a well-crystallized illite (WCI) (Meunier and Velde 2004). In this second scenario, the PCI is actually an I/S, while the WCI consists of ~100% nonexpandable illite layers. It might be appropriate to recall Elliott and Matisoff's (1996) calculations of kinetic rates of the burial-induced transformation of smectite into illite that provide different models of kinetic rates depending on the studied sedimentary basins, pointing to the fact that the rates strictly depend on the evolution of the studied sedimentary sequences.

The sequential reaction, as applied to a two-step process of mineral transformation, is analogous to the case of radioactive decay, where a radioactive parent decays to the first radioactive daughter that in turn decays to the second stable daughter. In a one-step first-order process, the decrease in the abundance of an expandable phase (S) may be represented by the following equation:

$$\mathrm{S} = \mathrm{S}_0 \exp(-kt),$$

where S_0 is the initial abundance, k is a first-order rate constant (time $^{-1}$), and t is time. The quotient S/S_0 is the fraction remaining. If this first-order reaction law is obeyed then the value of the rate constant k may be determined from a plot of ln S/S_0 against t. Whitney and Northrop (1988) demonstrated in experiments of 120-day duration at temperatures ranging from 250 to 400° C that the reaction progress is not linear during the first 30 days and it may be interpreted as a first-order mechanism only after the first 30 days. The effect of temperature on reaction rate is usually represented by the Arrhenius-type relationship between the reaction rate constant (k), the temperature (T in kelvin), and the activation energy of the reaction (E), which is assumed constant and independent of temperature:

$$\ln k_2/k_1 = (E/R)(1/T_1 - 1/T_2),$$

where R is the gas constant (1.987 cal/mol/K, or 8.1345 J/mol/K). In this format, the equation describes a change in ln k as a result of the temperature increase from T_1 to T_2.

As temperature increases with depth, the reaction rate also increases, and its increase is determined by the activation energy. High-temperature experimental determinations of the smectite-to-illite reaction rates (temperature above 250° C; e.g., Eberl and Hower 1976, Roberson and Lahann 1981, Whitney and Northrop 1988, Huang et al. 1993) have reported representative activation energy values of about 30 and 28 kcal/mol. Small (1993) reported an activation energy of 26.3 kcal/mol for direct precipitation of illite from solution, between 200 and 240° C. These are high values of activation energy in comparison to those resulting from reactions such as silicate mineral dissolution rates at Earth's surface temperatures.

Lower values of the activation energy, ranging from 3.0 to 3.5 kcal/mol, were reported in experiments conducted between 150 and 250° C (Howard and Roy 1985), and values of 7.2 to 7.5 kcal/mol were reported in experiments conducted between 120 and 200° C (Cuadros and Linares 1996). The latter authors pointed out that these low values for the activation energy of illitization indicate a solid-state transformation mechanism rather than dissolution and precipitation. In other words, illitization occurring at relatively low temperatures could preferentially be a solid-state transformation reaction, but illitization that occurs at relatively high temperatures would be a solution–precipitation reaction. Determinations based on natural occurrences from wells at seven locations (Texas, California, Japan, Paris Basin) give a range of E from 3.8 to 12 kcal/mol (Velde and Vasseur 1992). However, these authors have also shown that the choice of the reaction rate model may greatly affect the activation energy value obtained from it, as was discussed in the case of the study of the Lower Triassic sandstones from Paris Basin. In a two-step model, they obtained $E_1 = 16.7$ kcal/mol and $E_2 = 8.9$ kcal/mol. Renac and Meunier (1995) hypothesized that in addition to the two first-order reaction steps (smectite to PCI to WCI), a third step in which illite layers precipitate on preexisting crystals also occurs, for which the activation energy would be lower than the $E_2 = 8.9$ kcal/mol of the second first-order step in the chain reaction. A value of $E = 7$ kcal/mol, at approximately the midpoint of the cited range from 3 to 12 kcal/mol, would increase the reaction rate parameter k by a factor of 5, in a temperature interval from 60 to 120° C that corresponds to a depth interval from ~1.5 to 3.5 km.

In conclusion, a wide range in the activation energy values derived from the smectite-to-illite transition in sediment cores of different geologic ages and from sequences subjected to varied subsidence rates have been reported, as have a variety of kinetic models describing the reaction. Measured profiles show that the fraction of the expandable phase may decrease by a factor of 2 over a period of 10^6 to 10^7 years. A decrease occurring over a period of time shorter than 1 to 10 Ma can be viewed as an illitization pulse, which can relate to a punctuated (hydro)-thermal event.

ILLITIZATION AND K-AR DATING

The progress of illitization with increasing depth is related to the ^{40}Ar/^{40}K atomic ratios in the clays, in which ^{40}Ar forms from the radioactive decay of ^{40}K. The K-Ar ages of the fine-size clay fractions derived from the analytically determined ^{40}Ar/^{40}K ratios variably reflect the clay–mineral diagenesis because of the differences in the physical and chemical behavior of K and Ar in clay–mineral lattices and in the pore space filled with a solution and/or gas. The ^{40}Ar/^{40}K ratios and K-Ar ages of K-containing clays often decrease with increasing depth and increasing stratigraphic age in sediment, although an opposite relationship was reported by Elliott et al. (1991) in Upper Cretaceous bentonite of the Denver Basin. In a closed system in which no K or ^{40}Ar is added to or released from the clay particle, the K-Ar age increases with time or the sediment stratigraphic age, which in turn reflects the age of the sediment source terranes when no syn-sedimentary diagenesis occurred.

The often-encountered decrease in the K-Ar age with increasing stratigraphic age can be interpreted to represent a loss of ^{40}Ar from the minerals, an addition of K to the clay particles that carries ^{40}K with it, or a combination of the two processes (Clauer and Lerman 2009), as shown in Figure 19. This figure shows calculated changes with time in the K-Ar age of a clay particle with an initial K-Ar age of 55 Ma under the conditions of (1) a closed system, (2) a ^{40}Ar loss, (3) a K gain, (4) a combination of the two processes, and (5) a simulation of a nearly constant K-Ar age profile with depth. The K-Ar initial age reflects the age of formation of the clay material in the source terrane, which accounts for the old ages relative to the time of deposition. Lerman and Clauer (2005) and Lerman et al. (2007) applied a theory to analyze the loss of radiogenic ^{40}Ar and the addition of K to clay size fractions of several sedimentary sequences of different stratigraphic intervals and from varied places. More recently, Clauer and Lerman (2009) strengthened the mathematical model by comparing the calculated addition and diffusion rates with analytical data that were published independently. In other words, they showed how the K-Ar ages of illite-rich size fractions extracted from progressively buried sedimentary rocks can be explained by the rates of combined K addition and radiogenic ^{40}Ar release from clay-particle mixtures. This model simulates well the changes with depth from 1800 to 4200 m of the K-Ar ages of fine (<0.1-µm) to coarse (>2-µm) clay fractions, showing that K addition and ^{40}Ar loss are changing with depth in sedimentary sequences in the Texas Gulf Coast, North Sea Bergen High section, and the Mahakam formations offshore the Island of Borneo. Clauer and Lerman (2009) also showed that some of the

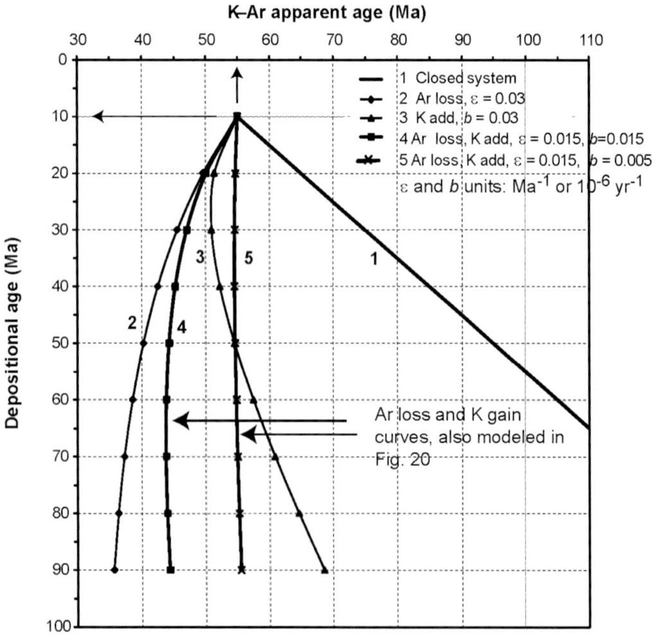

FIG. 19.—Model-calculated changes in a theoretical K-Ar age, from an initial value of 55 Ma at a sediment depth of depositional age 10 Ma, due to loss of radiogenic ^{40}Ar, gain of K, and both processes operating simultaneously. Changes in the K-Ar age with time should be compared to the change in a closed system (line 1), in which no Ar or K is lost or gained. Slower and faster rates are shown in curve 2 for ^{40}Ar loss, curve 3 for K gain, and curve 4 for both processes. Curve 5 simulates a nearly constant K-Ar age with increasing depth (after Clauer and Lerman 2009). Vertical scale represents time since deposition. Note the curves for Ar loss and K gain (curves 4 and 5), also plotted in Figure 20.

observed trends of the K-Ar age with increasing depth, over stratigraphic time intervals of 10^6 to 10^7 years, can be simulated by variable diagenetic processes as ^{40}Ar loss by diffusion out of spherical or flake-shaped particles, ^{40}Ar loss by a first-order flux out of the particles, and/or ^{40}K addition to the particles. Thus, the choice of the model describing the changes in the $^{40}Ar/^{40}K$ ratio and K-Ar age with time needs to consider not only the age trends but also the concentrations of the ^{40}K and the radiogenic ^{40}Ar in the separated clay fractions. The interpretation of the K-Ar ages with depth of the coarser and finer clay fractions in the Neogene Mahakam shale is shown in Figure 20.

The analysis of the Mahakam shale sequence is based on a model of continuous loss of ^{40}Ar as a first-order process with a rate constant ε (Ma^{-1}) and gain of K at a constant linear rate b (Ma^{-1}). The rate of change in the ^{40}Ar in the clay fraction is the difference between the K-gain rate, $^{40}K_0(1+bt)e^{-\lambda t}$, and the ^{40}Ar mass loss rate, $-\varepsilon\,(^{40}Ar)$. In the four studied sections, the K and ^{40}Ar concentrations of the fine clay fractions increase with increasing depth. The increase in the K content is consistent with the progress of illitization. However, the increase in the ^{40}Ar content with depth is smaller than would be observed in a closed system, and, therefore, the $^{40}Ar/^{40}K$ ratios and K-Ar ages both decrease with increasing depth. An unexpected conclusion of this analysis is that the Ar loss and K gain can be described by constant rate parameters that are independent of sediment depth, and therefore independent of temperature. The order of magnitude of the rate parameters of ^{40}Ar loss ε and K gain b is in the range from 10^{-3} to 10^{-2}%/Ma, and 10^{-1} to 10^{0}%/Ma, respectively. These magnitudes are similar to that of the two rate constants of the two-step illitization in the Upper Triassic–Lower Jurassic section studied in the borehole BA1 near Balazuc in southeastern France (Renac and Meunier 1995; C. Renac's data, as cited in Meunier and Velde [2004]): in the stratigraphic interval between ~1100 and 1500 m, where the temperature increase may be as little as 14° C (from the geothermal gradient of 35° C/km), the rate constant k_1 for the reaction from smectite to PCI is essentially constant with depth at 0.026%/Ma (range, 0.021–0.032), and the second rate constant k_2 for the reaction PCI to WCI doubles in value over the 400-m interval, from 0.007 to 0.015%/Ma. This similarity, perhaps fortuitous, of the rate constants of these two very different models approximates a complex process by conceptually simple kinetic relationships.

The decrease with depth in the K-Ar age of the coarser fractions can be bracketed by calculated values of the K addition rate and the ^{40}Ar release rate in a range from 0.2 to 3.5%/Ma and from 0.5 to 4.5%/Ma, respectively (Clauer and Lerman 2009). For the fine-size fractions, the K-gain rate varies from 3.5 to 6%/Ma and the ^{40}Ar release rate from 0.7 to 6%/Ma. When the rates of K addition and ^{40}Ar escape are small and similar, each about 1%/Ma, the model shows that the mean K-Ar age of fine illite-rich size fractions remains constant with depth over several hundreds of meters, demonstrating that such relationships between the K-Ar ages and depth are not representative of a steady state but rather of a continuing dynamic diagenetic process with changing rates of K addition and radiogenic ^{40}Ar release. These results provide a preliminary kinetic formulation of the combined K and ^{40}Ar behavior during illitization in size fractions containing mixtures of authigenic and detrital particles of sedimentary sections extending over stratigraphic intervals of 10^3 m or 10^1 Ma.

CONCLUSIONS

Illitization is a widely described process in sedimentary sequences that is attributed to varied crystal-chemical processes, including solid-state reactions, dissolution of detrital illite–smectite mixed layers combined to precipitation of new illite, as well as direct precipitation from interstitial fluids. Whatever the process, illitization is an indicator of the postdepositional history of clays: burial-induced diagenetic

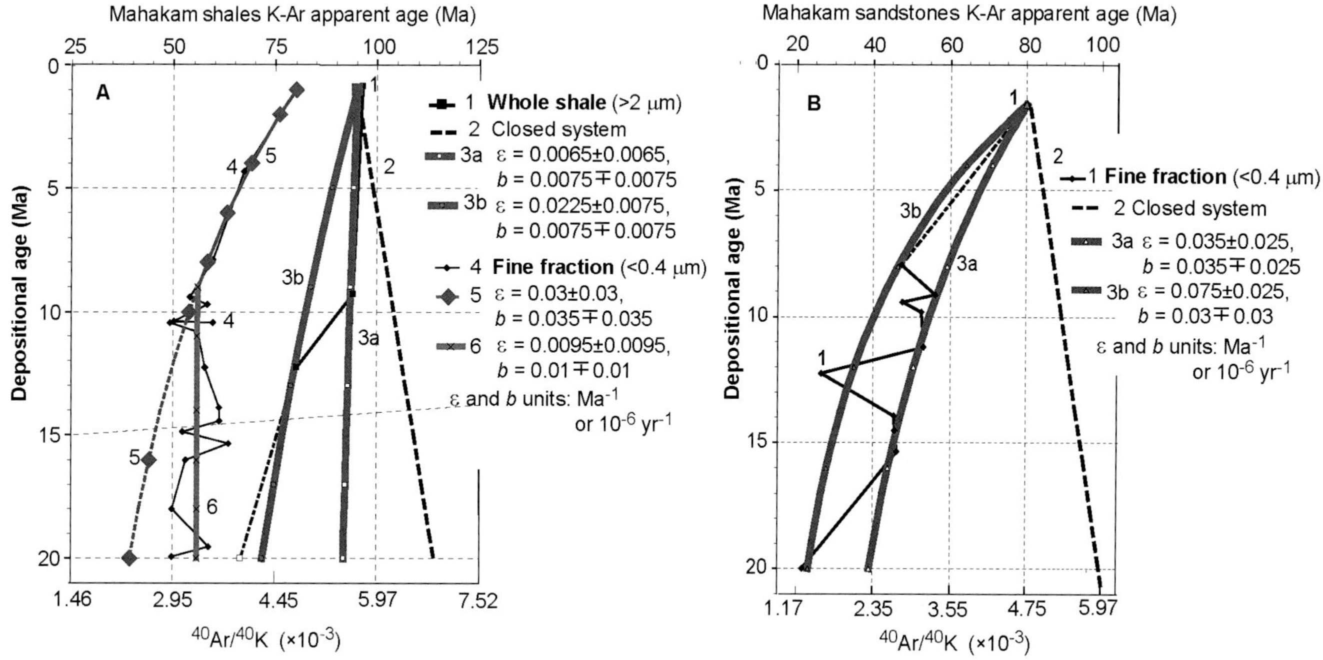

FIG. 20.—(A) Model-calculated changes in the K-Ar ages with depth in whole rocks and fine clay fractions of Mahakam shales (Eastern Borneo) by a two-process model of ^{40}Ar loss and K gain. The section from about 10 to 20 Ma of stratigraphic age is approximated by a starting mean K-Ar age of 56 Ma and is simulated by the two-process model, with variable parameter values for ^{40}Ar release (ε) and K gain (b). The whole shale data (1) are bracketed by the model-calculated curves 3a and 3b. Fine fraction data are best fit in the upper part by a simultaneous Ar loss and K gain shown by curve 5 and in the lower part by curve 6 that approximates constant age of 56 Ma for the fraction with depth. (B) Simulation of the K-Ar age changes with depth in the fine clay fraction of Mahakam sandstones (Eastern Borneo) by the two-process model. The data are bracketed by the curves 3a and 3b, showing simultaneous Ar loss and K gain (line 1). The calculated K-Ar ages for a theoretical closed-system evolution are shown in both cases (after Clauer and Lerman 2009; data from Furlan et al. 1996 and Clauer et al. 1999).

changes, for instance, are often identified by a change in the clay-sized mineralogy from smectite-rich to illite-rich. Generally, the changes are attributable to a temperature-induced increase of the K content in smectite and illite–smectite mixed layered structures that can be progressive during burial or instantaneous and episodic during interactions of migrating, hot, and K-rich hydrothermal pore fluids with their host rocks.

Comprehension of illitization in sedimentary sequences is also germane to the potential use of authigenic illite as oxygen and hydrogen isotope geothermometric records of the thermal history of sedimentary basins, even if this geothermometric use is still debatable. Because illite crystallization can result either from a progressive burial or a sudden and episodic precipitation from migrating hot fluids initiated by regional tectonic activities, especially in porous sedimentary lithologies, stable isotope geochemistry can be helpful in distinguishing these two potential scenarios. Determination of the maximum paleotemperature is also an important step in the reconstruction of the thermal history of a sedimentary basin, but it is not the only important step; knowledge of when this maximum temperature occurred is also essential for the interpretation. The application of stable and radiogenic isotope studies to illitization in varied rock lithologies of the sedimentary basins of Northern Germany, Paris, East Slovakia, and the Mahakam Delta shows, on the basis of combined mineralogical, morphological, and isotopic databases of illite and associated materials, that the isotopic approach of illitization is of discriminating help in the reconstruction of sedimentary basins.

Oxygen and hydrogen isotope geothermometry and radiogenic isotope dating (K-Ar or other methods) of authigenic illite in sedimentary rocks represent an appropriate approach to complement the reconstruction of the thermal history of sedimentary rocks, but they need to take into account some prerequisites. For instance, a reliable application needs an adequate mineral separation, which is a serious technical challenge, and a precise identification of the authigenic illite particles, including XRD and SEM or TEM observation. One also needs to be aware of drawbacks in the isotopic methods, such as approximate isotope fractionation rates of oxygen in illite, because of the difficulty they pose in terms of evaluation in the low-temperature domain.

ACKNOWLEDGMENTS

We express our sincere thanks to N.B. Harris (University of Alberta, Edmonton, Canada) and K. Peters (US Geological Survey, Denver, Colorado), the editors of this issue on "Thermal history analysis of sedimentary basins: Methods and applications," for their kind invitation to contribute. We are particularly grateful to E. Eslinger (Eric Geoscience, Inc., Glenmont, New York) and N.B. Harris for their extremely detailed and questioning review and for the editorial work on a previous version of the manuscript draft. We also thank the second anonymous reviewer for his improving inputs and comments. This research was supported in part by the Arthur L. Howland Fund of the Department of Earth and Planetary Sciences, Northwestern University.

REFERENCES

Aja SU. 1995. Thermodynamic properties of some 2:1 layer clay minerals from solution-equilibration data. *European Journal of Mineralogy* 7:325–333.

Altaner SP, Hower J, Whitney G, Aronson JL. 1984. Model for K-bentonite formation: Evidence from zoned K-bentonites in the Disturbed Belt, Montana. *Geology* 12:412–425.

Altaner SP, Ylagan RF. 1997. Comparison of structural models of mixed-layer illite/smectite and reaction mechanisms of smectite illitization. *Clays and Clay Minerals* 45:517–533.

Amouric M, Olives J. 1991. Illitization of smectite as seen by high-resolution transmission electron microscopy. *European Journal of Mineralogy* 3:831–835.

Árkai P, Tóth NM. 1983. Illite crystallinity: Combined effects of domain size and lattice distorsion. *Acta Geologica Hungarica* 26:341–348.

Arnórsson S, Stefánsson A. 1999. Assessment of feldspar solubility constants in water in the range 0 to 350 degrees C at vapor saturation pressures. *American Journal of Science* 299:173–209.

Aronson JL, Hower J. 1976. Mechanism of burial metamorphism of argillaceous sediments: 2. Radiogenic argon evidence. *Geological Society of America Bulletin* 87:738–744.

Awwiller DN. 1994. Geochronology and mass transfer in Gulf Coast mudrocks (south-central Texas, U.S.A.): Rb-Sr, Sm-Nd and REE systematics. *Chemical Geology* 116:61–84.

Bechtel A, Hoernes S. 1990. Oxygen isotope fractionation between oxygen of different sites in illite minerals; A potential single-mineral thermometer. *Contribution to Mineralogy and Petrology* 104:463–470.

Berthelsen A. 1992. Mobile Europe. *In* Blundell D, Freeman R, Mueller S (Editors). *A Continent revealed. The European Geotraverse*: Cambridge University Press, UK. p. 11–32.

Bottinga Y, Javoy M. 1973. Comments on oxygen isotope geothermometry. *Earth and Planetary Science Letters* 20:250–265.

Bottinga Y, Javoy M. 1975. Oxygen isotope partitioning among the minerals in igneous and metamorphic rocks. *Review of Geophysical and Space Physics* 13:401–418.

Bruce CH. 1984. Smectite dehydration—its relationship to structural disposition and hydrocarbon migration in the Gulf of Mexico. *American Association of Petroleum Geologists Bulletin* 68:673–683.

Brunet MF, Le Pichon X. 1982. Subsidence of the Paris Basin. *Journal of Geophysical Research* 87:8547–8560.

Burkhard M. 1988. L'Hélvétique de la bordure occidentale du Massif de l'Aar; Evolution tectonique et métamorphique. *Eclogae Geologicae Helvetiae* 81:63–114.

Burst JF. 1959. Post diagenetic clay mineral-environmental relationships in the Gulf Coast Eocene in clays and clay minerals. *Clays and Clay Minerals* 6:327–341.

Cathelineau M, Fourcade S, Clauer N, Buschaert S, Rousset D, Boiron MC, Meunier A, Lavastre V, Javoy M. 2004. Dating multistage paleofluid percolations: A K-Ar and ^{18}O study of fracture illites from altered Hercynian plutonites at the basement/cover interface (Poitou High, France). *Geochimica et Cosmochimica Acta* 68:2529–2542.

Chaudhuri S, Środoń J, Clauer N. 1999. K-Ar dating of illitic fractions of Estonian "Blue Clay" treated with alkylammonium cations. *Clays and Clay Minerals* 47:96–102.

Chiba H, Chacko T, Clayton RN, Goldsmith JR. 1989. Oxygen isotope fractionations involving diopside, forsterite, magnetite, and calcite; Application to geothermometry. *Geochimica et Cosmochimica Acta* 53:2985–2995.

Clauer N. 2006. Towards an isotopic modeling of the illitization process based on data of illite-type fundamental particles from mixed layered illite–smectite. *Clays and Clay Minerals* 54:119–130.

Clauer N, Chaudhuri S. 1995. *Clays in Crustal Environments: Isotope Dating and Tracing*: Springer Verlag, New York. 359 p.

Clauer N, Chaudhuri S. 1997. Isotopic dating of very low-grade metasedimentary and metavolcanic rocks: Techniques and methods. *In* Frey M, Robinson D (Editors). *Low-Grade Metamorphism*: Blackwell Science, Oxford, UK. p. 202–226.

Clauer N, Cocker JD, Chaudhuri S. 1992. Isotopic dating of diagenetic illites in reservoir sandstones: Influence of the investigator effect. *In* Houseknecht DW, Pittman ED (Editors). *Origin, Diagenesis and Petrophysics of Clay Minerals in Sandstones*, Special Publication 47: Society of Economic Paleontologists and Mineralogists, Tulsa, Oklahoma. p. 5–12.

Clauer N, Lerman A. 2009. Modeling a concomitant K gain and radiogenic Ar loss in clay fractions during burial illitization. *In* 46th Annual Meeting Abstracts, Billings, Montana, June 6–11. Clay Minerals Society, Chantilly, VA. (abstract).

Clauer N, Liewig N, Pierret MC, Toulkeridis T. 2003. Crystallization conditions of fundamental particles from mixed-layers illite–smectite of bentonites based on isotopic data (K-Ar, Rb-Sr and $\delta^{18}O$). *Clays and Clay Minerals* 51:664–674.

Clauer N, O'Neil JR, Furlan S. 1995. Clay minerals as record of temperature conditions and duration of thermal anomalies in the Paris Basin, France. *Clay Minerals* 30:1–13.

Clauer N, O'Neil JR, Furlan S, Mossmann JR. 1996a. Clay minerals as records of temperature conditions and duration of thermal anomalies in the Paris Basin, France: A reply to the discussion by Spötl et al. (1996). *Clay Minerals* 31:209–215.

Clauer N, Rinckenbach T, Weber F, Sommer F, Chaudhuri S, O'Neil JR. 1999. Diagenetic evolution of clay minerals in oil-bearing Neogene sandstones and associated shales from Mahakam Delta Basin (Kalimantan, Indonesia). *American Association of Petroleum Geologists Bulletin* 83:62–87.

Clauer N, Rousset D, Środoń J. 2004. Modeled shale and sandstone burial diagenesis based on the K-Ar systematics of illite-type fundamental particles. *Clays and Clay Minerals* 52:576–588.

Clauer N, Środoń J, Francú J, Šucha V. 1997. K-Ar dating of illite fundamental particles separated from illite–smectite. *Clay Minerals* 32:181–196.

Clauer N, Zwingmann H, Chaudhuri S. 1996b. Isotope (K-Ar and oxygen) constraints on the extent and importance of the Liassic hydrothermal activity in Western Europe. *Clay Minerals* 31:301–318.

Cuadros J, Linares J. 1996. Experimental kinetic study of the smectite-to-illite transformation. *Geochimica et Cosmochimica Acta* 60:439–453.

Curnelle R, Dubois P. 1986. Evolution mésozoïque des grands basins sédimentaires français: Basins de Paris, d'Aquitaine et du Sud-Est. *Bulletin de la Société Géologique de France* 8:529–546.

Dayal R. 1975. Clay-sea water interaction at elevated pressures [PhD dissertation]: Dalhousie University, Halifax, Nova Scotia, Canada, 234 p.

Delgado A, Reyes E. 1996. Oxygen and hydrogen isotope compositions in clay minerals: A potential single-mineral geothermometer. *Geochimica et Cosmochimica Acta* 60:4285–4289.

Demars C, Pagel M. 1994. *Paléotempératures et Paléosalinités dans les Grès du Keuper du Bassin de Paris: Inclusions Fluides dans les Minéraux Authigènes*, Vol. 319, II: Comptes Rendus de l'Académie des Sciences, Paris. p. 427–434.

Drong HJ. 1979. Diagenetische Veränderung in den Rotliegend Sandsteinen im NW-deutschen Becken. *Geologische Rundschau* 68:1172–1183.

Dudek T, Środoń J, Eberl DD, Elsass F, Uhlik P. 2002. Thickness distribution of illite crystals in shales. I: X-ray diffraction vs. high-resolution transmission electron microscopy measurements. *Clays and Clay Minerals* 50:562–577.

Dunoyer de Segonzac G. 1970. The transformation of clay minerals during diagenesis and low-grade metamorphism: A review. *Sedimentology* 15:281–346.

Dunoyer de Segonzac G, Bernoulli D. 1976. Diagenèse et métamorphisme des argiles dans le Rhétien Sud-Alpin et Austro–Alpin Lombardi et Grison. *Bulletin de la Société Géologique de France* 18:1283–1293.

Eberl DD, Drits V, Środoń J, Nuesch R. 1996. MUDMASTER: A program for calculating crystallite size distributions and strain from the shapes of X-ray diffraction peaks: US Geological Survey Open-File Report, Vol. 96-171, 45 p.

Eberl DD, Hower J. 1976. Kinetics of illite formation. *Geological Society of America Bulletin* 87:1326–1330.

Eberl DD, Środoń J. 1988. Ostwald ripening and interparticle-diffraction effects from illite crystals. *American Mineralogist* 73:1335–1345.

Eberl DD, Środoń J, Kralik M, Taylor BE, Peterman ZE. 1990. Ostwald ripening of clays and metamorphic minerals. *Science* 248:474–477.

Eberl DD, Velde B. 1989. Beyond the Kübler index. *Clay Minerals* 24:571–577.

Elliott WC, Aronson JL. 1987. Alleghanian episode of K-bentonites illitization in the southern Appalachian Basin. *Geology* 15:735–739.

Elliott WC, Aronson JL, Matisoff G, Gautier DL. 1991. Kinetics of the smectite to illite transformation in the Denver Basin; Clay mineral, K-Ar data and mathematical model results. *American Association of Petroleum Geologists Bulletin* 75:436–462.

Elliott WC, Matisoff G. 1996. Evaluation of kinetic models for the smectite to illite transformation. *Clays and Clay Minerals* 44:77–87.

Eslinger EV, Savin SM. 1973. Oxygen isotope geothermometry of the burial metamorphic rocks of the Precambrian Belt Supergroup, Glacier National Park, Montana. *Geological Society of America Bulletin* 84:2549–2560.

Essene EJ, Peacor DR. 1995. Clay mineral thermometry—a critical prospective. *Clays and Clay Minerals* 43:540–553.

Faure G, Mensing TM. 2005. *Isotopes. Principles and Applications*, 3rd ed.: Wiley & Sons, Hoboken, New Jersey. 897 p.

Fournier RO, Truesdell AH. 1973. An empirical Na-K-Ca geothermometer for natural waters. *American Journal of Sciences* 37:1255–1275.

Francú J, Muller P, Šucha V, Zatkaliková V. 1990. Organic matter and clay minerals as indicators of thermal history in the Transcarpathian Depression (East Slovakian Neogene Basin) and the Vienna Basin. *Geologica Carpathica* 41:535–546.

Freed RL, Peacor DR. 1992. Diagenesis and the formation of authigenic illite-rich crystals in Gulf Coast shales: TEM study of clay separates. *Journal of Sedimentary Petrology* 62:220–234.

Frey M. 1987. Very low-grade metamorphism of clastic sedimentary rocks. *In* Frey M (Editor). *Low Temperature Metamorphism*: Blackie and Son, Ltd, Glasgow, Scotland. p. 9–58.

Frey M, Teichmüller M, Teichmüller R, Mullis J, Künzi B, Breitschmid A, Gruner U, Schwizer B. 1980. Very low-grade metamorphism in external parts of the Central Alps: Illite crystallinity, coal rank and fluid inclusion data. *Eclogae Geologicae Helvetiae* 73:173–203.

Furlan S. 1994. Transferts de matière au cours de la diagenèse d'enfouissement dans le bassin du delta de la Mahakam (Indonésie). Un nouveau concept pour le mécanisme de illitisation [PhD thesis]: Université Louis Pasteur, Strasbourg, 224 p.

Furlan S, Clauer N, Chaudhuri S, Sommer F. 1996. K transfer during burial diagenesis in the Mahakam Delta Basin (Kalimantan, Indonesia). *Clays and Clay Minerals* 44:157–169.

Gaupp R, Matter A, Platt J, Ramseyer K, Walzerbruck J. 1993. Diagenesis and fluid evolution of deeply buried Permian (Rotliegend) gas reservoirs, Northwest Germany. *American Association of Petroleum Geologists Bulletin* 77:1111–1128.

Gautier DL, Bird KJ, Colten-Bradley VA. 1987. Relationship of clay mineralogy, thermal maturity, and geopressure in wells of the Point Thomson area. *In* Bird KJ, Magoon LB (Editors). *Petroleum Geology of the Northern Part of the Arctic National Wildlife Refuge, Northeastern Alaska*, Vol. 1778: US Geological Survey Bulletin. p. 199–207.

Girard JP, Barnes DA. 1995. Illitization and paleothermal regimes in the Middle Ordovician St Peter Sandstone, Central Michigan Basin, United States: K-Ar, oxygen isotopes, and fluid inclusion data. *American Association of Petroleum Geologists Bulletin* 79:49–69.

Girard JP, Savin SM. 1996. Intracrystalline fractionation of oxygen isotopes between hydroxyl and non-hydroxyl sites in kaolinite measured by thermal dehydroxylation and partial fluorination. *Geochimica et Cosmochimica Acta* 60:469–487.

Glasmann JR, Larter S, Briedis NA, Lundegard PD. 1989. Shale diagenesis in the Bergen area, North Sea. *Clays and Clay Minerals* 37:97–112.

Glennie KW, Mudd GC, Nagtegaal PJ. 1978. Depositional environment and diagenesis of Permian Rotliegend sandstones in Leman Babkand Sole Pit areas of the UK southern North Sea. *Journal of the Geological Society* 135:25–34.

Guilhaumou N. 1993. Paleotemperatures inferred from fluid inclusions in diagenetic cements: Implications for the thermal history of the Paris Basin. *European Journal of Mineralogy* 5:1217–1226.

Guilhaumou N, Gaulier JM. 1991. *Détermination de Paléotempératures dans les Roches-Mères du Bassin de Paris: Etude d'Inclusions Fluides et Implications pour l'Histoire Thermique du Basin*, Vol. 313, II: Comptes Rendus de l'Académie des Sciences, Paris. p. 773–780.

Gustafson LB, Curtis LW. 1983. Post Kombolgie metasomatism at Jabiluka, Northern Territory, Australia, and its significance in the formation of high grade uranium mineralization in lower Proterozoic rocks. *Economic Geology* 35:26–56.

Halliday AN, Ohr M, Mezger K, Chesley JT, Nakai S, Dewolf CP. 1991. Recent developments in dating ancient crustal fluids. *Reviews in Geophysics* 29:577–584.

Hamilton PJ, Giles MR, Ainsworth P. 1992. K-Ar dating of illites Brent Group reservoirs: A regional perspective. *In* Morton AC, Haszeldine RS, Giles MR, Brown S (Editors). *Geology of the Brent Group,* Special Publication 61: Geological Society of London, UK. p. 377–400.

Hamilton PJ, Kelley S, Fallick AE. 1989. K-Ar dating of illite in hydrocarbon reservoirs. *Clay Minerals* 24:21–31.

Hamza MS, Epstein S. 1980. Oxygen isotopic fractionation between oxygen of different sites in hydroxyl-bearing silicate minerals. *Geochimica et Cosmochimica Acta* 44:173–182.

Harris NB. 1992. Burial diagenesis of Brent sandstones: A study of Statfjord, Hutton and Lyell fields: *In* Morton AC, Haszeldine RS, Giles MR, Brown S (Editors). *Geology of the Brent Group,* Special Publication 61: Geological Society of London, UK. p. 351–375.

Hoeve J, Quirt D. 1987. A stationary redox front as a critical factor in the formation of high grade unconformity-type uranium ores in the Athabasca basin, Saskatchewan, Canada. *Bulletin de Minéralogie* 110:157–171.

Honty M, Uhlik P, Šucha V, Caplovicová M, Francú J, Clauer N, Biroń A. 2004. Smectite to illite alteration in salt-bearing bentonites; The East Slovak Basin. *Clays and Clay Minerals* 52:533–551.

Howard JJ, Roy DM. 1985. Development of layer charge and kinetics of smectite alteration. *Clays and Clay Minerals* 33:81–88.

Hower J, Eslinger EV, Hower M, Perry EA. 1976. Mechanism of burial metamorphism of argillaceous sediments: 1. Mineralogical and chemical evidence. *Geological Society of America Bulletin* 87:725–737.

Huang WL, Longo JM, Pevear DR. 1993. An experimentally derived kinetic model for smectite-to-illite conversion and its use as a geothermometer. *Clays and Clay Minerals* 41:162–177.

Hunziker JC, Frey M, Clauer N, Dallmeyer RD, Friedrichsen H, Flehmig W, Hochstrasser K, Roggwiller P, Schwander H. 1986. The evolution of illite to muscovite: Mineralogical and isotopic data from the Glarus Alps, Switzerland. *Contribution to Mineralogy and Petrology* 92:157–180.

Inoue A, Kohyama N, Kitagawa R, Watanabe T. 1987. Chemical and morphological evidence for the conversion of smectite to illite. *Clays and Clay Minerals* 35:111–120.

James AT, Baker DR. 1976. Oxygen isotope exchange between illite and water at 22° C. *Geochimica et Cosmochimica Acta* 40:235–239.

Klug HP, Alexander LE. 1974. *X-Ray Diffraction Procedures*, 2nd ed.: Wiley, New York. 992 p.

Kotzer TG, Kyser TK. 1995. Petrogenesis of the Proterozoic Athabasca basin, Northern Saskatchewan, Canada, and its relation to diagenesis, hydrothermal uranium mineralization and paleohydrogeology. *Chemical Geology* 120:45–89.

Kübler B. 1966. La cristallinité de l'illite et les zones tout à fait supérieures du métamorphisme. *In Étages Tectoniques*: Colloque de Neuchâtel, La Baconnière, Suisse. p. 105–122.

Kübler B. 1997. Concomitant alteration of clay minerals and organic matter during burial diagenesis. *In* Paquet H, Clauer N (Editors). *Soils and Sediments*: Springer Verlag, New York. p. 327–362.

Kübler B, Goy-Eggenberger D. 2001. La cristallinité de l'illite revisitée: Un bilan des connaissances acquises ces trente dernières années. *Clay Minerals* 36:143–157.

Kulla JB, Anderson TF. 1978. Experimental oxygen isotope fractionation between kaolinite and water. *In* Zartman RE (Editor). *Short Papers of the 4th International Conference of Geochronology, Cosmochronology and Isotope geology*: US Geological Survey, Open File Report. p. 234–235.

Lampe C, Person M, Nöth S, Ricken W. 2001. Episodic fluid flow within continental rift basins: Some insights from field data and mathematical models of the Rhinegraben. *Geofluids* 1:42–52.

Land LS, Dutton SP. 1978. Cementation of a Pennsylvanian Deltaic Sandstone: Isotopic data. *Journal of Sedimentary Petrology* 48:1167–1176.

Lanson B, Besson G. 1992. Characterization of the end of smectite-to-illite transformation: Decomposition of X-ray patterns. *Clays and Clay Minerals* 40:40–52.

Lee MC. 1984. Diagenesis of the Permian Rotliegendes Sandstone, North Sea: K/Ar, O^{18}/O^{16} and petrologic evidence [PhD Thesis]: Case Western Reserve University, Cleveland, Ohio.

Lee MC, Aronson JL, Savin SM. 1989. Timing and conditions of Permian Rotliegendes sandstone diagenesis, southern North Sea: K/Ar and oxygen isotope data. *American Association of Petroleum Geologists Bulletin* 73:195–215.

Lerman A, Clauer N. 2005. Losses of radiogenic ^{40}Ar in sediment fine-clay size fractions of sediments. *Clays and Clay Minerals* 53:233–248.

Lerman A, Ray BM, Clauer N. 2007. Radioactive production and diffusional loss of radiogenic ^{40}Ar in clays in relation to its flux to the atmosphere. *Chemical Geology* 243:205–224.

Liewig N, Clauer N. 2000. K-Ar dating of varied microtextural illite in Permian gas reservoirs, northern Germany. *Clay Minerals* 35:271–281.

Liewig N, Clauer N, Sommer F. 1987. Rb-Sr and K-Ar dating of clay diagenesis in Jurassic sandstone reservoirs. *American Association of Petroleum Geologists Bulletin* 71:1467–1474.

Merriman RJ, Roberts B, Peacor DR. 1990. A transmission electron microscope study of white mica crystallite size distribution in mudstone to slate transitional sequence, North Wales. *Contribution to Mineralogy and Petrology* 106:27–40.

Meunier A, Velde B. 2004. *Illite*: Springer, Berlin. 286 p.

Milliken KL. 1992. Chemical behavior of detrital feldspars in mudrocks versus sandstones, Frio Formation (Oligocene), South Texas. *Journal of Sedimentary Petrology* 62:790–801.

Moore DM, Reynolds RC. 1997. *X-Ray Diffraction and Identification of Clay Minerals,* 2nd ed.: Oxford University Press. 378 p.

Mossmann JR, Clauer N, Liewig N. 1992. Dating thermal anomalies in sedimentary basins: The diagenetic history of clay minerals in the Triassic sandstones of the Paris Basin (France). *Clay Minerals* 27:211–226.

Nadeau PH, Wilson MJ, McHardy WJ, Tait JM. 1984. Interstratified clays as fundamental particles. *Science* 225:923–925.

National Institute of Standards and Technology (NIST). 2005. Chemistry Web Book, Thermodynamic Properties of Fluid Systems. http://webbook.nist.gov/chemistry/.

O'Neil JR, Kharaka YK. 1976. Hydrogen and oxygen isotope exchange reactions between clay minerals and water. *Geochimica et Cosmochimica Acta* 40:241–246.

Osborne MJ, Swarbrick RE. 1999. Diagenesis in North Sea HPHT clastic reservoirs—consequences for porosity and overpressure prediction. *Marine and Petroleum Geology* 16:337–353.

Pagel M, Meyer A. 1991. A fluid inclusion and fission track study in the Sancerre-Crouy deep drill hole: ECROFI XI [abstract], 10–12 April, Florence. 1 p.

Patrier P, Beaufort D, Laverret E, Bruneton P. 2003. Dickite and $2M_1$ illite in deeply buried sandstones from the middle Proterozoic Kombolgie Formation (Northern Territory, Australia). *Clays and Clay Minerals* 51:102–116.

Peacor DR. 1992. Diagenesis and low-grade metamorphism of shales and slates. *Reviews in Mineralogy and Geochemistry* 27:335–380.

Person M, Raffensperger JP, Ge S, Garven G. 1996. Basin-scale hydrogeologic modeling. *Reviews in Geophysics* 34:61–87.

Platt J. 1993. Controls on the clay mineral distribution and chemistry in the early Permian Rotliegend of Germany. *Clay Minerals* 28:393–416.

Pollastro RM. 1993. Considerations and applications of the illite/smectite geothermometer in hydrocarbon-bearing rocks of Miocene to Mississippian age. *Clays and Clay Minerals* 41:119–133.

Powers MC. 1959. Adjustment of clays to chemical change and the concept of the equivalence level. *Clays and Clay Minerals* 6:309–326.

Renac C, Kyser TK, Durocher K, Dreaver G, O'Connor T. 2002. Comparison of diagenetic fluids in the Proterozoic Thelon and Athabasca basins, Canada: Implications for protracted fluid histories in stable intracratonic basins. *Canadian Journal of Earth Sciences* 39:113–132.

Renac C, Meunier A. 1995. Reconstruction of paleothermal conditions in a passive margin using illite–smectite mixed-layer series (BA1 scientific deep drill-hole, Ardeche, France). *Clay Minerals* 30:107–118.

Reynolds RC Jr. 1985. *NEWMOD©: A Computer Program for the Calculation of One-Dimensional Diffraction Patterns of Mixed-Layered Clays*: R.C. Reynolds, Jr., Hanover, New Hampshire.

Rieken R. 1988. *Lösungs-Zusammensetzung und Migrationsprozesse von Paläo-Fluidsystemen in Sedimentgesteinen des Norddeutschen Beckens (Mikrothermometrie, Laser-Raman-Spektroskopie und Isotopen–Geochemie)*: Arbeiten der Geologie und Paläontologie, Universität Göttingen.

Roberson HE, Lahann RW. 1981. Smectite to illite conversion rates: Effect of solution chemistry. *Clays and Clay Minerals* 29:129–135.

Rosenberg PE. 2002. The nature, formation and stability of end-member illite: A hypothesis. *American Mineralogist* 87:103–107.

Savin SM. 1967. Oxygen and hydrogen isotope ratios in sedimentary rocks and minerals [PhD dissertation]: California Institute of Technology, Pasadena.

Savin SM, Lee M. 1988. Isotopic studies of phyllosilicates. *Reviews in Mineralogy and Geochemistry* 19:189–224.

Shang C, Rice JA, Eberl DD, Lin JS. 2003. Measurement of illite particle thickness using a direct Fourier transform of small-angle X-ray scattering data. *Clays and Clay Minerals* 51:293–300.

Sharp ZD. 1997. Application of stable isotope geochemistry to low-grade metamorphic rocks. *In* Frey M, Robinson D (Editors). *Low-Grade Metamorphism*: Blackwell Science Ltd, Oxford, UK. p. 227–260.

Sheppard SMF. 1986. Characterization and isotopic variations in natural. *In* Valley JW, Taylor HP, O'Neil JR (Editors). *Stable Isotopes in High Temperature Geological Processes*, Reviews in Mineralogy 16: Mineralogical Society of America, Chelsea. p. 165–183.

Sheppard SMF, Gilg HA. 1996. Stable isotope geochemistry of clay minerals. *Clay Minerals* 31:1–24.

Small JS. 1993. Experimental determination of the rates of precipitation of authigenic illite and kaolinite in the presence of aqueous oxalate and comparison to the K/Ar ages of authigenic illite in reservoir sandstones. *Clays and Clay Minerals* 41:191–208.

Spötl C, Worden RH, Walgenwitz F. 1996. Clay minerals as records of temperature conditions and duration of thermal anomalies in the Paris Basin, France: Discussion. *Clay Minerals* 31:203–208.

Środoń J. 1995. Reconstruction of maximum paleotemperatures at present erosional surface of the Upper Silesia Basin, based on the composition of illite/smectite in shales. *Studia Geologica Polonica* 108:9–22.

Środoń J. 1999. The use of clay minerals in reconstructing geological processes: Current advances and some prospectives. *Clay Minerals* 34:27–37.

Środoń J, Eberl DD. 1987. Illite. *Reviews in Mineralogy and Geochemistry* 13:495–544.

Šucha V, Kraus I, Gerthofferová H, Peteš J, Sereková M. 1993. Smectite to illite conversion in bentonites and shales of the East Slovak Basin. *Clay Minerals* 28:243–253.

Taylor HP Jr, Epstein S. 1962. Relationship between O^{18}/O^{16} ratios in coexisting minerals of igneous and metamorphic rocks, Part 1. *Geological Society of America Bulletin* 73:461–480.

Timar-Geng Z, Fügenshuh B, Schaltegger U, Wetzel A. 2004. The impact of the Jurassic hydrothermal activity on zircon fission track data from the southern Upper Rhine Graben area. *Schweizerische Mineralogische und Petrographische Mitteilungen* 84:257–269.

Ustinov VI, Grinenko VA. 1985. Intrastructural isotope distribution during mineral formation. *Geochemistry International* 22:143–149.

Velde B. 1985. *Clay Minerals: A Physico-Chemical Explanation of their Occurrence*, Developments in Sedimentology 40: Elsevier, Amsterdam. 426 p.

Velde B (Editor). 1995. *Origin and Mineralogy of Clays*: Springer Verlag, Heidelberg. 334 p.

Velde B, Vasseur G. 1992. Estimation of the diagenetic smectite to illite transformation in time-temperature space. *American Mineralogist* 77:967–976.

Warren BE, Averbach BL. 1950. The effect of cold-work distortion on X-ray patterns. *Journal of Applied Physics* 21:595–599

Weaver CE. 1957. The clay petrology of sediments. *Clays and Clay Minerals* 6:154–187.

Weaver CE. 1961. Clay minerals of the Ouachita structural belt and adjacent foreland. *In Ouachita System,* University of Texas Publication, 6120: Bureau of Economic Geology, Austin, Texas. p. 147–160.

Whitney G, Northrop HR. 1988. Experimental investigation of the smectite to illite reaction: Dual reaction mechanisms and oxygen–isotope systematics. *American Mineralogist* 73:77–90.

Yates DM, Rosenberg PE. 1996. Formation and stability of endmember illite: I.

Solution equilibration experiments at 100–250°C and $P_{v,soln}$. *Geochimica et Cosmochimica Acta* 60:1873–1883.
Zheng YF. 1993. Calculation of oxygen isotope fractionation in hydroxyl-bearing silicates. *Earth and Planetary Science Letters* 120:247–263.
Ziegler K, Longstaffe FJ. 2000. Clay mineral authigenesis along a mid-continental scale fluid conduit in Palaeozoic sedimentary rocks from southern Ontario, Canada. *Clay Minerals* 35:239–260.
Zwingmann H, Clauer N, Gaupp R. 1999. Structure-related geochemical (REE) and isotopic (K-Ar, Rb-Sr, $\delta^{18}O$) characteristics of clay minerals from Rotliegend sandstone reservoirs (Permian, Northern Germany). *Geochimica et Cosmochimica Acta* 63:2805–2823.

APPENDIX.—Calculation of $^{18}O/^{16}O$ fractionation in illite-water system as a function of water–solid ratio.

This section presents the equations of the $^{18}O/^{16}O$ equilibrium between water and illite at different water–mineral volume ratios. Starting with an initial condition of nonequilibrium in a sediment between pore water and illite, the equations define the final equilibrium isotopic values in the two phases.

1. The molar mass of oxygen ($^{16}O + {}^{18}O$) in H_2O, $m_{O,l}$, is given by

$$m_{O,l} = V_l \rho_l / M_{H_2O} \qquad (1)$$

where V_l is the solution volume in the bulk sediment, assumed as an approximation to be the volume of H_2O, ρ_l is the solution (water) density, and M_{H_2O} is the gram-formula mass of H_2O. ρ_{H_2O} (NIST 2005) at 60° C: 100 bar, 0.98748; 150 bar, 0.98960 g/cm^3 (hydrostatic pressure at about 1–1.5 km depth) at 150° C: 300 bar, 0.93286; 450 bar, 0.94030 g/cm^3 (hydrostatic pressure at about 3–4.5 km depth) M_{H_2O} molecular weight of H_2O = 18.01528 g/mol ($M_H = 1.00794$, $M_O = 15.9994$)

2. The molar mass of oxygen ($^{16}O + {}^{18}O$) in illite, $m_{O,s}$, of stoichiometric composition $K_{0.6}Al_{2.2}Si_{3.7}O_{10}(OH)_2$ is

$$m_{O,s} = V_s \rho_s N_O / M_s \qquad (2)$$

where V_s is the solid mineral volume in the bulk sediment, ρ_s is the mineral density, N_O is the number of O atoms in the stoichiometric formula that exchange $^{18}O/^{16}O$ with H_2O in the solution, and M_s is the gram-formula mass of the mineral illite.

$\rho_{illite} = 2.7$ g/cm^3 (range 2.6–2.9)

$N_O = 2$ or 12 (2 hydroxyl oxygens exchanging or all 12 oxygens exchanging with water)

M_s molecular weight of $K_{0.6}Al_{2.2}Si_{3.7}O_{10}(OH)_2 = 380.7434$ g/mol

3. The ratio of the oxygen molar masses in the liquid and solid phases, R_{ls}, is from Eqs. 1 and 2, thus:

$$R_{ls} \equiv \frac{m_{O,l}}{m_{O,s}} = \frac{\rho_l}{\rho_s N_O} \times \frac{M_s}{M_{H_2O}} \times \frac{V_l}{V_s}. \qquad (3)$$

In Eq. 3, the variable of interest is the volume quotient water–mineral, V_l/V_s. The effects on water density, ρ_l, of a higher temperature (60–150° C, considered in this article) and higher hydrostatic pressure at depths of about 1.5 to 4 km are small, and they insignificantly affect the results based on the value of $\rho_l = 1$ g/cm^3. Using the parameter values given above,

$R_{ls} = 3.914 V_l/V_s$ for $N_O = 2$ when only 2 O atoms in illite are exchanging with water, and

$R_{ls} = 0.652 V_l/V_s$ for $N_O = 12$ when all of the O atoms in illite are exchanging with water.

4. The number of moles of ^{18}O in H_2O initially is the product of the moles of oxygen, $m_{O,l}$, and the mole fraction of ^{18}O, $n_{l,0}$, where subscript 0 denotes the initial state: $m_{O,l}\, n_{l,0}$. After a fraction x of ^{18}O in water is withdrawn by isotope equilibrium exchange with illite, the remaining number of moles in solution is

$$m_{O,l} n_{l,0}(1-x)$$

and the remaining mole fraction is

$$\frac{m_{O,l} n_{l,0}(1-x)}{m_{O,l}} = n_{l,0}(1-x). \qquad (4)$$

The ^{18}O mole fraction is a small number, of an order of 0.002 in natural waters:

$$n_l = \frac{^{18}O}{^{18}O + {}^{16}O} \approx \frac{^{18}O}{^{16}O}.$$

Small changes in the $^{18}O/^{16}O$ ratio due to exchange of ^{18}O with ^{16}O are not considered to affect the mean molecular weight or gram-formula mass of H_2O, and the quantity $m_{O,l}$ is taken as constant.

5. Illite, in an initial state of nonequilibrium with the solution, may contain $m_{O,s}\, n_{s,0}$ moles ^{18}O, where $n_{s,0}$ is the initial mole fraction of ^{18}O in the illite oxygen, and it will gain $m_{O,l}\, n_{l,0}\, x$ moles ^{18}O by exchange with the solution. Thus, the total number of moles ^{18}O in illite after the exchange with the solution is

$$m_{O,s} n_{s,0} + m_{O,l} n_{l,0} x$$

and the ^{18}O mole fraction in illite is:

$$\frac{m_{O,s} n_{s,0} + m_{O,l} n_{l,0} x}{m_{O,s}} = n_{s,0} + R_{ls} n_{l,0} x. \qquad (5)$$

6. The equilibrium fractionation factor for $^{18}O/^{16}O$ exchange between illite and water is the quotient of the isotope ratios, equivalent to their mole fractions, in the two phases. Also, in the notation of $\delta^{18}O$ (fraction or $\delta^{18}O$ ‰/1000), the fractionation factor α is

$$\alpha = \frac{(^{18}O/^{16}O)_{illite}}{(^{18}O/^{16}O)_{liq}} = \frac{1+\delta^{18}O_s}{1+\delta^{18}O_l}. \qquad (6)$$

For a system of water–mineral volume ratio V_l/V_s, the ratio of the mole fractions is from Eqs. 5 and 6, as follows:

$$\alpha = \frac{n_{s,0} + R_{ls}n_{l,0}x}{n_{l,0}(1-x)} = \frac{n_{s,0}/n_{l,0} + R_{ls}x}{1-x}$$
$$= \frac{(1+\delta_{s,0})/(1+\delta_{l,0}) + R_{ls}x}{1-x}. \tag{7}$$

In Eq. 7, the amount of ^{18}O that is transferred either from the solution to illite ($x > 0$) or from the illite to solution ($x < 0$) at equilibrium depends on α, which is a function of temperature, the initial nonequilibrium isotope concentrations in the solid ($\delta_{s,0}$) and solution ($\delta_{l,0}$), and the water–mineral volume ratio in the term R_{ls}:

$$x = \frac{\alpha - (1+\delta_{s,0})/(1+\delta_{l,0})}{\alpha + R_{ls}}. \tag{8}$$

8. Finally, the isotopic composition of the solution (δ_l) and illite (δ_s) is obtained from Eqs. 4, 6, 7, and 8:

$$\delta_l = (1+\delta_{l,0})(1-x) - 1; \tag{9}$$

$$\delta_s = (1+\delta_l)\alpha - 1. \tag{10}$$

In Figure 10, the calculated values of δ_l and δ_s are shown at two temperatures and a wide range of the water–mineral volume ratios.

APPLICATION OF GAS ISOTOPES TO THE THERMAL HISTORY ANALYSIS OF BASINS

XINYU XIA AND YONGCHUN TANG
Power Environmental Energy Research Institute, 738 Arrow Grand Circle, Covina, California 91722, USA
e-mail: tang@peeri.org

ABSTRACT: Carbon isotope composition can be used to derive natural gas generation temperature and source rock maturity by applying a rigorous parallel first-order reaction model based on quantum chemistry and microkinetics. In gas generation reactions, the breaking of $^{13}C–^{12}C$ and $^{12}C–^{12}C$ bonds has different frequency factors and activation energies. After the optimizing of these kinetic parameters with laboratory experiments, we can use the model to derive natural gas formation temperature and source rock maturity with given carbon isotope composition data. This method is useful for analyzing the thermal history of a basin along with petroleum generation and accumulation history of reservoirs, especially for strata that lack vitrinite for reflectance measurements.

KEY WORDS: natural gas, carbon isotope, thermal history, basin analysis, kinetics

GEOCHEMICAL PRINCIPLES UNDERLYING THE METHOD

Hydrocarbons in natural gas (methane, ethane, propane, and isomers of butane and pentane) are predominantly derived from kerogen and oil cracking (Tissot and Welte 1984). The main geochemical indices of natural gas are the fractions of the different gas components (hydrocarbon gases and major non-hydrocarbon gases such as CO_2 and N_2) and isotopic composition of these components. With increasing maturity, the methane content of gas generated from a source rock, or gas dryness, increases (Schoell 1983), although this variation may be small, according to hydrous pyrolysis studies by Lewan (1997) and Schimmelmann et al. (1999). Because of secondary processes, in particular fractionation during migration, the relation between gas dryness and maturity is often hard to quantify and apply. Instead, application of isotope composition is more common, where the carbon isotope composition ($\delta^{13}C$) of each hydrocarbon component in natural gas is defined by

$$\delta^{13}C_i = \left[\frac{(^{13}C/^{12}C)_i}{(^{13}C/^{12}C)_r} - 1\right] \times 1000, \quad (1)$$

where the subscript i indicates the hydrocarbon molecule with i carbon atoms and where r indicates the reference sample of carbon isotope composition, that is, the international Pee Dee belemnite standard with a generally accepted absolute $^{13}C/^{12}C$ ratio of 0.0112372 (Werner and Brand 2001).

The reaction of kerogen cracking is kinetically controlled and has a kinetic isotope effect (KIE). Therefore, just as with vitrinite reflectance (R_o), the isotopic composition of natural gas is a geochemical indicator of the thermal history of sedimentary basins. Under some circumstances, gas isotopes can supplement R_o in thermal history analysis, especially for strata that lack vitrinite.

Natural gas formation is generally accepted as a result of a set of parallel first-order reactions (e.g., Burnham et al. 1987, Behar et al. 1992), meaning that a group of substances reacts simultaneously to yield one or more groups of products (Lakshmanan and White 1994): for example, the decomposition of kerogen to hydrocarbons, NSO compounds, and asphaltene is a set of parallel reactions. Each reaction in the set of parallel reactions has a unique activation energy. Kerogen can be regarded as a mixture of the reacting precursors of these reactions. A first-order reaction means that the reaction rate is proportional to the concentration of only one of the reactants of the reaction and is independent of the other reactants (if they exist). This proportional coefficient, or rate constant, is expressed in the Arrhenius form as

$$k_a = A\exp(-E_a/RT), \quad (2)$$

and the reaction rate is expressed as

$$dc/dt = -k_a c, \quad (3)$$

where k_a is the rate constant of any of the parallel reactions; A and E_a are the frequency factor and activation energy, respectively; R is the gas constant; T is temperature; and c is the concentration of the precursor that generates hydrocarbon in each of the reactions.

The hydrocarbon generation process involves numerous complex reactions; however, the rate-limiting step in all of these reactions is the rupture of C-C bonds, which determines the A and E_a of the overall reaction. Carbon isotopes are an influential factor in this step. Because there are two stable carbon isotopes in organic matter (^{12}C and ^{13}C, with ^{12}C accounting for >98% of carbon atoms), there are consequently two types of C-C bonds: ^{12}C-^{12}C and ^{12}C-^{13}C (the ^{13}C-^{13}C bond has a rather small probability of occurrence and can thus be omitted). Breakage of these different bonds has different A and E_a values, resulting in different reaction rates during hydrocarbon generation. The carbon isotope composition of hydrocarbon products will therefore change with varying temperatures and reaction extents. This KIE is the chemical basis for analyzing thermal histories with the gas isotope method.

Some empirical relations have been established between carbon isotope composition and thermal maturity (Stahl and Carey 1975, Stahl et al. 1977, Schoell 1983), but accurate and generalized relationships between carbon isotope composition and thermal history rely on theoretical kinetic studies (Cramer et al. 1998, 2001; Tang et al. 2000). Tang et al. (2000) investigated the difference of k_a and E_a using quantum chemistry calculations for the above two types of C-C bonds; each C-C bond is adjacent to different functional groups that possibly exist in kerogens. The results show that the frequency factor of ^{13}C-^{12}C cracking is about 1.02 times larger than that of ^{12}C-^{12}C cracking. The difference between the activation energies of these two types of C-C bonds is not constant; it changes with the activation energy itself and can be expressed in the following formula:

Analyzing the Thermal History of Sedimentary Basins: Methods and Case Studies
SEPM Special Publication No. 103, Copyright © 2012
SEPM (Society for Sedimentary Geology), ISBN Print ISBN 978-1-56576-315-9, CD/DVD ISBN 978-1-56576-317-3, p. 147–152.

$$\Delta E_a = \beta_L + (\beta_H - \beta_L) f(E_a), \quad (4)$$

where ΔE_a is the difference between E_a for a ^{13}C-^{12}C bond and a ^{12}C-^{12}C bond; β_H and β_L are two constants standing for the minimum and maximum probable value of ΔE_a, respectively; and $f(E_a)$ is the cumulative Gaussian distribution as a function of E_a of ^{13}C-^{12}C bonds. Examples of ΔE_a changing with E_a are shown in Figure 1. Typically the value of ΔE_a for natural gas generation is between $\beta_L = 0$ cal/mol and $\beta_H = 70$ cal/mol.

By substituting the difference between kinetic parameters for ^{12}C-^{12}C and ^{13}C-^{12}C bonds into Eq. 2, the yield of hydrocarbon products with and without ^{13}C atoms can be estimated for a given frequency factor and distribution of activation energy.

Eqs. 1 through 4 reflect the relationship between carbon isotope composition and hydrocarbon formation temperature. Therefore, the thermal history of a gas-producing basin can be obtained from the gas isotope composition if the gas generation model is calibrated.

An insight into Eqs. 1 and 3 reveals more details about natural gas isotope composition. The timing and amount of trapped natural gas affect the cumulative isotope composition of an accumulation. Natural gas generation is a continuous process, and the instantaneous carbon isotope ratios of the generated gas change continuously. Suppose a hydrocarbon generation "kitchen" is presently at high maturity; there are two cases for gas accumulation history, as follows: (1) all gas generated from low to high maturity is trapped in a single reservoir, or (2) gas generated at different maturity stages is trapped in different reservoirs (or the gas from early maturity stage is not trapped). We refer to the first case "cumulative," since the reservoir has accumulated all of the gas generated from the source rock over the complete maturation history. In this case, the gas isotope composition is a weighted average of gas from low to high maturity components. We refer to the second case "instantaneous," because each reservoir trapped gas generated at one instant in time (more realistically, over a short period of time), and the gas isotope composition reflects only the maturity of the source rock at the time the gas was generated.

PRACTICAL APPLICATION OF THE METHOD

For real application, measurements of carbon isotope composition of hydrocarbons in natural gas are needed, and a suitable kinetic model of hydrocarbon generation from kerogen should be calibrated. Figure 2 offers an example that presents the activation energy distribution of methane generation from n-octadecane ($C_{18}H_{38}$) cracking, including the distribution of ΔE_a, which changes with E_a.

Figure 3 shows the value of $\delta^{13}C$ of methane generated at two geologically reasonable heating rates, using the above activation energy distribution; the frequency factor for ^{12}C-^{12}C bond cracking is $A_{12} = 9.37 \times 10^{11}\ s^{-1}$, and $A_{13} = A_{12} \times 1.02$ for ^{13}C-^{12}C bond cracking. This theoretical example illustrates the relationship between the values of gas isotope composition and the formation temperature of natural gas. Figure 3 shows that an instantaneous gas reservoir generally has a heavier carbon isotope composition (more positive $\delta^{13}C$ values) for methane compared with a cumulative gas reservoir, except for gases formed at low source rock maturity ($T < 150°$ C). The heating rate, a

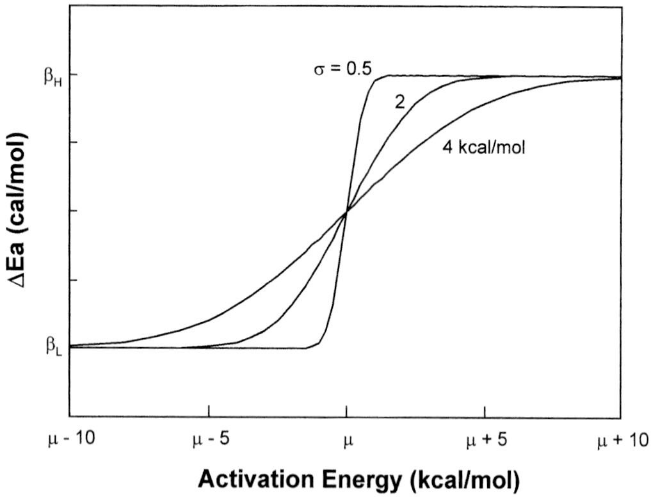

FIG. 1.—Examples of ΔE_a as a cumulative Gaussian distribution function of variance from E_a (Tang et al. 2000). μ and σ represent the mean and variance of the Gaussian distribution, respectively.

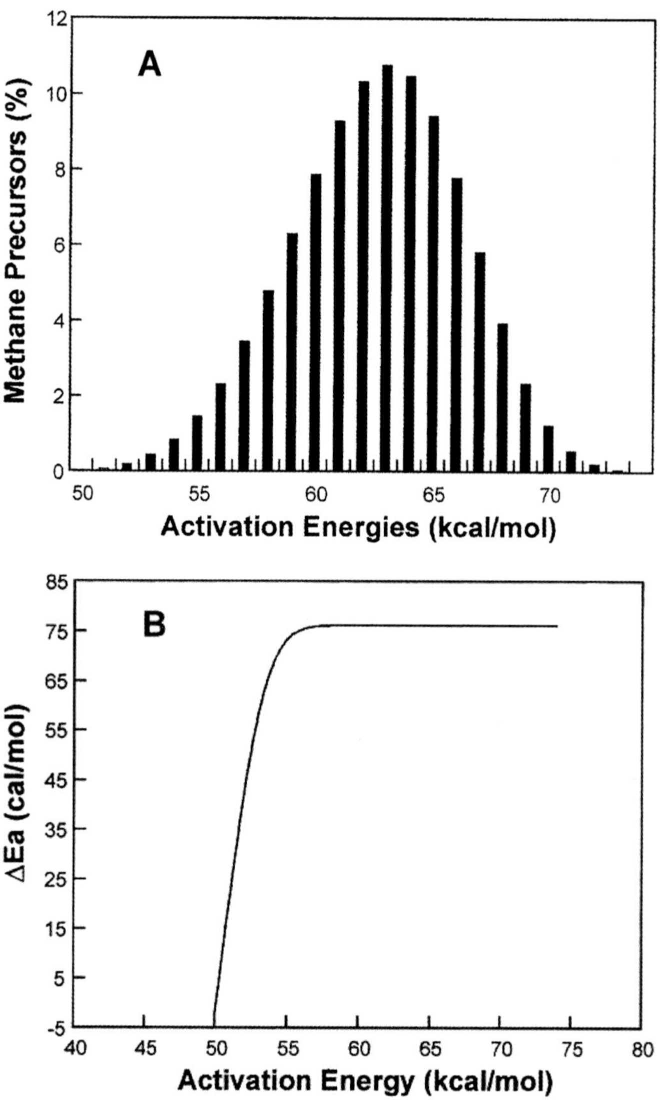

FIG. 2.—(A) Calibrated distribution of activation energies calculated for methane generation from thermal cracking of n-octadecane in the laboratory; (B) Calibrated distribution of activation energy difference (ΔE_a) between ^{12}C-^{13}C and ^{13}C-^{12}C bonds as a function of activation energy (Tang et al. 2000).

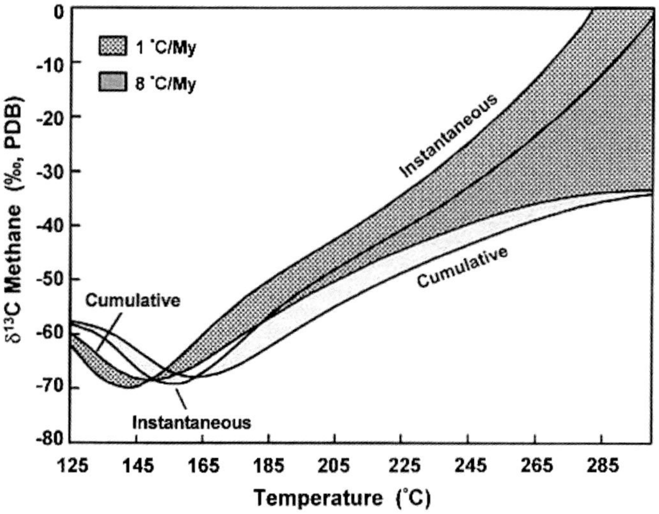

FIG. 3.—Plot of predicted isotopic composition vs. temperature for instantaneous and cumulative methane generated from pyrolysis of n-octadecane. The dark shaded zone = heating rate 1° C/my; the light shaded zone = heating rate 8° C/my. Each zone is bracketed with the cumulative curve and instantaneous curve. It should be noted that the $\delta^{13}C$ value for the instantaneous curve of each heating rate is more positive (heavier) than that for the cumulative curve, except at low temperatures. The curves for 1 and 8° C/my bracket the range of heating rates applicable to most sedimentary basins (Tang et al. 2000).

significant aspect of thermal history, also affects isotope composition, with higher heating rate associated with more negative $\delta^{13}C$ values.

PITFALLS AND UNCERTAINTIES IN APPLICATION OF THE METHOD

The relationship between gas isotopes and basin thermal history is based on rigorous derivation of chemical kinetics; however, the complexity of geological processes results in some pitfalls and uncertainties in the practical application of this method.

Migration of Natural Gas

Natural gas may have migrated long vertical distances from the source rock through faults or fracture systems, but it will nonetheless reflect the thermal history of the source rock instead of the reservoir strata. Therefore, determination of the source rock depth is essential for thermal analysis using gas isotopes. For example, the natural gas in Tertiary reservoirs in the Gippsland basin, offshore Australia, has migrated from deeper Cretaceous source rocks (James 1983). However, change of isotope composition of natural gas due to diffusive fractionation is generally thought to be negligible (<1 per mil for methane; Fuex 1980), although Prinzhofer and Huc (1995) suggested that leakage from a reservoir affects the difference between $\delta^{13}C_1$ and $\delta^{13}C_2$.

Complexity of Mathematical Model

To describe natural gas generation process with isotope fractionation, dozens of parallel reactions of various activation energies may be involved; each reaction is described by an ordinary differential equation (ODE). The simultaneous solving of an ODE group of this size requires relatively complicated mathematical computation. This refers to computational difficulty, which can be largely addressed with modern computer-aided approaches, but it also implies that multiple thermal history solution may be possible for individual cases.

Variation of Kerogen Composition

Different types of kerogen may have different initial carbon–isotope compositions and gas-generation kinetics, which influence the relationship between isotope composition and formation temperature of natural gas. Therefore, for an accurate model the hydrocarbon gas should be correctly correlated to its source rock.

Other Mechanisms of Hydrocarbon Gas Generation

In addition to the conventional mechanisms of gas generation from kerogen pyrolysis and oil cracking, there are other mechanisms. For example, methane can originate from bacterial degradation of organic matter, which results in light isotope composition. The source rock maturity or gas formation temperature can therefore be underestimated by using a conventional thermal model if a gas reservoir has a significant component of methane from this source. Recently it was found that the hydrogen isotope compositions of methane can become lighter at high maturity in some natural gas fields, such as the Appalachian Basin (Burruss and Laughrey 2010). Although the mechanism for this is not totally clear, this effect may indicate water involvement in hydrocarbon generation at high maturity. The relationship between isotope ratios and thermal history for these mechanisms remains an open question.

CASE STUDY

We present a case study that employs gas isotopes to determine the thermal conditions for natural gas formation in the Kuqa Depression of the Tarim Basin, China. The Kuqa Depression, a foreland basin, is located in the northern Tarim Basin south of the Tianshan Mountains in northwestern China. It contains Mesozoic–Cenozoic, nonmarine sedimentary strata up to 12 km thick, including the Lower and Middle Jurassic coal-bearing sequence, over 8000 m of Cretaceous and Tertiary clastic red beds, and Eocene and Miocene evaporates that serve as cap rocks (Liang et al. 2003). Several gas fields have been discovered since 1993, with over 500 billion cubic meters of proven gas reserves.

The most important gas source rock in the Kuqa Depression includes Jurassic coal and carbonaceous mudstone (Liang et al. 2003). To calibrate the kinetic parameters of the isotopic fractionation model for the source rocks, pyrolysis experiments on a Jurassic coal sample from the Yinan-2 Well (depth: 4316.8 m; for location see Liang et al. 2003) were performed, and the methane yield and its carbon isotope composition were measured. The sample contained type-III organic matter with vitrinite reflectance of 0.88% and total organic carbon (TOC) of 76.5%. Rock-Eval analysis showed S1 = 6.34 mg/g and S2 = 112.8 mg/g and reflected a hydrogen index of 147.39 mg/g TOC. Kerogen was extracted with methanol/acetone/benzene (=2:3:5 v/v/v) from a 100-mesh–sieved fraction of the coal sample.

Anhydrous pyrolysis of kerogen was performed in a closed system in order to take into account natural gas generated from the secondary cracking. The reactor is a gold tube with a length of 40 mm and a base diameter of 4.5 mm. The prepared kerogen samples of 20 to 50 mg were sealed into the tube under argon atmosphere; the pressure was maintained at 50 MPa bar with an autoclave. The samples were heated up to different temperature, from 300 to 600° C, with two heating rates (2° C/hr and 20° C/hr). The experimental temperature error was less than 1° C, and the variation of pressure was less than 0.1 MPa.

After each of the tubes reached the planned temperature point, it was removed from the autoclave and the gas products were transferred to a volumetrically calibrated system for yield and isotope analysis. Gas composition of the products was analyzed with an HP 5890-II gas chromatograph, with an error of less than 1%. The carbon isotope composition of methane was analyzed with a VG Isochrom GC-IRMS, with an error of less than 0.3 per mil. The experiments were carried out in the Guangzhou Institute of Geochemistry, China.

Figure 4 shows the methane yield and methane carbon isotope composition from two series of pyrolysis experiments on a Jurassic coal sample at two different heating rates (2° C/hr and 20° C/hr). Methane generation was considered as a group of first-order parallel reactions from n precursors with different activation energies; the decomposition of each precursor can be written from Eqs. 1 and 2, thus:

$$\frac{dy_i}{dT} = -\frac{A}{v}\exp\left(-\frac{E_i}{RT}\right)y_i, \tag{5}$$

where A indicates the frequency factor; v indicates the heating rate; E_i indicates the activation energy of the ith precursor of methane; and y_i indicates the conversion of the ith precursor of methane (the initial value is assumed to be 1), which changes with time. The overall yield of methane changing with temperature is

$$y_{Methene} = \frac{1}{n}\sum_{i=1}^{n}[1 - y_i(T)]. \tag{6}$$

This equation is fitted to the data in Figure 4A; the best fit provides the optimized parameters of A and E_i of Eq. 4. To make the model more rigorous (that is, to reduce the number of parameters needed to fit), we applied a rectangular (uniform) distribution of the precursors with different activation energy (for a physical explanation of a rectangular distribution, see Rudzinski and Panczyk [2000]).

The yield of $^{13}CH_4$ can also be expressed in the form of Eqs. 5 and 6, using $A_{13} = A \times 1.02$, and $E_{13,I} = E_i + \Delta E_a$, where ΔE_a is expressed as Eq. 3. Then the theoretical isotope composition of methane product can be obtained using Eq. 1 after the $^{12}CH_4$ and $^{13}CH_4$ is calculated for each time point. By fitting the theoretical results to Figure 4B, the optimized parameters for kinetic fractionation are obtained. All of the parameters are shown in Table 1.

Using the model with the above optimized parameters, the relationship between gas isotopes and thermal parameters under geological conditions can be derived. The Kuqa Depression experienced rapid sedimentation in the Neogene, with a maximum burial rate of nearly 600 m/my. With a maximum probable thermal gradient of 25° C/km (Wang et al. 2003), the maximum possible heating rate is about 15° C/my, much higher than is common in sedimentary basins.

To examine the effect of uncertainty in the thermal history model, we compared the methane isotope compositions corresponding to two heating rates: 2° C/my for a common burial rate and 15° C/my for the most rapid burial rate, as in this example. The two different heating rates results in greater than a 10° C difference in the estimated

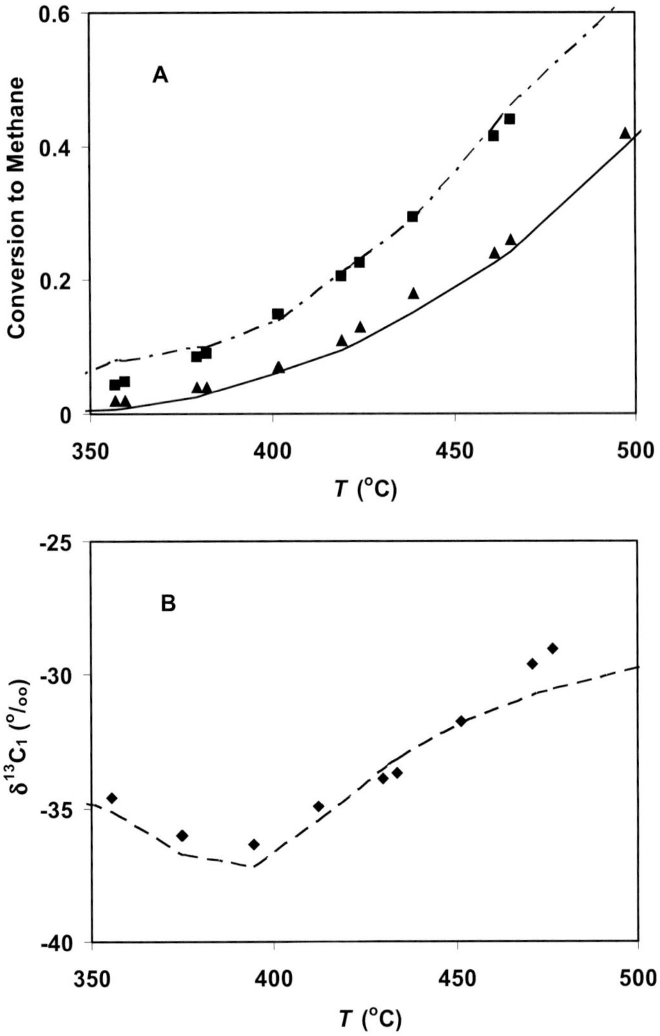

FIG. 4.—Best fit of the isotope kinetics model based on the Jurassic coal pyrolysis experiments. (A) Yield of methane, triangles and solid line: experimental and calculated data with heating rate of 20° C/hr, respectively; squares and dashed line: experimental and calculated data with heating rate of 2° C/hr, respectively; (B) carbon isotope composition of methane from pyrolysis, dots and line: experimental and calculated results, respectively. Experimental data from Li et al. (2005). For optimized parameters, see Table 1.

TABLE 1.—Kinetic parameters for methane generated from Jurassic coal in the Kuqa Depression.

$\delta^{13}C_{initial}$ (per mil)	A_{12} (s^{-1})	A_{13}/A_{12}	$E_{12,min}$ (kcal/mol)	$E_{12,max}$ (kcal/mol)	β_L (cal/mol)	β_H (cal/mol)	μ_μ (kcal/mol)	σ (kcal/mol)
−26.0	1.0×10^{14}	1.02	53.6	73.8	4	80	54.9	1.39

Note: $\delta^{13}C_{initial}$:$\delta^{13}C$ of precursors of methane; A_{12} and A_{13}: frequency factors of ^{12}C-^{12}C breaking and ^{13}C-^{12}C breaking, respectively; $E_{12,min}$ and $E_{12,max}$: the minimum and maximum values of a uniform distribution of activation energy to break ^{12}C-^{12}C bonds, respectively; β_L and β_H: the minimum and maximum values of activation energy difference between ^{12}C-^{12}C and ^{13}C-^{12}C, respectively; σ and μ: the mean and variance of a cumulative Gaussian function to describe the activation energy difference between ^{12}C-^{12}C and ^{13}C-^{12}C, respectively. Also see Eq. 4.

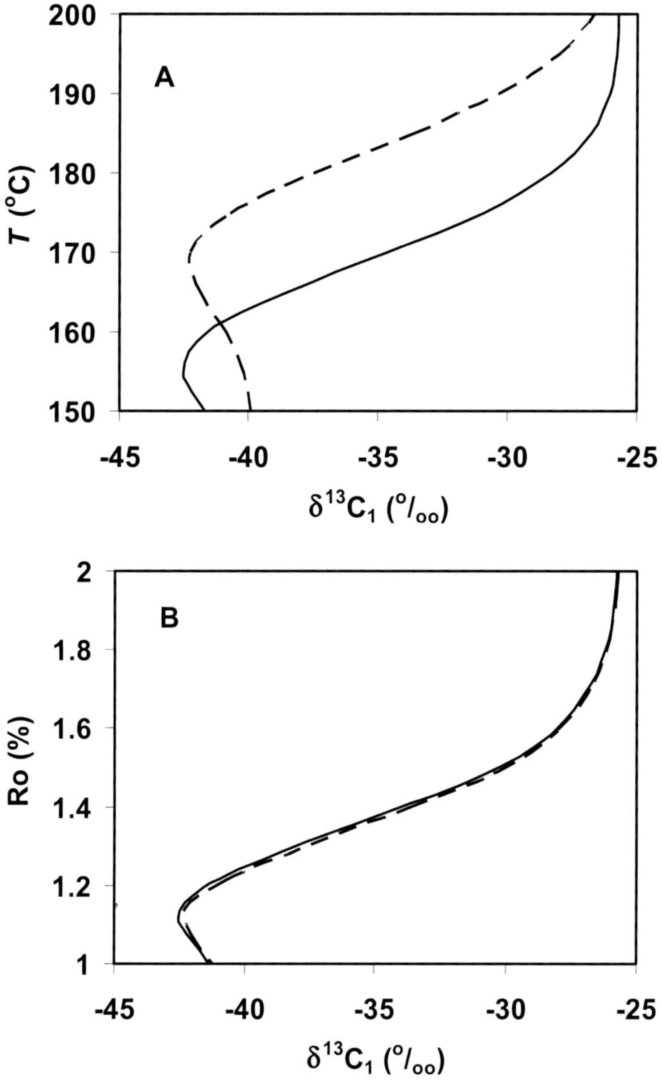

FIG. 5.—Relationship between isotope composition of methane (instantaneous case) and formation temperature (A) or vitrinite reflectance (B) of natural gas. Solid lines = heating rate of 15° C/my; dashed lines = heating rate 2° C/my.

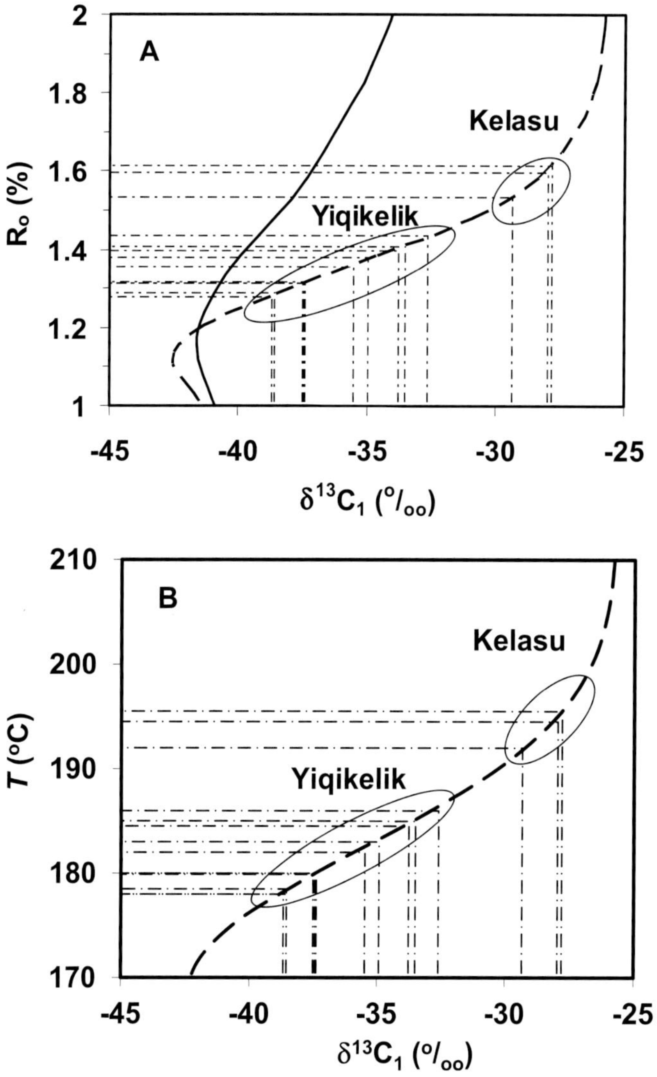

FIG. 6.—Derivation of R_o (A) and temperature (B) of gas formation from methane isotope composition. Solid line = cumulative trapping; dashed lines = instantaneous trapping; dotted-dash lines show the R_o or T values predicted from field $\delta^{13}C_1$ data ($\delta^{13}C_1$ from Liang et al. [2003]).

temperature of gas formation for a given $\delta^{13}C_1$ value; however, the difference for the estimated R_o at the time of gas generation (calculated using the Easy-Ro method; Sweeney and Burnham 1990) is negligible (Fig. 5). Therefore, gas isotope composition can be used to determine the maturity of the source rock and the corresponding R_o for the reconstruction of thermal history.

Figure 6A shows the derivation of R_o at the time of gas formation based on $\delta^{13}C_1$ data for some gas samples from the Kelasu–Yiqikelik zone in the Kuqa Depression (data from Liang et al. 2003). R_o–$\delta^{13}C_1$ curves for both cumulative and instantaneous models are presented; the plot indicates that the gas reservoir did not form as a cumulative case; otherwise the R_o values derived from $\delta^{13}C_1$ would be high, much greater than 2.0% for the samples from Kelasu Zone and some samples from the Yiqikelike zone, contradicting available geological and thermal data. Therefore, instantaneous trapping is the more reasonable scenario, from which R_o of ~1.6% is derived for maturity of the Jurassic source rock in the Kelasu Zone, and from which R_o of ~1.4% is derived for the maturity of the source rock in the Yiqikelik Zone. Using the temperature–$\delta^{13}C_1$ curve for instantaneous conditions and a heating rate of 15° C/my, the temperature of gas generation is estimated to have been close to 200° C.

CONCLUSIONS

Carbon isotope composition can be applied as an approach to derive natural gas generation temperature and source rock maturity based on a rigorous parallel first-order reaction model, considering the differences between kinetic parameters (including activation energies and frequency factors) for a ^{13}C–^{12}C bond and a ^{12}C–^{12}C bond breaking. After the kinetic parameters of the model are optimized with laboratory experiments, we can use them to derive natural gas formation temperature and source rock maturity with given carbon isotope composition data. This method is useful for analyzing the thermal history of a basin along with petroleum generation and accumulation history of reservoirs.

ACKNOWLEDGMENTS

We are thankful for the review by Nick Harris and Kenneth Peters, which greatly improved the article.

REFERENCES

Behar F, Kressmann S, Rudkiewicz JL, Vandenbroucke M. 1992. Experimental simulation in a confined system and kinetic modelling of kerogen and oil cracking. *Organic Geochemistry* 19:173–189.

Burnham AK, Braun RL, Gregg HR, Samoun AM. 1987. Comparison of methods for measuring kerogen pyrolysis rates and fitting kinetic parameters. *Energy Fuels* 1:452–458.

Burruss RC, Laughrey CD. 2010. Carbon and hydrogen isotopic reversals in deep basin gas: Evidence for limits to the stability of hydrocarbons. *Organic Geochemistry* 41:1285–1296.

Cramer B, Faber E, Gerling P, Krooss BM. 2001. Reaction kinetics of stable carbon isotopes in natural gas—insights from dry, open system pyrolysis experiments. *Energy Fuels* 15:517–532.

Cramer B, Krooss BM, Littke R. 1998. Modelling isotope fractionation during primary cracking of natural gas: A reaction kinetic approach. *Chemical Geology* 149:235–250.

Fuex AA. 1980. Experimental evidence against an appreciable isotopic fractionation of methane during migration. *Physics and Chemistry of the Earth* 12:725–732.

James AT. 1983. Correlation of natural gas by use of carbon isotopic distribution between hydrocarbon components. *American Association of Petroleum Geologists Bulletin* 67:1176–1191.

Lakshmanan CC, White N. 1994. A new distributed activation energy model using Weibull distribution for the representation of complex kinetics. *Energy Fuels* 8:1158–1167.

Lewan MD. 1997. Experiments on the role of water in petroleum formation. *Geochimica et Cosmochimica Acta* 61:3691–3723.

Li X, Xiao X, Mi J, Tang Y, Xiao Z, Liu D, Shen J, Liu J. 2005. Kinetic parameters of methane generated from source rocks and its application in the Kuqa Depression of the Tarim Basin. *Acta Geologica Sinica* 79:133–142. (in Chinese)

Liang D, Zhang S, Chen J, Wang F, Wang P. 2003. Organic geochemistry of oil and gas in the Kuqa Depression, Tarim Basin, NW China. *Organic Geochemistry* 34:873–888.

Prinzhofer AA, Huc AY. 1995. Genetic and post-genetic molecular and isotopic fractionations in natural gases. *Chemical Geology* 126:281–290.

Rudzinski W, Panczyk T. 2000. Kinetics of isothermal adsorption on energetically heterogeneous solid surfaces: A new theoretical description based on the statistical rate theory of interfacial transport. *Journal of Physical Chemistry B* 104:9149–9162.

Schimmelmann A, Lewan MD, Wintsch RP. 1999. D/H isotope ratios of kerogen, bitumen, oil, and water in hydrous pyrolysis of source rocks containing kerogen types I, II, IIS, and III. *Geochimica et Cosmochimica Acta* 63:3751–3766.

Schoell M. 1983. Modeling thermogenic gas generation using carbon isotope ratios of natural gas hydrocarbons. *American Association of Petroleum Geologists Bulletin* 67:2225–2238.

Stahl WJ, Carey BD Jr. 1975. Source-rock identification by isotope analyses of natural gases from fields in the Val Verde and Delaware basins, west Texas. *Chemical Geology* 16:257–267.

Stahl WJ, Wollanke G, Boigk H. 1977. Carbon and nitrogen isotope data of Upper Carboniferous and Rotliegend natural gases from North Germany and their relationship to the maturity of the organic source material. *In* Campos R, Gonji J (Editors). *Advances in Organic Geochemistry*: Enadisma, Madrid, Spain. p. 539–559.

Sweeney JJ, Burnham AK. 1990. Evaluation of a simple model of vitrinite reflectance based on chemical kinetics. *American Association of Petroleum Geologists Bulletin* 74:1559–1570.

Tang Y, Perry JK, Jenden PD, Schoell M. 2000. Mathematical modeling of stable carbon isotope ratios in natural gases. *Geochimica et Cosmochimica Acta* 64:2673–2687.

Tissot BP, Welte DH. 1984. *Petroleum Formation and Occurrence*, 2nd ed.: Springer, New York, 669 p.

Wang L, Li C, Liu S, Li H, Xu M, Wang Q, Ge R, Jia C, Wei G. 2003. Geotemperature gradient distribution of Kuqa Foreland Basin, North of Tarim, China. *Chinese Journal of Geophysics* 46:403–407. (in Chinese)

Werner RA, Brand WA. 2001. Referencing strategies and techniques in stable isotope ratio analysis. *Rapid Communications in Mass Spectrometry* 15:501–519.

EVALUATION OF THE HEAT FLOW IN THE SOUTHERN ALPS DURING MESOZOIC EXTENSION: IMPLICATIONS FOR HYDROCARBON EXPLORATION IN THE PO PLAIN FOREDEEP

PAOLO SCOTTI AND ROBERTO FANTONI

Eni S.p.A.–Exploration and Production Division, Via Emilia 1, 20097 San Donato Milanese, Milan, Italy

e-mail: paolo.scotti@eni.com

ABSTRACT: We reconstructed the thermal history, and therefore the evolution of heat flow in the southern Alps region (northern Italy), using organic matter (OM) maturity data obtained from samples of sedimentary units outcropping along the entire mountain chain. Regional OM maturity patterns are largely controlled by a high geothermal gradient and differential burial during the Norian–Liassic extensional phase. Good results for thermal history calibrations from several locations were obtained using a large database of measured OM maturity, consisting of different maturity parameters, often all in agreement, for a wide range of lithostratigraphic sequences.

One-dimensional thermal modeling was applied to selected successions that were not overprinted by alpine evolution. Resulting heat-flow values were high and relatively uniform (85 to 105 mW/m^2) throughout the southern Alps during the Lias–Early Dogger, and they progressively decreased after the Bajocian–Bathonian to values similar to the present-day heat flow (50 to 55 mW/m^2) by the end of the Cretaceous. This heat-flow reconstruction is consistent with the known tectonic evolution of Mesozoic extension in the southern Alps, characterized by a rifting stage lasting to the Lias, followed by a drifting stage beginning in the Middle Jurassic.

Elevated heat flow in the southern Alps has important implications for hydrocarbon exploration. Upper Triassic source rocks in basinal successions attained high maturity during the Jurassic, and this is even more likely for the deeper Middle Triassic source rocks.

Improved definition of the heat-flow peak in the Jurassic helps to define the hydrocarbon charge risk associated with leads and prospects in the Po Plain subsurface, where traps formed in the Cenozoic. In areas characterized by little Rhaetian–Liassic burial, the source rocks retained their original petroleum potential prior to strong Neogene–Quaternary burial and heating.

KEY WORDS: thermal history, heat flow, organic matter maturity, southern Alps, hydrocarbon generation

INTRODUCTION

Maturity data obtained from the analysis of organic matter are excellent indicators of the thermal evolution of sedimentary basins. Most maturity parameters are sensitive to temperatures between 60° C and 200° C.

The most commonly used maturity parameter is vitrinite reflectance (R_o). R_o measurements are commonly corroborated by T_{max} from Rock-Eval pyrolysis, as well as certain biomarker ratios and TAI (thermal alteration index of palynomorphs). These parameters record heating events experienced by organic matter contained in the rocks, and they can help to reconstruct the thermal history of sedimentary successions. Measured data, such as vitrinite reflectance, can be used to check the consistency of geological models that may incorporate sedimentary/tectonic burial and heat-flow histories. Integration with geothermobarometers from fluid inclusion studies, and other inorganic parameters sensitive to thermal evolution (apatite fission-track analysis, clay mineralogy), may improve the reconstruction of the geological/thermal model.

A good reconstruction of the geological model requires the definition of all elements that can influence temperature. This model should contain (1) the most probable burial history, (2) a correct interpretation of the lithology of the different formations, which controls the compaction and thermal conductivity of the rock, and (3) changes in paleolatitude and water depth affecting the temperature at the water–sediment interface. Calibration of thermal history is more reliable when most of the measured maturity data refer to a thick lithostratigraphic interval. When the heat-flow history is known, the sedimentary and/or tectonic burial history, and also the amount of erosion at unconformities can be reconstructed. Alternatively, when the burial history is known, the heat flow at identified geological time intervals that are characterized by certain types of geodynamic settings (e.g., rifting) can be evaluated. In these ways, higher or lower maturity gradients and discontinuities in the trend of maturity can be explained. Modeling, specifically the calculation of organic matter maturity as a comparison to measured data, has been substantially improved by the application of sophisticated software that has been developed by petroleum companies and research institutes over the last two decades.

The aim of our work is to assess the most probable heat-flow evolution of the southern Alps (northern Italy) during Mesozoic extensional phases by analyzing organic matter (OM) maturity and modeling selected successions that were not overprinted by the Alpine orogeny. Various thermal histories have been described over the last few years (Fantoni and Scotti 2003; including references therein). This has resulted in a significant case history (Iseo Lake succession, western-side outcrops) related to the progressive updating of measured OM maturity data (Scotti 2005). It is very important to have reliable maturity data from key successions in the burial history in order to reconstruct a robust evolution of the burial and heat flow. We used the modeled heat-flow history to make predictions for oil exploration in the Po Plain foredeep.

MESOZOIC EXTENSION IN THE SOUTHERN ALPS

The structural pattern of the southern Alps is characterized by E-W– and WSW-ENE–trending compressional structures in the western and eastern sectors, respectively. Mesozoic extension, related to the tectonic movements between Adria and Europe, resulted in the creation of a N–S half graben, bounded by E- and W-dipping master normal faults (section A-A′ in Fig. 1). Tectonic activity started during the Norian. After relative quiescence during the Rhaetian, renewed extension took place during the Liassic. Extension shifted westward to the Ligurian–Piedmont area starting in the Late Jurassic, where oceanic crust formed. From this time until the Early Cretaceous, the southern

Analyzing the Thermal History of Sedimentary Basins: Methods and Case Studies
SEPM Special Publication No. 103, Copyright © 2012
SEPM (Society for Sedimentary Geology), Print ISBN 978-1-56576-315-9, CD/DVD ISBN 978-1-56576-317-3, p. 153–165.

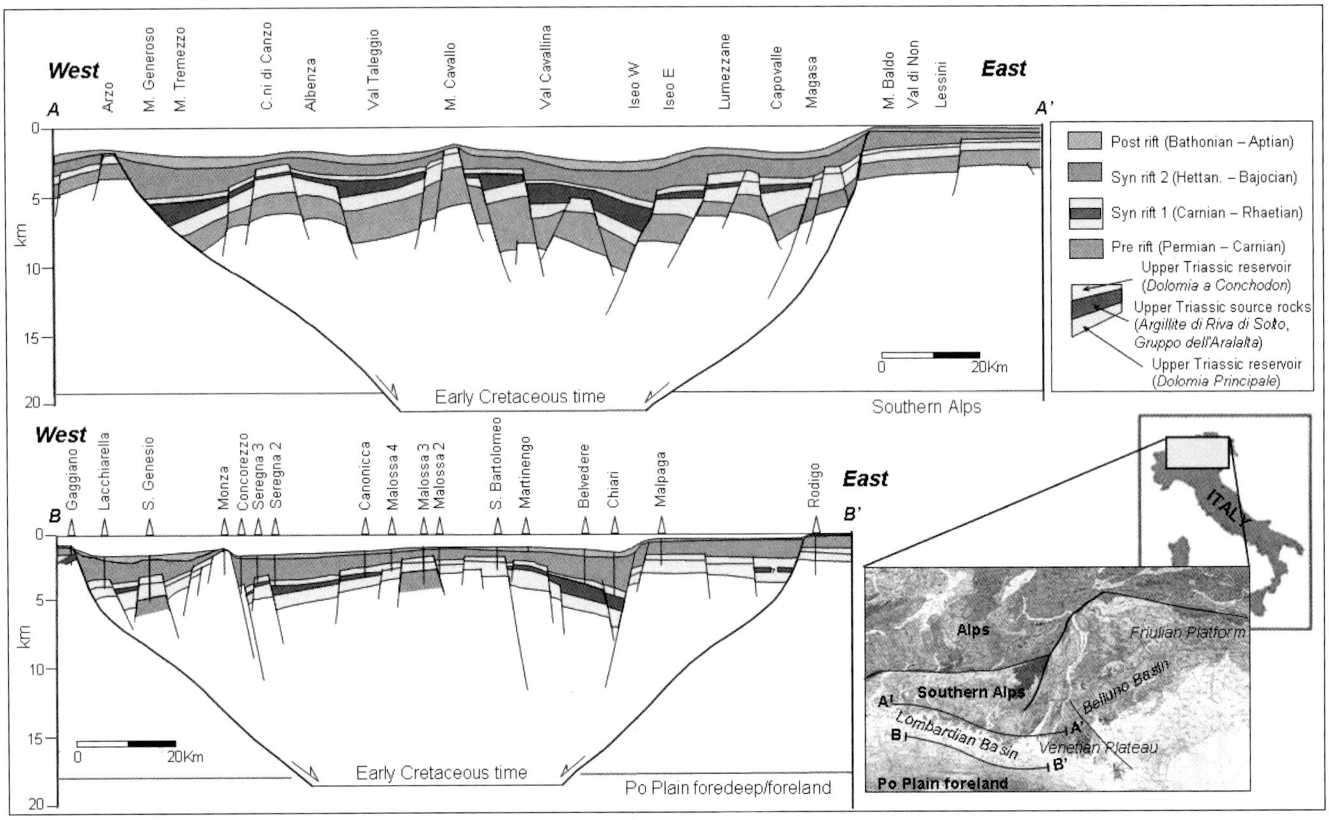

FIG. 1.—Synextensional Mesozoic successions of the southern Alps (A-A′) and the Po Plain foredeep/foreland (B-B′).

Alps underwent postrift thermal subsidence (Bertotti et al. 1993, and references therein). The same structural architecture is present in the Po Plain foreland (section B-B′ in Fig. 1).

ORGANIC MATTER MATURITY IN THE MESOZOIC SUCCESSION OF THE SOUTHERN ALPS

The thermal history was reconstructed using organic matter (OM) maturity values obtained from samples collected from Permian to Cretaceous sedimentary units outcropping along the whole chain. Organic matter maturity data in Table 1 represent samples taken from the western–central sector of the southern Alps (Fig. 2). Data affected by anomalous thermal perturbations were previously excluded; e.g., significant tectonic thickening due to overthrusts for which hanging wall ramps had been subsequently eroded; deposition of thick foredeep clastics subsequently eroded during the recent alpine cycles; thermal perturbation due to nearby magmatic intrusions.

Almost all OM maturity data are from intervals with fair to good total organic carbon (TOC) values (most samples have from 0.4 to 2% TOC). In some localities, the presence of several "good TOC" levels provides a vertical set of data along a thick sedimentary succession, resulting in a better definition of the maturity gradient. Vitrinite reflectance (R_o %) values from the western–central sector of the southern Alps (from Biella area to Trento Plateau) are described in Fantoni and Scotti (2003), Calabrò et al. (2003), Bersezio et al. (2005), and Scotti (2005). The kerogen in Upper Triassic and Liassic formations of southern Alps is often rich in inertinite and poor in vitrinite. Comparisons of the different maturity parameters (described herein) and the relationship of increasing maturity with depth improve confidence in some data; e.g., vitrinite reflectance data in samples where less than 20 measurements of R_o are available.

The lowest OM maturity values occur on structural highs having thin overlying sedimentation. For example, on the Gozzano and Arzo highs (western side of section A-A′ in Figure 1 and locations Monfenera, Besano, and Porto Ceresio in Figure 2 and Table 1), T_{max} from Rock-Eval pyrolysis for Middle Triassic units is <425° C (immature kerogen), suggesting maximum temperatures lower than 60 to 70° C (using a heating rate of 3° C/Ma). Samples from the Trento Plateau (eastern side of section A-A′ in Figure 1, and projected locations Cavedine and Coredo 1 well in Figure 2 and Table 1) also show low OM maturity (~0.5% R_o and 435° C T_{max}) for Upper Triassic units. Increased OM maturity can primarily be attributed to a thicker synrift sedimentary succession. The highest maturity values correspond to the depocenters of synrift successions, such as the Upper Triassic units of the M. Tremezzo/M. Galbiga (M. Generoso Basin) and Iseo Lake successions (Iseo Basin) (see section A-A′ in Fig. 1; locations in Fig. 2; geochemical data in Table 1). As an example, very high OM maturity was reached by Norian units (R_o ~3.5% at the base of the Calcare di Zorzino Formation) in the Iseo depocenter.

THERMAL MODELING

Vitrinite reflectance measured data were used for thermal model calibration using a proprietary Eni S.p.A. one-dimensional mathematical model that simulates the degree of maturity reached by kerogen. The model input data are: (1) formation thickness, (2) formation lithology (in order to define the compaction curve and thermal conductivity of the rock matrix), (3) ages of sedimentation and paleobathymetry for each stratigraphic event (all of these data define

TABLE 1.—Maturity parameters of the samples with individual values, number of measurements, and average values of vitrinite reflectance (R_o), thermal alteration index (TAI), T_{max} from Rock-Eval pyrolysis, and biomarker isomerization ratios. Mean values of source rock properties are also reported: total organic carbon (TOC), free hydrocarbons (S1), residual petroleum potential (S2), CO_2 from kerogen (S3), hydrogen index (HI), and oxygen index (OI).

Locations of thermal modelling (Figure 2)	Sample point symbol (Figures 9-13)	Area	Locality	Formation	Age	Kerogen Type immature samples (1)	Optical analyses *: R_o % and R_o % equivalent from bitumen* individual sample measurements	number of measurements	R_o % mean (x) = number of samples	TAI (1-5 scale)	TOC and Rock-eval **: T_{max} °C	TOC %	S1 mg HC/g rock	S2 mg HC/g rock	S3 mg CO_2/g rock	HI mg HC/g TOC	OI mg CO_2/g TOC	Biomarker ratios: sterane isomer. 20S/(20S+20R)	hopane isomer. 22S/(22S+22R)
	△	Gozzano high	Monfenera (P.te S.Quirico and Pissone)	"Dolomie Stratificate"	Upper Anisian	III	0.41-0.35-0.33-0.33	15-18-19-10	**0.35** (4)	1.6 (1)	**423** (5)	0.73 (5)	0.03 (5)	1.24 (5)	0.51 (5)	132 (5)	87 (5)	-	-
	▲	Arzo high	Besano (old "oil shales" / fossils quarry)	Besano Fm	Anisian-Ladinian bound.	II	-	-	-	-	**422** (24)	12.18 (24)	2.05 (24)	64.8 (24)	1.98 (24)	499 (24)	50 (24)	**0.29** (5)	**0.58** (5)
	▲	Arzo high	Besano (Rio Ponticelli)	Besano Fm (upper portion)	Anisian-Ladinian bound.	II/I	0.43-0.41-0.37-0.36	9-8-13-16	**0.39** (4)	**1.5** (7)	**420** (8)	1.54 (8)	0.27 (8)	10.6 (8)	0.58 (8)	670 (8)	46 (8)	-	-
	■	Arzo high	Porto Ceresio (Rio Vallone)	Calcare di Meride	Ladinian	II	-	-	-	-	**411** (28)	2.22 (28)	0.82 (28)	13.4 (28)	0.67 (28)	527 (28)	44 (28)	**0.11** (4)	**0.27** (4)
	●	M.Generoso Basin	Valsolda	Calcare di Zu	Rhaetian	III	0.92-0.63	14-7	**0.77** (2)	2.2 (1)	**447** (11)	0.52 (12)	0.03 (12)	0.17 (12)	0.36 (11)	29 (12)	67 (11)	-	-
	●	M.Generoso Basin	Bene Lario	Calcare di Zu	Rhaetian	III	1.92-1.49 (1.83)*	3-11 (5)*	**1,71** (2)	-	**490** (2)	1.17 (2)	0.00 (2)	0.09 (2)	0.32 (2)	7 (2)	26 (2)	-	-
	○	M.Generoso Basin	M. Galbiga (CMo lower portion)	Calcare di Moltrasio	Lower Lias	II/III	1.13-1.03-1.00-0.94 (1.00-0.96)*	37-26-10-18 (7-11)*	**1.03** (4)	-	**461** (4)	0.71 (6)	0.05 (6)	0.13 (6)	0.45 (6)	24 (6)	55 (6)	-	-
	○	M.Generoso Basin	M. Galbiga (CMo upper portion)	Calcare di Moltrasio	Lower Lias	II/III	0.84-0.77-0.75-0.69 (0.72)*	3-9-14-3 (1)*	**0.76** (4)	-	438 (3)	0.42 (3)	0.08 (3)	0.33 (3)	0.15 (3)	92 (3)	34 (3)	**0.54** (1)	**0.57** (1)
	○	M.Generoso Basin	Osteno	As above ("Livello Osteno")	Sinemurian	II/III	-	-	-	-	**438** (1)	0.38 (1)	0.34 (1)	0.80 (1)	0.22 (1)	211 (1)	58 (1)	**0.58** (1)	**0.56** (1)
	○	M.Generoso Basin	Ponna	Calcare di Moltrasio	Lower Lias	II/III	0.70	6	0.70 (1)	-	**436** (1)	0.37 (1)	0.21 (1)	1.13 (1)	0.10 (1)	305 (1)	27 (1)	**0.53** (1)	**0.57** (1)
	○	M.Generoso Basin	Laglio	Calcare di Moltrasio	Lower Lias	II/III	-	-	-	-	**435** (1)	0.58 (1)	0.15 (1)	1.59 (1)	-	274 (1)	-	**0.31** (1)	**0.60** (1)
	○	M.Generoso Basin	Monte S. Zeno	Calcare di Moltrasio	Lower Lias	II/III	0.53	6	0.53 (1)	-	**431** (2)	0.41 (2)	0.12 (2)	1.31 (2)	0.10 (2)	319 (2)	25 (2)	**0.42** (2)	**0.60** (2)
	○	M.Generoso Basin	M.Generoso (Sud)	Calcare di Moltrasio	Lower Lias	II/III	0.52-0.49	2-9	**0.51** (2)	1.7 (2)	**420** (3)	0.58 (3)	0.04 (3)	1.46 (3)	0.13 (3)	222 (3)	17 (3)	**0.11** (1)	**0.45** (1)
	○	M.Generoso Basin	M.Generoso (Nord)	Calcare di Moltrasio	Lower Lias	II/III	0.60-0.55-0.54	5-6-7	**0.56** (3)	**1.5** (2)	**427** (7)	0.35 (7)	0.03 (7)	0.66 (7)	0.08 (7)	179 (7)	22 (7)	**0.29** (2)	**0.57** (2)
2	○	M.Generoso Basin	Val Breggia (CMo upper portion)	Calcare di Moltrasio	Lower Lias	II/III	0.55-0.51-0.49-0.48-0.46-0.43	7-11-9-4-6-7	**0.48** (6)	1.8 (4)	**423** (5)	0.43 (5)	0.03 (5)	0.87 (5)	0.11 (5)	194 (5)	30 (5)	**0.10** (1)	**0.38** (1)
	●	Albenza Plateau	M. Cornizzolo A	Calcare di Zu	Rhaetian	III	-	-	-	-	**435** (25)	0.59 (28)	0.02 (28)	0.56 (28)	0.47 (28)	65 (28)	89 (28)	-	-
	●	Albenza Plateau	M. Cornizzolo B	Calcare di Zu	Rhaetian	III-III/II	-	-	-	**1.7** (4)	438 (2)	0.42 (2)	0.04 (2)	0.76 (2)	0.27 (2)	179 (2)	67 (2)	-	-
	●	Albenza Plateau	M. Cornizzolo C	Calcare di Zu	Rhaetian	III-III/II	-	-	-	-	**434** (3)	0.50 (3)	0.04 (3)	1.05 (3)	0.30 (3)	185 (3)	70 (3)	-	-
	●	Albenza Plateau	Corni di Canzo	Calcare di Zu	Rhaetian	III	0.46	7	**0.46** (1)	**1.2** (4)	**436** (19)	0.54 (23)	0.07 (23)	0.63 (23)	0.72 (23)	79 (23)	171 (23)	-	-
	●	Albenza Plateau	Valmadrera (S. Tommaso)	Calcare di Zu	Rhaetian	III-III/II	0.67-0.62 (0.66-0.63)*	2-1 (14-8)*	**0.65** (2)	-	**442** (5)	0.44 (4)	0.06 (4)	0.40 (4)	0.42 (3)	89 (4)	98 (3)	-	-
	○	Albenza Plateau	Canzo (Fontana Gajum)	Calcare di Moltrasio	Lower Lias	II/III	-	-	-	2.1 (1)	**432** (3)	0.47 (3)	0.11 (3)	1.33 (3)	0.15 (3)	284 (3)	32 (3)	-	-
	○	Albenza Plateau	Asso	Calcare di Moltrasio	Lower Lias	II/III	-	-	-	-	**434** (5)	0.48 (5)	0.13 (5)	1.17 (5)	0.13 (5)	221 (5)	29 (5)	-	-
	□	M.Cavallo high	Zandobbio	Calcare di Moltrasio	Lower Lias	II/III	0.68-0.65-0.58-0.53-0.52-0.48-0.48	4-3-6-6-11-11-9	**0.56** (7)	-	**431** (4)	0.64 (4)	0.34 (4)	1.47 (4)	0.12 (4)	229 (4)	25 (4)	0.47 (4)	**0.55** (4)
1	■	Iseo Basin	Zorzino (CZo lower portion)	Calcare di Zorzino	Norian	II	3.46-3.32-3.18-2.92 (3.08-2.62)*	3-15-6-9 (5-3)*	**3.0 - 3.5** (4)	3.9 (9)	-	1.14 (11)	0.03 (11)	0.01 (11)	0.92 (11)	2 (11)	127 (11)	-	-
1	■	Iseo Basin	Zorzino (CZo upper portion)	Calcare di Zorzino	Norian	II	2.47-2.36 (2.47-2.25)*	12-12 (3-2)*	**2.5 - 3.0** (2)	4.0 (17)	-	0.76 (16)	0.02 (16)	0.01 (16)	0.80 (16)	2 (16)	135 (16)	-	-
1	◆	Iseo Basin	Riva di Solto (ARS lower portion)	Argillite di Riva di Solto	Norian	III/II	2.17-2.08	1-4	2.1 - 2.5 (2)	3.6 (13)	-	0.66 (30)	0.01 (30)	0.01 (30)	0.72 (30)	1 (30)	103 (30)	-	-
1	■	Iseo Basin	Zu (CZu Top)	Calcare di Zu	Rhaetian	III	1.11-1.10-1.00 (1.05-1.03-0.95)*	7-5-2 (5-1-4)*	**1.07** (3)		**456** (1)	0.43 (3)	0.05 (3)	0.14 (3)	0.03 (3)	32 (3)	8 (3)	-	-
1	□	Iseo Basin	Tavernola (CMo lower portion)	Calcare di Moltrasio	Lower Lias	II/III	0.98-0.93-0.92-0.91-0.90 (1.04-1.00-0.99)*	11-14-15-10-11 (18-2-5)*	**0.93** (5)	-	463 (3)	0.55 (6)	0.09 (6)	0.24 (6)	0.07 (6)	37 (6)	12 (6)	-	-
1	□	Iseo Basin	Tavernola (CMo upper portion)	Calcare di Moltrasio	Lower Lias	II/III	0.70-0.63	10-9	0.67 (2)	-	450 (2)	0.52 (8)	0.04 (8)	0.22 (8)	0.05 (8)	38 (8)	10 (8)	-	-
1	△	Iseo Basin	Tavernola (CDo middle portion)	Calcare di Domaro	Pliensbachian	II/III	0.78-0.75-0.69	4-2-3	0.74 (3)	-	-	0.29 (3)	0.04 (3)	0.10 (3)	0.02 (3)	36 (3)	6 (3)	0.46 (1)	0.59 (1)
1	△	Iseo Basin	Tavernola (CDo upper portion)	Calcare di Domaro	Pliensbachian	II/III	0.54-0.50	7-2	0.52 (2)	-	442 (2)	0.53 (2)	0.19 (2)	1.60 (2)	0.58 (2)	302 (2)	106 (2)	**0.51** (2)	**0.58** (2)
	△	Transition	Provaglio d'Iseo	Calcare di Domaro	Pliensbachian	II/III	-	-	-	-	**427** (2)	0.68 (2)	0.03 (2)	1.78 (2)	0.13 (2)	260 (2)	19 (2)	**0.12** (1)	**0.52** (1)
	◆	Idro/Mag./Lum. Basins	Capovalle	Argillite di Riva di Solto ?	Norian	III/II or II	1.47 (1.52)*	12 (40)*	1.52* (1)	-	505 (7)	1.53 (9)	0.07 (9)	0.18 (9)	0.65 (9)	17 (9)	59 (9)	-	-
	◆	Idro/Mag./Lum. Basins	Passo Tremalzo	Argillite di Riva di Solto	Norian	III/II	-	-	-	-	464 (3)	0.61 (5)	0.04 (5)	0.11 (5)	0.43 (5)	17 (5)	76 (5)	-	-
	■	Idro/Mag./Lum. Basins	Lumezzane (Agnosine-Casale)	"Facies Eteropiche"	Norian	II (S)	(1.55-1.34)*	(20-30)*	1.45* (2)	-	-	0.74 (1)	0.01 (1)	0.02 (1)	0.91 (1)	3 (1)	123 (1)	-	-
	■	Idro/Mag./Lum. Basins	Magasa (Sp 9 km 23)	"Facies Eteropiche"	Norian	II (S)	(1.53)*	(5)*	1.53* (1)	-	-	2.57 (5)	0.03 (5)	0.03 (5)	1,82 (5)	1 (5)	112 (5)	-	-
	■	Trento Plateau	Cavedine (M. Grattacul)	"Calcari Neri"	Norian	II	0.52	10	**0.52** (1)	**1.5** (2)	**435** (11)	0.87 (11)	0.14 (11)	3.75 (11)	0.41 (11)	378 (11)	45 (11)	**0.29** (3)	**0.57** (3)
	▲	Trento Plateau	Coredo 1 well (1250 m)	Raibl	Carnian	III	0.47-0.47-0.44	16-20-20	0.46 (3)	2.0 (1)	435 (3)	0.36 (3)	0.17 (3)	0.63 (3)	-	175 (3)	-	0.50 (3)	0.58 (3)

Symbols: the above symbols are used in the figures.
. *White (black outline)* : Liassic formations
. *Black* : Upper Triassic formations (Type III-III/II kerogen)
. *Grey (black outline)* : Upper Triassic formations (Type II kerogen)
. *Light grey (grey outline)* : Lower Triassic formations

Localities ordered starting from the western sector and by oldest formation
(1) = Rough determination of kerogen type of immature samples, or after assessment of HI before maturation via optical determination of organic matter type, sedimentary environment and comparison with same/analogous formation at different maturity (see trends HI vs T_{max} in Fig. 10).
Range of HI° adopted for kerogen type definition: Type IV (HI° < 50 mg HC/g TOC); Type III (50-200); Type II/III (200-350); Type II (350-650); Type I (> 650).

(x) = number of valid samples
* = vitrinite reflectance equivalent from bitumen (in brackets)
** = only samples with TOC > 0.25 % have been taken into account
Note : average TOC and HI values refer only to samples with TOC>0.25 %.
Source rock properties are the mean values for the sample population, but not of the formation.
Most reliable data (R_o, TAI, T_{max} and biomarker, in **bold type**)

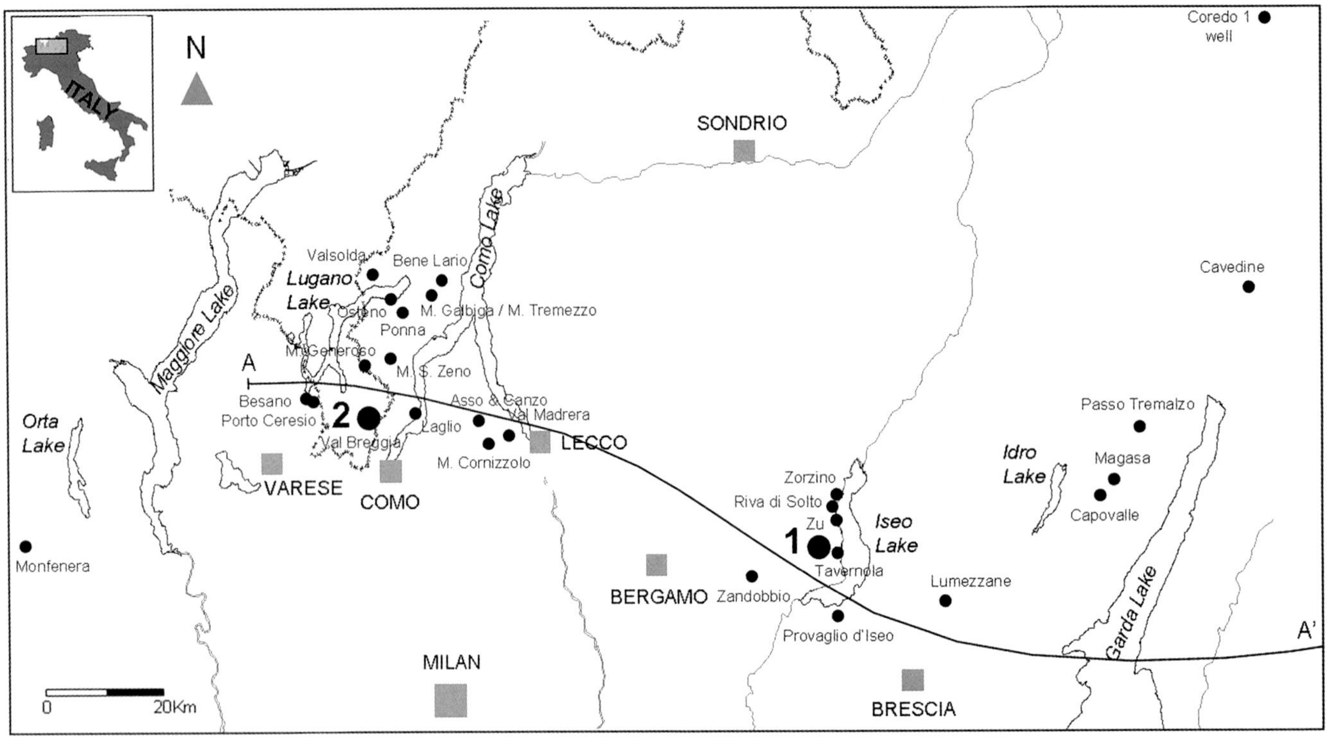

FIG. 2.—Locations of the thermal history calibrations (large dots; 1 = Iseo Lake western-side outcrops; 2 = Val Breggia) and the main sample localities (small dots) for this study.

the burial history), (4) heat-flow values through time, and (5) paleolatitude, which is used to convert paleosurface temperature to sea bottom temperature according to paleobathymetry. Finally, the input for the kinetic model has to be defined (e.g., kinetics of reactions governing vitrinite reflectance—EasyR_o method of Sweeney and Burnham 1990).

Heat flow is the most important unknown variable and is related to the tectonic setting/geodynamic evolution of the area. Novelli et al. (1987), in accordance with the geodynamic evolution of the southern Alps (Bertotti et al. 1993, and references therein), although lacking calibration points, hypothesized high heat-flow values of ~75 mW/m^2 in the northwestern Po Basin during the Early Lias. Greber et al. (1997) and Bertotti et al. (1999) suggested a higher heat flow for the Lombardian Basin, but predicted its peak value at ~90 mW/m^2 at Anisian–Ladinian and Carnian–Norian ages, respectively. Bersezio and Bellantani (1997) interpreted the thermal perturbation as Liassic (and pre-Toarcian) in age for the same basin.

The Eni S.p.A. database was recently enhanced with a large amount of OM maturity data for Mesozoic sedimentary successions throughout the southern Alps region. Therefore, constraints for the reconstruction of the thermal history were available. Recent thermal modeling indicates that the last probable heat-flow peak was later than the Lias (although other significant earlier peaks cannot be excluded), when the Upper Triassic and Liassic formations were subject to deep burial. Also, the maximum heat flow of this late peak has been better quantified (Calabrò et al. 2003; Ronchi et al. 2003; Fantoni and Scotti 2003; Zattin et al. 2003, 2006).

Best-fit solutions between calculated and measured OM maturity for some Mesozoic successions (i.e., Arzo; M. Generoso; M. Tremezzo; Iseo; Magasa; Val di Non areas—see Fig. 1) and other simulation points in the eastern southern Alps were obtained by assuming increased heat flow throughout the rifting stage, with a maximum during Bajocian–Bathonian time. Resulting heat-flow values are quite high and relatively uniform across the southern Alps (85 to 105 mW/m^2), suggesting that this kind of thermal regime was widespread (Fantoni and Scotti 2003).

Areal fluctuations in peak heat flow could be due to differential crustal thinning or to other causes, such as radiogenic heat within the crust, hydrothermal fluid circulation, or underplating phenomena. Heat flow progressively decreased after the Bajocian–Bathonian to values similar to present day (for present-day heat flow values, see Novelli et al. 1987, Della Vedova et al. 2001) by the end of the Cretaceous. This reconstruction is consistent with the known tectonic evolution of Mesozoic extension in the southern Alps, characterized by a rifting stage up to the Lias, followed by a drifting stage (westward opening of the Ligurian–Piedmont ocean) and thermal subsidence from the Middle Jurassic.

An update of the previous model was made to improve the heat-flow history at the most significant locality (Iseo Lake, western-side outcrops; Scotti 2005). In this paper, detailed thermal modeling of two locations (Iseo Lake and Val Breggia; see locations in Fig. 2) will be described. Implications for oil exploration in the Po Plain foredeep (Scotti and Fantoni 2008) are discussed later in this paper.

Details of the Iseo Lake case history (location 1 in Fig. 2) are shown in Figures 3–8, with Figures 3–5 presenting an earlier geological model in terms of Cretaceous–Tertiary burial histories, and Figures 6–8 presenting a more recent model. Figures 3 and 6 also show the work flow for the thermal regime assessment: The burial history combined with the heat-flow evolution (a and b) determines the thermal history of all the stratigraphic events. The thermal evolution of the base of Calcare di Zorzino and Argillite di Riva di Solto formations is shown as an example in Figures 3c and 6c. By linking a suitable kinetic model (EasyR_o) to the thermal history, the maturity evolution of all stratigraphic events can be evaluated; the maturity of the base of the Calcare di Zorzino (CZo) and Argillite di Riva di Solto (ARS) formations is shown as an example in Figures 3d and 6d. The maturity

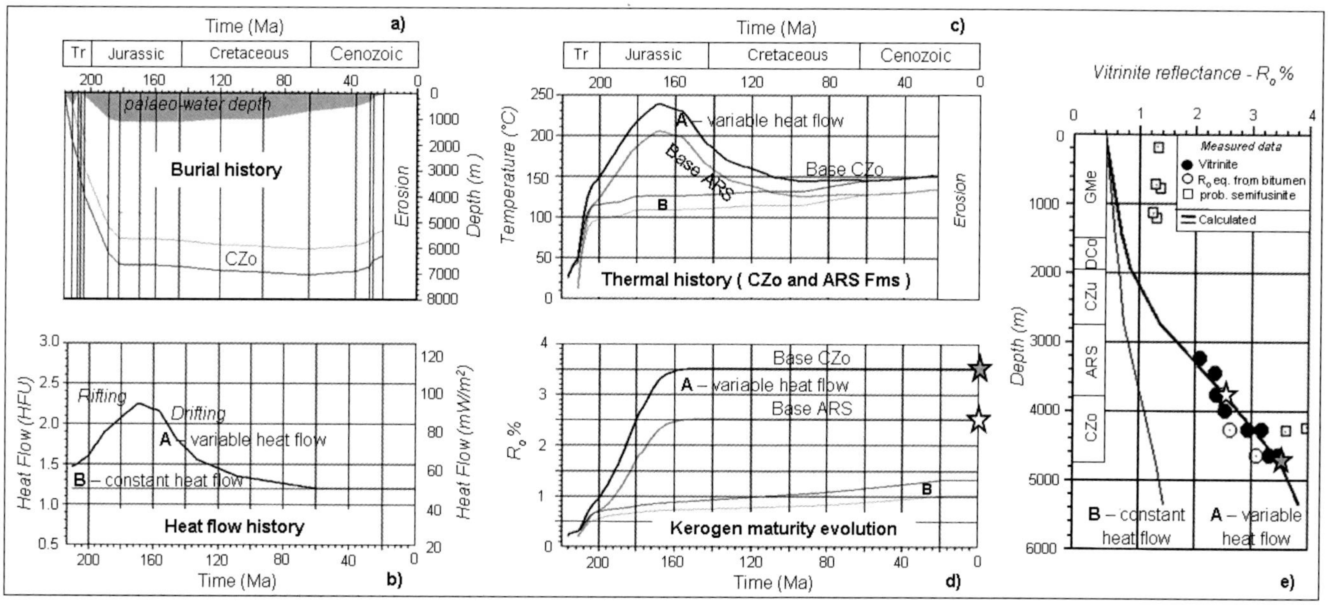

FIG. 3.—Thermal modeling performed in 2003 (Fantoni and Scotti) for Iseo Lake western-side outcrops (lithostratigraphic units: CZo = Calcare di Zorzino; ARS = Argillite di Riva di Solto; CZu = Calcare di Zu; DCo = Dolomia a Conchodon; GMe = Gruppo del Medolo): Burial history (a) and heat-flow history (b) during the extensional Mesozoic phase and later, and comparison between variable (A) and constant heat flow (B); thermal history (c); organic matter (OM) maturity history (d); and measured and calculated OM maturity (vitrinite reflectance and EasyR_o, respectively) vs. depth (e).

level is also plotted against depth: Note the different R_o trends calculated by using a constant heat flow through time (equal to the present day—thin line), or a variable heat flow according to the geodynamic evolution of the area (bold line) in Figures 3e and 6e.

Figure 3 shows the modeling performed by Fantoni and Scotti (2003). Note the high maturity gradient within the Upper Triassic succession (CZo and ARS formations) and the lack of reliable vitrinite reflectance data in the Liassic succession (GMe—Gruppo del Medolo;

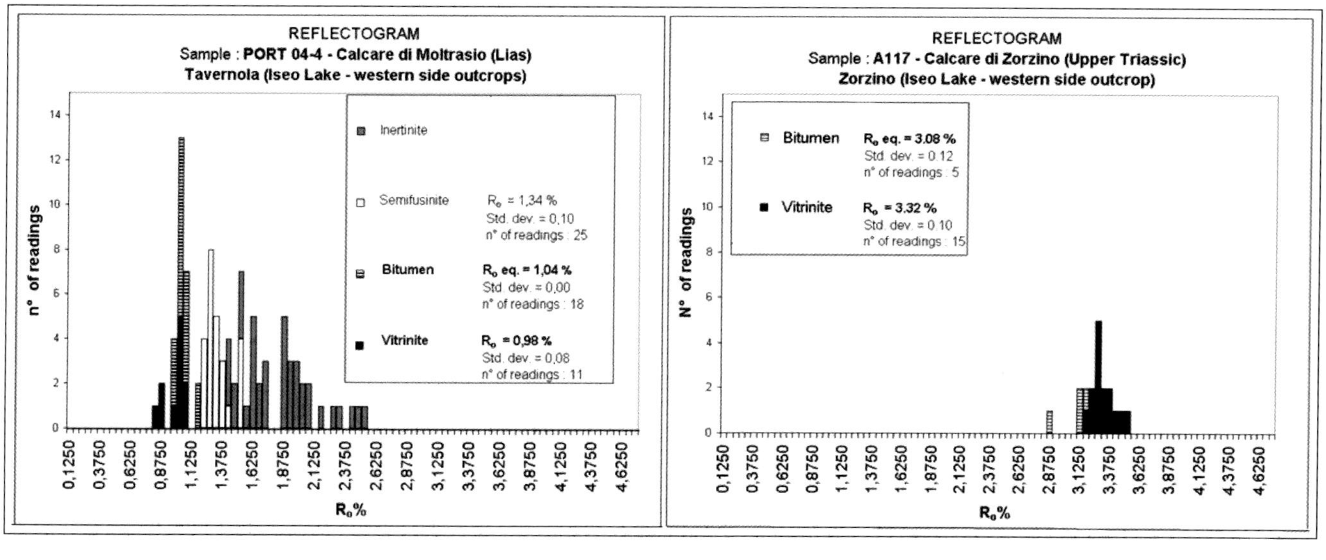

FIG. 4.—Reflectograms for two samples of Iseo Lake western-side outcrops. (a) A complete reflectogram for a sample of Calcare di Moltrasio (Lower Lias). This sample contains abundant inertinite, interpreted as semifusinite (mean value of $R_o = 1.34\%$), and fusinite (mean value of $R_o > 1.6\%$). Vitrinite particles have a mean value of $R_o = 0.98\%$. Bitumen is also present, with an equivalent $R_o = 1.04\%$. (b) A reflectogram for a sample of Calcare di Zorzino (Norian) for only R_o % and R_o % equiv. from bitumen (inertinite data are not shown). The number of measurements is low (15 particles), but strongly anisotropic vitrinite is easily recognizable. The standard deviation of these measurements is low. In addition, these data for the sample at the base of the formation are in good agreement with the general maturity trend of this succession.

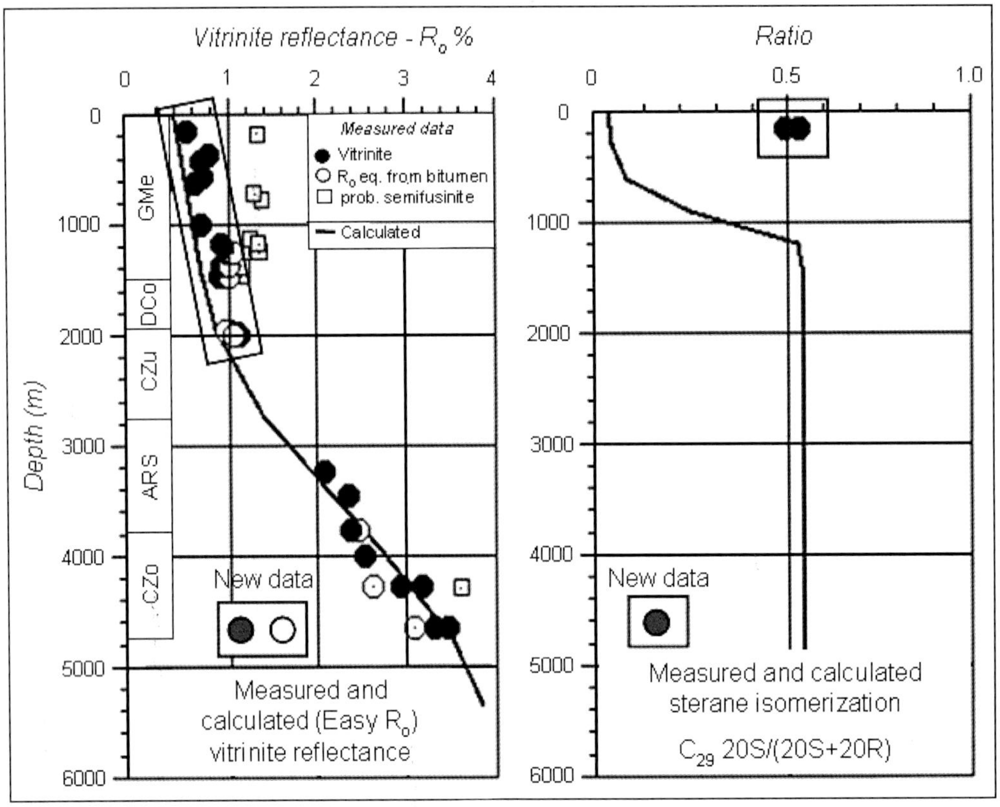

FIG. 5.—Iseo Lake western-side outcrops. New maturity data for the upper portion of the succession became available in 2005. Comparison between previously calculated maturity profiles and new data (vitrinite reflectance and sterane isomerization) suggested the need for a review of the thermal history.

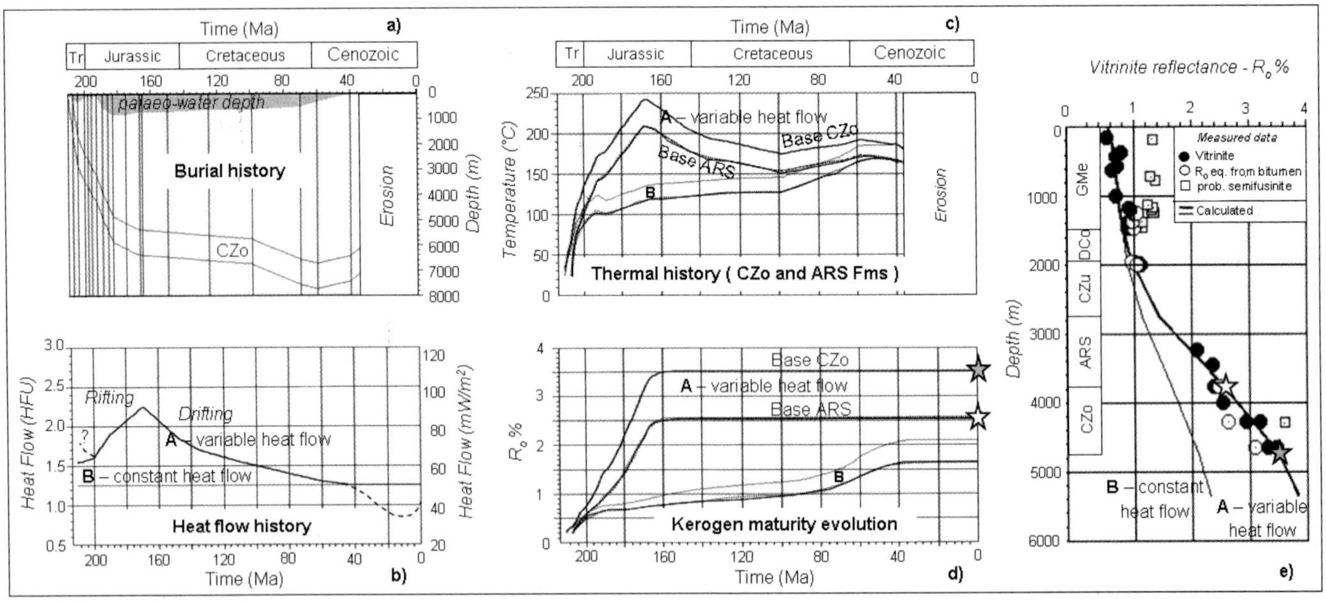

FIG. 6.—Updated modeling (Scotti 2005, Scotti and Fantoni 2008) of Iseo Lake western-side outcrops: Burial history (a) and heat-flow history (b) during the extensional Mesozoic phase and later, and comparison between variable (A) and constant heat flow (B) (dashed line for Po Plain foredeep simulation points); thermal history (c); organic matter (OM) maturity history (d); and measured and calculated OM maturity (vitrinite reflectance and Easy R_o, respectively) vs. depth (e).

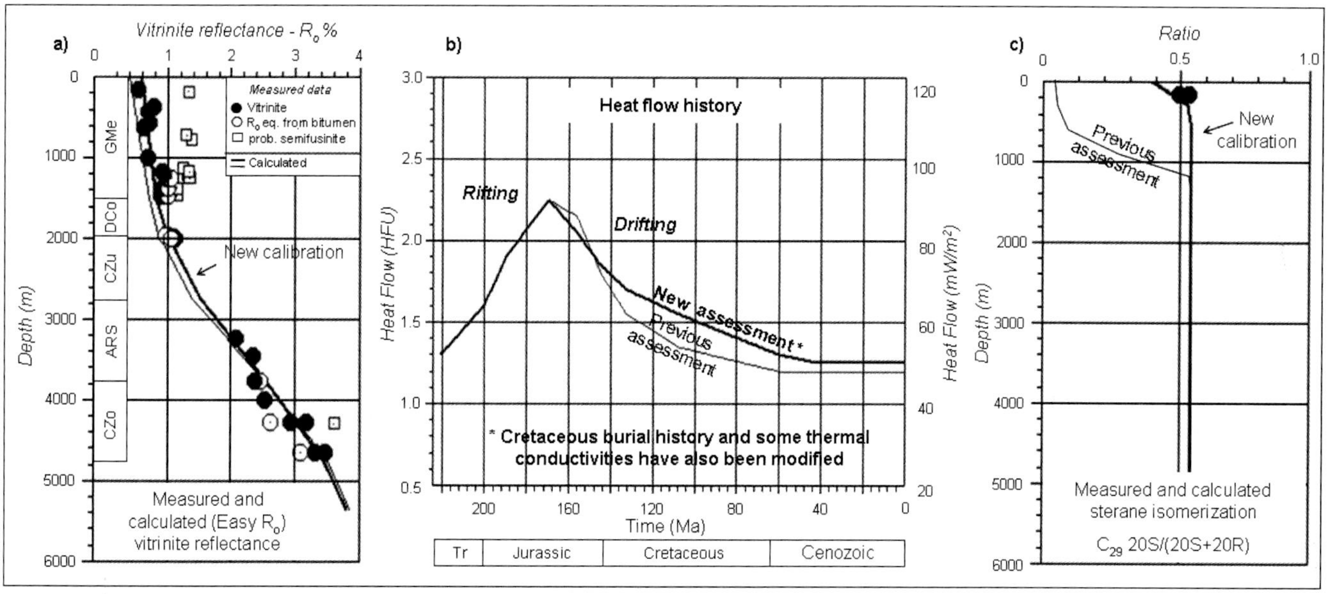

FIG. 7.—Iseo Lake western-side outcrops. (a) The latest assessment (bold type), together with a modification of burial history in Cretaceous times, results in a good fit between the calculated curves and the measured data (R$_o$). (b) Comparison of two different assessments of the heat-flow history. (c) The latest assessment (bold type) results in a good fit also for sterane isomerization.

Fig. 3e). Within this Liassic succession, the vertical maturity trend is based on reflective maceral fragments that likely consist of semifusinite. The maturity gradient within the Upper Triassic succession is, however, useful enough to assess the Middle Jurassic heat-flow peak (as previously discussed).

New thermal maturity data from the shallower portion of the sedimentary succession became available in 2005, following detailed sampling that focused on the scarce, thin, more argillaceous mudstones interbedded in the succession. These samples, although predominantly containing inertinite, contain rare but fairly reliable vitrinite fragments (Fig. 4). Hence, an improved thermal history reconstruction for Late Jurassic–Early Tertiary times was created (Scotti 2005). Using the thermal histories of each event previously assessed, the calculated maturity within the shallower portion of the sequence did not match the

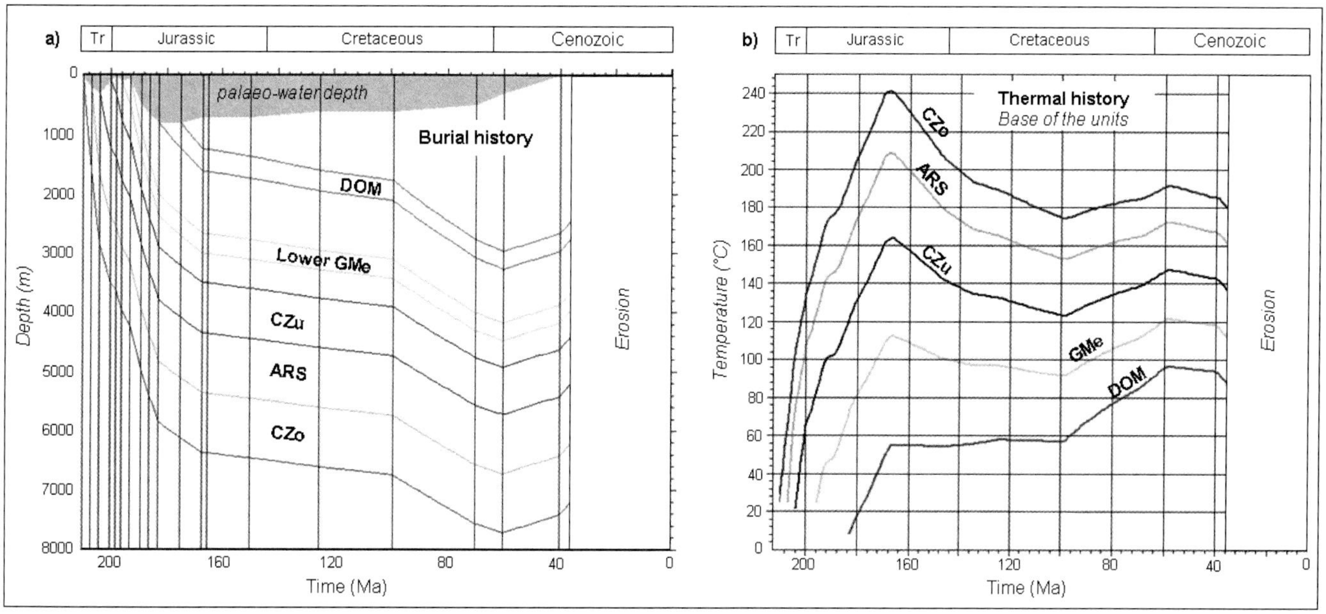

FIG. 8.—Iseo Lake western-side outcrops. (a) Burial histories for some of the lithostratigraphic units: CZo—Calcare di Zorzino; ARS—Argillite di Riva di Solto; CZu—Calcare di Zu; GMe—Gruppo del Medolo; Dom—Domaro Formation. (b) Thermal histories of the base of the lithostratigraphic units.

new vitrinite reflectance data (Fig. 5). Additional data from biomarker maturities were also available that gave results similar to those using the new vitrinite reflectance data. As shown in Figure 5, calculated R_o and the biomarker ratio, sterane isomerization, underestimate the new measured data. Thus, modification of the post-Jurassic thermal history was required.

The new maturity data suggested that the upper portion of the Medolo Group was subjected to higher temperatures than previously interpreted. The vitrinite reflectance maturity gradient within the Medolo Group is very low (Figs. 5, 6e), indicating a low geothermal gradient at the time of maximum heating. Only the paleotemperatures of the Medolo Group needed to be significantly modified in order to obtain a calibration match. This reconstruction was possible by changing the Cretaceous–Tertiary burial history (units now eroded), while maintaining a low geothermal gradient. In this way, higher temperatures for the Medolo Group were obtained (the highest of its thermal history).

The maximum thickness that can reasonably be assigned to the eroded portion of the Cretaceous–Paleocene succession is about 2200 m (including ~1700 m of Turonian–Campanian Flysch Lombardo and Maastrichtian–Paleocene Scaglia Formation) (Figs. 6a, 8a). This is similar to the thickness of the depocenter in the southern foredeep of the Eo Alpine chain (Bersezio et al. 1993). By assuming this (eroded) thickness, the resulting temperature increase was insufficient to match the modeled maturity to the measured data, so the heat flow needed to be increased in Cretaceous–Early Tertiary time, compared to the original model. This new assessment of the thermal regime evolution (Scotti 2005) is shown in Figure 6. The model presented in Figure 6 also includes some slight modifications to thermal conductivity, paleobathymetry, and absolute ages of the Upper Triassic–Liassic formations (Gradstein et al. 2004; updating of Remane 2000).

The R_o trend vs. depth (Fig. 6e), calculated by using a constant heat flow of 52 mW/m^2 (equal to the present day—thin line), shows a low maturity gradient, even after assuming deep Cretaceous burial. Figure 7 compares the two thermal models for the variable heat-flow scenario.

Increasing burial during the Cretaceous in a normal thermal regime (e.g., 52 to 58 mW/m^2) also produced an increase in temperature for the oldest formations, but not enough to significantly increase their maturity. The oldest formations experienced higher temperatures during the Dogger, when burial was significant and the geothermal gradient was very high. As Figure 8 demonstrates, only units younger than the Calcare di Zu Formation were submitted to the highest temperatures during the Late Cretaceous–Early Tertiary.

The validity of this heat-flow history, mostly during Late Cretaceous–Early Tertiary times, can be verified using successions that have a well-known Cretaceous–Tertiary burial history and have not undergone significant post-Cretaceous heating due to Tertiary burial. The Jurassic–Cretaceous Val Breggia succession near Mendrisio (Switzerland; location 2 in Fig. 1) is suitable for this purpose. The available maturity data for the Val Breggia succession are of good quality (Bersezio et al. 2002, 2005). Vitrinite reflectance, T_{max} from Rock-Eval pyrolysis, and biomarker maturity data are in agreement. The selected biomarker ratios are suitable for thermal calibration because at these low levels of maturity, they have not reached end-point values.

In general, comparisons of the different maturity parameters can help to identify and reject anomalous data, and thus define the most probable maturity stage at each sample location. This helps with the selection of modeling locations by taking into account the measured data available from several localities at different stages of maturity in the southern Alps (see Table 1). A quick check of the reliability of the available maturity data can be carried out by observing Figures 9, 10, and 11, and also by analyzing the biomarker maturity in Figures 12 and 13. Well-defined maturity trends for similar formations with constant kerogen type can be recognized, therefore improving the confidence

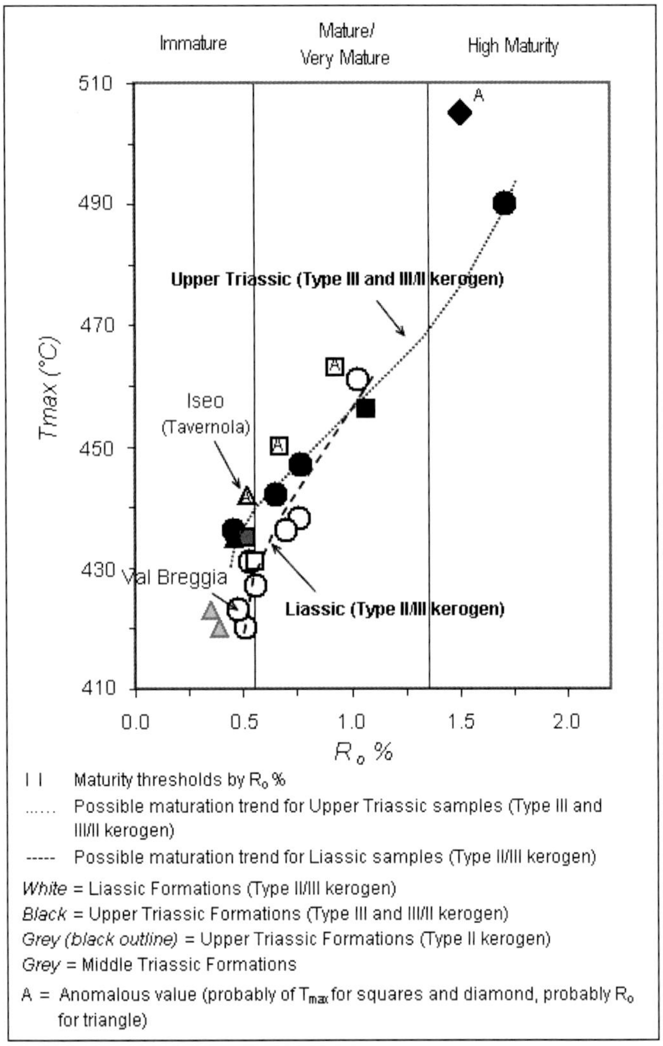

FIG. 9.—Correlations between average values of R_o and T_{max} (Table 1) for some formations sampled in the southern Alps (focusing on Iseo and Monte Generoso Basins and nearby areas). Some anomalous (A) T_{max} and R_o values are present.

assigned to some data (e.g., R_o data with less than 20 measurements per sample, or localities with few samples).

The relationship between T_{max} and vitrinite reflectance (for average values at several locations, see Table 1) for two different kerogen types is shown in Figure 9. The black symbols represent Upper Triassic formations with kerogen type III and III/II. The relationship is good, with no significant scattering, as is often shown in literature. Only the diamond point is off trend: This location (Capovalle–Lumezzane Basin) has several samples from the Argillite di Riva di Solto Formation with an average TOC of 1.5% (significantly higher than the typical average TOC of this formation). It is likely that these samples come from the basal part of the formation, at the boundary with the richer Calcare di Zorzino Formation, or they may in fact come from this older formation. In either case, it means that kerogen quality has changed to type II, and by changing the kinetics, the correlation with vitrinite reflectance could also change. The most reliable maturity indicator may be the vitrinite reflectance (~1.5%), which is in agreement with R_o equivalent from bitumen. If we were to convert T_{max}

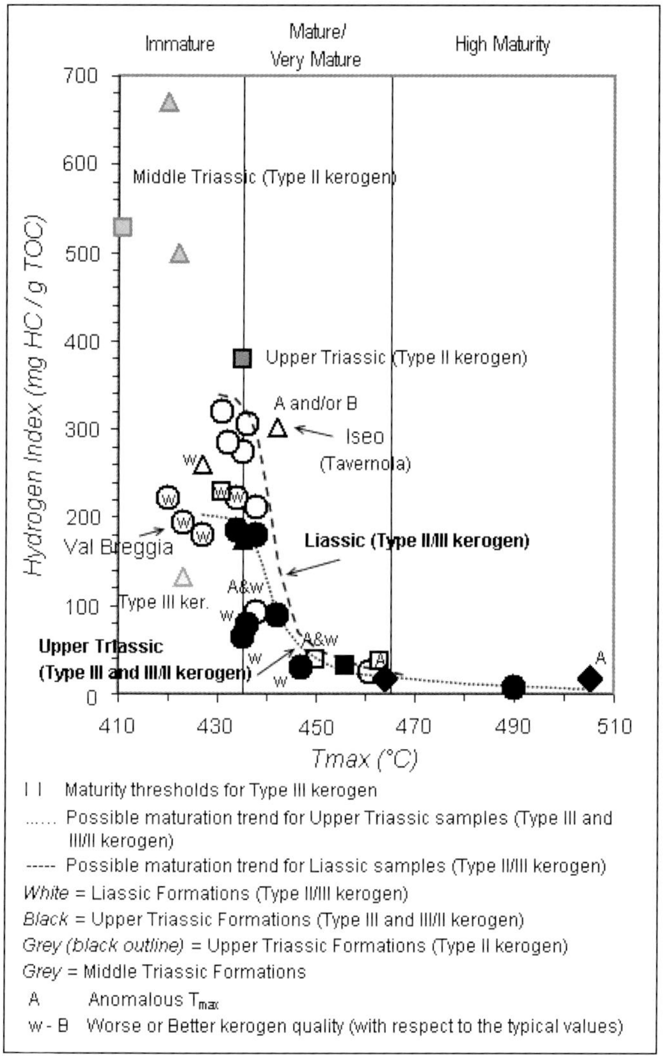

FIG. 10.—Correlation between hydrogen index and T_{max} for different sample populations. Some anomalous (A) T_{max} values are present but are not as clear as in Figure 9. Some Liassic immature samples have poor kerogen quality (w) due to their high inertinite content, but they have apparently reliable T_{max} values.

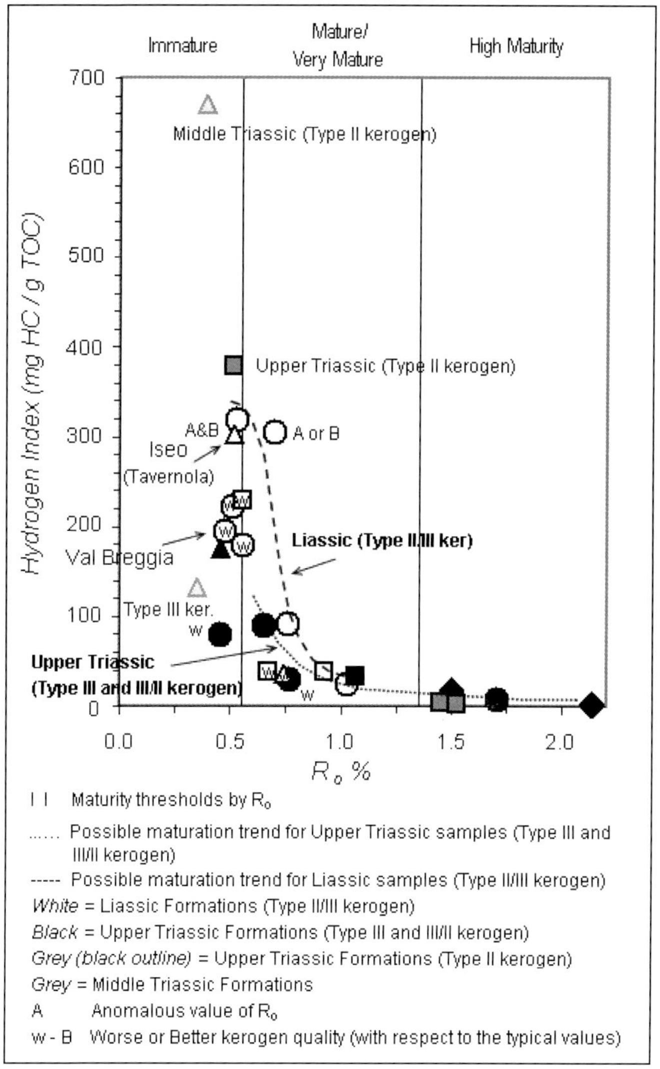

FIG. 11.—Correlation between hydrogen index and vitrinite reflectance for different sample populations. Some R_o anomalies (A) may be present; the symbol marked "A or B" may relate to better kerogen quality.

(505° C) into R_o equivalent using the possible maturation trend for type III and III/II kerogen (dashed fine line in Fig. 9), the resulting value would be over 2%.

Liassic formations are problematic because small quantities of marine organic matter are mixed with terrestrial organic matter (mainly inertinite: fusinite and semifusinite), resulting in a kerogen (in terms of original Hydrogen Index, HI) that can be considered as type II/III. The relationship between vitrinite reflectance and T_{max} is not very well defined. The samples from the Calcare di Moltrasio Formation from the Monte Generoso Basin and Albenza Plateau (Table 1; white circles in Fig. 9) show a good correlation (thick dashed line), which is dissimilar to the Upper Triassic samples. Two samples from the Tavernola location (Iseo Basin–Iseo Lake succession) have low vitrinite reflectance (0.50 and 0.54%) compared to a T_{max} of over 440° C (Table 1; Fig. 9—white triangles represent average values). Vitrinite reflectance is probably unreliable (only two and seven measurements were made for the first and the second sample, respectively), and it is likely to be underestimated. If we believe the T_{max} and the maturation trend for Liassic samples in Figure 9, R_o should be ~0.75%. Comparing vitrinite reflectance and a biomarker ratio (sterane isomerization—Fig. 13), and trusting the biomarker ratio data (free hydrocarbons seem to be indigenous—average production index is 0.11) and the kinetics of both maturity parameters, R_o must be higher than 0.65%. To complicate the maturity data assessment, a different kerogen quality could be interpreted with a higher original HI with respect to typical values for Liassic formations (see Fig. 10, where residual HI is high compared to the T_{max} value). Therefore, a different T_{max}–R_o correlation curve could be interpreted. Finally, the accepted thermal modeling of the Iseo Lake succession suggests that an overall good fit between calculated and measured vitrinite reflectance in the Liassic succession has been achieved if we consider that the measured vitrinite reflectance at the Tavernola location (the shallowest R_o value in Fig. 6e) is slightly underestimated with respect to the likely correct value (calculated EasyR_o ~0.6%). The samples from the lower and

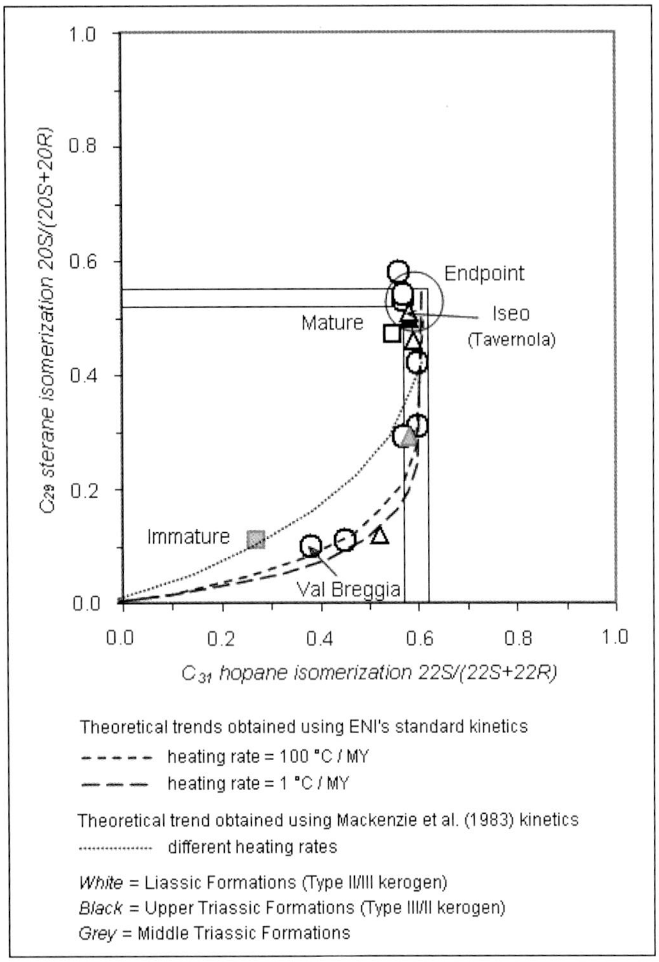

FIG. 12.—Biomarker maturity parameters (sterane and hopane isomerization—mean values; sample locations in Table 1). Note the fairly good agreement between measured data and theoretical trends using different kinetic parameters: Eni's standard kinetics and Mackenzie et al. (1983). The Mackenzie et al. (1983) sterane and hopane isomerization kinetics have the same activation energy and differ only in their frequency factor; hence, the relationship between the two parameters is independent of heating rate.

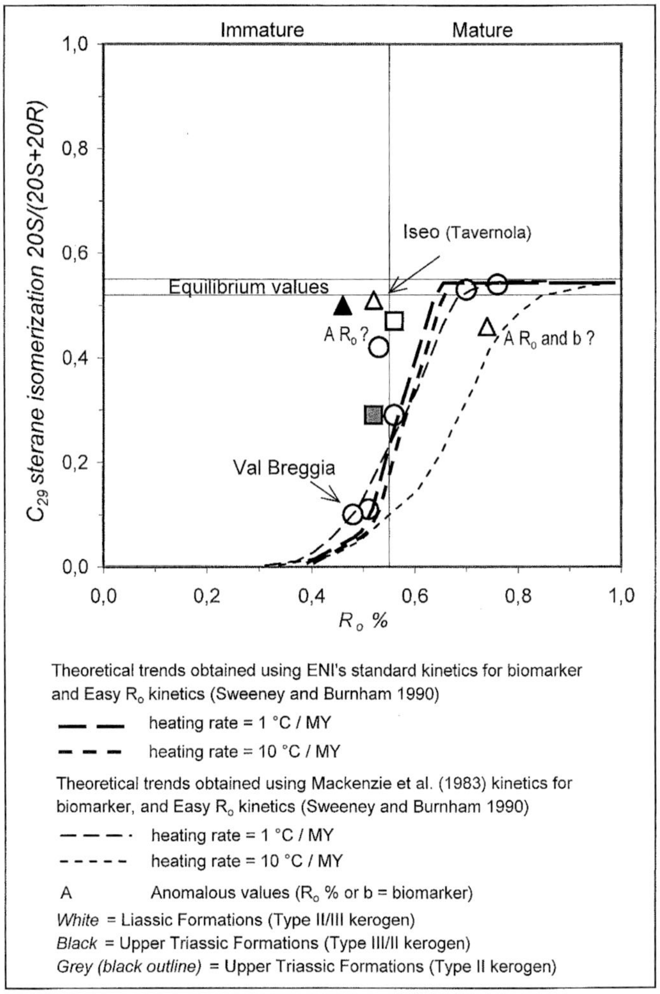

FIG. 13.—Biomarker maturity parameters (sterane isomerization) and vitrinite reflectance (mean values; sample locations in Table 1). Note the fairly good agreement between measured data and theoretical trends using different kinetic parameters. The relationship between the two parameters is not very dependent on heating rate using Eni's sterane isomerization kinetic parameters, although these consider different activation energies. Using Mackenzie et al. (1983) biomarker kinetics, the relationship between the two parameters strongly depends on heating rate.

upper parts of the Calcare di Moltrasio at the Tavernola location (white square symbols in Fig. 9) exhibit anomalously high T_{max} values compared to vitrinite reflectance (463° C with 0.93%, and 450° C with 0.67%, respectively). It is not clear if the particular kerogen composition (small quantity of marine organic matter with a significant amount of semifusinite) is responsible for these anomalously high values, as the vitrinite measurements seem to be correct. This case also shows that thermal modeling can help in understanding the reliability of maturity data by suggesting that the measured vitrinite reflectance can be considered more realistic, and that T_{max} cannot be easily converted into R_o equivalent.

Returning to the Val Breggia thermal modeling, the good fit between calculated and measured maturity data is shown in Figure 14. The maximum heat flow is slightly lower than that used in the Iseo Lake simulation (heat-flow peak in the Middle Jurassic is 2.1 Heat Flow Units (HFU), about 88 mW/m^2 as calibrated by Bersezio et al. 2005). The small overestimation of biomarker maturity for the Val Breggia succession (d and e in Fig. 14), assuming kinetic parameters are correct, can easily be corrected using a lower heat flow during the Tertiary, or with slightly thinner Cretaceous–Tertiary events. However, the use of these biomarker maturity parameters, which assume simple precursor–product relationships (oversimplified), is still recommended only as a rough constraint on thermal history (Peters et al. 2005). Nevertheless, in some cases (e.g., Val Breggia and Iseo Lake successions), maturity information from biomarkers can be useful. In particular, for the thermal modeling of the Iseo Lake succession (regarding the thermal history assessment of the Upper Liassic), the biomarker maturities are more accurate than R_o, as the measured vitrinite reflectance values are scattered.

The Val Breggia modeling seems to provide us with a constraint on the maximum possible heat flow in the Latest Cretaceous and in the

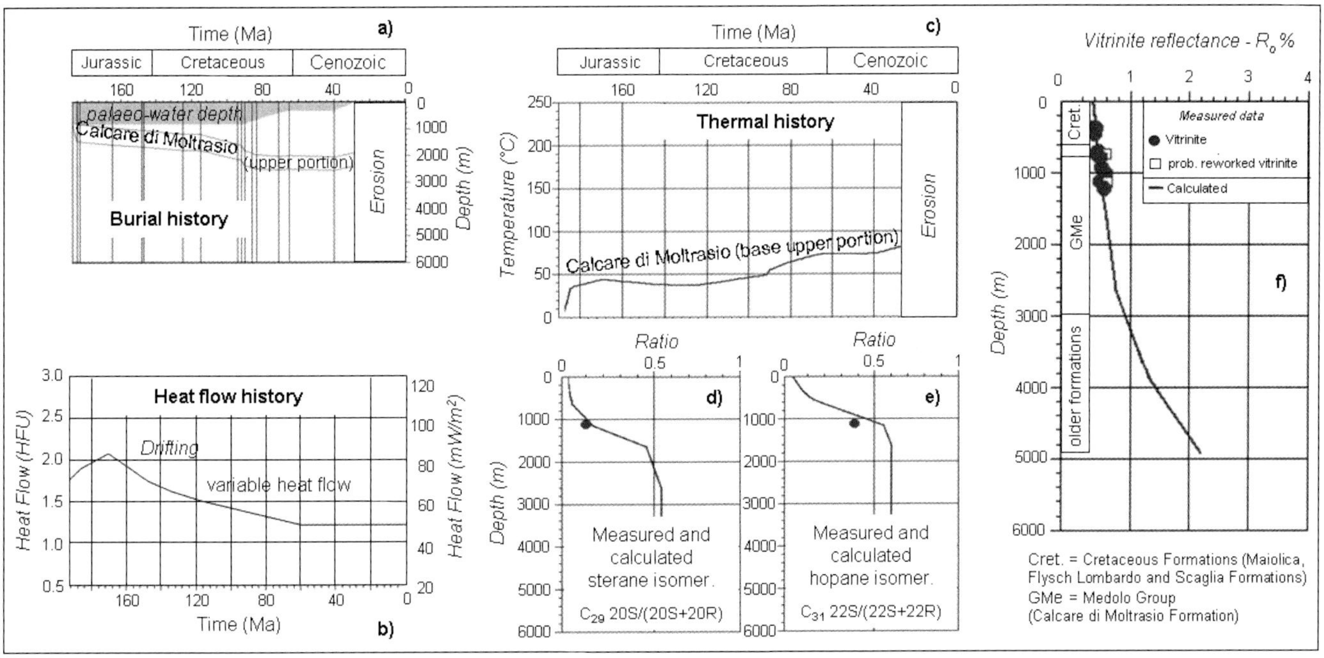

FIG. 14.—Thermal modeling for Val Breggia section: (a) Burial history of the upper portion (thickness 500 m) of the Calcare di Moltrasio and (b) adopted heat-flow history during the extensional Mesozoic phase and later. (c) Thermal history of the upper portion of Calcare di Moltrasio Formation. Results in (d), (e), and (f) compare well between calculated and measured maturity data (R_o and biomarker maturity values).

Lower Tertiary (between 70 and 30 Ma ago). This is because the paleotemperature at the base of the upper portion of the Calcare di Moltrasio (depth ~1200 m in Fig. 14f) must be lower than 75–80° C (Fig. 14c); otherwise, the calculated maturity will be overestimated. This was obtained using a heat flow of 1.2 HFU (~50 mW/m^2). Considering small errors in the burial history (as previously mentioned) and in the thermal model (i.e., overestimation of paleosurface temperature, underestimation of water depth), and assuming that regional heat flow was homogeneous between Val Breggia and Iseo Lake (same geological context and nearby; see Fig. 1), the heat flow at Iseo Lake location must be ~1.2 HFU (in any case, not higher than 1.3–1.4 HFU; 54.5–58.5 mW/m^2). This seems to confirm that in order to increase the maturity (and therefore the temperature) for the Domaro Formation of the Iseo Lake succession ("DOM" unit in Fig. 8b), the only way to reach the required temperature (~90–100° C over a short time) is to consider a thick eroded succession, as shown in Figures 6a and 8a.

Cooling During Burial

Our thermal modeling indicates a cooling event that occurred during postrift deposition at the end of the Middle Jurassic–Early Cretaceous, because the hypothesized very low sedimentation rate cannot compensate for the decreasing heat flow. The decrease of temperature is particularly evident for the Upper Triassic formations in the depocenters (i.e., Iseo Lake succession; Figs. 6c, 8b). Therefore, it is suggested that this first cooling event was mainly due to decreased geothermal gradient (decreasing heat flow) rather than uplift and exhumation of the chain, which actually occurred only much later, for example, in the eastern sector of the southern Alps, during the Late Miocene (Zattin et al. 2006).

IMPLICATIONS FOR HYDROCARBON EXPLORATION IN THE PO PLAIN FOREDEEP

The regional distribution of OM maturity in the southern Alps appears to be primarily controlled by differential burial during the Norian–Liassic extensional phase, which was associated with a high geothermal gradient. This kind of thermal regime can have significant implications for hydrocarbon exploration, even when taking into consideration the earlier timing assumed (without calibration points) by Novelli et al. (1987) and Greber et al. (1997). A higher and later heat-flow peak (when burial was greater), as evaluated in more recent literature, can have stronger effects on hydrocarbon generation. Geochemical modeling of some basinal successions (for example, Iseo Lake), calibrated by measured thermal maturity profiles, suggests that the Upper Triassic source rocks may have attained high maturity levels during the Jurassic, and this is even more likely for the deeper Middle Triassic source rocks.

In the Po Plain foredeep, the same successions as in the southern Alps are present, yet they have only undergone mild Tertiary structural compression (Fig. 1). It is necessary to carry out detailed modeling (based on the same heat-flow history calibrated on the outcrop analogues) to predict the timing of hydrocarbon generation and expulsion. In areas of shallow Rhaetian–Liassic burial in the Po Plain subsurface, the source rocks retained their original petroleum potential (immature or low maturity) prior to major Neogene–Quaternary burial. The high late heating allowed generation/expulsion of hydrocarbons after trap formation (Novelli et al. 1987). Better definition of the heat-flow peak in the Jurassic (Scotti 2005, and references therein) helps to further refine the hydrocarbon charge risk for leads and prospects in the Po Plain subsurface, where traps formed in the Cenozoic (Fantoni et al. 2004, Scotti and Fantoni 2008), as shown in Figure 15.

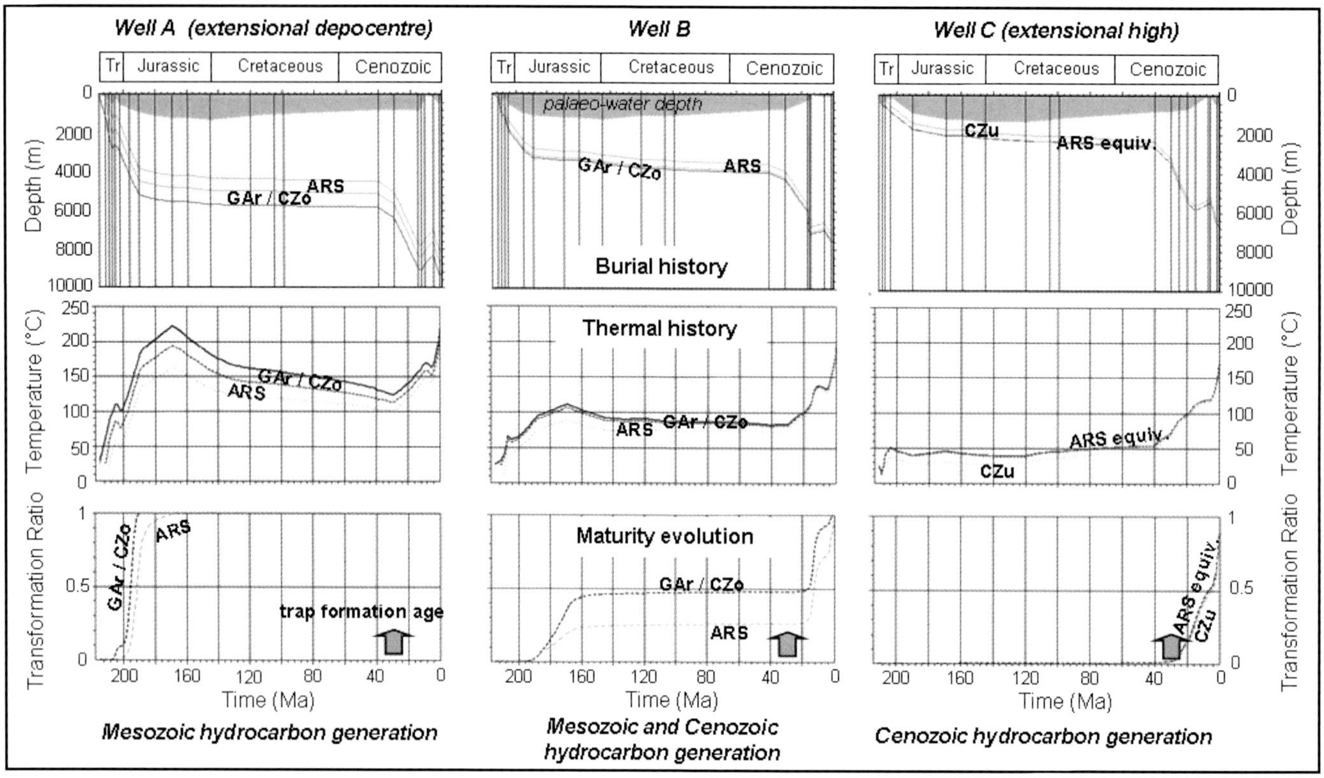

FIG. 15.—Burial history, thermal history, and timing of hydrocarbon generation (transformation ratio) for the main Upper Triassic source rocks in different tectonic settings of the Po Plain foreland. Type IIS (GAr—Gruppo Aralalta, and CZo—Calcare di Zorzino) and type III kerogen (ARS—Argillite di Riva di Solto, and CZu—Calcare di Zu) kinetic parameters were used (Eni S.p.A. proprietary database).

CONCLUSIONS

The heat-flow and burial histories of two sedimentary successions from the southern Alps were assessed using thermal maturity parameters from organic matter, including vitrinite reflectance, biomarker maturity, and T_{max} from Rock-Eval pyrolysis. It is important to have reliable maturity data for key successions to reconstruct burial and heat-flow histories.

Agreement between the different maturity parameters and the availability of measured maturity data for a thick lithostratigraphic sequence are critical to obtain reliable thermal reconstruction. Our reconstruction of the heat-flow evolution at the transition between rift and postrift phases (Middle Jurassic) was possible only because a clear maturity trend within the Upper Triassic sedimentary succession (Iseo section) was available.

The availability of additional thermal maturity data from the shallower portion of the sedimentary succession (Medolo Group—Liassic) allowed the reconstruction of a better thermal history for the Late Jurassic–Early Tertiary. In particular, we were able to better define the Cretaceous burial history in the Iseo Basin and evaluate the heat-flow history in the Late Jurassic–Early Tertiary for the southern Alps. The decreasing heat flow in the drifting phase was less drastic than the previous assessment. A 2200-m-thick eroded section of Cretaceous–Paleocene succession (mostly due to the Flysch Lombardo), instead of the previously adopted 1100 m, was used to obtain a good fit between the calculated and measured maturity data within the Liassic succession of the Iseo section.

In the southern Alps in general, OM maturity appears to have been primarily controlled by differential burial during the Norian–Liassic extensional phase and by high heat flow. Resulting heat-flow values, checked in several localities in the southern Alps, using recent literature, are quite high and relatively uniform (85 to 105 mW/m^2), suggesting that this thermal regime occurred throughout the southern Alps domain. This type of heat-flow history may have significant implications for hydrocarbon exploration. The modeled maturity profiles of some basinal successions suggest that the Upper Triassic source rocks could have already attained high maturity during the Jurassic, and this is even more likely for the deeper Middle Triassic source rocks.

Where traps in the Po Plain foredeep formed by Tertiary Alpine compression, accurate modeling is necessary to predict the time of hydrocarbon generation and expulsion. In the areas of low Rhaetian–Liassic burial, the source rocks retained their original petroleum potential prior to strong Neogene–Quaternary burial. High recent heating resulted in generation/expulsion of hydrocarbons after trap formation. Accurate definition of the Jurassic heat-flow peak can therefore help to better define hydrocarbon charge risk.

ACKNOWLEDGMENTS

We are grateful to Eni S.p.A., Exploration and Production Division, for their permission to publish the data. We thank reviewers Kenneth E. Peters, Nicholas B. Harris, Deborah Higley, and Catherine Hamblett for improving clarity and grammar and for their useful suggestions.

REFERENCES

Bersezio R, Bellantani G. 1997. The thermal maturity of the South Alpine

Mesozoic succession north of Bergamo by vitrinite reflectance data. *Atti Ticinensi di Scienza della Terra* 5:101–114.

Bersezio R, Erba E, Gorza M, Riva A. 2002. Berriasian–Aptian black shales of the Maiolica formation (Lombardian Basin, southern Alps, northern Italy): local to global events. *Palaeogeography, Palaeoclimatology, Palaeoecology* 180:253–275.

Bersezio R, Fornaciari M, Gelati R, Napolitano A, Valdisturlo A. 1993. The significance of the Upper Cretaceous to Miocene clastic wedges in the deformation history of the Lombardian southern Alps. *Gèologie Alpine* 69(1):3–20.

Bersezio R, Scotti P, Tonetti M. 2005. Maturity of organic matter in the Liassic succession of the Mt. Generoso rifted basin (southern Alps, Lombardy). *In* Corrado S, Di Giulio A, Zattin M, Aldega L (Editors). *The Record of Thermal History in Sedimentary Succession: Course on Methods and Examples from the Italian Region*: Atti Ticinensi di Scienza della Terra. Special Series 10, p. 71–76.

Bertotti G, Picotti V, Bernoulli D, Castellarin A. 1993. From rifting to drifting: tectonic evolution of the South Alpine upper crust from the Triassic to the Early Cretaceous. *Sedimentary Geology* 86(1/2):53–76.

Bertotti G, Seward D, Wijbrans J, Ter Voorde M, Hurford AJ. 1999. Crustal thermal regime prior to, during, and after rifting: a geochronological and modeling study of the Mesozoic South Alpine rifted margin. *Tectonics* 18(2):185–200.

Calabrò R, Ceriani A, Di Giulio A, Fantoni R, Scotti P. 2003. Reconstruction of thermal history of the syn-rift sequence between the Iseo Basin and Trento Plateau: results from the integrated study of organic matter and fluid inclusions. *In* Ceriani A, Di Giulio A (Editors). *The Record of Thermal History in Sedimentary Succession: Course on Methods and Examples from the Italian Region*: Atti Ticinensi di Scienza della Terra. Special Series 9, p. 88–91.

Della Vedova B, Bellani S, Pellis G, Squarci P. 2001. Deep temperatures and surface heat flow density distribution. *In* Vai GB, Martini P (Editors). *Anatomy of an Orogen: The Apennines and Adjacent Mediterranean Basins*: Kluwer Academic Publishers, Dordrecht, Netherlands. p. 65–76.

Fantoni R, Bersezio R, Forcella F. 2004. Alpine structure and deformation chronology at the southern Alps–Po Plain border in Lombardy. *Bollettino Società Geologica Italiana* 123:463–476.

Fantoni R, Scotti P. 2003. Thermal record of the Mesozoic extensional tectonics in the southern Alps. *In* Ceriani A, Di Giulio A (Editors). *The Record of Thermal History in Sedimentary Succession: Course on Methods and Examples from the Italian Region*: Atti Ticinensi di Scienza della Terra. Special Series 9, p. 96–101.

Gradstein FM, Ogg JG, Smith AG. 2004. *A Geologic Time Scale 2004*: Cambridge University Press, Cambridge, UK. 610 p.

Greber E, Leu W, Bernoulli D, Schumacher M, Wyss R. 1997. Hydrocarbon provinces in the Swiss southern Alps—a gas geochemistry and basin modelling study. *Marine and Petroleum Geology* 14(1):3–25.

Mackenzie AS, Wolff GA, Maxwell JR. 1983. Fatty acids in some biodegraded petroleums. Possible origins and significance. *In* Bjorøy M, Albrecht C, Cornford, C (Editors). *Advances in Organic Geochemistry, 1981*: John Wiley and Sons, New York. p. 637–649.

Novelli L, Chiaramonte MA, Mattavelli L, Pizzi G, Sartori L, Scotti P. 1987. Oil habitat in the northwestern Po Basin. *In* Doligez B (Editor). *Migration of Hydrocarbons in Sedimentary Basins*: Editions Technip, Paris. p. 27–57.

Peters KE, Walters CC, Moldowan JM. 2005. *Biomarkers and isotopes in petroleum exploration and Earth history. In The Biomarker Guide,* 2nd ed.: Cambridge University Press, Cambridge UK. p. 475-1155.

Remane J. 2000. International Stratigraphic Chart, with explanatory note. *In* 31st International Geological Congress, Brazil, 6-17 August, Rio de Janeiro 2000, Abstracts volume ICS, IUGS and UNESCO, Paris. p. 16.

Ronchi P, Imperial E, Scotti P, Ceriani A, Jadoul F, Ricchiuto T. 2003. The thermal and diagenetic history of the Dolomia Principale in the Iseo Basin (southern Alps, Italy). *In* Ceriani A, Di Giulio A (Editors). *The Record of Thermal History in Sedimentary Succession: Course on Methods and Examples from the Italian Region*: Atti Ticinensi di Scienza della Terra. Special Series 9, p. 83–87.

Scotti P. 2005. Thermal constraints suggested by the study of the organic matter and thermal modelling strategies—A case history from southern Alps. *In* Corrado S, Di Giulio A, Zattin M, Aldega L (Editors). *The Record of Thermal History in Sedimentary Succession: Course on Methods and Examples from the Italian Region*: Atti Ticinensi di Scienza della Terra. Special Series 10, p. 21–35.

Scotti P, Fantoni R. 2008. Thermal modelling of the extensional Mesozoic succession of the southern Alps and implications for oil exploration in the Po Plain foredeep. *In* 70th EAGE Conference and Exhibition; June 9–12, 2008; Rome, Italy, Extended abstract, A0016, 5p.

Sweeney JJ, Burnham AK. 1990. Evaluation of a simple model of vitrinite reflectance based on chemical kinetics. *American Association of Petroleum Geologists Bulletin* 74:1559–1570.

Zattin M, Cuman A, Fantoni R, Martin S, Scotti P, Stefani C. 2003. Thermochronological evolution of the eastern southern Alps from vitrinite reflectance and fission-track data. *In* Ceriani A, Di Giulio A (Editors). *The Record of Thermal History in Sedimentary Succession: Course on Methods and Examples from the Italian Region*: Atti Ticinensi di Scienza della Terra. Special Series 9, p. 92–95.

Zattin M, Cuman A, Fantoni R, Martin S, Scotti P, Stefani C. 2006. From Middle Jurassic heating to Neogene cooling: the thermochronological evolution of the southern Alps. *Tectonophysics* 414:191–202.

INTRODUCTION

In the scope of the Sub-Thrust Reservoir Appraisal (SUBTRAP) project (e.g., Roure et al. 2005, 2010), IFP New Energies (formerly IFP) and the Mexican Petroleum Institute (IMP) initiated a long-term collaborative study of fluid–rock interactions and reservoir quality prediction in Cretaceous carbonates of the Córdoba Platform in Eastern Mexico. This area constitutes the southernmost extent of the Laramide thrust front and can be traced along the eastern edge of the North American Cordillera, from Canada in the north to the Gulf of Mexico in the south (Fig. 1). This study identified numerous episodes of pressure–solution fracturation and hydraulic brecciation in the platform carbonates related to increasing burial and tectonic stress (layer parallel shortening), as well as exposure events and coeval paleokarst development (Ferket et al. 2000, 2003, 2004, 2006; Ferket 2004, 2006). During the first phase of the SUBTRAP project, two-dimensional (2D) basin modeling was also performed to reconstruct the paleoburial history of individual reservoir and source rock horizons using the software Thrustpack (Divies and Sassi 1996; Sassi and Rudkiewicz 1999, 2000). This modeling coupled kinematic evolution of the thrust belt with erosion and deposition of synorogenic sediments. The modeling was also applied to reconstruct the structural evolution of three regional sections extending from the Cordilleran foothills in the west to the Veracruz Basin in the east from their Middle Cretaceous pre-orogenic configuration until present day (Ortuño et al. 1999, 2003). The Thrustpack modeling enabled us to (1) reconstruct the thermal evolution of these cross sections using bottom-hole temperature (BHT) data and maturity ranks of the organic matter (i.e., vitrinite reflectance [R_o] and Rock-Eval [T_{max}]) for calibration and (2) simulate the maturity evolution of source rocks from the Late Jurassic onward. These preliminary models and current petroleum occurrences indicate that a petroleum charge was generated in the adjacent Veracruz Basin (Figs. 1, 2) during post-Laramide subsidence, which migrated westward, filling traps in frontal duplexes of the Cordilleran thrust belt.

This article summarizes results from the second phase of the SUBTRAP project, which focused on remaining uncertainties relating to (1) the past attitude of the basement, (2) the timing and amount of exhumation and unroofing of the foothills, and (3) the timing and pathways of regional fluid flow and hydrocarbon migration. Following a short summary of the regional geology and current understanding of the petroleum systems, the article focuses on fluid inclusion (FI) studies that were employed to better constrain estimates of the eroded thicknesses. Fluid inclusion analysis is a paleothermo-barometer that is independent of the other thermal calibrations, which are based on

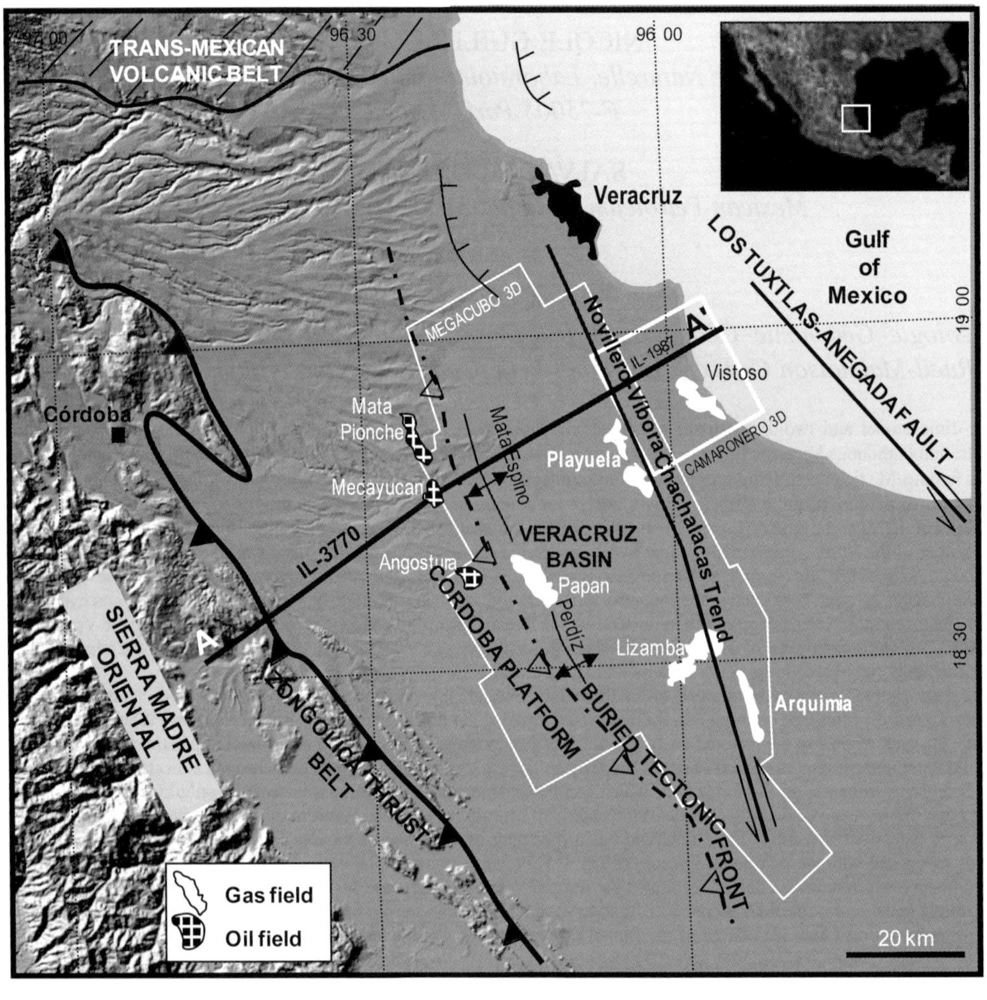

FIG. 1.—Location map of the Veracruz Basin, Córdoba Platform, and Zongolica thrust belt, outlining as well the main oil and gas fields of the platform. AA′ indicates the trace of the modeled transect.

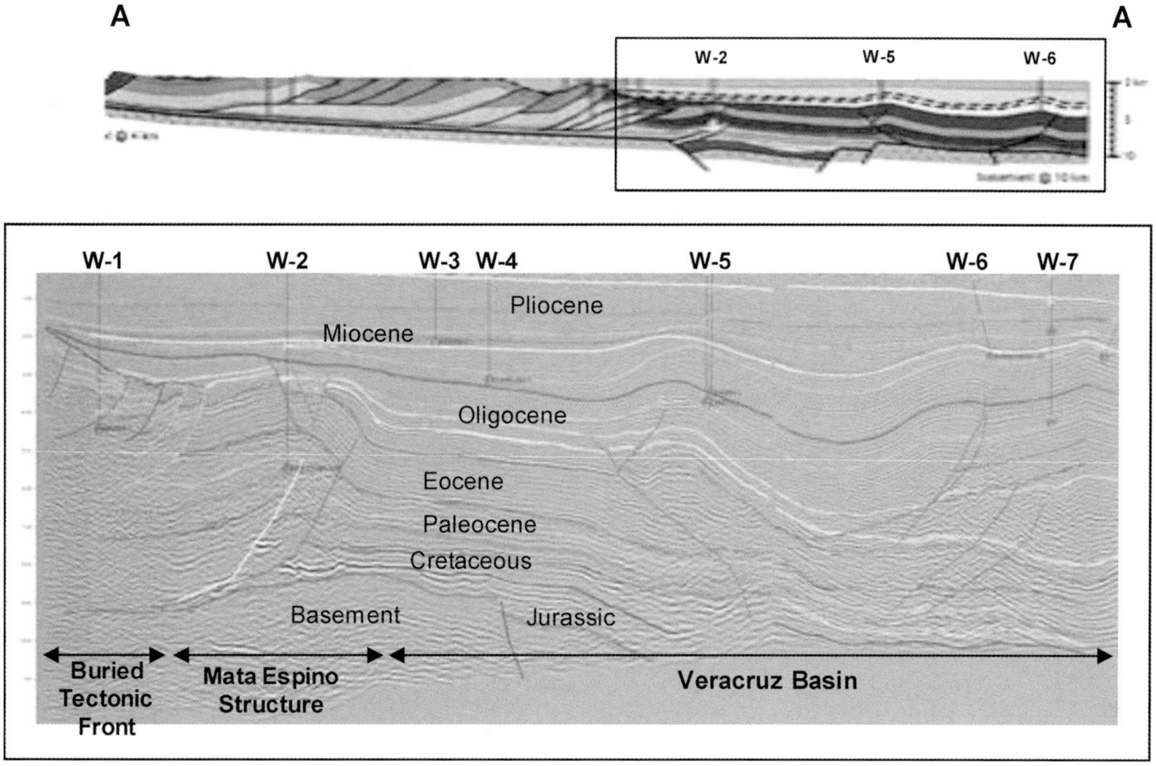

FIG. 2.—Top: Structural section AA'. Bottom: Seismic profile imaging the transition between the Córdoba Platform and thin-skinned Laramide thrust front in the west and basement-involving inverted structures of the Veracruz Basin in the east.

organic matter maturity ranks (see Roure et al. 2010). A direct consequence of the FI study was to provide more realistic control on the paleoburial of the Cretaceous platform carbonates and on the former configuration of the underlying basement. Subsequently, we used the inferred structural evolution of an existing regional cross section as input data for a basin modeling experiment using the CERES2D software, a dedicated basin modeling tool, developed for tectonically complex areas (see Appendix). This software was used to simulate the regional fluid flow evolution and hydrocarbon migration.

GEOLOGICAL SETTING AND CURRENT KNOWLEDGE OF THE PETROLEUM SYSTEMS OF THE CÓRDOBA PLATFORM AND VERACRUZ BASIN

The Córdoba Platform is located south of the Trans-Mexican volcanic axis (Fig. 1) and belongs to the Zongolica thrust belt (Ortuño 1991; Meneses-Rocha et al. 1997; Ortuño et al. 1999, 2003; Rojas 1999; Fig. 2). The Córdoba Platform is predominantly made up of Cretaceous carbonates and constitutes the southernmost portion of the Laramide thrust belt system. This thrust belt developed along the eastern edge of the North American Cordillera during Late Cretaceous to Eocene, from the Canadian Rockies in the north to the Sierra Madre Oriental in Mexico in the south (Fig. 1). To the east, the foreland autochthon consists of the Veracruz Basin (Fig. 1), where thick Oligocene and Neogene siliciclastics derived from post-Laramide erosion of the thrust belt accumulated (Jennette et al. 2003a, 2003b). A short summary of the geodynamic and paleogeographic evolution, lithostratigraphy, and petroleum systems of this onshore part of Eastern Mexico is provided below.

Geodynamic Evolution of Eastern Mexico

The tectonic evolution of eastern Mexico is directly linked to the history of the Gulf of Mexico. The Jurassic events that eventually formed the Gulf of Mexico began with a rifting episode during Late Triassic to Jurassic, leading to continental break up and progressive oceanic accretion between the Maya block in the west and the actual North American Platform (Campa and Coney 1983, Salvador 1991, Pindell 1994). At the western margin of the Gulf of Mexico, rift-related crustal thinning resulted in the formation of two basins: the Veracruz Basin in the east and the Zongolica Basin in the west (not shown). In these basins, euxinic conditions and source rock deposition prevailed during the Late Jurassic. These two domains were still characterized by pelagic carbonate sedimentation during Cretaceous episodes of post-rift thermal subsidence, whereas the intervening Córdoba Platform, comprised of ancient horsts, was characterized by shallow-water sedimentation during the Early and Middle Cretaceous. Cordilleran deformation was first recorded in the study area by local uplift and karstification of the carbonate platform during the Campanian in response to tectonic loading of the hinterland. A foredeep with calcareous flysch deposition formed between the forebulge and the tectonic front at that time. Platform exposure was followed by drowning of ancient platform domains and onset of Maastrichtian deep-water clastic sedimentation in the various units of the Zongolica thrust belt. These events thus record a short episode of foreland flexuration.

During the Late Cretaceous and Paleocene, the Mesozoic series of the Zongolica Basin and Córdoba Platform were tectonically detached from their crystalline basement and thrusted eastward toward the autochthonous foreland. The thrust front was located at the former platform-to-basin transition, where major deformation of carbonaceous sediments and stacking of numerous duplexes occurred. These duplexes formed traps for major oil pools in Middle and Upper Cretaceous carbonate reservoirs (Fig. 2; Roure 2008). Thrusting resulted in a rapid exhumation and unroofing of the Mesozoic carbonate units, with a massive supply of clastic material toward the still-subsiding Veracruz Basin.

Basement faults of the Veracruz Basin were ultimately reactivated by transpression during the Late Miocene, resulting in the development of wide anticlinal traps, such as the Novillero–Vibora–Chachalacas (NVC) anticline and the NW-trending Mata Espino and Perdiz structures (Fig. 1), where biogenic gas accumulations occur in Neogene sandstone reservoirs (Espinoza and Toriz 2005).

Lithostratigraphic Description

Pre-Rift and Syn-Rift Sequences: The crystalline basement of the Córdoba Platform and Veracruz Basin is made up of Permo-Triassic metamorphic rocks (Jacobo 1986). In the former basinal domains, the basement is overlain by Triassic–Jurassic (?) continental red beds of the Cahuasas and Rosario formations (Figs. 2, 3).

Passive Margin Sequence: Late Jurassic calcareous black shales of the Tepexilotla Formation (Figs. 2, 3) crop out along the exposure of the sole thrust of the Sierra de Zongolica in the west and have been drilled in one well in the southern part of the Veracruz Basin. The thickness of these rocks measures approximately 250 m (Vázquez 2004).

Overlying Cretaceous sediments comprise a succession of limestones (mudstones to grainstones) that display drastic lateral thickness and facies variations between platform and basinal domains. The Lower Cretaceous interval is 1000 to 1500 m thick and consists of partially dolomitized micritic limestones typical of a lagoonal environment. These rocks grade upward into dolomites and partially dolomitized limestone with evaporate intercalations. The overlying Middle Cretaceous Orizaba Formation is 1500 to 2000 m thick in the central and western parts of the platform (Ferket 2004) but gradually thins to the east. In the SW portion of the platform, the basal units of this formation consist of dolomites and evaporites, which commonly form decollement horizons during Laramide deformation. The lower and middle parts of the Orizaba Formation are partly dolomitized, while the upper part is dominated by light-gray, miliolid-rich, bioclastic limestone.

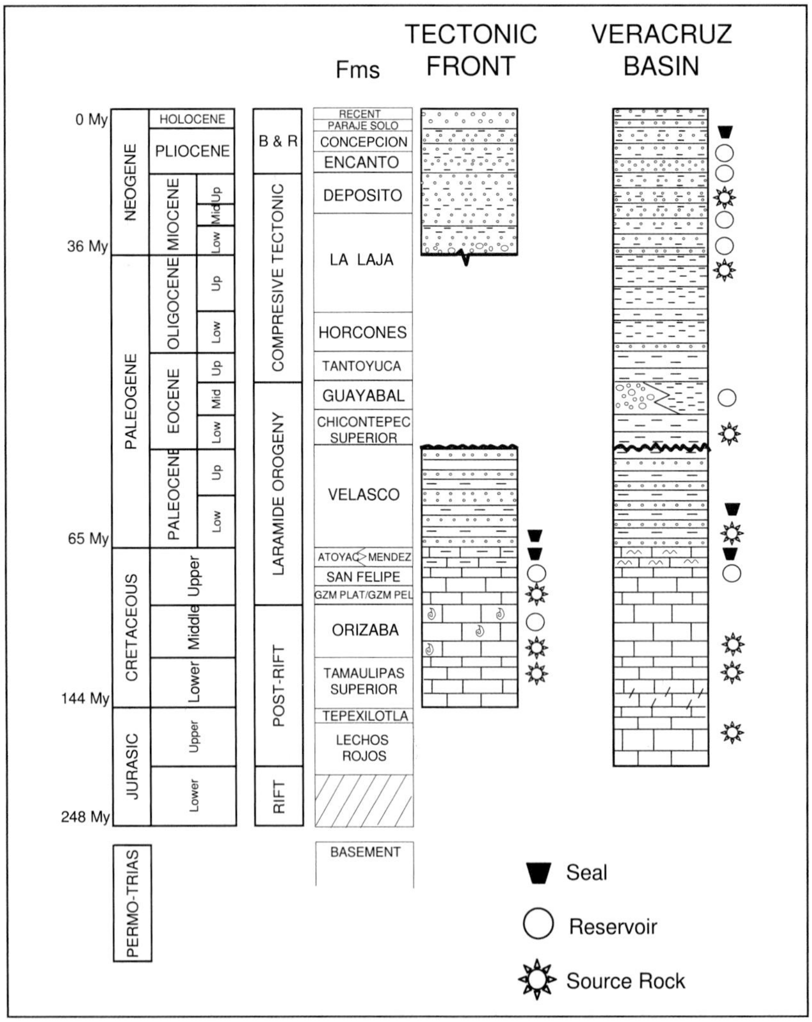

FIG. 3.—Stratigraphic column for the Córdoba Platform and Veracruz Basin, showing potential source rock, reservoir, and seal horizons.

The Upper Cretaceous sequence is divided in five formations, some of which are coeval but which record different paleo-environments. The Guzmantla Formation is 1500 m thick and comprises Lower Turonian and Upper Coniacian–Santonian units (Figs. 2, 3). The lower unit (Guzmantla "pelagica") is composed of well-bedded argillaceous limestones deposited during a maximum flooding phase. The upper unit consists of partly dolomitized limestone, typical of an inner platform with isolated reefs, shoals, and lagoons. The Maltrata Formation is 100 to 400 m thick, consists of argillaceous limestones deposited in deeper anoxic water around the platform, and is considered to be time-equivalent to the lower Guzmantla Formation.

In the Zongolica belt, the Campanian is represented by a depositional hiatus and local paleokarst development; i.e., subaerial exposure. At the same time along the eastern platform edge, slope breccias and intercalated pelagic limestones of the San Felipe Formation were deposited. The Atoyac Formation of Maastrichtian age overlies the Campanian unconformity only in the easternmost outcrops of the Zongolica thrust belt and constitutes the last platform sedimentation in the area. No complete section of the Atoyac Formation has been preserved as a result of post-Laramide erosion, leaving only minimum estimates of the original thickness (300–400 m). The Upper Cretaceous Méndez Formation comprises deep-water shaly limestones (Ferket 2004) that are believed to have originally covered the whole passive margin, but their residual thickness after erosion locally amounts to a maximum of 300 m.

Laramide Flexural Sequence: Erosional remnants in the Zongolica foothills are preserved in a sequence of Laramide-age Late Cretaceous and Paleocene siliciclastic turbidites, and few well records are available with which to document the Paleocene lithofacies in the adjacent Veracruz Basin (Berman et al. 1995).

In contrast, the Eocene sequence of the Veracruz Basin is represented by up to 1400 m of conglomerate, sandstone, and shale derived from the erosion of the Laramide thrust belt (Figs. 2, 3). Seismic and well data indicate that near-source debris-flow sediments close to the tectonic front grade into thick packages of turbidites in the basin, resulting in a basinward thickening of the sequence (Jennette et al. 2003a, 2003b).

Post-Laramide Sequences: Outcrops and wells provide direct control on Oligocene lithofacies in the Veracruz Basin. Nevertheless, interpretations of lithologies and depositional systems are largely based on seismic data. In a few locations where the Oligocene has been drilled it consists of fine sediments deposited in deep-water environments.

The Miocene series unconformably overlies the Oligocene. Lower Miocene is made up of conglomerates with subordinate beds of sandstone and shales that constitutes the upper part of the La Laja Formation (Figs. 2, 3). These sediments were probably transported through large submarine canyons that served as important conduits for the transport of erosional products from the thrust belt to the Gulf of Mexico (Berman et al. 1995). In the Middle Miocene, the basin was still fed through a complex of channels, also filling the former canyons created during the Early Miocene. During the Late Miocene (Depósito Formation; Fig. 3), uplifted and eroded areas to the NW and SW of the Córdoba Platform continued to feed basin-floor fans and complex channel systems that carried a great quantity of sand and mud toward the basin. Simultaneously, during the Late Miocene, reactivation of basement faults in the Veracruz Basin resulted in strata onlapping onto the flanks of localized antiforms, accounting for a number of stratigraphic traps that host major gas fields (e.g., Arquimia field located along the southeastern flank of the NVC structure; Fig. 1).

The Pliocene sequence consists of clayey–sandy sediments of the Encanto and Depósito formations (Fig. 3) that onlap a gently eastward-dipping erosional surface in the foothills (Fig. 2) and rapidly thickens east of the thrust front. Its overall thickness varies from 1000 m near the thrust front in the west to 2000 m or more in the southeastern part of the Veracruz Basin. Large clinoforms prograde to the southeast and have their principal supply source in the northwest. They contribute to form great submarine fans constituted by a complex of interbedded channels. Sandstone layers of Lower Pliocene form clinoforms that host important gas fields in the Veracruz Basin (e.g., Lleida field situated north of Veracruz; Fig. 1).

TABLE 1.—Source rock data of petroleum system in Veracruz Basin (Gonzalez-Mercado 2007).

	Upper Jurassic	Lower and Middle Cretaceous	Upper Cretaceous	Paleocene to Miocene
Kerogen	II	II	II	III
TOC (%)	2	3.50	2	1
IH	475	511–531	300–600	
S_2 (mg HC/g)	2.12	7.5 and 10.3	7.36	
T_{max} (° C)	>450	426–470	410–434	

TOC = total organic compound; IH = hydrogen index; S2 = second peak during Rock-Eval pyrolysis; Tmax = maximal temperature at S2 peak.

Current Knowledge on the Petroleum Systems in the Córdoba Platform and Veracruz Basin

The pattern of oil and bitumen occurrences indicates that two petroleum systems are present (Guzman-Vega et al. 2001, Prost and Aranda-Garcia 2001). The first of these includes bitumen and heavy oil in Cretaceous platform carbonates of the Cordillera allochthon: for example, the Peñuela quarry near the city of Córdoba. These hydrocarbon accumulations are believed to record early petroleum migration and trapping, followed by biodegradation in anticlines that formed in the Paleogene, and are located quite far from the productive oil fields of the frontal duplexes. Ferket et al. (2010) suggested that oil shows near the city of Córdoba are related to short vertical migration from undocumented intraplatform, organic-rich sediments. Another possibility for these oil occurrences is a longer-range eastward migration from the Jurassic source rocks of the former Zongolica Basin. This should have occurred prior to thrusting and unroofing of this basinal domain eastward on top of the Córdoba Platform.

Oil fields in the frontal complexes are almost certainly related to a different petroleum system, directly associated with the Veracruz Basin (Guzman-Vega et al. 2001). Only this latter petroleum system is discussed further in this article.

Source Rocks: Source rocks associated with hydrocarbons in the frontal complexes include the Upper Jurassic Tepexilotla Formation, Lower Cretaceous Tamaulipas Superior Formation, and Middle Cretaceous Orizaba Formation (Vázquez 2004, 2008) and intervals in the Paleocene to Miocene section. Details regarding source rock quality, character, and interpretation are provided in Table 1.

a. The Upper Jurassic (Tithonian) is the principal source rock (Vázquez 2004) and consists of shaly limestones of the Tepexilotla Formation. The net thickness of the source interval is about 150 m (Vázquez 2004). The organic matter is algal-derived Type II kerogen, with total organic carbon (TOC) content averaging 2% (Guzman-Vega et al. 2001); a hydrocarbon index (HI) of 475; S_2 of 2.12 mg hydrocarbons (HC)/g rock, and a T_{max} of >450° C. These analyses correspond to mature source rocks; the original TOC for Jurassic source rocks is 3.5 to 4% (Vázquez 2008). Biomarker analyses on Jurassic oils indicate a calcareous marine siliciclastic environment (Berman et al. 1995).

b. Middle and Lower Cretaceous argillaceous and evaporitic limestones also are considered to be source rocks for liquid and gas hydrocarbons (Vázquez 2004). The net thickness of the source interval is 130 m. Pyrolysis and organic petrography indicate a mean TOC value of 3.5% (Guzman-Vega et al. 2001), Type II kerogen, an HI ranging from 511 to 531, S_2 between 7.5 and 10.3 mg HC/g rock, and T_{max} values between 426 and 470° C. These parameters correspond to a relatively immature rock and represent good source rock potential (Vázquez 2008). Biomarkers on such oils point out a calcareous–evaporitic marine family.

c. The Upper Cretaceous (Turonian) contains argillaceous limestones with a net generative thickness ranging from 50 to 80 m. According to Vázquez (2004), the average TOC value is 2%; HI is 300 to 600; S_2 is 7.36 mg HC/g rock, and T_{max} ranges from 410 to 434° C. The kerogen is Type II, and biomarkers indicate an anoxic calcareous marine source rock.

d. Some units within the Paleocene to Miocene sequence have been characterized as source rocks for biogenic gas based on Rock-Eval analyses. Their kerogen is Type III, and TOC amounts to 1% (Vázquez 2008). The next table summarizes the source rock parameters.

A consequence of the geologic complexity of the Veracruz Basin is that the same source rocks can present different degrees of maturity. Thus, according to Table 1, the T_{max} of the Upper Jurassic rocks from the basin depocenter indicates that this source rock is currently overmature, while the same rocks located along the Zongolica thrust belt (Fig. 1) are still into the oil generation window.

According to their T_{max} data, the Middle and Lower Cretaceous source rocks are currently within the window of generation, in contrast to the Upper Cretaceous source rock from the Platform of Córdoba, which is currently immature.

At the moment, for the Tertiary source rocks (Paleocene and Eocene) located in the Veracruz Basin, thermal conditions for oil generation are present; however, the organic matter contained in these rocks is Type III, which generates gas.

Reservoirs: Oil reservoirs in the Córdoba Platform are hosted in limestone and dolomites of the Middle Cretaceous Orizaba Formation, as well as in limestone of the Upper Cretaceous Guzmantla and San Felipe formations (Berman et al. 1995). Some limestones within the Guzmantla and San Felipe formations exhibit breccia textures. The Middle and Upper Cretaceous reservoirs are characterized by dual porosity–permeability systems, with contributions from both matrix and fracture porosity.

Hydrocarbon production in the Veracruz Basin is mainly thermogenic and biogenic gas trapped in clastics sequences throughout the entire Miocene and in the Lower Pliocene (Vázquez 2008). The only oil field in the Veracruz Basin is located in conglomerates of Eocene age (Geologic report of well Perdíz-1, unpublished data, 2000).

Seals: The main seals for reservoirs located in duplexes at the Laramide tectonic front are the Upper Cretaceous Méndez Formation and the Paleocene shales of the Velasco Formation (Fig. 3). Reverse faults form efficient lateral seals (Espinoza and Toriz 2005; Fig. 2). In the Veracruz Basin, the seals are essentially Pliocene shale varying from 100 m thick near the thrust front to 2500 m thick in the eastern part of Veracruz Basin (AIV-Pemex 2007, 2010).

Timing of Hydrocarbon Generation, Migration, and Trapping: Organic geochemistry data such as biomarkers and diamandoids (Vázquez 2004, 2008) indicate multiple sources and migration pathways, revealing a complicated petroleum system. At least two hydrocarbon migration episodes of oils of differing maturity were recorded in Cretaceous reservoirs at the buried tectonic front. Oil migration occurred westward from sources in the foreland toward the thrust belt and is generally believed to have occurred mainly from the Miocene onward (e.g., Ortuño et al. 2003 [and references therein]). Two types of gases are observed in the oil fields located in the thrust belt cretaceous rocks. The first type of gas is found in association with oil, whereas the second one is a more evolved dry gas, identified through isotopic analysis, chromatography, and biomarker analysis. We propose that both gases come from the same source rock, the evolved gas originating from a second phase of maturation, probably after the source rock was affected by the buried caused by the Laramide Orogeny. Gases in the Veracruz Basin often represent mixtures between thermogenic and biogenic gases (Vázquez 2004, 2008), but because direct evidence concerning the timing and identification of migration pathways is lacking, basin modeling is required.

TABLE 2.—Tectonic events and associated heat flow data used for the thermal field modeling (Gonzalez-Mercado 2007).

Age	Stage	Tectonic event	Heat flow
416	crustal part	pre-rift	65
248	basement		60
144.2	Jurassic	rift	55
112.2	Early Cretaceous		50
93.5	Middle Cretaceous	post-rift	50
83.5	Santonian		50
65	Maastrichtian	Laramide stage	50
50	Early Paeocene		50
41.3	Upper Paleocene to lower Eocene		50
37.1	Middle Eocene		47
33.8	Upper Eocene	main phase of shortening	35
24.3	Oligocene		30
16.4	Lower Miocene		30
11.7	Middle Miocene		30
5.7	Upper Miocene		30
0	Pliocene to present day	basin and range stage	30

CONSTRUCTION OF THE STRUCTURAL SECTION AND QUANTIFICATION OF ERODED THICKNESSES

Present-Day Structural Section

The 120-km-long section is located in the north of Veracruz. It extends NE–SW from the Tezonapa-1 well at the western edge of the section, to the Tortuguilla-1 well in the east of the Veracruz Basin (Fig. 1). From west to east the section includes the Zongolica thrust belt; the buried, folded Cordoba Platform; and the Veracruz Basin (Fig. 2). Seismic data are lacking for the western part of the modeled section (Córdoba Platform). Structural interpretations and reconstructions in this area are therefore derived from field observations and available well data. In the eastern part of the Veracruz Basin, data from two three-dimensional (3D) seismic volumes were available to reconstruct the transect (Figs. 1, 2). Calibration data along the entire transect were provided by a large number of exploration wells.

Fluid Inclusion Data, PVT Modeling, and Paleoburial Estimates

The amount of erosion of platform carbonates in the thrust belt was first estimated based on platform reconstructions and thickness variations of the carbonate formations throughout the platform and basin, as recorded in several exploration wells (Ferket 2004, 2006). This exercise was relatively straightforward for the Orizaba and Guzmantla formations but much more ambiguous for laterally restricted formations, such as the Atoyac, Maltrata, and San Felipe formations. Nevertheless, the latter formations are thin compared to the main platform formations, and possible errors in estimates of their original thicknesses therefore do not engender significant errors in reconstructing paleoburial and erosion.

In order to further estimate paleoburial depths and amount of erosion in the thrust belt zone, where several stratigraphic sequences have been stacked and where syntectonic flysch has accumulated in basins, several generations of crosscutting veins were studied from outcrops and exploration wells. As much as possible, these diagenetic phases were placed in a relative timescale and, where possible, also in an absolute timescale. Some minerals contain fluid inclusions that captured relics of ancient formation fluids. Microthermometric study of fluid inclusions gives information on paleotemperature and composition of the fluids preserved in them. The measured homogenization temperatures (Th) give minimum trapping temperatures. In order to determine the real trapping conditions, a pressure correction is needed. In some cases paleopressures can be calculated or estimated from crosscutting relationships between the water isochors and an independent paleothermometer or from cogenetic aqueous and hydrocarbon-filled fluid inclusions. Furthermore, it is not always clear whether a fluid inclusion formed from a fluid in equilibrium with the host rock, from a hydrothermal fluid at higher temperature, or from a cooler fluid. Although exact temperature values might not be available, the diagenetic events and minimum temperatures can be valuable in unraveling the burial history of complex areas.

A data set with useful diagenetic markers has been compiled over the last 12 years for the Córdoba–Veracruz area. The vein filling minerals have been characterized by petrography, stable oxygen and carbon isotopes and Sr isotope analyses, microstructural characteristics, and fluid inclusion data. When fluid inclusion characteristics, such as vapor:liquid ratio and Th-Tm spreads, show relatively homogeneous populations, then it is reasonable to assume that the fluids were in thermal equilibrium with the host rock. Any fluid inclusion population representing either necking down, extreme variation in Th, or variable characteristics that could not be assigned to a specific generation was left out of the data set. Under these conditions one can use the Th values as control points on the burial curve to test kinematic scenarios, always keeping in mind that they represent minimum trapping temperatures. Depending on the isochors (composition), the needed correction is expected to be smaller or higher. In the Córdoba area, fluid inclusions were studied in several minerals (dolomite, fluorite, quartz, and calcite) that precipitated during the burial history before, coeval with, and after hydrocarbon migration, from two Mesozoic levels: the Orizaba and Guzmanta formations (Ferket 2006, Ferket et al. 2010; Figs. 4, 5).

The information inferred from the fluid inclusion data set (Pressure and Temperature of entrapment) was used to calibrate one-dimensional (1D) basin modeling. In two cases, fluid inclusions provided crucial information that could be used to calculate the amount of burial. Quartz–cement in dolomites of the Orizaba Formation hosted four different generations of water–hydrocarbon fluid inclusions (both mixed water and hydrocarbon fluid inclusions, and water-free fluid inclusions). The first two generations are cogenetic primary water–hydrocarbon fluid inclusions, and there are also two sets of secondary fluid inclusions (Fig. 4). In this case, a combination of microthermometric and Synchroton Fourier Transform Infra-Red analyses (SFTIR; see Guilhaumou and Dumas 2005 for explanation) enabled us to perform Pressure and Temperature (PT)-modeling and to calculate paleopressures (Ferket et al. 2010; Fig. 4). These techniques are described in detail in Ferket et al. (2010). The isochors for the hydrocarbon phase were constructed in a PT diagram based on modeling of the progressive volume changes of the vapor bubble in the hydrocarbon fluid inclusion during heating (Fig. 6) and based on the SFTIR results (see Guilhaumou et al. 2004 for explanation). Combination of these isochors with the water isochors gives an estimate of the real trapping pressure. Elsewhere, calcite cement in limestone from the Guzmantla Formation at the Peñuela paleoseep (Ferket et al. 2003, 2010; Fig. 5) appeared to contain pure-oil fluid inclusions. This points to an already-differentiated reservoir with well-individualized water, oil, and gas levels, rather than to primary hydrocarbon migration. Pressure and Temperature (PT)-modeling of these fluid inclusions (Fig. 5) also gave minimum estimates for the paleopressure.

Based on these two cases, it could be derived that an original Late Cretaceous to Paleocene flysch cover of at least 4.5 km must have been present in the western part of the Córdoba Platform. This is particularly well expressed from the modeled fluid inclusion for the Orizaba Formation, with an entrapment pressure of more than 250 bars (Fig. 4). Thermal maturity (T_{max}) in the Córdoba area (Ferket et al. 2006) also fits this kinematic evolution, even better than do the classical burial scenarios (Fig. 7). This thickness of eroded sediments should have been present originally on top of the Peñuela paleoseep, which is situated in an upper thrust sheet. This degree of missing section cannot be explained by tectonic loading on top of this thrust unit and cannot be restored in kinematic reconstructions. Such deposition can only be explained by the development of a local foredeep that received erosional products of the Zongolica thrust belt (Ferket et al. 2010). This foredeep developed just after a short episode of foreland flexuring, documented by uplift and local karstification of the carbonate platform (forebulge) during the Late Cretaceous. A minor portion of the calculated 4.5-km overburden may also have been related to tectonic stacking (assuming an extension of the allochthonous nappes with relics of the former Zongolica Basin eastward for more than 16 km).

TWO-DIMENSIONAL RECONSTRUCTION OF INTERMEDIATE GEOMETRIES AND PALEOBURIAL: TOWARD A KINEMATIC SCENARIO

Starting from the present-day structural section, a back-stripping restoration of all lithostratigraphic units was carried out that accounted for paleobathymetry, deformation, erosion, and compaction. The geologic concepts for this restoration were taken from SUBTRAP (Ortuño et al. 1999, 2003) and Ferket (2006). A summary of the main kinematic steps is given below.

Paleozoic to Early Cretaceous

The oldest rocks of the section correspond to crystalline basement of Paleozoic–Triassic age (radiometric ages of 248.2 Ma; Pemex-Schlumberger 2000), partly eroded during the pre-rift period. The first tectonic episode relates to Late Jurassic rifting and opening of the Gulf of Mexico. During the Upper Jurassic, continental red beds as well as marine calcareous sediments and clays (Tithonian source rocks) were deposited in half-graben structures. The subsequent Early Cretaceous post-rift stage is marked by regional carbonate sedimentation in platform and basin domains on former horsts and grabens (Fig. 8a).

Albian–Cenomanian

This stage corresponds to the principal development of the Córdoba Platform, characterized by thick accumulation of bioclastic limestone.

Cement phase	Petrography	Stable isotopes (‰PDB)	FI	Interpretation
Dolomite D	Zoned with cloudy core and transparent rims; in (breccia-) veins; CL: dull red luminescent	$\delta^{18}O$ -3.5 to -6.0 $\delta^{13}C$ +3.7 to +0.2	Primary FI H_2O-NaCl T_h (center) 50-60°C T_h (rim) 70-80°C	Burial cement
Fluorite F	Cubic crystals with a bitumen-coating on free faces, often corroded before precipitation of calcite; in (breccia-) veins; CL: extinguishing blue to non-luminescence		Secondary FI Non-fluorescent H_2O-NaCl T_h 110-155°C T_m -5.7 to -9.1°C	Hot saline fluid preceding HC trapping in reservoirs i.e. peak Laramide burial
Quartz Q	Euhedral cement in breccia vein, postdating D and F and preceding C1, intergrown with solid bitumen and containing blue fluorescent H_2O – hydrocarbon (HC) FI		Primary FI i) HC-(H_2O) T_h 90-105°C 35-40mol% CH_4 C7-C8 alkane equivalent ii) H_2O-NaCl-(CH_4-HC) T_h 140-150°C T_m -8.1 to -8.3°C Secondary FI iii) HC T_h 75-85°C 35-55mol% CH_4 C6-C7 alkane equivalent iv) H_2O-NaCl T_h 70-80°C T_m -0.6 to -1.7°C	Trapping and early post-trapping conditions at high temperature and pressure (200 bar < P < 350 bar), i.e. peak Laramide burial Intergrown bitumen relates to thermochemical sulfate reduction (TSR)
Calcite C1	Replacive after anhydrite, neomorphic, megacrystals after corrosion phase; associated with sulfides and pyrobitumen; CL: non-luminescent	$\delta^{18}O$ -16.7 to -5.0 $\delta^{13}C$ -18.0 to -2.5	At least partly secondary H_2O-NaCl T_h 71-110°C and 45-75°C T_m -3.3 to +0.0°C	Related to thermochemical sulfate reduction (TSR)
Calcite C2	Common cement phase in veins; after corrosion phase; CL: sector-zoned dull luminescent	$\delta^{18}O$ -11.9 to -3.5 $\delta^{13}C$ -2.4 to +1.9	Primary FI H_2O-NaCl T_h 45-66°C T_m ~0.0°C	Post-Laramide denudation and cooling phase

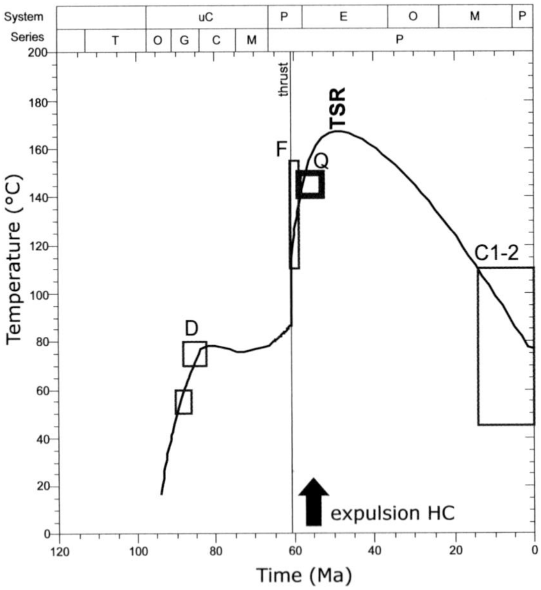

FIG. 4.—Diagenetic data used to calibrate the burial history of the Orizaba Formation in the Córdoba area. The burial curve was modeled by the Genex-1D code (IFP) making use of a kinematic reconstruction in line with inferred eroded thicknesses based on paleopressure calculations. CL = cathodoluminescence; FI = fluid inclusions; HC = hydrocarbons; for location of samples, see Figure 7 (see Ferket et al. [2010] for details).

Cement phase	Petrography	Stable isotopes (‰PDB)	FI	Interpretation
Calcite C3	Common cement phase in breccia-veins that postdate burial stylolites and precede tectonic stylolites CL: non/dull luminescent	$\delta^{18}O$ -5.0 to -13.0 $\delta^{13}C$ +1.0 to +2.5	Primary FI H_2O-NaCl T_h 40-65°C T_m -0.8 to -1.9°C	Early syn-Laramide
Calcite C4	Veins containing biodegraded oil and yellow fluorescent HC FI postdating tectonic stylolites	$\delta^{18}O$ -6.0 to -8.6 $\delta^{13}C$ -0.9 to +1.6	Primary-pseudosecondary HC (no water!) T_h 55.5-89°C 10-17mol% CH4 C9-C12 alkane equivalent	Syn- or early post-Laramide

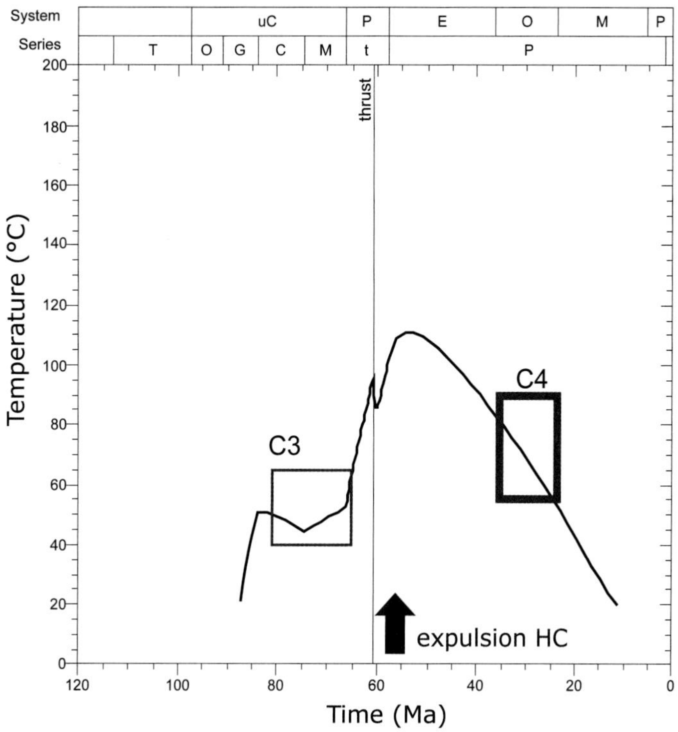

FIG. 5.—Diagenetic data used to calibrate the burial history of the Guzmantla Formation at the Peñuela paleoseep. The burial curve was modeled by the Genex-1D code (IFP) making use of a kinematic reconstruction in line with inferred eroded thicknesses based on paleopressure calculations. CL = cathodoluminescence; FI = fluid inclusions; HC = hydrocarbons; for location of samples, see Figure 7 (see Ferket et al. [2010] for details).

The thickness of the Cretaceous deposits decreases basinward. At this stage, the carbonated sequence displays an eastward dip controlled by the thermal subsidence of the Gulf of Mexico (Fig. 8b).

Santonian–Campanian

During the Santonian, stable platform conditions prevailed, characterized by a subhorizontal configuration and continuous deposition. Under the influence of tectonic load in the internal zones of the Cordillera (Zongolica fold and thrust belt), a flexure of the lithosphere developed in the foreland. These conditions led to local emergence and karstification of the platform in a forebulge domain (Ferket et al. 2000), leading to minor erosion and to the formation of a foredeep (Fig. 8c). In this foredeep, at least 4.5 km of argillaceous carbonated sediments (if not partly related to tectonic stacking) were deposited. It is assumed that the basement had developed a westward tilt by that time (Fig. 8c).

Maastrichtian

This period marks the beginning of the Laramide Orogeny. The first argillaceous carbonates filling the flexural basin(s) were deposited during the Maastrichtian. These sediments probably were particularly well expressed in the foredeep in the western part of the section and thinned to the east (Fig. 8d).

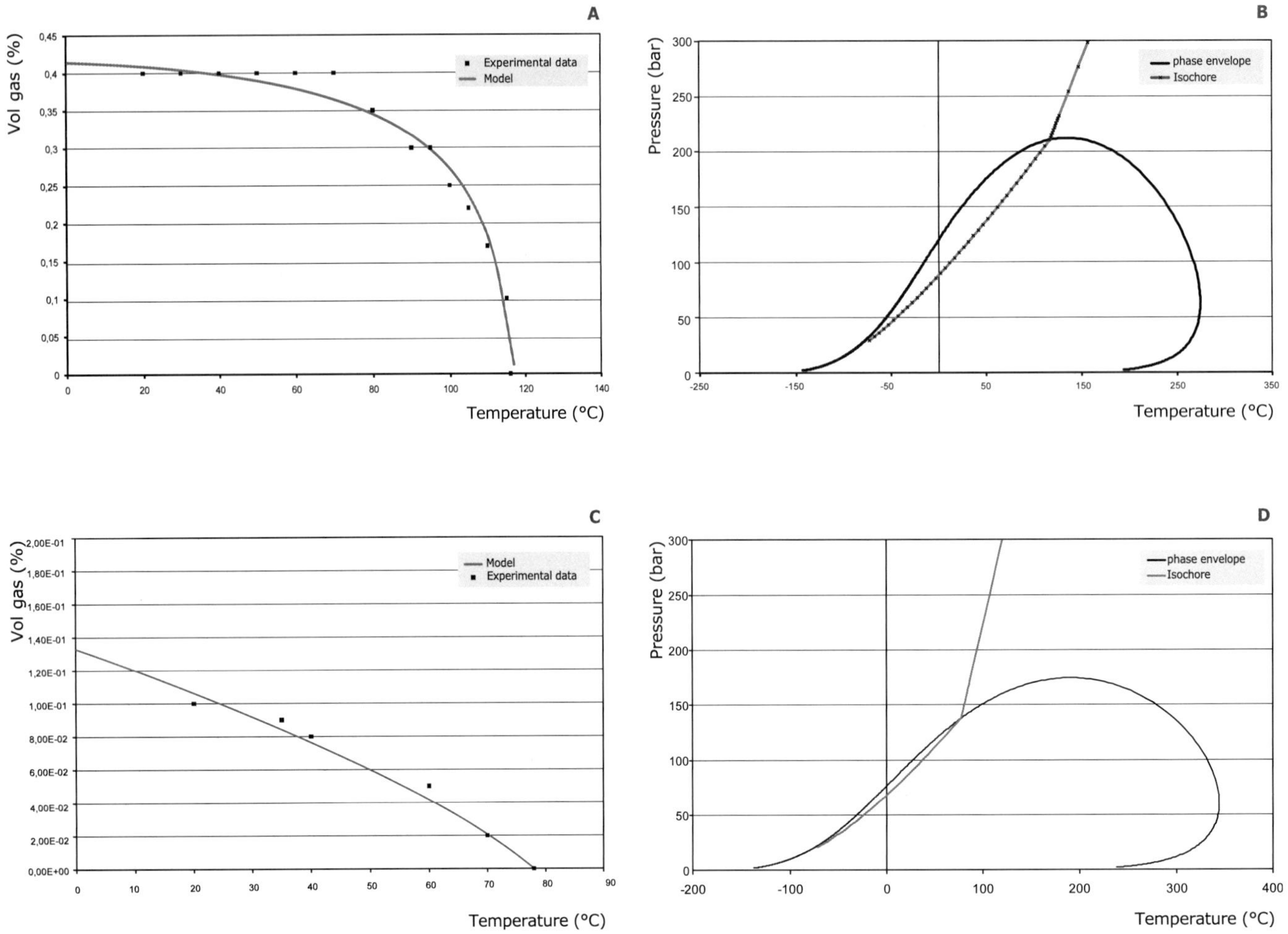

FIG. 6.—Pressure-Volume-Temperature (PVT) modeling for fluid inclusions found in the Córdoba area, respectively, in quartz from a well close to Córdoba (A and B) and in calcite from the Peñuela paleoseep (C and D). The progressive shrinking of the vapor bubble (% volume of gas) of two-phase hydrocarbon inclusions (A and C) is used, together with results from SFTIR analyses, to calculate the liquid-vapor curves of the fluid system entrapped and isochors relative to Th measured (B and D), from which pressure and temperature of the fluid inclusion entrapment were calculated (see Ferket et al. [2010] for details).

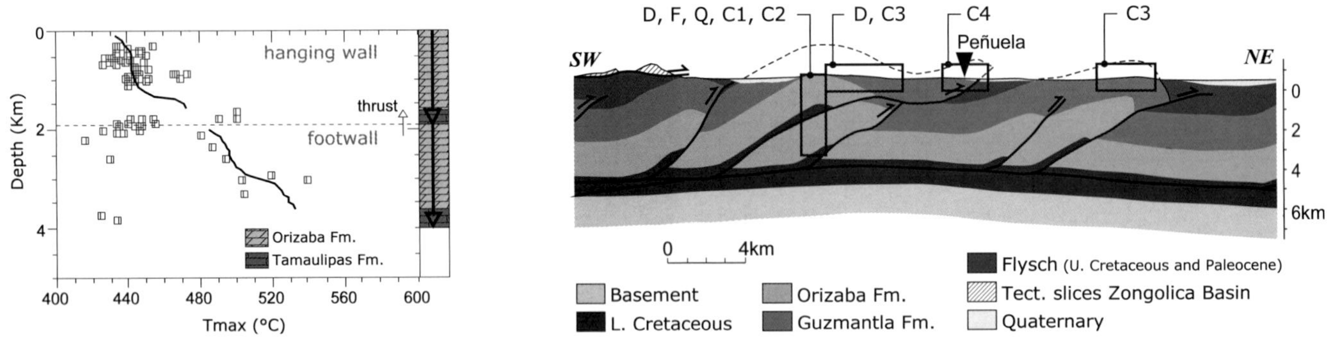

FIG. 7.—Maturity data (T_{max}, squares) used to calibrate thermal modeling (continuous line) in the Córdoba area. The IFP GenTect 1D basin modeling tool is used to model the temperature–depth curve (see detail in Ferket et al. [2010]). The represented model corresponds to a kinematic scenario taking into account an eroded sequence of 4.5 km, as derived from fluid inclusion data and Pressure-Volume-Temperature (PVT) modeling. The present model fits the maturity data much better than do classical concepts that consider higher heat flow and lesser erosion. The cross section gives the location of samples used for diagenetic studies. The letter codes refer to specific minerals that were identified and in which useful fluid inclusions were found (D = dolomite; Q = quartz; C = calcite; F = fluorite).

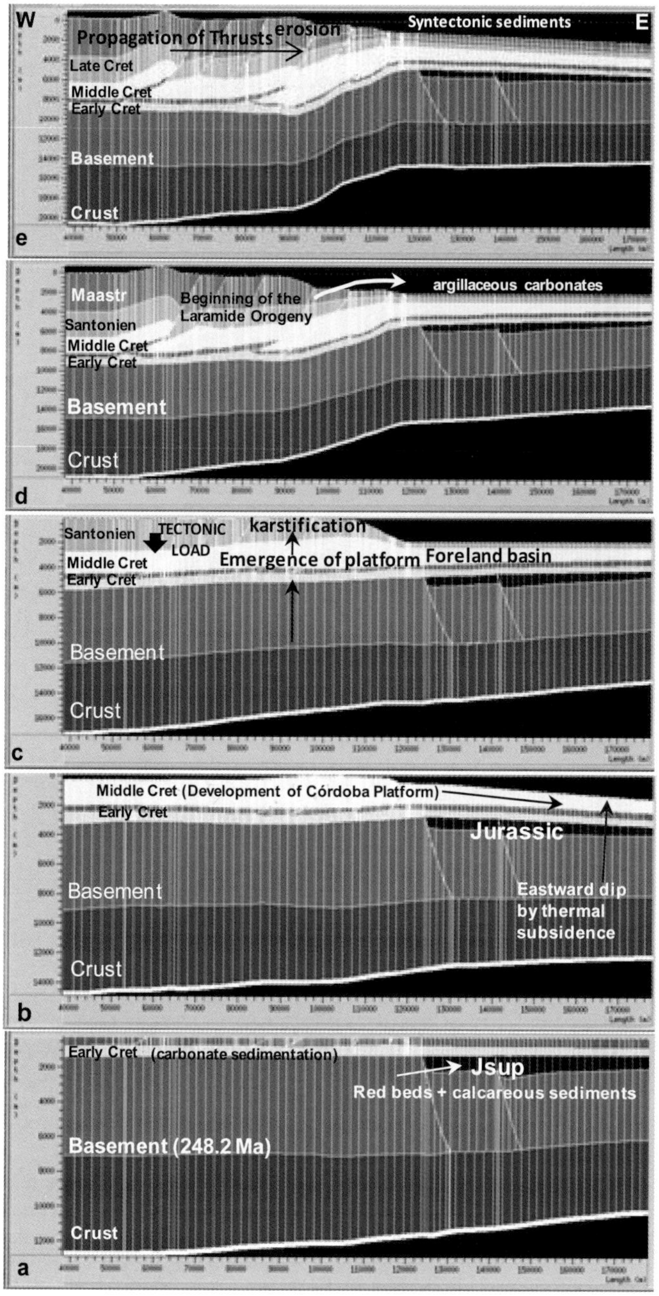

FIG. 8.—Main kinematic steps of section AA', obtained from backstripping restoration of all lithostratigraphic units. (a–e) Early to Late Cretaceous stages; (f–k) Paleocene to present day.

Paleocene

The Laramide Orogeny continued in the western part of the study area, represented by eastward propagation of thrusts and overthrusting of the Orizaba Platform and Zongolica Basin domains upon the Córdoba Platform. This period is marked by deposition of syntectonic sediments in structural lows and in the Veracruz Basin (Fig. 8e). This period is also characterized by a change in sedimentation in the region, from calcareous sediments to more siliciclastic deposits.

Late Paleocene–Early Eocene

This period is characterized by differences in the thickness of sediments deposited in the folded western border of the Veracruz Basin, recording the onset of Laramide deformation in this zone. Thick accumulation of Lower Eocene sediments in the western part of the basin indicates that the principal phase of deformation in this zone occurred at that time. From this time onward, subsidence caused the basement below the Veracruz Basin to tilt eastward again (Fig. 8f).

Middle Eocene–Late Eocene

Local emergence of the presently buried thrust front occurred during the Late Eocene to Early Oligocene (Ferket 2004, 2006). The last thrust deformation in this zone probably took place around the Middle Eocene. A major change in sedimentation was initiated in the Veracruz Basin starting in the Late Eocene, when the eastern part of the basin subsided faster than the western portion, resulting in an increased eastward tilt of the basement and thicker sedimentary sequences to the east (Fig. 8g).

Oligocene

The Oligocene is characterized by eastward thickening of the sedimentary sequence and continued eastward tilting of the basement below the Veracruz Basin. The Chattian unconformity clearly imaged in seismic lines resulted from a major sea-level drop that marks the end of Oligocene deposition (hiatus) in the basin. The Mata Espino antiformal structure is also interpreted to have initiated during the Oligocene (Fig. 8h).

Early Miocene–Middle Miocene

The Early Miocene is characterized by isopachous strata throughout the Veracruz Basin, indicating that tilting of the basement had decreased by that time. Small variations in sedimentary thickness can be attributed to early transpressional deformation of the basin. By the end of the Middle Miocene, sediments are folded and locally faulted, giving rise to the NVC and Mata Espino structures presented on the modeled section (Fig. 8i).

Upper Miocene

The Upper Miocene series onlaps the antiformal structures and almost fills the basin with strata that thicken to the east. The original westward extent of Miocene sediments over the thrust belt structures is unclear (Fig. 8j).

Pliocene–Recent

Late Miocene to Pliocene reactivation and uplift of frontal structures (possibly in response to basement faults reactivated by transpression) led to uplift of the Laramide Front with erosion of part of the Upper Miocene strata and dragging and steepening of Lower to Middle Miocene strata. A Pliocene piggyback basin formed to the west of the buried tectonic front zone. The Veracruz Basin was filled up further with sediments that started to develop prograding clinoforms (Fig. 8k).

A total tectonic shortening of 25 to 45% was proposed by Amoco–Pemex–IMP (1995) for the Córdoba thrust belt for the Laramide event. Post-Laramide uplift and erosion account for more than 4.5 km of erosion in the Córdoba foothills, as well as farther north in the Sierra Madre Oriental (Gray et al. 2001; Alzaga-Ruiz et al. 2008a, 2008b). Post-Laramide basement uplift is also associated with extensional structures in the hinterland and strong erosion in the foothills farther

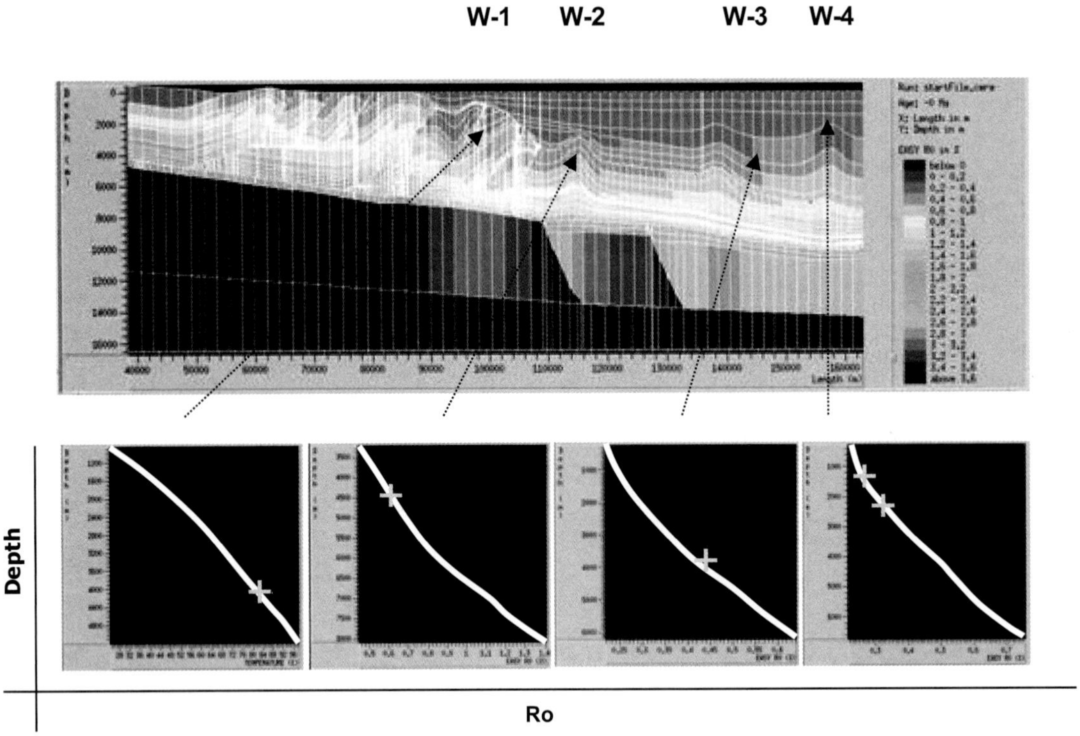

FIG. 9.—Thermal calibration of the Ceres model by means of R_o (location of the wells used for calibration is indicated by labels W1 to W4, indicated by small white symbols).

north in the Cordillera, similar to the Basin and Range province in the United States and to the Canadian Rockies (Hardebol et al. 2007, 2009; Roure et al. 2010). Unlike other thrust belts, however, the basement below the foothills is currently tilted toward the foreland. These uplift and unroofing processes can be explained by post-Laramide asthenospheric upwelling beneath the entire length of the North American Cordillera, a Pacific subduction-related "corner flow" thinning the continental crust and forcing uplift of the Córdoba thrust belt relative to the more stable Veracruz area, which continued to subside (Hyndman et al. 2005; Alzaga-Ruiz et al. 2008a, 2008b; Roure 2008; Roure et al. 2010).

THERMAL CALIBRATION AND RESULTS OF THE THERMAL MODELING

Thermal Calibration

Calibration of the thermal model was based on well data along the modeled section, including BHT, geothermal gradient, and core porosity data from wells located in Veracruz Basin and from T_{max} and R_o (vitrinite reflectance in oil) trends. Multiple simulations of 1D well models were initially carried out to constrain the thermal model. Additional temperature data from fluid inclusion studies was used to run a new set of 1D thermal models with the aim of constraining past heat flow values, which fit both the present-day temperature and maturity data and the FI burial (Ferket 2006, Ferket et al. 2010). The result of this calibration is a basal heat flow history starting with values of up to 50 mW/m^2 for the pre-Laramide periods and decreasing to 30 mW/m^2 for the post-Laramide period (Table 2). Several simulations employing both hydrostatic and nonhydrostatic conditions were carried out making use of these heat flow values for the calibration of the 2D basin models. A satisfactory agreement between well data and calibration was obtained from CERES2D simulations (Fig. 9).

Thermal Maturity Evolution

Lower and Middle Cretaceous source rocks located in the Córdoba Platform entered the oil window as early as 80 My, whereas the gas window was reached at 62 My (Fig. 10). This early maturation resulted from the early deep burial in this area. In contrast, Upper Jurassic source rocks in the Veracruz Basin entered the oil window at 45 My and the gas window at 39 My (Fig. 10). The increase of heat flow boundary conditions to ensure a reasonable fit with the temperature and vitrinite data led to an earlier generation of oil than was forecasted by the Genex modeling (see Fig. 4), but the timing of expulsion of the hydrocarbon from the source rocks is unchanged, as it depends on the expulsion saturation threshold, which is still reached during the Early Tertiary. Thus, the coupling between deformation and migration is still properly modeled.

Lower and Middle Cretaceous source rocks of the Veracruz Basin reached the oil window from 42 My onward, whereas the gas window was reached at 24 My. For the Paleocene source rocks in the basin, the CERES2D calculations indicate that they entered the oil and gas windows at 29 My and 1 My, respectively (Fig. 10). Upper Paleocene and Lower Eocene source rocks entered the oil window from 15 My onward (Fig. 10), whereas Middle Eocene and Upper Eocene source rocks entered the oil and gas windows at 11 My and 9 My, respectively (Figs. 10, 11). Tertiary source rocks have not yet reached the gas window as a result of lack of sufficient burial (Figs. 10, 11), although they may have produced biogenic gas. Finally, the potential Oligocene source rock is presently immature, except in the east (Fig. 10), where source rocks entered the oil window at 3 My (Figs. 10, 11).

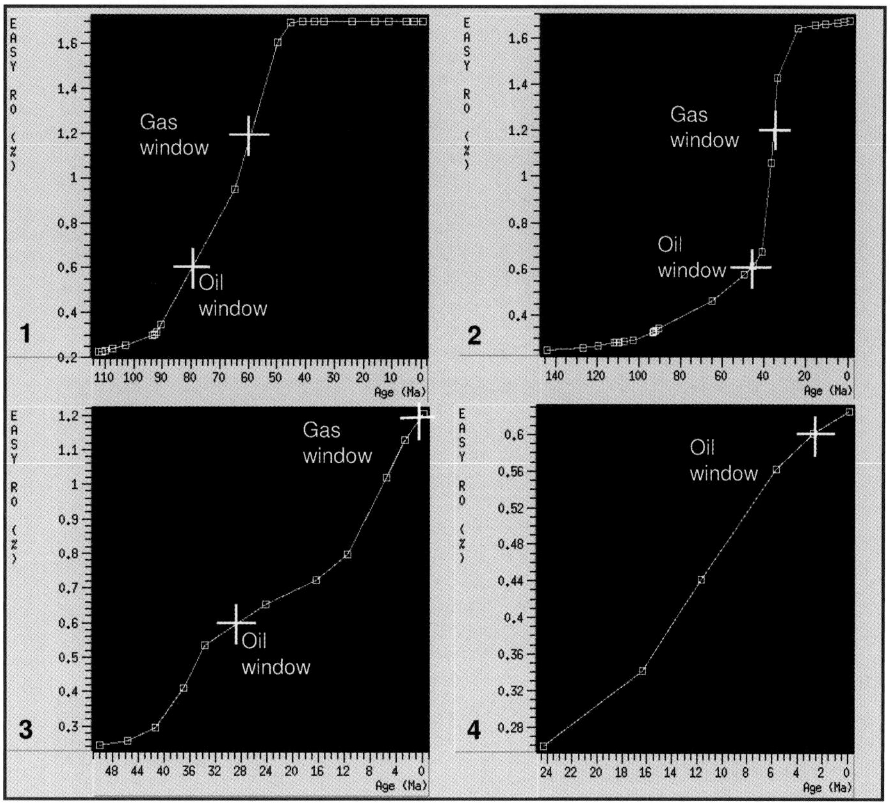

FIG. 10.—Diagrams showing the time at which different source rocks entered the oil and gas windows in the Córdoba Platform and Veracruz Basin: (1) Cretaceous source rocks from the Córdoba Platform; (2) Jurassic source rocks; (3) Lower Paleocene source rocks; (4) Oligocene source rocks from the Veracruz Basin. Large plus symbols indicate the time at which the source rock entered the oil or gas window.

Transformation Ratio

The "transformation ratio" (TR) parameter indicates the evolution stage of the kerogen, specifically addressing what proportion of the hydrocarbons has been produced with respect to the maximum potential hydrocarbon yield. These calculations are based on the kinetic parameters of Behar et al. (1997). Upper Jurassic and Lower to Middle Cretaceous source rocks located in the deepest parts of the Veracruz Basin have TR values measuring higher than 85%, indicating that these intervals have generated nearly all the possible hydrocarbons. Lower and Middle Cretaceous source rocks, located at the former platform-to-basin transition, also appear to have produced hydrocar-

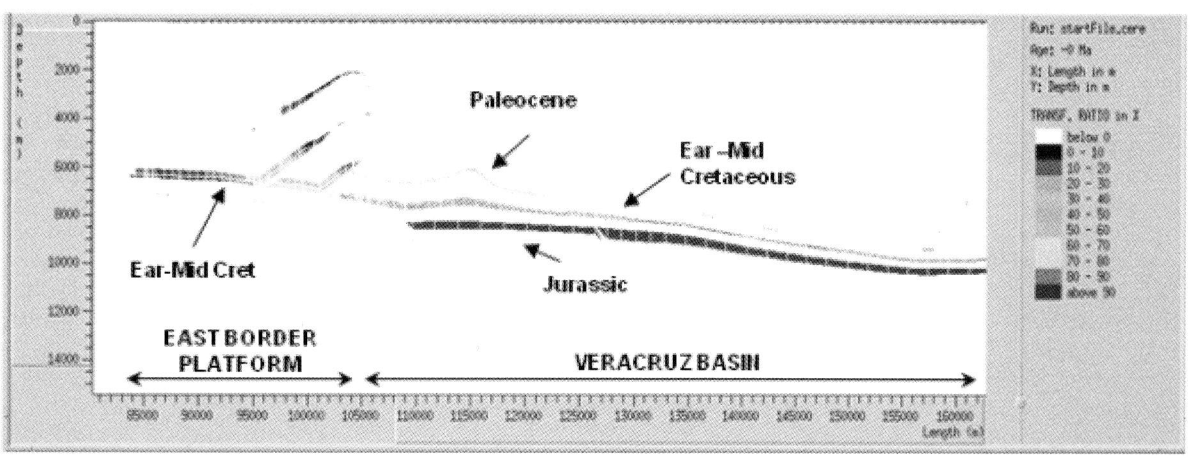

FIG. 11.—Present-day distribution of transformation ratio (ratio between the source rock potential and the effectively produced hydrocarbons) for different source rocks, according to their location.

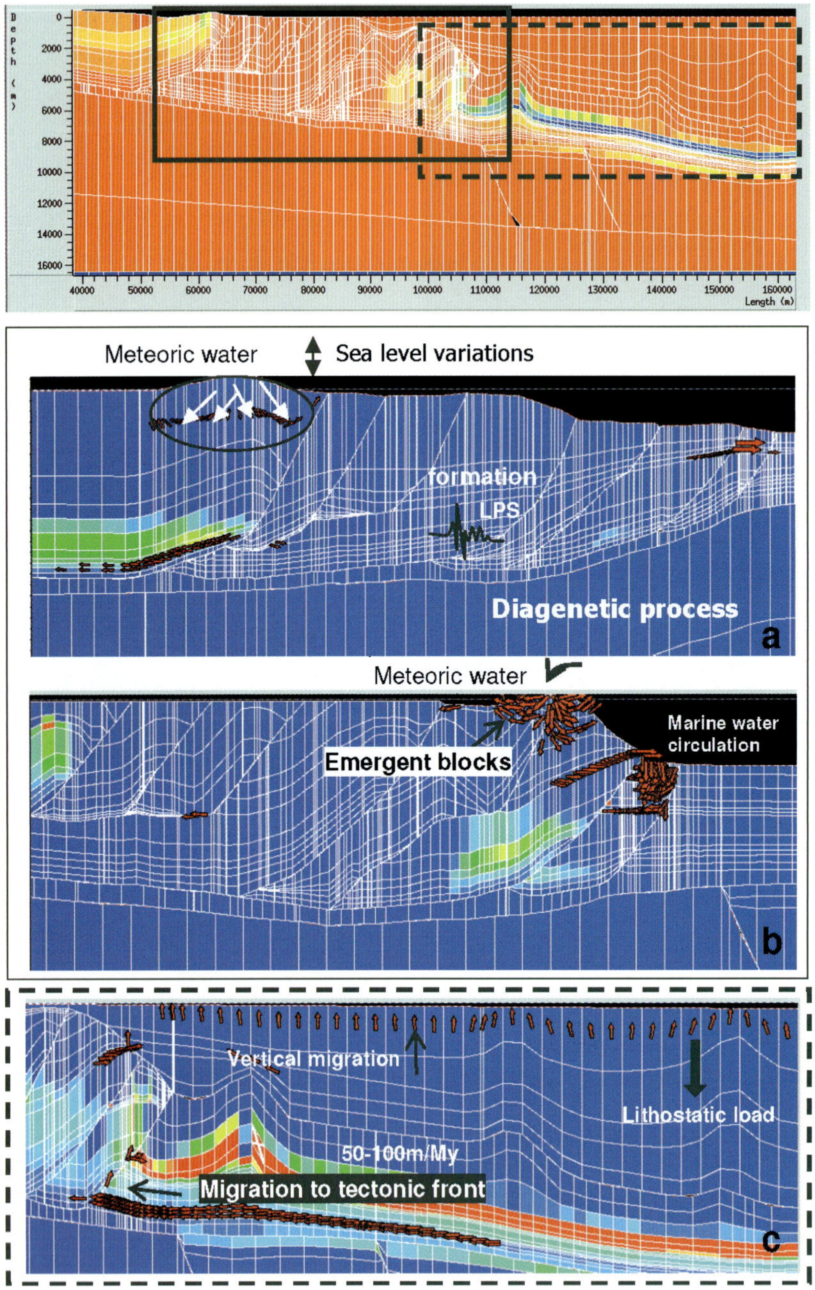

FIG. 12.—Top: Present-day architecture of the thrust front, with a rectangle localizing the two bottom inserts. Bottom: fluid flow simulations: (a) past architecture of this part of the section during the Campanian showing three flow systems: meteoric influx due to karstification of the emerged forebulge, squeegee flow toward the east and westward flow in isolated thrust compartments; (b) Early Eocene stage, showing (1) meteoric influx in locally emerged structures at the tectonic front, (2) westward flow toward the tectonic front from shale dewatering in the basin, and (3) eastward squeegee flow from the front toward the basin; (c) Late Miocene stage showing flow in the Veracruz Basin. Vertical fluid escape due to increased lithostatic load is the predominant mechanism with a continuous escape of basinal fluids. If a Cretaceous source rock is assumed, the model also predicts the onset of hydrocarbon migration from the basin up-dip to the tectonic front, with velocities ranging from 50 to 100 m/My.

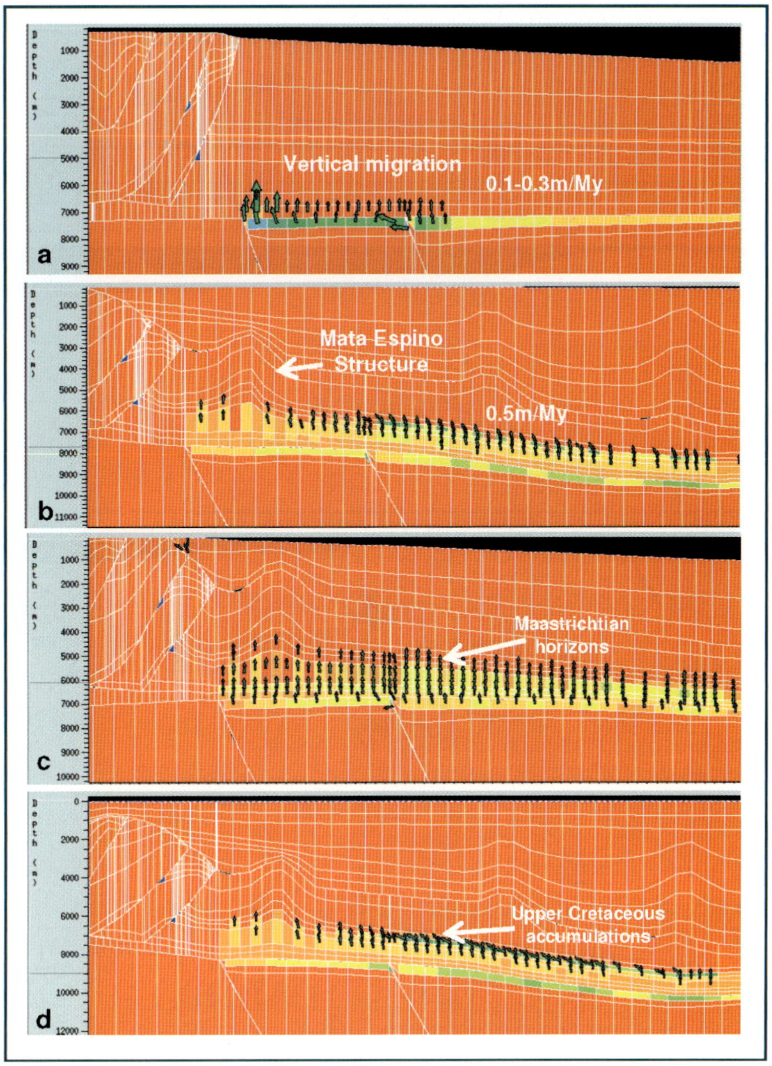

FIG. 13.—Hydrocarbon migration simulated for a Jurassic source rock located in the Veracruz Basin. (a) Late Eocene situation, (b) Early Miocene, (c) Late Miocene, (d) present day (see text for explanation).

bons. Similar source rocks located at the eastern border of the Córdoba Platform are characterized by average TR values below 60%, which implies that they may still be actively producing late oil or gas condensate. Such low maturity is a result of early uplift of the Laramide thrust belt (Fig. 11).

FLUID FLOW RECONSTRUCTION AND HISTORY OF HYDROCARBON MIGRATION

Scenario of Ancient Fluid Circulations and Comparison with the Data

The calculations carried out by CERES2D show a history of fluid migration, starting from the Maastrichtian in the western part of the section (Fig. 12a). Concurrently with the eastward propagation of the thrust belt, expulsion of fluids took place in the then-undeformed parts of the foreland, which recorded an early episode of layer parallel shortening (cf. Ferket et al. 2004). Formation fluids migrated up-dip along the foreland flexure, in the eastern part of the section (squeegee flow; Oliver 1986, Machel and Cavell 1999). Tectonic and burial compaction led to overpressures and expulsion of fluids, resulting in hydrofracturing and brecciation followed by precipitation of minerals with a different diagenetic signature (from stable isotope data) than is associated with the host rock (Ferket et al. 2003, 2004, 2006, 2010; Ferket 2004, 2006).

Ultimately, folding of the strata created additional fracture porosity (Fig. 12b). Locally, porosity was further enhanced by fluid circulations related to thermochemical sulfate reduction (in the deeply buried anhydrite-bearing intervals) or to karst development (in emerged structures). In contrast to the early eastward-directed squeegee flow, fluid flow in the already-compartmentalized thrust blocks in the western part of the section occurred from east to west, from the high-pressured zone of the ramp folds to their passive limbs. Such flow was focused at the base of the Middle Cretaceous dolomites and reached a mean velocity of 10 m/My.

During the Middle Eocene, the tectonic front had reached the eastern platform border and locally emerged in this easternmost zone. Meteoric water could thus enter the reservoirs and cause karstification below the presently buried Laramide unconformity (Fig. 12b). Simultaneously,

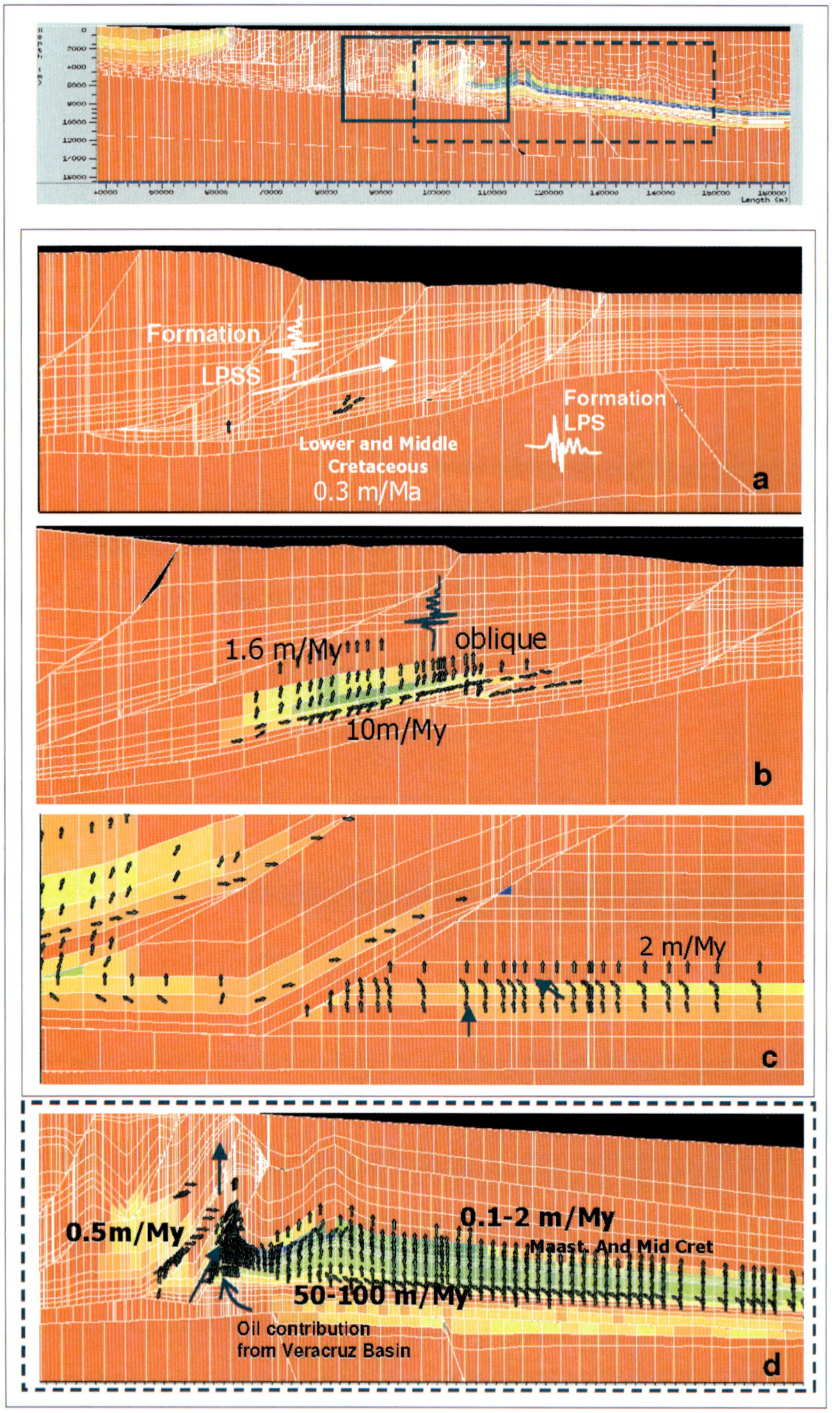

FIG. 14.—Hydrocarbon migration simulated for a model with two source rock horizons, in the Jurassic and Cretaceous series, respectively. Top: Present-day architecture of the thrust front, with a rectangle localizing the bottom inserts; Bottom: (a) Paleocene stage, (b) Early Eocene, (c) Late Eocene, (d) Early to Middle Miocene (see text for explanation). LPS = layer parallel shortening and associated stylolites; CS = compaction-related stylolites.

marine water circulation also took place in two directions, between the Córdoba Platform and the Veracruz Basin, associated with shale dewatering and continued squeegee flow (Fig. 12b).

During and following the Late Eocene, vertical fluid escape due to the increased lithostatic load became predominant in the Veracruz Basin (Fig. 12c). If a Cretaceous source rock is assumed, the model also predicts the onset of hydrocarbon migration from the basin up-dip to the tectonic front, with velocities ranging from 50 to 100 m/My (see

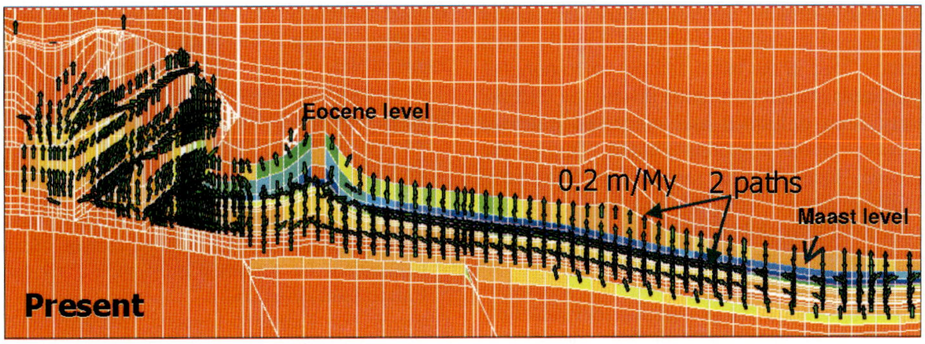

FIG. 15.—Hydrocarbon migration at present time. Note that oil has reached the Eocene reservoirs in the Veracruz Basin, with a velocity decrease down to 0.2 m/My.

below). This migration accelerated (~100 m/My) by the Late Miocene and was directed toward the tectonic front and upward along faults. Presently, the westward migration up to the tectonic front at the base of the Middle Cretaceous still represents the dominant flow pattern. Migration velocities of approximately 60 m/My are presently associated with vertical dewatering of the Tertiary series (Fig. 12c).

Hydrocarbon Migration History

As mentioned above, the Cretaceous platform carbonates in the easternmost frontal thrust zone produce oil and gas, whereas Tertiary clastic rocks in the Veracruz Basin produce only gas. As a result of current numerical limitations of CERES2D, migration paths could only be calculated for two-phase fluid flow (i.e., water and oil), thus excluding natural gas. The distribution and role of different source rocks in the eastern tectonic front and westernmost Veracruz Basin area are controversial, because the oils lack a typical Jurassic signature, and the existence of Cretaceous intraplatform source rocks is still questioned (Ferket 2004, 2006; Vázquez 2004). In order to respond to these uncertainties, two models have been constructed, and various simulations were run to test the impact of source rock distribution and other source rock characteristics. The first model considered only Upper Jurassic source rocks present in the Veracruz Basin. The second model assumes both Jurassic source rocks in the Veracruz Basin and Cretaceous source rocks located within the Veracruz Basin and in the eastern part of the platform. The model output defines water saturations, which are complimentary to hydrocarbon (in this case oil) saturation.

In the first model, which considers only Jurassic source rocks, oil migration starts during the Late Eocene (~37 Ma) in the westernmost part of the basin. Migration takes place vertically and the velocity is low (0.1–0.3 m/My; Fig. 13a). By the Early Miocene (~16 My) the velocity increased to 0.5 m/My and the migration front moved up to the Upper Cretaceous strata, filling the Mata Espino structure (Fig. 13b). A very minor westward migration toward the tectonic front occurred at this time. During the Late Miocene, the migration front has reached the Maastrichtian horizons (Fig. 13c), and velocity remained constant at ~0.5 m/My. At present, the hydrocarbon migration from Jurassic source rocks is vertical, and there is no significant migration toward the tectonic front. Therefore, traps are situated mainly in the Upper Cretaceous of the Veracruz Basin (Fig. 13d).

In the second model, hydrocarbon migration started at 65 My for source rocks located in the eastern part of the platform. Flow was concentrated at the Lower to Middle Cretaceous level and directed to the east with a velocity of 0.3 m/My (Fig. 14a). By the Early Eocene (~50 My; Fig. 14b) a complex migration pattern prevailed in the thrust belt. An eastward-directed migration of 10 m/My occurred at the Lower to Middle Cretaceous level following faults, whereas shallower levels exhibited vertical migration of hydrocarbons with a slower velocity of about 1.6 m/My (Fig. 14b). The same pattern continued during the Late Eocene (~37 My), but at this time hydrocarbon migration also started in the Veracruz Basin, with vertical and westward-directed flow of less than 2 m/Ma (Fig. 14c). By the Oligocene (~34 My), the oil migrated further upward along faults in the tectonic front, and a larger contribution (~5 m/My) from the Veracruz Basin helped charge the reservoirs in this zone. The oil migration front reached the Santonian level in the tectonic front by that time. At the beginning of the Miocene (~24 My), hydrocarbon migration became more rapid, with velocities around 50 m/My, and was directed both eastward along faults located in the eastern part of the platform and vertically in the Veracruz Basin. The velocities of upward migration in the Veracruz Basin were low (0.1–2 m/My). The Middle Miocene (~16 My; Fig. 14d) was characterized by slower eastward and upward migration (0.5 m/My). Nevertheless, westward migration from source rocks in the Veracruz Basin toward and through the tectonic front continued to contribute to the oil charge. This flow was concentrated mainly at the Maastrichtian and Middle Cretaceous levels (Fig. 14d). Presently (0 My; Fig. 15), the flow patterns in the Córdoba Platform remain the same, whereas the Veracruz Basin is characterized by a complex migration pattern. Below the Maastrichtian level flow is directed to the tectonic front in the west. Above this level a vertical migration predominates, with low velocities around 0.2 m/My. The model predicts that the oil migration currently has reached the Eocene level (Fig. 15).

CONCLUSIONS

Fluid inclusion data in the Córdoba Platform are consistent with the occurrence of a former Late Cretaceous–Paleocene flexural sequence that was up to 4.5 km thick, prior to uplift and removal by erosion. Accordingly, the present east-dipping attitude of the basement is best interpreted as the result of post-Laramide upwelling of the mantle beneath the North American Cordillera. This change in the dipping attitude of the basement was coeval with a rapid transfer of siliciclastic erosional products from the thrust belt toward the Veracruz Basin. This resulted in an increase in the source rock burial, which then entered the oil window in the basin.

The 2D fluid flow simulations presented here considered both Jurassic and Cretaceous series from the Veracruz Basin as potential source rocks. Flow simulation results are consistent with the known petroleum occurrences in this region. The source rocks located in the platform entered the oil window at 80 My, whereas the source rocks in the basin entered the oil window at 45 My. The first pulses of migration started at roughly the end of the Late Cretaceous (65 My). Migration paths changed through time. Initially migration was dominantly vertical, with rates that increased gradually from 0.3 to 2 m/My. With the increase in the lithostatic load in the basin and the corresponding increase in pore pressure, fluids started to migrate toward the tectonic front, which had already been deformed and contained potential structural traps when hydrocarbon migration started. During the Eocene and the Oligocene, hydrocarbons migrated with a mean velocity of 50 m/My (with velocities potentially reaching as high as 100 m/My from the basin toward the folded chain, as a result of the progressive inversion of the tilting). The speed of migration decreased to 0.5 m/My during the Lower Miocene.

As a result of numerical limitations, the present model could only simulate the fluid flow of both water and oil but not of gas. Nevertheless, it shows that oil could not reach stratigraphic levels shallower than the Eocene within the Veracruz Basin itself. This finding is consistent with the observations that in the Veracruz Basin, the Miocene series only host gas reservoirs, and the few oil wells in the basin produce from Eocene reservoirs. The fluid flow modeling is consistent with the various stages of water displacement recorded in the various stages of cementation.

ACKNOWLEDGMENTS

IMP provided the logistics for the fieldwork, and Pemex provided access to subsurface data. The authors acknowledge the work done by the CERES2D development and maintenance teams during many years: Isabelle Faille, Catherine Marquer (IFP New Energies), and Hervé Devoitine (Tech-Advantage). We are grateful to Pascal Mougin for the thermodynamic modeling and to Biliana Gasharova and Y.L. Mathis at the Institute for Synchrotron Radiation (ANKA beamline; Karlsruhe) for helpful assistance at the FTIR beam line. We thank E. Hiemstra and Editor N. Harris for the thorough reviews and improvement of the manuscript.

REFERENCES

Alzaga-Ruiz H, Granjeon D, Lopez M, Séranne M, Roure F. 2008a. Gravitational collapse and Neogene sediment transfer across the western margin of the Gulf of Mexico: Insights from numerical models. *Tectonophysics* doi:10.1016/j.tecto.2008.06.017.

Alzaga-Ruiz H, Lopez M, Roure F, Séranne M. 2008b. Interactions between the Laramide foreland and the passive margin of the Gulf of Mexico: Tectonics and sedimentation in the Golden Lane area, Veracruz State, Mexico. *Marine and Petroleum Geology* doi:10.1016/j.marpetgeo.2008.03.009.

Amoco–Pemex–IMP. 1995. Tertiary stratigraphy, basin evolution and economic potential of the Veracruz Basin, East–Central Mexico. Pemex Internal Report, 78 p.

Behar F, Espitalie J, Marquis F, Tang Y, Vandenbroucke M. 1997. Thermal cracking of kerogen in open and closed systems: Determination of kinetic parameters and stochiometric coefficients for oil and gas generation. *Organic Geochemistry* 26:321–339.

Berman AE, Pérez-Cruz G, Cantu-Chapa C. 1995. Tertiary stratigraphy, basin evolution and economic potential of the Veracruz Basin, East-Central Mexico. Pemex Internal Report, 78 p.

Bessis F, Burrus J, Chenet PY, Doligez B, Ungerer P. 1990. Basin evaluation by integrated two-dimensional modelling of heat transfer, fluid flow, hydrocarbon generation and migration. *American Association of Petroleum Geologists Bulletin* 74:309–335.

Campa MF, Coney PJ. 1983. Tectonostratigraphic terrane and mineral resources distribution in Mexico. *Canadian Journal of Earth Sciences* 13:1040–1051.

Chenet PY, Doligez B, Schmerber G. 1986. The Themis model: A new tool for the hydrocarbon exploration: Integrated study of the sedimentary basin from the sedimentation to the hydrocarbon accumulation. *Hidrocarburos, 1° Congreso Latino–Americano*, Buenos Aires, May 4–11 1986. 10:5187–5202.

Divies R, Sassi W. 1996. Foldis: Forward modelling of folding and compaction. *Annales Geophysicae* 14:1, C106.

Doligez B (Editor). 1987. *Migration of Hydrocarbons in Sedimentary Basins*: Editions Technip, Paris. 712 p.

Espinoza NM, Toriz GJ. 2005. Structural styles in the Veracruz area. *Review of the Mexican Association of Exploration Geophysicists* 4:3–20 (in Spanish).

Faille I, Nataf F, Schneider F, Willien F. 1998. Domain decomposition methods for fluid flows in porous medium. *In* Proceedings of the ECMOR 6th European Conference on the Mathematics of Oil Recovery; September 8–11, 1998; Peebles. Paper B-06, 6 p.

Ferket H. 2004. Sedimentology, diagenesis and fluid flow reconstruction in the Laramide fold-and-thrust belt of eastern Mexico (Córdoba Platform): Implications for petroleum exploration [PhD doctoral thesis]: Katholieke Universiteit Leuven, 281 p.

Ferket H. 2006. Kinematic evolution, diagenesis and fluid flow reconstruction in the Laramide fold-and-thrust belt of eastern Mexico (Córdoba Platform and Veracruz Basin): Implications for petroleum exploration. Institut Français du Pétrole Report, 123 p.

Ferket H, Guilhaumou N, Roure F, Swennen R. 2010. Insights from fluid inclusions, thermal and PVT modelling for paleoburial and thermal reconstruction of the Córdoba petroleum system (NE Mexico). *Marine and Petroleum Geology* 28:936–958.

Ferket H, Ortuño S, Swennen R, Roure F. 2003. Diagenesis and fluid flow history in reservoir carbonates of the Cordilleran fold- and thrust- belt: The Córdoba Platform. *In* Bartolini C, Burke K, Buffler R, Blickwede J, Burkart B (Editors). *Mexico and the Caribbean Region: Plate Tectonics, Basin Formation and Hydrocarbon Habitats*, Memoir 79: American Association of Petroleum Geologists, Tulsa, Oklahoma. p. 283–304.

Ferket H, Roure F, Swennen R, Ortuño S. 2000. Fluid migration placed into the deformation history of fold-and-thrust belts: An example from the Veracruz Basin (Mexico). *Journal of Geochemical Exploration* 69–70:275–279.

Ferket H, Swennen R, Ortuño-Arzate S, Cacas MC, Roure F. 2004. Hydrofracturing in the Laramide foreland fold-and-thrust belt of Eastern Mexico. *In* Swennen R, Roure F, Granath J (Editors). *Deformation, Fluid Flow and Reservoir Appraisal in Foreland Fold-and-Thrust Belts*, Hedberg Series, Memoir 1: American Association of Petroleum Geologists, Tulsa, Oklahoma. p. 133–156.

Ferket H, Swennen R, Ortuño-Arzate S, Roure F. 2006. Fluid flow evolution in petroleum reservoirs with a complex diagenetic history: An example from Veracruz, Mexico. *Journal of Geochemical Exploration* 89:108–111.

Gonzalez-Mercado GE. 2007. Reconstruction des circulations de fluides et de la migration des hydrocarbures (modélisation Ceres 2D) le long d'un transect à travers la Plateforme de Córdoba et le Bassin de Véracruz, Mexique. DES Université Pierre et Marie Curie. Institut Français du Pétrole Report 60112, 40 p.

Gray GG, Pottorf RJ, Yurewicz DA, Mahon KI, Pevear DR, Chuchla RJ. 2001. Thermal and chronological record of syn- and post-Laramide burial and exhumation, Sierra Madre Oriental, Mexico. *In* Bartolini C, Buffler RT, Cantu-Chapa A (Editors). *The Western Gulf of Mexico Basin: Tectonics, Sedimentary Basins and Petroleum Systems*, Memoir 75: American Association of Petroleum Geologists, Tulsa, Oklahoma. p. 159–181.

Guilhaumou N, Benchilla L, Mougin P, Dumas P. 2004. Advances in hydrocarbon fluid inclusions microanalysis and PVT modelling: Diagenetic history, P-T and fluid flow reconstruction, a case study in the North Potwar Basin, Pakistan. *In* Swennen R, Roure F, Granath J (Editors). *Deformation, Fluid Flow and Reservoir Appraisal in Foreland Fold-and-Thrust Belts*, Hedberg Series, Memoir 1: American Association of Petroleum Geologists, Tulsa, Oklahoma. p. 5–21.

Guilhaumou N, Dumas P. 2005. Synchrotron FTIR hydrocarbon fluid inclusions microanalysis applied to diagenetic history and fluid flow reconstruction in reservoir appraisal. *Oil and Gas Science Technology Revue de l'Institut Français du Pétrole* 60(5):1–12.

Guzman-Vega ME, Castro OL, Roman RJR, Medrano ML, Valdéz LC, Vàzquez CE, Ziga RG. 2001. Classification and origin of petroleum in the Mexican Gulf Coast Basin: An overview. *In* Bartolini C, Buffler RT, Cantu-Chapa A (Editors). *The Western Gulf of Mexico Basin: Tectonics, Sedimentary Basins and Petroleum Systems*, Memoir 75: American Association of Petroleum Geologists, Tulsa, Oklahoma. p. 127–142.

Hardebol N, Callot JP, Bertotti JL, Faure JL. 2009. Sedimentary and tectonic burial history appraisal and consequent temperature and organic maturation evolution in thrust-belt systems: A study on the SE Canadian Cordillera. *Tectonics* 28:TC3003, doi:10.1029/2008TC002335.

Hardebol NJ, Callot JP, Faure JL, Bertotti G, Roure F. 2007. Kinematics of the SE Canadian foreland fold and thrust belt: Implications for the thermal and organic maturation history. *In* Lacombe O, Lavé J, Roure F, Vergés J (Editors). *Thrust Belts and Foreland Basins*: New Frontiers in Earth Sciences: Springer-Verlag, Heidelberg, Germany. p. 179–202.

Hyndman RD, Currie CA, Mazzotti SP. 2005. Subduction zone backarcs, mobile belts, and orogenic heat. *Geological Society of America Bulletin* 15(2):4–10.

Jacobo AJ. 1986. The basement of the Poza Rica district and its controls on hydrocarbon generation. Internal review of the Mexican Petroleum Institute 18(1):5–24 (in Spanish).

Jennette DC, Fouad K, Wawrzyniec T, Dunlap D, Muñoz R, Meneses-Rocha J. 2003a. Slope and basin-floor reservoirs from the Miocene and Pliocene of the Veracruz Basin, south-eastern Mexico. *Marine and Petroleum Geology* 20:587–600.

Jennette DC, Wawrzyniec T, Fouad K, Dunlap D, Meneses-Rocha J, Grimaldo F, Muñoz R, Barrera D, Williams-Rojas CT, Escamilla-Herrera A. 2003b. Traps and turbidite reservoir characteristics from a complex and evolving tectonic setting, Veracruz Basin, south-eastern Mexico. *American Association of Petroleum Geologists Bulletin* 87:1599–1622.

Machel HG, Cavell PA. 1999. Low-flux, tectonically-induced squeegee fluid flow ("hot flash") into the Rocky Mountain foreland basin. *Bulletin of Canadian Petroleum Geologists* 47:510–533.

Meneses-Rocha JJ, Rodriguez-Figueroa D, Toriz-Gama J, Banda-Herandez J, Herandez-de La Fuente R, Valdivieso-Ramos V. 1997. *Field Guide to the Geologic Field Trip to the Zongolica Fold-and-Thrust Belt*: Associacion Mexicana de Geologos Petroleros-AMPG.

Ortuño AS. 1991. Contribution de la sédimentologie et de la télédétection à la reconstitution du bassin de Zongolica (Mexique centro-oriental). Place dans la dynamique du couple Golfe du Mexique-Système Cordillérain [PhD thesis]: University of Pau, France, 289 p.

Ortuño AS, Ferket H, Cacas MC, Swennen R, Roure F. 2003. Late Cretaceous carbonate reservoirs in the Córdoba Platform and Veracruz Basin (Eastern Mexico). *In* Bartolini C, Burke K, Buffler R, Blickwede J, Burkart B (Editors). *Mexico and the Caribbean Region: Plate Tectonics, Basin Formation and Hydrocarbon Habitats*, Memoir 79: American Association of Petroleum Geologists, Tulsa, Oklahoma. p. 476–514.

Ortuño AS, Roure F, Cacas MC, Swennen R, Ferket H. 1999. Subtrap Mexican transects across the Córdoba Platform. Institut Français du Pétrole-SUB-TRAP (SUBThrust Reservoir APpraisal) Report 45635–2.

Pemex-Schlumberger. 2000. Petroleum map of the Veracruz Basin. Pemex.

Pindell JL. 1994. Evolution of the Gulf of Mexico and the Caribbean. *In* Donovan SK, Jackson TA (Editors). *Caribbean Geology: An Introduction*: University of West-Indies Press, Kingston, Jamaica. p. 13–39.

Prost G, Aranda-Garcia M. 2001. Tectonics and hydrocarbon systems of Veracruz Basin, Mexico. *In* Bartolini C, Buffler RT, Cantu-Chapa A (Editors). *The Western Gulf of Mexico Basin: Tectonics, Sedimentary Basins and Petroleum Systems*, Memoir 75: American Association of Petroleum Geologists, Tulsa, Oklahoma. p. 271–291.

Rojas L. 1999. The tectonic evolution of the Zongolica flood-thrust belt and the Veracruz Basin, eastern Mexico [Master thesis]: Royal Holloway, University of London, 72 p.

Roure F. 2008. Foreland and hinterland basins: What controls their evolution? *Swiss Journal of Earth Sciences* 101(1):5–29. doi:10.1007/s00015–008-1285-x.

Roure F, Andriessen P, Callot JP, Faure JL, Ferket H, Gonzales E, Guilhaumou N, Lacombe O, Malandin J, Sassi W, Schneider F, Swennen R, Vilasi N. 2010. The use of paleo-thermo-barometers and coupled thermal, fluid flow and pore fluid pressure modelling for hydrocarbon and reservoir prediction in fold and thrust belts. *In* Goffey GP, Craig J, Needham T, Scott R (Editors). *Hydrocarbons in Contractional Belts*, Special Publication 348: Geological Society, London. p. 87–114.

Roure F, Alzaga H, Callot JP, Ferket H, Granjeon D, Gonzalez GE, Guilhaumou N, Lopez M, Mougin P, Ortuño S, Séranne M. 2009. Long lasting interactions between tectonic loading, unroofing, post-rift thermal subsidence and sedimentary transfers along the Western margin of the Gulf of Mexico: Some insights from integrated quantitative studies. *Tectonophysics* 475:169–189.

Roure F, Swennen R, Schneider F, Faure JL, Ferket H, Guilhaumou N, Osadetz K, Robion P, Vandeginste V. 2005. Incidence and importance of tectonics and natural fluid migration on reservoir evolution in foreland fold-and-thrust belts. *Oil and Gas Science and Technology Review Institut Français du Pétrole* 60:67–106.

Salvador A. 1991. Origin and development of the Gulf of Mexico basin. *In* Salvador A (Editor). *The Gulf of Mexico Basin*, The Geology of North America, Vol. J: Geological Society of America, Boulder, Colorado. p. 389–444.

Sassi W, Rudkiewicz JL. 1999. THRUSPACK Version 6.2: 2D Integrated maturity studies in thrust areas. Institut Français du Pétrole Report 45372.

Sassi W, Rudkiewicz JL. 2000. Computer modelling of petroleum systems along regional cross-sections in foreland fold-and-thrust belts: EAGE (European Association of Geoscientists & Engineers) Conference Malta 2000, extended abstract.

Schneider F. 2003. Basin modeling in complex are: Examples from Eastern Venezuelan and Canadian foothills. *Oil and Gas Science and Technology Revue de l'Institut Français du Pétrole* 58(2):313–324.

Schneider F, Devoitine H, Faille I, Flauraud E, Willien F. 2002. Ceres2D: A numerical prototype for HC potential evaluation in complex area. *Oil and Gas Science and Technology Revue de l'Institut Français du Pétrole* 57(6):607–619.

Schneider F, Faille I, Flauraud E, Willien F. 2000. A 2D basin modeling tool for HC potential evaluation in complex area: 62nd EAGE (European Association of Geoscientists & Engineers); May 29–June 2; Glasgow (Abstract).

Schneider F, Pagel M, and Hernandez E. 2004. Basin modeling in complex area: Example from Eastern Venezuelan Foothills. *In* Swennen R, Roure F, Granath J (Editors). *Deformation, Fluid Flow and Reservoir Appraisal in Foreland Fold-and-Thrust Belts*, Hedberg Series, Memoir 1: American Association of Petroleum Geologists, Tulsa, Oklahoma. p. 357–369.

Tissot B. 1969. Premières données sur les mécanismes et la cinétique de la formation du pétrole dans les sédiments: Simulation d'un schéma réactionnel sur ordinateur. *Revue de l'Institut Français du Pétrole* 24:470–501.

Ungerer P. 1990. State of the art of research in kinetic modelling of oil formation and expulsion. *Organic Geochemistry* 16:1–25.

Ungerer P, Burrus J, Doligez B, Chénet PY, Bessis F. 1990. Basin evaluation by integrated two dimensional modelling of heat transfer, fluid flow, hydrocarbon generation and migration. *American Association of Petroleum Geologists* 74:309–335.

Ungerer P, Doligez B, Chenet PY, Burrus J, Bessis F, Lafargue E, Giroir G, Heum R, Eggen R. 1987. A 2D model of basin scale petroleum migration by two-phase fluid flow: Application to some case studies. *In* Doligez B (Editor). *Migration of hydrocarbons in sedimentary basins.* Editions Technip, Paris, France. p. 415–456.

Vázquez CE. 2004. Final report on "generator systems of the Basin Veracruz. Pemex Report (in Spanish).

Vázquez CE. 2008. The Veracruz gas province: Origin and distribution of the gas and petroleum perspectives [Master's thesis]: Instituto Politécnico Nacional, México, 138 p. (in Spanish).

APPENDIX.—Two-dimensional (2D) basin modeling in a complex setting: The CERES2D software specificities.

HISTORICAL PERSPECTIVE

Basin modeling aims at reconstructing the time evolution of a sedimentary basin in order to make quantitative predictions of geological phenomena leading to pressure generation and hydrocarbon accumulations. It accounts for porous medium deformation, heat transfer, hydrocarbon formation, and multiphasic fluid migration (e.g., Schneider et al. 2000).

Since the pioneer works on the modeling of hydrocarbon formation (Tissot 1969) have been available, several generations of basin model have been built. Initially, 1D models such as *Genex* were used to simulate the temperature evolution and the maturation of the organic matter. In the sedimentary basin context, the 1D approach is satisfactory for these purposes because the thermal transfers are mainly vertical. Indeed, thermal convective transfers are most frequently negligible. Then 2D model, such as *Temispack* (Chenet et al. 1986; Doligez 1987; Bessis et al. 1990; Ungerer 1990; Ungerer et al. 1987, 1990), were built. These 2D models provide the possibilities of performing an evaluation of the pressure history and appreciating hydrocarbon migration and reservoir filling. However, these evaluations are only qualitative because fluid (water, oil, and gas) migration is mainly convective and therefore sensitive to 3D geometry and anisotropy. For these reasons, a generation of 3D basin model was designed (e.g., *Temis3D*; Schneider et al. 2000).

The previous sedimentary basin models are able to handle relatively simple geometries resulting from deposition, erosion, and vertical compaction. However, since the exploration is now focusing in area such as foothills, a new generation of models able to handle faults was needed.

CERES2D BASIN MODEL

The CERES2D software used is the current study is composed of several modules managed by a study browser. The main modules include a section editor, a restoration module, and a forward simulation simulator. The other modules include a chronostratigraphy editor, a lithology editor, a kerogen editor, fluid editors, a mesh editor, a run editor, and visualization modules. The main steps required to conduct a case study are (1) the edition of the initial section, (2) the restoration of the section, and (3) the forward simulations.

Initial Section: The initial section is composed of a 2D cross section, which can be edited directly from scratch or can be imported from other software. At this stage, it is recommended that one use structural tools to balance the section. The geometry of the section is defined first. The geological attributes are then assigned, including horizons and faults labeling, decollement level definition, and finally the lithology distribution, which may evolve spatially but not in time.

At this stage, the blocks, which represent the smallest kinematics units, are defined and meshed with their own grids, with no constraint coming from the other blocks. The faults grids are created dynamically during the forward simulation. The initial section holds the upper mantle, the ductile lower crust, the brittle upper crust, and the considered sedimentary part.

Kinematic Scenario: Once the edition of the initial section is terminated, the past stages of the sections are created following a backward process, which includes kinematic restoration, backstripping, and thickness modification.

For each restoration scenario, the backward process is composed of different steps (Schneider et al. 2002). The first step is the editing of the eroded parts if erosion has occurred during the considered period. Importing templates may help with this edition. The second step is an automatic backstripping of the section. Once the erosion and the sedimentation have been accounted for, the resulting section is uncompacted using porosity–depth relationships. This section is then restored from a kinematics point of view. At this stage, the displacements along faults are accounted for, using translations, vertical shear, and/or flexural slip mechanisms.

The last step of the backward simulation requires the correction of local inconsistencies in the computed thicknesses, which result from the use of the vertical shear mode of deformation. This step allows accounting for salt or mud tectonics. Correction of the edition of the erosion may be done at this stage. These steps should be performed for each layer initially defined in the section at present day.

Forward Simulations: In these complex geometries, faults cut the basin into blocks that naturally define computational subdomains, using Domain Decomposition Methods (e.g., Faille et al. 1998). In each block, the model accounts for the porous medium compaction, erosion, heat transfer, hydrocarbon formation, and migration (Schneider et al. 2002, 2004; Schneider 2003). The equations are mass conservation of solid and fluids (water, oil, and gas) coupled with Darcy's law and a compaction law. The faults have a constant thickness, and their permeability may evolve with time. The prototype allows using three permeability models for the faults: Faults can either be assumed to be pervious (i.e., first option) or impervious (i.e., second option), at which point they are considered a flow barrier. Alternatively, their permeability can also evolve through time according to the neighboring lithologies (i.e., third option). Permeability can also change with the strain rate. Whichever option is chosen, the faults are considered to be inactive when their velocities are lower than the defined speed limit of 50 m/Ma.

USING GEOCHEMICAL ANALYSIS TO IDENTIFY A SPATIAL MATURITY ANOMALY—AN EXAMPLE FROM THE UPPER RHINE GRABEN, GERMANY

CAROLYN LAMPE

ucon geoconsulting, Piusstr. 22a, 50823 Köln, Germany

e-mail: c.lampe@ucon4d.com

AND

LORENZ SCHWARK

Institute of Geoscience, Kiel University, Ludewig-Meyn Str. 10, 24118 Kiel, Germany

Abstract: Measurements of dispersed vitrinite in several exploration wells within the Tertiary and Quaternary rift sediments of the northern Upper Rhine Graben indicate a complex thermal history. While most wells show "normal" increasing maturity trends with depth, some lack any obvious trend. One of the investigated wells, Nordheim-1, even features a bell-shaped downhole anomaly having an inverse maturation trend. Abnormal maturation effects, such as oxidation, reworking, or deposition of previously coalified material from the hinterland, are ruled out as a result of the relative position of the well with respect to the graben shoulders or active fault systems. Thus, secondary maturation caused by focused, lateral hydrothermal fluid flow has been proposed.

To verify the measured maturity anomaly organic (Rock-Eval pyrolysis, methylphenanthrene distribution), geochemical analyses were used to obtain independent data for comparison. T_{max} values from Rock-Eval pyrolysis and various ratios of methylphenanthrene isomers indicate abnormally high maturities at shallow depth. Like the vitrinite reflectance values, these techniques show maturities that are inconsistent with their present burial depth but support enhanced convective hydrothermal heat flow in these strata. The consistency of results obtained using independent techniques verifies a hydrothermal origin of the maturity anomaly and excludes other modes of alteration.

Key words: Upper Rhine Graben, maturity anomaly, convective heat flow, methylphenanthrene index, transient heat flow effects

INTRODUCTION

Groundwater flow in sedimentary basins can have a profound effect on thermal maturity levels (Willett and Chapman 1989, Person and Garven 1992, Hullen et al. 1994, Person et al. 1995). Theoretical numerical models by Person et al. (1995) indicate that vertical flow rates of as little as 1 mm/yr can shift the oil-window upward by as much as 1000 m in discharge compared to recharge areas. Hydrodynamic processes may also enhance petroleum generation at shallow depths (~500 m) in hydrothermal discharge areas, as investigated by Kvenvolden and Simoneit (1990) in the Guaymas rift basin. In addition, groundwater flow may play an important role in long-range secondary oil migration (Garven 1989, Bethke et al. 1991, Garven et al. 1993). In particular, a topographically driven groundwater flow, enhanced during tectonic activity, may flush petroleum from structural reservoirs and drive it for hundreds of kilometers across a basin (Oliver 1986, Toth and Otto 1989, Toupin et al. 1997, Kennard et al. 1999).

Convective thermal anomalies induced by regional groundwater flow occur in many basins throughout the world, including the Rio Grande Rift, USA (Barroll and Reiter 1990, Mailloux et al. 1999); the Williston Basin, USA/Canada (Majorowicz et al. 1986); the Alberta Basin, Canada (Bachu 1995, 1997); the Eromanga Basin, Australia (Habermehl 1982, Toupin et al. 1997); and the Upper Rhine Graben (Clauser 1989, Person and Garven 1992, Aquilina et al. 1997, Lampe 2001, Lampe et al. 2001). In several basins, these hydrothermal disturbances of the conductive background temperature regime resulted in anomalies that affected the temperature and maturity of organic matter. Maturity anomalies such as those observed in the northern Upper Rhine Graben represent a comparably narrow time window in the geologic history of a basin, as overturned, inverse maturity trends indicate short-lived transient temperature conditions (Ziagos and Blackwell 1986). Such anomalies are reported from only a few basins, such as the Sub-Vulcan Basin, Australia (O'Brien et al. 1996, 1999; Yassir and Otto 1997; Kennard et al. 1999; Lisk et al. 1999). Maturity anomalies reported from these locations are typically associated with spatially confined increases in maturity featuring abnormally high maturity at shallow depth and/or inverse maturity with depth. While vitrinite reflectance anomalies, such as that in the Nordheim-1 well, are rare, they are consistent with maturation data from other sedimentary basins (Summer and Verosub 1987, Duddy et al. 1994, Goodhue and Clayton 1999, Parnell et al. 1999).

Vitrinite reflectance data can place time constraints on the relative duration of the fluid flow events and thus allow estimates of the factors that produce the anomaly (i.e., duration of flow, flow velocity, temperature of fluid; Lampe 2001, Lampe et al. 2001). However, proof is needed to verify the proposed anomaly and the geologic/hydrologic history behind the event.

The purpose of this work is to substantiate a thermal anomaly based on a bell-shaped vitrinite reflectance profile in the Nordheim-1 well in the northern Upper Rhine Graben using various geochemical measurements, including Rock-Eval pyrolysis, T_{max}, and molecular maturity parameters based on methylphenanthrene distribution.

STUDY AREA

The approximately 300-km-long and 30-km-wide Upper Rhine Graben is part of a continental Tertiary trough that trends SSW–NNE from Basel to Mainz (Fig. 1A). Its flanks are bordered by deep basement faults (Fig. 1B). Subsidence of the graben along these faults resulted in accumulation of ~3000 m of sediment (Illies and Mueller 1970, Illies and Fuchs 1974, Pflug 1982).

Rifting and deposition in the northern Upper Rhine Graben started in the lower Oligocene. The graben subsided continuously, with the exception of a ~9-my hiatus between the JT I and JT II formations (upper Miocene to lower Pliocene) (Fig. 2). During this time, considerable shoulder uplift and subsequent erosion of the graben strata, on the order of 1000 to 2000 m, may have affected the western

Analyzing the Thermal History of Sedimentary Basins: Methods and Case Studies
SEPM Special Publication No. 103, Copyright © 2012
SEPM (Society for Sedimentary Geology), Print ISBN 978-1-56576-315-9, CD/DVD ISBN 978-1-56576-317-3, p. 187–198.

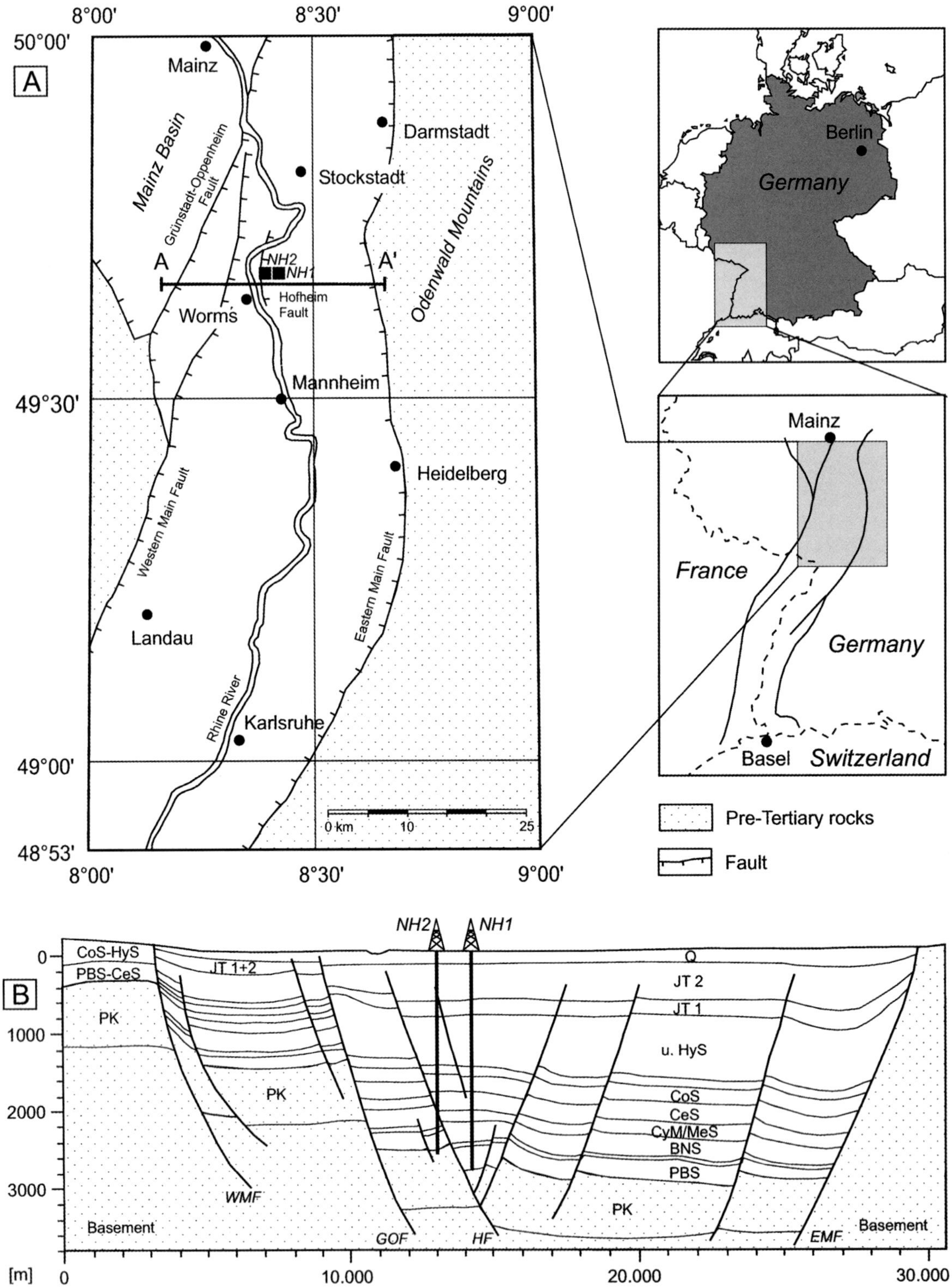

FIG. 1.—Map of the northern Upper Rhine Graben with major fault systems and key wells (A) and position of cross section A-A′ (B). The cross section shows the stratigraphic units and the major fault zones (GOF = Grünstadt–Oppenheim Fault; WMF = Western Main Fault; HF = Hofheim Fault; EMF = Eastern Main Fault). Wells Nordheim-1 and Nordheim-2 are projected into the profile. For information about the stratigraphy refer to Table 1 and Figure 2.

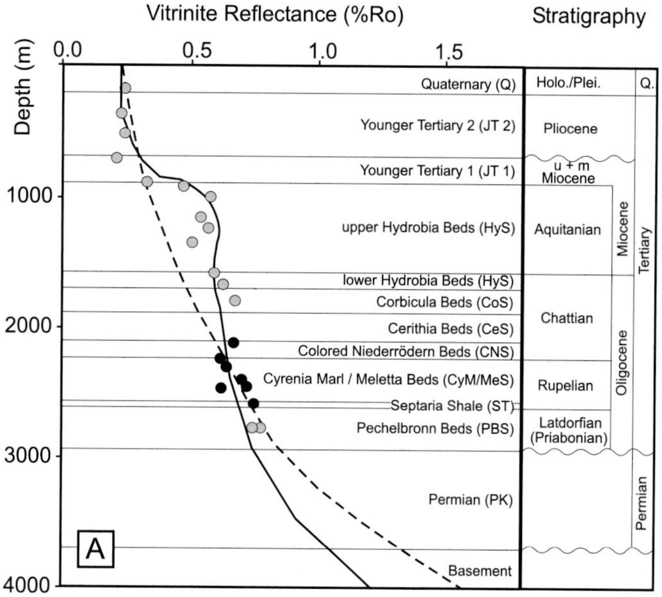

FIG. 2.—Stratigraphy of the northern Upper Rhine Graben and measured vitrinite reflectance for the Nordheim-1 and –2 wells (shaded and black dots, respectively) together with computed vitrinite reflectance profiles (dashed and solid lines). The dashed line shows Ro calculated using a conductive basal heat flow, while the solid line is based on a basal heat flow combined with a single lateral heat pulse of 50,000 years with a velocity of 1.0 m/yr and a temperature of 130° C.

rift margin, as evidenced by reduced sediment thickness and interpreted numerical models (Lampe 2001); however, the central part of the graben seems to have been largely unaffected by erosion.

The Tertiary graben sediments mainly comprise fine-grained limnic to brackish-marine mixed lithologies of marl, claystone, and limestone. Evaporitic, coarse-grained sandstones and calcarenites occur as interbeds in most of the Oligocene and lower Miocene formations. The Younger Tertiary and the overlying Quaternary largely consists of coarse-grained alluvial, fluvial, and eolian sediments with interbedded fine-grained horizons and lenses of clay and silt. These sediments unconformably overlie Permian basement, which consists of fluvial conglomerates and interbedded volcanics. The Permian basement, as well as the coarse-grained Younger Tertiary and Quaternary sediments, have sufficient permeability to allow groundwater circulation. Permeable alluvial fan deposits along the major faults, such as the Western Main Fault, the Grünstadt–Oppenheim Fault, the Hofheim Fault, and the Eastern Main Fault (Fig. 1), together with phases of tectonic reactivation of the fault systems make the faults possible fluid pathways for groundwater, connecting the deep and the shallow groundwater systems across the lower permeable Tertiary strata.

Hydrodynamic conditions within the Rhine Graben are complex and consist of a combination of flow patterns: gravity/topography-driven flow (more or less vertical, along the main fault escarpments and the associated alluvial fan sediments) and, possibly, compaction-driven flow (±horizontal, within the low-permeability central Tertiary sediments). A regional redistribution of heat associated with a large-scale flow pattern has been proposed for the Rhine Graben (Doebl and Teichmüller 1979, Clauser 1989, Clauser and Villinger 1990, Person and Garven 1992). Evidence for a deep groundwater circulation system is provided by chemical analysis of the groundwater, which indicates that it originated from the Permian basement (Sonne 1972, Meier 1989). Groundwater must have passed upward through the Tertiary rocks based on high contents of sulfur derived from the pyrite-rich Tertiary strata and halite derived from the evaporitic Oligocene CoS formation (Scharpf 1977, Schmitt 1992, Kaupe 1995).

THE THERMAL ANOMALY

Measurements of dispersed vitrinite in the Nordheim-1 well show an unusual vertical maturity trend (Fig. 2). Above and below an 800-m-thick Miocene sedimentary unit, the vitrinite reflectance data indicate a relatively undisturbed basal heat flow. However, within this unit vitrinite reflectance is elevated and cannot be explained assuming only conductive heat flow (Lampe et al. 1999). This 800-m-thick maturity anomaly was interpreted to indicate that the thermal history of the northern Upper Rhine Graben is not only the result of basal conductive heat flow but was also greatly influenced by circulating deep groundwater (Lampe et al. 1999, Lampe and Person 2000). The thermal anomaly that caused this maturation anomaly must have been hydrothermal in origin and persisted long enough to elevate the vitrinite reflectance values within this unit, yet it must have dissipated before the thermal perturbation would have altered the organic matter below and above the unit.

Numerical modeling studies substantiate that the vitrinite reflectance anomaly was caused by a transient thermal inversion induced by episodic, lateral flow of hot (130–160° C) groundwater along conductive fractures and bedding planes (Lampe et al. 2001). Inflow temperature, horizontal velocity, duration and frequency of flow, and thickness of the aquifer were varied in these models and showed that a thermal maturity anomaly could have only been produced by a rather restrictive set of hydrothermal conditions. The observed vitrinite reflectance anomaly could have been caused by a single hydrothermal flow event. Figure 2 shows the measured vitrinite reflectance (dots) and two calculated maturity trends derived from two-dimensional numerical modeling studies (Lampe 2001, Lampe et al. 2001). The dashed line represents the vitrinite reflectance profile developed in response to a conductive basal heat flow, while the solid line is based on a basal

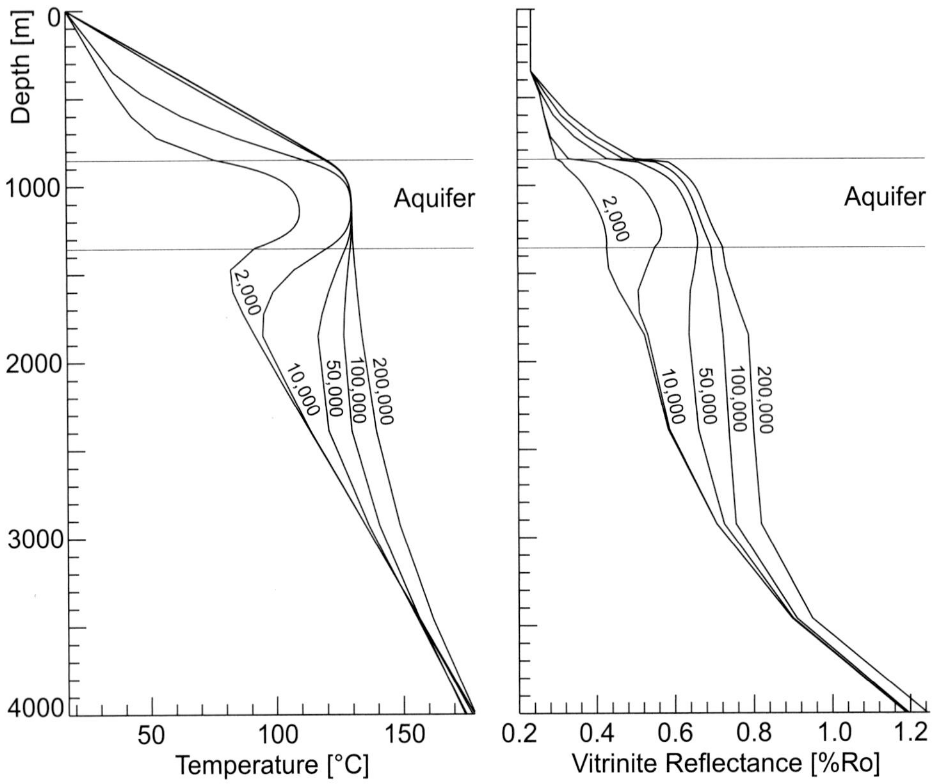

FIG. 3.—Temperature and vitrinite reflectance profiles along a generic well 500 m from a presumed feeding fault. This idealized model assumes a constant fluid flow of 130° C at a velocity of 1 m/yr. Numbers indicate years after initiation of fluid flow. Note the pronounced overturn in vitrinite reflectance after just 10,000 years of flow. The system has reached steady-state conditions after 200,000 years with respect to both temperature and vitrinite reflectance. After Lampe (2001).

heat flow combined with a single lateral heat pulse of 50,000 years with a velocity of 1.0 m/yr and a temperature of 130° C.

After roughly 50,000 years both the thermal and the vitrinite reflectance profile dissipate (Fig. 3), changing from a bell-shaped curve (indicating transient conditions) to a dogleg-shaped curve (approaching steady-state conditions). If the flow event lasted longer than 100,000 years, then the maturation anomaly disappeared as the lower confining unit approached steady-state thermal conditions (Fig. 3).

The groundwater flow regime in the northern Upper Rhine Graben indicates that flow of heated groundwater is directed upward along deep faults, which serve as conduits connecting the porous Permian aquifers with the younger, shallower strata. Subsequently, the groundwater flows laterally into the strata (aquifer) where the anomaly is observed. Flow rates that develop under topography-driven groundwater flow are too low to preserve the hot temperatures of deep-seated upward-migrating brines. The higher flow rates that are needed may be provided by episodic, short-term changes in fault permeability (Rojstaczer and Wolf 1992, Muir-Wood 1994). Tectonic forcing that triggered high discharge fluid flow during phases of active rifting (possibly associated with earthquakes) seems to be the only feasible explanation for the short-term transient fluid flow events necessary to cause such a spatial temperature and maturity anomaly (Lampe 2001, Lampe et al. 2001).

METHODS

In order to gain insight regarding the temperature development of the Upper Rhine Graben, temperature-sensitive calibration parameters were investigated. These include vitrinite reflectance and T_{max} values, which are derived from Rock-Eval pyrolysis and extrapolated from various methylphenanthrene indices.

Vitrinite Reflectance

Vitrinite reflectance is one of the most widely used methods for assessing the thermal maturity of sedimentary rocks. Measurements were carried out on dispersed vitrinite on polished mounts derived from exploration well cuttings (Table 2). Unlike measurements performed on coals, reflectance data derived from dispersed vitrinite are less uniform (Scheidt and Littke 1989), resulting in fairly high standard deviations (0.03–0.34; Table 2). The often relatively small particles make valuable measurements difficult, resulting in few measurements per sample and/or high standard deviations. Model-based calculated maturity (Figs. 2, 3, 6) was performed using the kinetic EASY %Ro approach based on the algorithm introduced by Sweeney and Burnham (1990), including assumptions about heat flow, surface temperature, and depositional history, such as erosion events or hiatuses.

Total Organic Carbon/Total Carbon

The amounts of total carbon (TC) and total organic carbon (TOC) contained in the various formations of the northern Upper Rhine Graben are basic parameters with which to evaluate source rock quality (Table 3). Total carbon and TOC were determined by combustion using a LECO CS212 elemental analyzer.

TABLE 1.—Tertiary stratigraphy of the northern Upper Rhine Graben. The bold-type abbreviations are used throughout the text.

<table>
<tr><th colspan="3">Abbreviation</th><th colspan="3">Formation Name</th><th colspan="3">Translation</th></tr>
<tr><td colspan="3">JT 2</td><td colspan="3">Jungtertiär 2</td><td colspan="3">Younger Tertiary 2</td></tr>
<tr><td colspan="3">JT 1</td><td colspan="3">Jungtertiär 1</td><td colspan="3">Younger Tertiary 1</td></tr>
<tr><td colspan="3">HyS</td><td colspan="3">Hydrobienschichten</td><td colspan="3">Hydrobia Beds</td></tr>
<tr><td colspan="3">CoS</td><td colspan="3">Corbiculaschichten</td><td colspan="3">Corbicula Beds</td></tr>
<tr><td colspan="3">CeS</td><td colspan="3">Cerithienschichten</td><td colspan="3">Cerithia Beds</td></tr>
<tr><td colspan="3">BNS</td><td colspan="3">Bunte Niederröderner Schichten</td><td colspan="3">"colored" Niederrödern Beds</td></tr>
<tr><td colspan="2">CyM</td><td rowspan="4">GS</td><td colspan="2">Cyrenenmergel</td><td rowspan="4">Graue Serie</td><td colspan="2">Cyrenia Marl</td><td rowspan="4">Gray Series</td></tr>
<tr><td colspan="2">MeS</td><td colspan="2">Melettaschichten</td><td colspan="2">Meletta Beds</td></tr>
<tr><td rowspan="2">ST</td><td>FiS</td><td rowspan="2">Septarienton</td><td>Fischschiefer</td><td rowspan="2">Septaria Shale</td><td>Fish Shale</td></tr>
<tr><td>FoM</td><td>Foraminiferen-mergel</td><td>Foraminifera Marl</td></tr>
<tr><td colspan="3">PBS</td><td colspan="3">Pechelbronner Schichten</td><td colspan="3">Pechelbronn Beds</td></tr>
</table>

TABLE 2.—Vitrinite reflectance data for Nordheim-1 and Nordheim-2 wells.

Sample No.	Sample type	Formation[1]	Depth (m)	Vitrinite reflectance (%Ro)	Standard deviation	No. of measurements
NH1-S5	cuttings	Q	160.00	0.24	0.04	40
NH1-S11	cuttings	JT 2	350.00	0.23	0.04	41
NH1-S14	cuttings	JT 2	500.00	0.24	0.04	20
NH1-S21	cuttings	JT 1	700.00	0.21	0.08	6
NH1-S26	cuttings	JT 1	875.00	0.33	0.07	13
NH1-S28	cuttings	u. HyS	915.00	0.47	0.18	38
NH1-S30	cuttings	u. HyS	990.00	0.58	0.21	23
NH1-S35	cuttings	u. HyS	1155.00	0.54	0.20	23
NH1-S37	cuttings	u. HyS	1230.00	0.57	0.27	19
NH1-S41	cuttings	u. HyS	1340.00	0.51	0.25	35
NH1-S48	cuttings	u. HyS	1580.00	0.59	0.22	46
NH1-S50	cuttings	l. HyS	1675.00	0.63	0.26	50
NH1-S53	cuttings	CoS	1790.00	0.68	0.22	33
*		PBS	2767.90	0.75	0.04	20
*		PBS	2768.40	0.77	0.07	49
NH2-S1	cuttings	CeS	2105.00	0.67	0.26	10
NH2-S4	cuttings	CyM–MeS	2220.00	0.61	0.27	35
NH2-S8	cuttings	CyM–MeS	2285.00	0.63	0.18	27
NH2-S12	cuttings	ST	2380.00	0.69	0.23	22
*		PBS	2430.00	0.72	0.07	31
NH2-B15	core	PBS	2441.28	0.72	0.34	52
*		PBS	2443.00	0.72	0.03	46
NH2-B7	core	PBS	2453.40	0.61	0.08	55
*		PBS	2562.00	0.75	0.06	20

[1] The formation names, ages, and abbreviations are described in Table 1 and Figure 2.
* Measured by Rückheim (1989).

TABLE 3.—Total carbon and total organic carbon (TC/TOC) for Nordheim-1 and Nordheim-2 wells.

Sample No.	Type	Formation[1]	Depth (m)	TC (wt. %)	TOC (wt. %)
NH1-S7	cuttings	JT 2	250.00	2.63	1.37
NH1-S8	cuttings	JT 2	275.00	2.35	0.05
NH1-S13	cuttings	JT2	450.00	0.37	0.07
NH1-S17	cuttings	JT2	585.00	3.24	0.05
NH1-S22	cuttings	JT1	735.00	4.74	0.40
NH1-S27	cuttings	u. HyS	885.00	6.55	3.05
NH1-S28	cuttings	u. HyS	915.00	6.44	4.52
NH1-S32	cuttings	u. HyS	1080.00	6.19	2.94
NH1-S38	cuttings	u. HyS	1275.00	7.11	3.00
NH1-S44	cuttings	u. HyS	1450.00	9.55	1.80
NH1-S52	cuttings	CoS	1730.00	5.49	1.76
NH2-S1	cuttings	CeS	2105.00	5.41	0.93
NH2-S6	cuttings	CyM–MeS	2250.00	4.24	0.15
NH2-S11	cuttings	ST	2360.00	5.21	1.69
NH2-S15	cuttings	PBS	2439.00	3.65	0.75
NH2-S17	cuttings	PBS	2500.00	4.27	1.07
NH2-B5	core	PBS	2560.04	2.42	2.45
NH2-S20	cuttings	PBS	2600.00	2.26	0.57

[1] The formation names, ages, and abbreviations are described in Table 1 and Figure 2.

Rock-Eval Pyrolysis

The temperature (° C) at which the pyrolytic yield of hydrocarbons from a rock sample reaches its maximum (T_{max}) is determined using Rock-Eval pyrolysis. This technique, which is based on the anhydrous pyrolysis of source rocks and was developed by the Institut Français du Pétrole (IFP), enables the determination of hydrocarbon generative potential. T_{max}, one of the measurements that results from a Rock-Eval analysis, is a kerogen-type dependent maturation parameter that can be correlated with vitrinite reflectance (Table 4) (Peters et al. 2005).

Rock-Eval pyrolysis was performed to gain information about the kerogen type and maturity. All measurements were performed in duplicate using a Rock-Eval II Plus Analyzer (Vinci Technologies) following standardized procedures (Espitalié et al. 1977, Bordenave et al. 1993). Calibration of the instrument was performed using the 55.000 IFP standard. Samples were decarbonated prior to Rock-Eval analysis to minimize the mineral matrix effect and to avoid any contribution of dissociating carbonates to the S_3 peaks.

Maturation and source quality parameters derived from Rock-Eval pyrolysis (Table 3) include the following: hydrogen index ($HI = [S_2/TOC] \times 100$, mg/g TOC), oxygen index ($OI = [S_3/TOC] \times 100$, mg/g TOC), production index ($PI = S_1/[S_2 + S_2]$), and T_{max} (° C).

TABLE 4.—Rock-Eval pyrolysis and geochemical analysis for Nordheim-1 and Nordheim-2 wells. Bold values indicate strong evidence for lowered T_{max} due to contamination or migrated oil (Peters 1986).

Sample No.	Sample type	Formation[1]	Depth (m)	TOC cf[2] (wt. %)	S_1 (mg HC/g rock)	S_2 (mg HC/g rock)	S_3 (CO_2/g rock)	HI (mg HC/g TOC)
NH1-S7	cuttings	JT 2	250.00	1.56	0.21	0.51	1.36	32
NH1-S11	cuttings	JT 2	350.00	13.50	2.70	14.75	7.68	109
NH1-S27	cuttings	u. HyS	885.00	5.07	1.75	29.70	2.36	585
NH1-S28	cuttings	u. HyS	915.00	5.72	2.84	26.81	2.55	469
NH1-S32	cuttings	u. HyS	1080.00	4.28	2.00	14.51	2.10	339
NH1-S38	cuttings	u. HyS	1275.00	4.76	1.69	19.65	1.64	412
NH1-S44	cuttings	u. HyS	1450.00	5.14	4.11	21.38	1.58	415
NH1-S48	cuttings	u. HyS	1580.00	2.87	1.70	7.11	0.83	248
NH1-S50	cuttings	l. HyS	1675.00	4.41	1.55	15.47	1.74	350
NH1-S52	cuttings	CoS	1730.00	3.00	0.78	9.42	0.74	314
NH2-S1	cuttings	CeS	2105.00	1.70	0.27	1.81	0.48	107
NH2-S11	cuttings	ST	2360.00	2.69	0.43	4.22	0.50	157
NH2-S15	cuttings	PBS	2439.00	0.97	0.19	0.89	0.47	91
NH2-S17	cuttings	PBS	2500.00	1.74	0.76	3.26	0.54	186
NH2-B5	core	PBS	2560.04	1.41	0.10	0.62	0.30	43

[1] The formation names, ages, and abbreviations are described in Table 1 and Figure 2.
[2] TOC cf = carbonate-free TOC; HC = hydrocarbon.

Methylphenanthrene Index

The relative distribution of phenanthrene and its methylated derivatives in rock extracts provides information on maturity and kerogen type. These biomarker-derived maturity parameters utilize ratios of phenanthrene and various methylphenanthrene isomers, distinguished by the position of the methyl group on the aromatic ring (Radke and Welte 1983; Radke et al. 1986, 1997; Radke 1987).

The methylphenanthrene index (MPI) parameter is highly dependent on depositional environment, which is both laterally and vertically variable throughout the Upper Rhine Graben. The MPI is defined for type III kerogen (Radke 1987), ideally contained continuously throughout the section, often mixed with type II or type I kerogen.

The concentration of 2- and 3-methylphenanthrenes increases in rocks as a result of increasing maturation. This is thought to be the result of rearrangement of 1- and 9-methylphenanthrenes, as the 2- and 3-isomers are thermodynamically favored at higher temperatures (Radke 1987). Relative thermal maturity was determined using the MPI equations after Radke (1987):

$$\text{MPI1} = 1.5(2\text{MP} + 3\text{MP})/(\text{P} + 1\text{MP} + 9\text{MP}), \quad (1)$$

$$\text{MPI3} = (2 - \text{MP} + 3 - \text{MP})/(1 - \text{MP} + 9 - \text{MP}), \quad (2)$$

$$MPR = 2 - \text{MP}/1 - \text{MP}, \quad (3)$$

where MP is methylphenanthrene, P is phenanthrene, and the hyphenated numbers depict the position of the methyl group on the aromatic ring.

To determine the concentration of the respective methylphenanthrenes and phenanthrenes, rocks were extracted using dichloromethane as an organic solvent. The bitumen obtained was then fractionated by preparative chromatography into saturated and aromatic hydrocarbons as well as nitrogen, sulfur, and oxygen–bearing components. Quantitative analysis of the phenanthrenes was performed using gas chromatography coupled with mass spectrometry (GC-MS). The total extracts were analyzed in single ion monitoring mode, recording masses m/z 178 (phenanthrene) and m/z 192 (methylphenanthrene).

TABLE 4.—Extended.

OI (mg CO_2/g TOC)	PI	T_{max} (° C)	MPI (%R_c)
87	**0.30**	**385**	0.71
57	**0.16**	**370**	0.62
46	0.05	426	0.63
44	0.10	415	0.65
49	0.12	401	0.71
34	0.08	401	0.69
30	**0.16**	**400**	0.70
29	**0.19**	**397**	0.64
39	0.10	406	0.70
25	0.08	422	0.68
28	0.12	420	0.67
18	0.09	422	0.75
48	**0.18**	**411**	0.72
31	**0.19**	**404**	—
20	0.14	454	0.74

Equivalent vitrinite reflectance values were calculated from the concentrations of the methylphenanthrenes and phenanthrenes, following Radke et al. (1991):

$$\%R_c = 0.6\ \text{MPI1} + 0.4(for\%\ \text{R} \leq 1.35\%\ ); \quad (4)$$

$$\%R_c = -0.6\ \text{MPI1} + 2.6(for\%\ \text{R} \geq 1.35\%\ ). \quad (5)$$

RESULTS AND DISCUSSION

Samples were chosen from the Nordheim-1 and Nordheim-2 wells for Rock-Eval pyrolysis in order to test whether the vitrinite reflectance maturity anomaly in the Nordheim-1 well is evident in other geochemical parameters, such as T_{max} or various ratios of methylphenanthrene isomers. To show a continuous downhole profile, the maturation data from the two Nordheim wells were combined because they are close to each other and were interpreted to have very similar thermal histories, based on their respective vitrinite reflectance profiles (Lampe 2001).

Some of the investigated wells contain reworked material as a result of their proximity to the active graben faults where sediment from the hinterland was brought into the graben; however, the Nordheim-1 well shows no petrographic evidence for reworked material, despite its relative position with regard to active faults.

Total organic carbon values vary between 0.05% (JT2 formation) and 4.52% (upper HyS formation; Table 4). Generally low values are found in the coarse clastic Quaternary and Younger Tertiary formations (JT2 and JT1). The fine-grained Tertiary graben fill (i.e., HyS to PBS formations) shows higher TOC, with a mean value of 1.41%. Although varying on a relatively small scale, the highest values are measured in the HyS formation (Nordheim-1 well), with a mean of 3.06%. Total organic carbon values commonly vary on a small scale (cm to dm), which reflects the heterogeneous depositional environments that dominate the Upper Rhine Graben stratigraphy (Rückheim 1989).

Figure 4A shows a general increase of T_{max} as a function of depth. Abnormally low T_{max} values (diamond symbol in Fig. 4A) represent contamination by drilling fluids or natural impregnation of migrated hydrocarbons (Peters 1986, Bordenave et al. 1993). The unusually high values of the Nordheim-1 samples at a depth of ~900 m (T_{max} = 426 and 415° C) coincide with the maturity anomaly observed in the measured vitrinite reflectance; deeper samples, between 1080 and 1675 m, have significantly lower T_{max} values, between 397 and 406° C. T_{max} increases in the deepest samples (depths from 1730 to 2560 m) to between 404 and 454"° C.

With the exception of the terrigenous sandstones (diamond symbol), Figure 4B shows a decreasing trend in TOC with increasing depth and temperature for the Nordheim wells, commonly associated with organic carbon being converted to oil and/or gas, assuming that the starting TOC content was approximately the same. Figure 4C, D indicate mainly type II kerogen with a standard maturation trend, while the terrigenous samples (Q and JT) indicate type III to IV kerogen. The contribution of different kerogen types from the diverse depositional environments within the northern Upper Rhine Graben becomes evident in Figure 4D. The mostly limnic to brackish-marine samples of the Tertiary graben fill plot around the type II line, more mature samples plotting closer to the origin (square symbol). The Quaternary and Younger Tertiary alluvial samples (diamond symbol), on the other hand, are below an HI of 110 mg hydrocarbon/g TOC and indicate type III to type IV kerogen.

Although an effect of different kerogen type and/or impregnation of strata by migrated hydrocarbons may be responsible for the unusual T_{max} values in the Nordheim-1 well, it is significant that the anomaly of high T_{max} values at comparably shallow depth coincides with the one

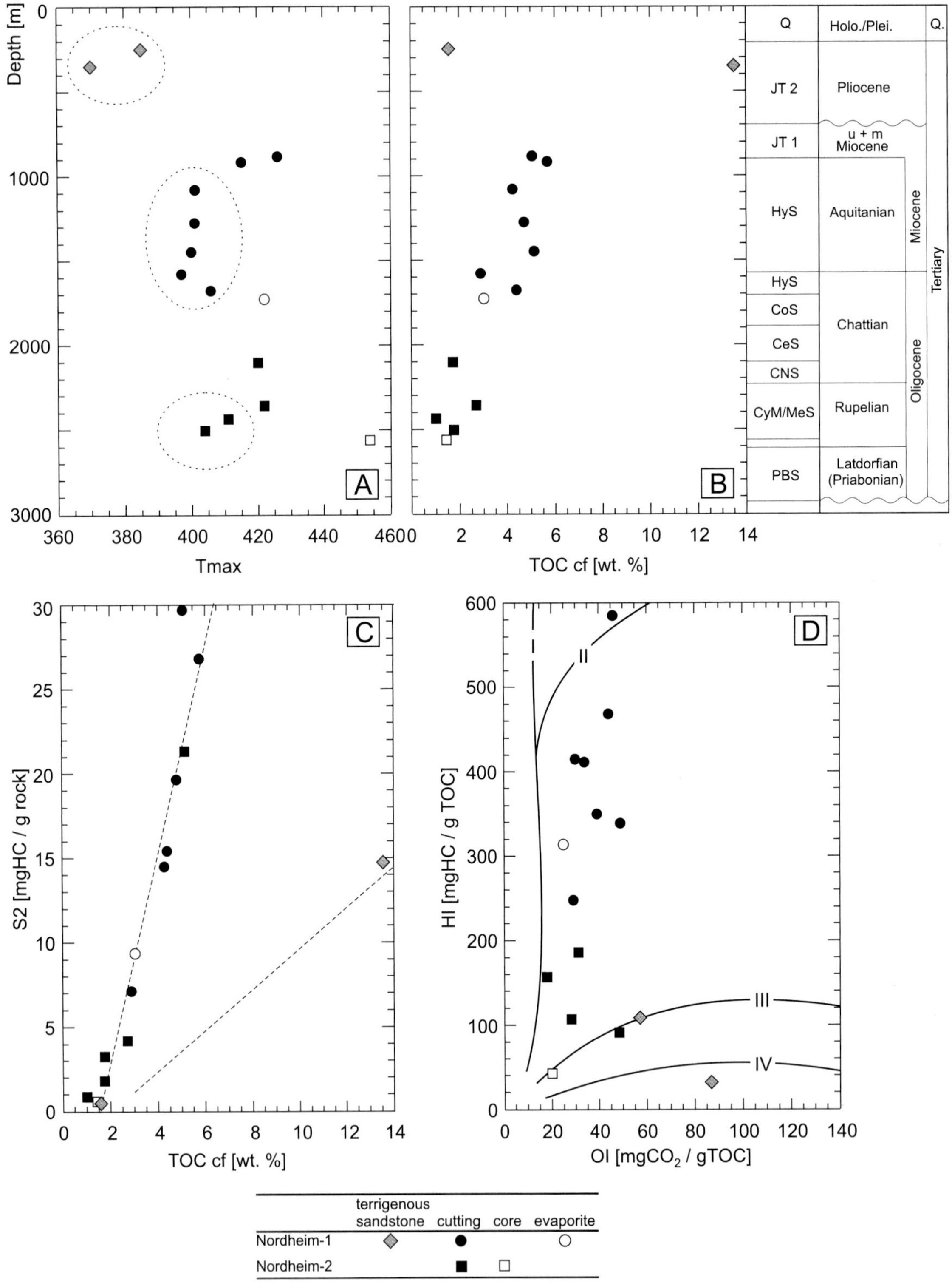

FIG. 4.—Rock-Eval pyrolysis plots showing (A) T_{max} values derived from the S_2 peak (samples in dashed circles show strong evidence for lowered T_{max} due to contamination by additives or migrated oil); (B) carbonate-free TOC; (C) carbonate-free TOC vs. S_2; and (D) classification of kerogen types based on a pseudo–van Krevelen diagram of HI vs. OI. The formation names and abbreviations are described in Table 1 and Figure 2.

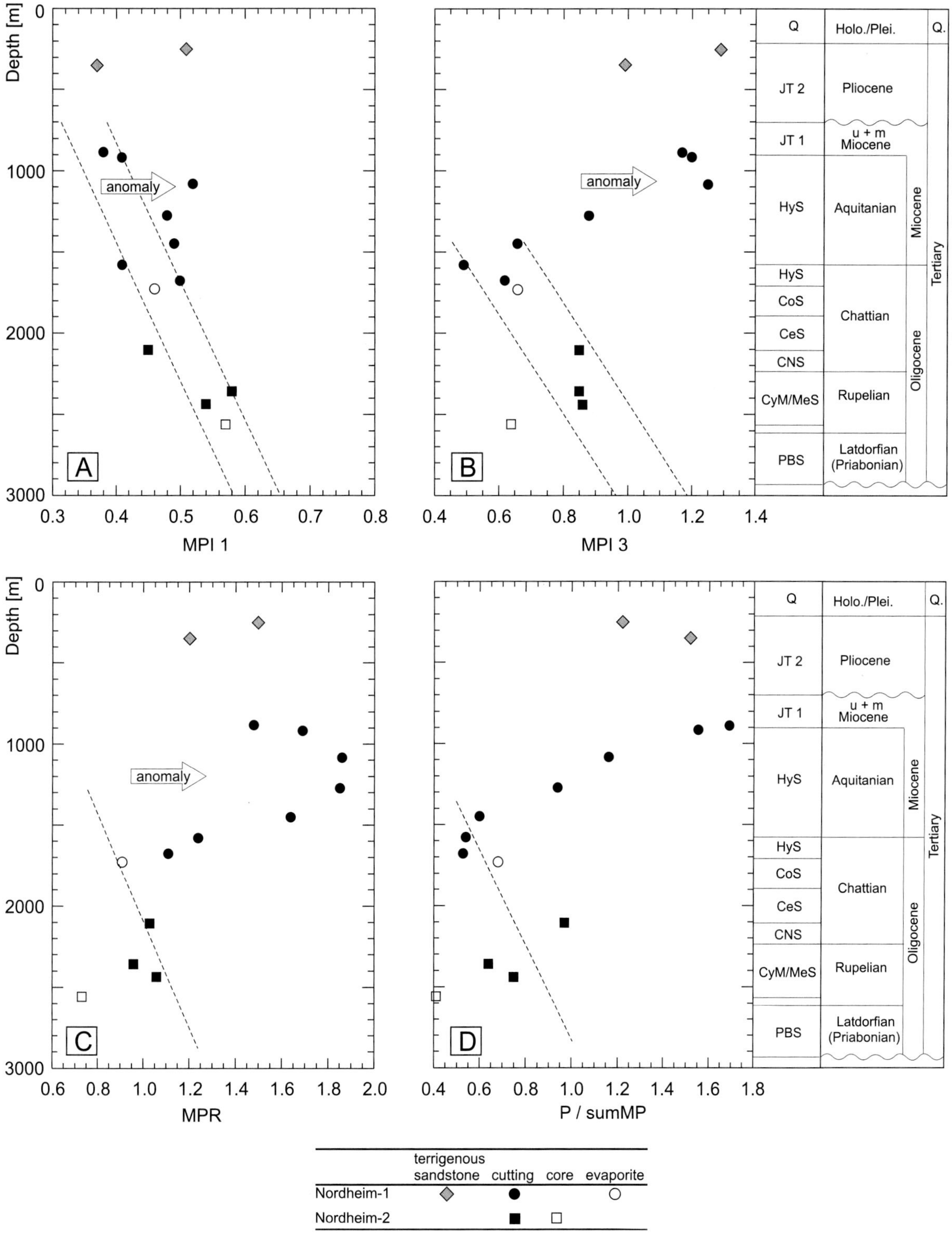

FIG. 5.—Methylphenanthrene index plots showing (A) MPI 1; (B) MPI 3; (C) MPR: ratio of 2- to 1-methylphenanthrene; and (D) phenanthrene concentration to the sum of methylphenanthrene. All four plots show unusual maturities in the strata below the JT formation (gray diamonds) down to a depth of ~1500 m. The dashed lines indicate hypothetical maturity trends. The formation names and abbreviations are described in Table 1 and Figure 2.

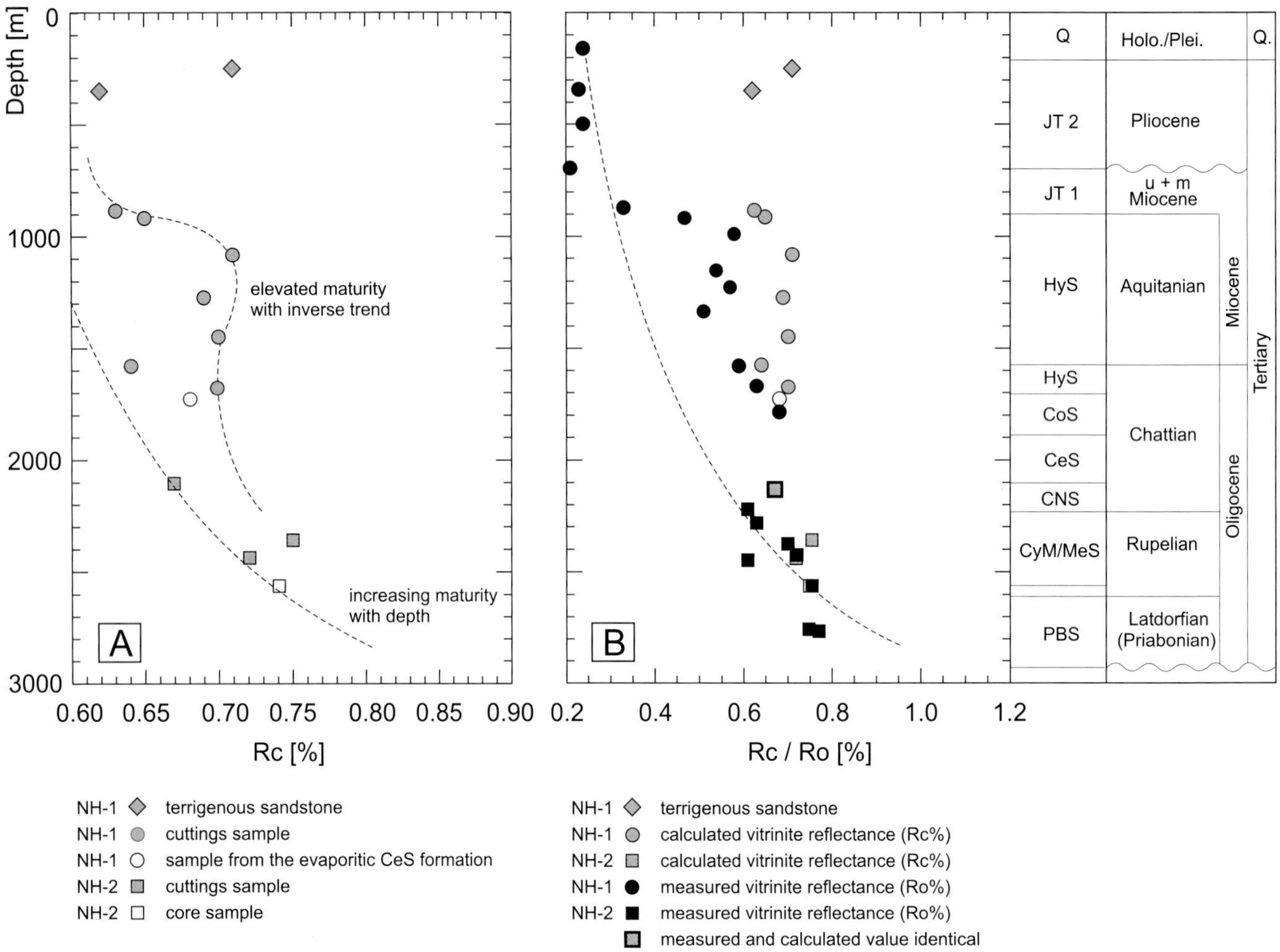

FIG. 6.—(A) Calculated vitrinite reflectance from MPI. The plot shows unusual maturities in the strata below the JT formation (gray diamonds) to a depth of ~1500 m. (B) Calculated vitrinite reflectance from MPI (black symbols) and measured vitrinite reflectance (gray symbols). Both profiles indicate elevated maturities in the strata below the JT formation (>850 m). The dashed line depicts a possible maturity trend. The formation names and abbreviations are described in Table 1 and Figure 2.

observed in the vitrinite reflectance data. Migrated hydrocarbons would not have affected the vitrinite data.

Gas chromatography-MS analysis shows that the samples contain sufficient phenanthrene and methylphenanthrene isomers to gain information on their maturation history. Additionally, saturate biomarker distribution determined by GC-MS indicates that most samples, in particular those from the Oligocene HyS, CoS, and CeS formations, were deposited under saline conditions. For example, very low pristine:phytane ratios of 0.15 to 0.25 for the HyS or CoS formations and of 0.35 to 0.5 for the CeS, ST, and PBS formations support hypersaline conditions (Peters et al. 2005). This organic facies interpretation is complemented by the high abundance of gammacerane, with gammacerane/$C_{30}\alpha\beta$-hopane ratios of 2.0 for the CoS formation, 0.63 to 1.12 for the HyS formation, and 0.31 to 0.92 for the CeS, ST, and PBS formations, respectively.

Figure 5 shows the MPI 1, MPI 3, and MPR parameters and the ratio of phenanthrene to the sum of the methylphenanthrenes (P/ΣMP) ratios. The two samples derived from the terrigenous Quaternary and Younger Tertiary sandstones (diamond symbol) show unusually high maturities with respect to their shallow burial depths. These younger sediments locally contain abundant coalified wood and may have also been impregnated by migrated hydrocarbons from depth. Thus, the apparent high maturity values of these samples may reflect a facies rather than a maturity effect (e.g., remnants of bitumen or a kerogen type differing from the other samples). Another unusual sample type is depicted by the open square symbol in Figure 5. This is the only sample derived directly from core, whereas the other samples were obtained from well cuttings.

All other samples (solid squares and circles and open circle depicting a sample derived from the evaporitic CeS formation) show significant shifts to higher maturity at a depth of ~850 m below the terrigenous sandstones of the Quaternary and Younger Tertiary formations. With increasing depth below 850 m, maturities show an inverse trend, indicating a lower relative maturity. This unusual vertical maturity trend is consistent with the inverse trend in vitrinite reflectance measured in the Nordheim-1 well.

The magnitude and trend of the MPI 1 values are consistent with a "normal" maturity, with the exception of the three data points between

1000 m and 1500 m (Fig. 5A). Although the apparent maturity of these samples could be derived from remnants of secondary migration, it is the authors' interpretation that the elevation is caused by secondary maturation; e.g., spatially confined hydrothermal events. Similar explanations can be found for the elevated maturities in the MPI 3 (Fig. 5B), the MPR (Fig. 5C), and the P/ΣMP (Fig. 5D) parameter plots. The observed maturities are not consistent with their present burial depth.

Plotting vitrinite reflectance calculated from the MPI 1 using Eq. 4 against measured vitrinite reflectance indicates a bell-shaped vertical profile (Fig. 6) indicating the same maturity anomaly (location and magnitude) observed in the measured calibration data (Fig. 6B). However, the calculated vitrinite reflectance values are generally higher than those derived from the optical measurements.

The MPI maturity data clearly show unusual downhole values and trends at relatively shallow depth. The lack of suitable samples of type II kerogen for the JT formation hinders reliable maturity assessment when applying these ratios. The expression of the maturation anomaly is relatively subtle in the MPI 1 data but is significantly more pronounced in the MPI 3, MPR, and P/ΣMP ratios (Fig. 5).

CONCLUSIONS

The heat flow density distribution measured in the Upper Rhine Graben indicates that it is not caused solely by an elevated basal conductive heat flux due to rift-related thinning of the lithosphere. The local variations in heat flow seem to be associated with redistribution of surface heat by hydrothermal fluid flow through the regional groundwater system.

It is normally not possible to evaluate the timescale of paleo-heating from vitrinite reflectance data because it only records the maximum temperature. However, in the case of transient heating, kinetics can be used to estimate the duration of the elevated temperatures (Duddy et al. 1994). To produce the observed anomaly, not only time and temperature but also width of aquifer and velocity of flow have to be taken into account.

The above results are based on a series of comparison or control data, typically used for calibration of thermal history models (Leischner 1994, Radke et al. 1997). These include parameters such as vitrinite reflectance, T_{max} from Rock-Eval pyrolysis, and molecular ratios of phenanthrenes and methylphenanthrenes (i.e., MPI 1, MPI 3, or MPR). Additional thermochronometric data (e.g., sufficient amount of apatite grains for fission track analysis or fluid inclusion data) were not available for the study area.

While each investigated maturity parameter on its own can easily be argued to be deficient, the sum of the parameters indicates that the unusual downhole thermal anomaly trend is due to additional thermal alteration rather than to other factors, such as different lithofacies or kerogen types or secondary migration of hydrocarbons through the Tertiary strata.

ACKNOWLEDGMENTS

We thank BEB Erdgas and Erdöl GmbH, Hannover, for allowing us to view their core material and to collect samples. We also thank Eric Michael, Ken Peters, and Nick Harris for their thorough and helpful reviews.

REFERENCES

Aquilina L, Pauwels H, Genter A, Fouillac C. 1997. Water–rock interaction processes in the Triassic sandstone and the granitic basement of the Rhine Graben: Geochemical investigation of a geothermal reservoir. *Geochimica et Cosmochimica Acta* 61:4281–4295.

Bachu S. 1995. Synthesis and model of formation-water flow, Alberta Basin, Canada. *American Association of Petroleum Geologists Bulletin* 79:1159–1178.

Bachu S. 1997. Flow of formation waters, aquifer characteristics, and their relation to hydrocarbon accumulations, northern Alberta Basin, Canada. *American Association of Petroleum Geologists Bulletin* 81:712–733.

Barroll MW, Reiter M. 1990. Analysis of the Socorro hydrogeothermal system: Central New Mexico. *Journal of Geophysical Research* 95:21949–21963.

Bethke CM, Reed JD, Oltz DF. 1991. Long-range petroleum migration in the Illinois basin. *American Association of Petroleum Geologists Bulletin* 75:925–945.

Bordenave ML, Espitalié J, Leplat P, Oudin JL, Vandenbroucke M. 1993. Screening techniques for source rock evaluation. *In* Bordenave ML (Editor). *Applied Petroleum Geochemistry*: Èditions Technip, Paris. p. 217–278.

Clauser C. 1989. Conductive and convective heat flow components in the Rheingraben and implications for the deep permeability distribution. *In* Beck AE, Garven G, Stegena L (Editors). *Hydrogeological Regimes and their Subsurface Thermal Effects*: American Geophysical Union, Washington, DC. p. 59–64.

Clauser C, Villinger H. 1990. Analysis of conductive and convective heat transfer in a sedimentary basin, demonstrated for the Rheingraben. *Geophysical Journal International* 100:393–414.

Doebl F, Teichmüller R. 1979. Zur Geologie und heutigen Geothermik im mittleren Oberrhein-Graben. *Fortschritte in der Geologie von Rheinland und Westfalen* 27:1–17.

Duddy IR, Green PF, Bray RJ, Hegarty KA. 1994. Recognition of the thermal effects of fluid flow in sedimentary basins. *In* Parnell J (Editor). *Geofluids: Origin, Migration and Evolution of Fluids in Sedimentary Basins*, Special Publication 144: Geological Society of London, London. p. 325–345.

Espitalié J, Laporte JL, Madec M, Marquis F, Leplat P, Paulet J, Boutefeu A. 1977. Méthode rapide de caractérisation des roches mères de leur potentiel pétrolier et de leur degré d'évolution. *Revue de l'Institut Francais du Pétrole* 32:23–42.

Garven G. 1989. Hydrogeologic model for the formation of the giant oil sands deposit of the Western Canadian Sedimentary Basin. *American Journal of Science* 289:105–166.

Garven G, Ge S, Person M, Sverjenski D. 1993. The genesis of stratabound ore deposits in the midcontinent basins of North America: 1. The role of regional groundwater flow. *American Journal of Science* 293:479–568.

Goodhue R, Clayton G. 1999. Organic maturation levels, thermal history and hydrocarbon source rock potential of the Namurian rocks of the Clare Basin, Ireland. *Marine and Petroleum Geology* 16:667–675.

Habermehl MA. 1982. The Eromanga Basin within the Great Artesian Basin. *In* Moore PS, Mount TJ (Editors). *Eromanga Basin Symposium*, Abstracts 5: Geological Society of Australia, Sydney. p. 384–398.

Hullen JB, Goff F, Bortz LC, Bereskin SR. 1994. Geology and geothermal origin of Grant Canyon and Bacon Flat oil fields, Railroad Valley, Nevada. *American Association of Petroleum Geologists Bulletin* 78:596–631.

Illies HJ, Mueller S. 1970. *Graben Problems*: E. Schweizerbart'sche Verlagsbuchhandlung, Stuttgart. 316 p.

Illies JH, Fuchs K. 1974. *Approaches to Taphrogenesis*: E. Schweizerbart'sche Verlagsbuchhandlung, Stuttgart. 460 p.

Kaupe M. 1995. Dreidimensionale Strömungs- und Stofftransportmodellierung in einem quartären Porenaquifer unter Einfluß von Uferfiltration [PhD thesis]: University of Mainz, 173 p.

Kennard JM, Deighton I, Edwards DS, Colwell JB, O'Brien GW, Boreham CJ. 1999. Thermal history modeling and transient heat pulses: New insights into hydrocarbon expulsion and "hot flushes" in the Vulcan Sub-basin, Timor Sea. *Australian Petroleum Production and Exploration Association Journal* 39:177–207.

Kvenvolden KA, Simoneit BRT. 1990. Hydrothermally derived petroleum: Examples from Guyamas Basin, Gulf of California, and Escanaba Trough, northeast Pacific Ocean. *American Association of Petroleum Geologists Bulletin* 74:223–237.

Lampe C. 2001. *The Effects of Hydrothermal Fluid Flow on the Temperature History of the Northern Upper Rhinegraben: Numerical Simulation Studies*, Vol. 8: Kölner Forum für Geologie und Paläontologie, Köln, Germany. 126 p.

Lampe C, Nöth S, Ricken W. 1999. Burial and thermal history of Cenozoic sediments in the northern Rhinegraben: A 1 D-simulation based on well

Nordheim-1. *Zentralblatt für Geologie und Paläontologie Teil I* 10–12:1375–1389.

Lampe C, Person M. 2000. Episodic hydrothermal fluid flow in the Upper Rhinegraben, Germany. *Journal of Geochemical Exploration* 69–70:37–40.

Lampe C, Person M, Nöth S, Ricken W. 2001. Episodic fluid flow within continental rift basins—some insights from field data and mathematical models. *Geofluids* 1:42–54.

Leischner K. 1994. Kalibration simulierter Temperaturgeschichten von Sedimentgesteinen [PhD thesis]: Berichte des Forschungszentrums Jülich, 2909, 309 p.

Lisk M, Brincat MP, O'Brien GW, Eadington PJ, Faiz M. 1999. Palaeohydrology of the Vulcan Sub-basin: Implications for trap integrity. *Australian Petroleum Production and Exploration Association Journal* 39:208–225.

Mailloux B, Person M, Strayer P, Hudleston PJ, Carter S, Dunbar N. 1999. Tectonic and stratigraphic controls on the hydrothermal evolution of the Rio Grande Rift. *Water Resources Research* 35:2641–2659.

Majorowicz JA, Jones FW, Jessop AM. 1986. Geothermics of the Williston Basin in Canada in relation to hydrodynamics and hydrocarbon occurrences. *Geophysics* 51:767–779.

Meier L. 1989. Ein Modell für die Tiefenstruktur und Kinematik im Bereich des Nördlichen Rheingrabens [PhD thesis]: University of Karlsruhe, 146 p.

Muir-Wood R. 1994. Earthquakes, strain-cycling, and the mobilization of fluids. *In* Parnell J (Editor). *Geofluids: Origin, Migration and Evolution of Fluids in Sedimentary Basins*, Special Publication 144: Geological Society of London, London. p. 85–98.

O'Brien GW, Lisk M, Duddy I, Eadington PJ, Cadman S, Fellows M. 1996. Late tertiary fluid migration in the Timor Sea: A key control on thermal and diagenetic histories? *Australian Petroleum Production and Exploration Association Journal* 36:399–427.

O'Brien GW, Lisk M, Duddy IR, Hamilton J, Woods P, Cowley R. 1999. Plate convergence, foreland development and fault reactivation: Primary controls on brine migration, thermal histories and trap breach in the Timor Sea, Australia. *Marine and Petroleum Geology* 16:533–560.

Oliver J. 1986. Fluids expelled tectonically from orogenic belts: Their role in hydrocarbon migration and other geologic phenomena. *Geology* 14:99–102.

Parnell J, Carey PF, Green P, Duncan W. 1999. Hydrocarbon migration history, West of Shetland: Integrated fluid inclusion and fission track studies. *In* Fleet AJ, Boldy SAR (Editors). *Petroleum Geology of Northwest Europe*: Proceedings of the 5th Conference, London. p. 613–625.

Person M, Toupin D, Eadington P. 1995. One-dimensional models of groundwater flow, sediment thermal history and petroleum generation within continental rift basins. *Basin Research* 7:81–96.

Person MA, Garven G. 1992. Hydrologic constraints on petroleum generation within continental rift basins; Theory and application to the Rhine Graben. *American Association of Petroleum Geologists Bulletin* 76:468–488.

Peters KE. 1986. Guidelines for evaluating petroleum source rock using programmed pyrolysis. *American Association of Petroleum Geologists Bulletin* 70:318–329.

Peters KE, Walters CC, Moldowan JM. 2005. *The Biomarker Guide*: Cambridge University Press, Cambridge, UK. 1155 p.

Pflug R. 1982. *Bau und Entwicklung des Oberrheingrabens*: Wissenschaftliche Buchgesellschaft, Darmstadt. 145 p.

Radke M. 1987. Organic geochemistry of aromatic hydrocarbons. *In* Brooks J, Welte DH (Editors). *Advances in Petroleum Geochemistry*: Academic Press, New York. p. 141–207.

Radke M, Horsfield B, Littke R, Rullkötter J. 1997. Maturation and petroleum generation. *In* Welte DH, Horsfield B, Baker DR (Editors). *Petroleum and Basin Evolution*: Springer, Stuttgart. p. 169–230.

Radke M, Welte DH. 1983. The methylphenanthrene index (MPI): A maturity parameter based on aromatic hydrocarbons. *In* Bjorøy M (Editor). *Advances in Organic Geochemistry*: Wiley, Chicester, UK. p. 504–512.

Radke M, Welte DH, Willisch H. 1986. Maturity parameters based on aromatic hydrocarbons: Influence of the organic matter. *Organic Geochemistry* 10:51–63.

Radke M, Welte DH, Willisch H. 1991. Distribution of alkylated aromatic hydrocarbons and dibenzothiophenes in rocks of the Upper Rhine Graben. *Chemical Geology* 93:325–341.

Rojstaczer S, Wolf S. 1992. Permeability changes associated with large earthquakes: An example from Loma Prieta, California. *Geology* 20:211–214.

Rückheim J. 1989. Migrations- und Akkumulationsgeschichte der Erdöle des nördlichen Oberrheingrabens und deren Beziehung zur Diagenese der klastischen Speichergesteine [PhD thesis]: RWTH Aachen, 282 p.

Scharpf H-J. 1977. Erläuterungen zur Geologischen Karte von Hessen 1:25000 Blatt 6316 Worms: Wiesbaden.

Scheidt G, Littke R. 1989. Comparative organic petrology of interlayered sandstones, siltstones, mudstones and coals in the Upper Carboniferous Ruhr Basin, Northwest Germany, and their thermal history and methane generation. *Geologische Rundschau* 78:375–390.

Schmitt M. 1992. Ursache und Ablauf der Versalzung der oberen Grundwasserleiter im nördlichen Oberrheingraben [PhD thesis]: University of Darmstadt (TH), Darmstadt.

Sonne V. 1972. Jungtertiäre Ablagerungen ("Aquitan") am Nordwestrand des Mainzer Beckens. *Mainzer Geowissenschaftliche Mitteilungen* 1:137–142.

Summer NS, Verosub KL. 1987. Maturation anomalies in sediments underlying thick volcanic strata, Oregon: Evidence for a thermal event. *Geology* 15:30–33.

Sweeney JJ, Burnham AK. 1990. Evaluation of a simple model of vitrinite reflectance based on chemical kinetics. *American Association of Petroleum Geologists Bulletin* 74:1559–1570.

Toth J, Otto CJ. 1989. Hydrogeology and oil deposits at Pechelbronn-Soultz, Upper Rhine Graben; Ramifications for exploration in intermontane basins. *In* Thanasuthipitak T, Ounchanum P (Editors). *Proceedings of the International Symposium on Intermontane Basins; Geology and Resources*: Chiang Mai University, Thailand. p. 59–76.

Toupin D, Eadington PJ, Person M, Morin P, Wieck J, Warner D. 1997. Petroleum hydrogeology of the Cooper and Eromanga basins, Australia: Some insights from mathematical modeling and fluid inclusion data. *American Association of Petroleum Geologists Bulletin* 81:577–603.

Willet SD, Chapman DS. 1989. Temperatures, fluid flow, and the thermal history in the Uinta Basin. *In* Beck AE, Garven G, Stegena L (Editors). *Hydrogeological Regimes and their Subsurface Thermal Effects*: American Geophysical Union, Washington, DC. p. 29–33.

Yassir N, Otto CJ. 1997. Hydrodynamics and fault seal assessment in the vulcan sub-basin, Timor Sea. *Australian Petroleum Production and Exploration Association Journal* 37:380–389.

Ziagos JP, Blackwell DD. 1986. A model for the transient temperature effects of horizontal fluid flow in geothermal systems. *Journal of Volcanology and Geothermal Research* 27:371–397.

THERMAL MATURATION HISTORY OF ARCTIC ALASKA AND THE SOUTHERN CANADA BASIN

DAVID W. HOUSEKNECHT AND W. MATTHEW BURNS
US Geological Survey, Reston, Virginia 20192 USA
e-mail: dhouse@usgs.gov

AND

KENNETH J. BIRD
US Geological Survey, Menlo Park, California 94025 USA

ABSTRACT: The emerging global focus on the oil and gas potential of the Arctic underscores the importance of understanding petroleum systems with limited data. Geohistory modeling of Arctic Alaska (including the Chukchi shelf) and the southern Canada basin indicates that regional patterns of thermal maturity and timing of petroleum generation reflect geologic processes associated with rift-opening of the Canada basin and collision orogenesis along the Brooks Range–Herald arch from Jurassic through Tertiary time. The base of the Cretaceous–Tertiary Brookian sequence provides a regional reference horizon because most oil generation occurred as the result of Brookian burial.

In Arctic Alaska, basal Brookian strata on the Beaufort rift shoulder grade from immature in the west to overmature in the east. From the crest of the rift shoulder, thermal maturity of basal Brookian strata increases southward into the oil window on the north flank of the Colville foreland basin and into the gas window in the foredeep. A >200-mile-wide area of immature to mature strata in the Chukchi Sea narrows eastward as the Brooks Range converges with the rift shoulder in the eastern North Slope. These patterns reflect generally low Jurassic to Tertiary sediment accommodation on the rift shoulder, large Cretaceous sediment accommodation in the Colville foredeep, and northward impingement of the Brooks Range onto the eastern part of the rift shoulder during the Tertiary.

Fewer geologic data in the Canada basin increases the uncertainty of modeling. Projection of stratigraphy from the rift shoulder, reconstruction of regional sediment dispersal patterns, and consideration of source rocks in Arctic Alaska and Canada indicate the potential for four source rocks in the Cretaceous and Paleogene. Model results indicate that all four source rocks are mature or overmature across much of the southern Canada basin. The highest thermal maturity occurs in depocenters immediately north of the rift shoulder and on the eastern margin of the study area, which is the distal Mackenzie delta. The lowest thermal maturity occurs at the northern limit of modeling, more than 200 miles north of the rift shoulder and on the western margin of the study area, adjacent to the Chukchi borderland. A potential source rock in the Lower Cretaceous likely matured during the Early Cretaceous in a western depocenter related to sediment by-pass of the Chukchi shelf, but maturation of all source rocks elsewhere occurred during the Paleogene when large volumes of sediment were shed from the Brooks Range and through the Mackenzie delta.

KEY WORDS: Arctic Alaska, Canada basin, thermal maturation, oil generation

INTRODUCTION

Arctic Alaska, its continental shelves, and adjacent parts of the Canada basin hold significant potential for future oil and gas discoveries. After more than a half century of exploration and the discovery of North America's largest oil field at Prudhoe Bay, a relatively small part of the region is thoroughly explored because of technical, economic, environmental, and policy limitations. Thus, large expanses of onshore Arctic Alaska and most of the adjacent offshore remain virtually unexplored.

As technological and economic barriers to exploration become less daunting, and as global policymakers consider environmental and sovereignty issues that will determine the future scope of petroleum exploration in the Arctic, regional petroleum systems have emerged as a critical factor that will influence both exploration and policy decisions. The objective of this article is to present the results of regional modeling of burial and thermal maturation history that influence the petroleum resource potential of this region.

This article is organized as follows: (1) We present a synthesis of the geology and tectonic evolution of the region; (2) We infer the stratigraphy and source-rock character of unexplored parts of the region; (3) We describe the methods used for thermal history modeling; (4) We analyze heat flow, including considerations of constraining data, temporal variation related to geologic history, and implications to regional modeling; (5) We model the thermal and hydrocarbon-generation history of exploration wells for which empirical thermal maturity data are available; (6) We model the thermal and hydrocarbon-generation history of pseudowells located in unexplored parts of the region; (7) We present the thermal maturation history of the entire region by integrating information from exploration wells and pseudowells; and (8) We draw conclusions regarding the hydrocarbon potential of the region based on this work.

GEOLOGIC SETTING

Stratigraphy and Geologic History

The study area includes Arctic Alaska north of the Brooks Range, the US Chukchi Sea, the southern and eastern parts of the Chukchi borderland, the US Beaufort Sea, and the southern part of the Canada basin (Fig. 1). The study area is divided into two geologically distinct provinces by the Beaufort rift shoulder (Fig. 1), which formed during Jurassic to Early Cretaceous opening of the Canada basin. The Arctic Alaska microplate (including Arctic Alaska and the Chukchi shelf) is thought to have rifted from Arctic Canada, probably by counterclockwise rotational opening of the Canada basin (Grantz et al. 1990b, 2007), although alternative interpretations exist. The rift shoulder is defined by high-standing acoustic basement that generally dips gently southward beneath Arctic Alaska and the Chukchi shelf and that steps northward across normal faults to great depths beneath the Beaufort

Analyzing the Thermal History of Sedimentary Basins: Methods and Case Studies
SEPM Special Publication No. 103, Copyright © 2012
SEPM (Society for Sedimentary Geology), Print ISBN 978-1-56576-315-9, CD/DVD ISBN 978-1-56576-317-3, p. 199–219.

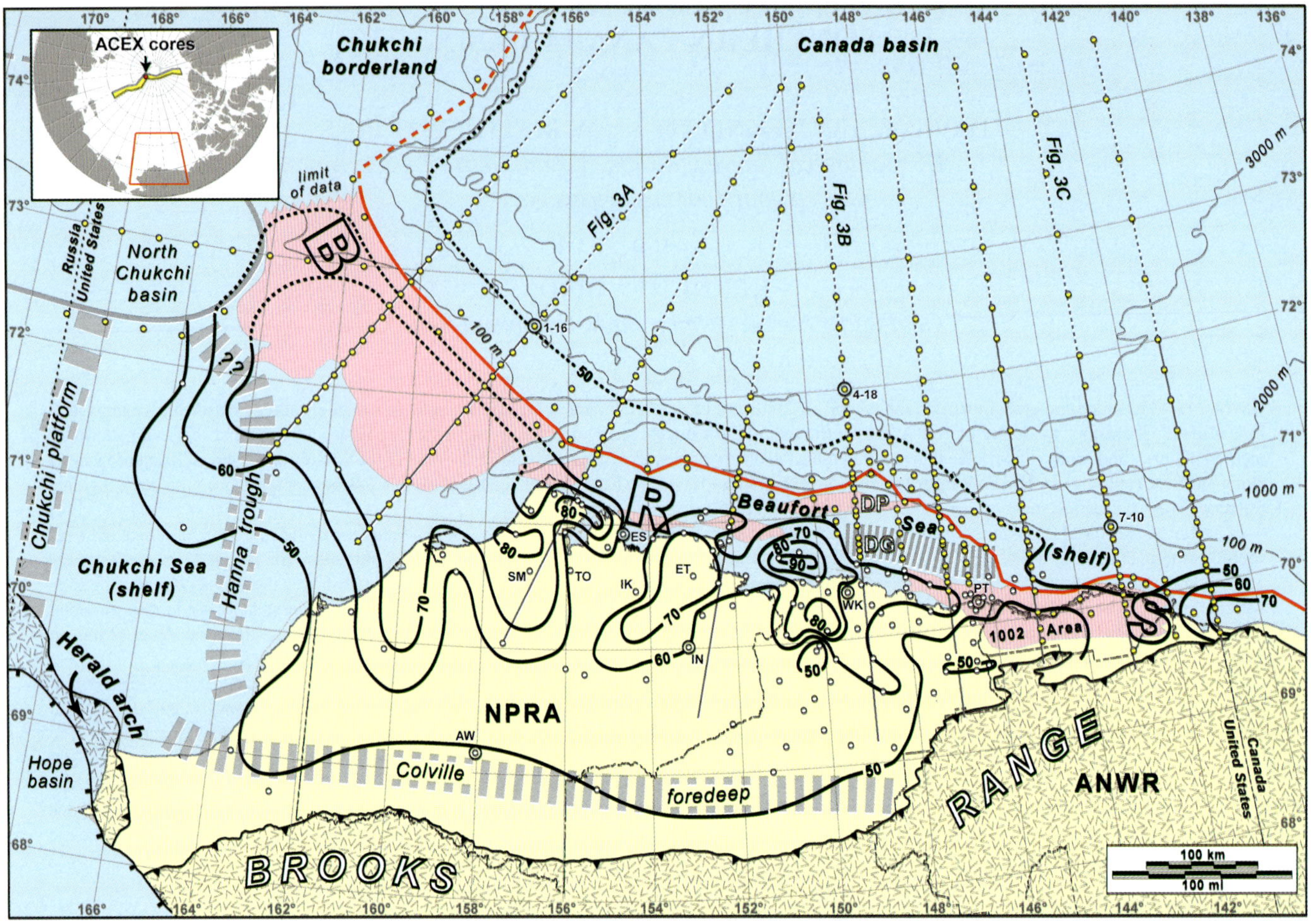

FIG. 1.—Map of Arctic Alaska and southern Canada basin showing main geologic elements and modeling parameters. Heavy red line, Beaufort hingeline; heavy dashed red line, inferred hingeline of Chukchi borderland; pink areas, Franklinian basement subcrops LCU; block letters "BRS," trend of the Beaufort rift shoulder defined by hingeline and Franklinian subcrop. Straight gray lines, locations of schematic cross sections used for modeling Beaufort rift shoulder and Canada basin, solid where constrained with seismic data, dashed where not, labeled lines shown in Figure 3; white dots with black rims, exploration wells used for modeling; yellow dots with black rims, pseudowells used for modeling; yellow and black contours, heat flow estimated from depth-VR data, dashed where inferred from geologic relations; thin black contours, bathymetry in meters. Circled and labeled wells shown in Figures 6 and 7 include the following: AW, Awuna #1; ES, East Simpson #1; IN, Inigok #1; PT, Pt. Thomson #1; WK, West Kuparuk #3-11-11. Circled and labeled pseudowells shown in Figure 10. Other labeled wells mentioned in text include the following: ET, East Teshekpuk #1; IK, Ikpikpuk #1; SM, South Meade #1; TO, Topagoruk #1. Broad dashed gray lines depict approximate trends of Chukchi platform, Hanna Trough, Colville foredeep, and Dinkum graben (DG). DP, Dinkum plateau; random V pattern, tectonic belts. ANWR, Arctic National Wildlife Refuge; 1002 Area, ANWR coastal plain; NPRA, National Petroleum Reserve in Alaska. Inset map shows location of study area in red polygon, Lomonosov Ridge (polygon with black outline and yellow fill and location of ACEX cores [red dot with black rim]).

shelf. The crest of the rift shoulder is commonly referred to as the Barrow arch. Acoustic basement of the Arctic Alaska microplate comprises mainly pre-Mississippian low-grade metamorphic rocks (Fig. 2, Franklinian sequence; Dumoulin 2001). Basement north of the rift shoulder likely comprises highly attenuated continental crust that grades northward to oceanic crust beneath the floor of the Canada basin (Grantz et al. 1990b, 2007).

Arctic Alaska displays the overall geometry of a foreland basin, which formed during Jurassic to Tertiary time in response to collision-induced tectonism along the Brooks Range and Herald arch thrust belts (Fig. 1; Bird and Molenaar 1992, Moore et al. 1994). Franklinian basement is overlain by Mississippian to Jurassic strata (Fig. 2) deposited prior to (Ellesmerian sequence) and during (Beaufortian sequence) rift-opening of the Canada basin. The older part of the Ellesmerian sequence locally includes thick graben-filling successions. The Beaufortian and Ellesmerian sequences are truncated progressively northward beneath the Lower Cretaceous unconformity (LCU; Fig. 2), which is considered to be the breakup unconformity formed during rift-shoulder uplift, and Franklinian basement subcrops the LCU in areas of maximum uplift (Fig. 1). Overlying the LCU, Lower Cretaceous to Tertiary foreland basin strata (Brookian sequence) generally thicken southward into the foredeep and thin northward by onlap against the rift shoulder. The thickest part of the foreland-basin fill grades from Early Cretaceous beneath the western Alaska North

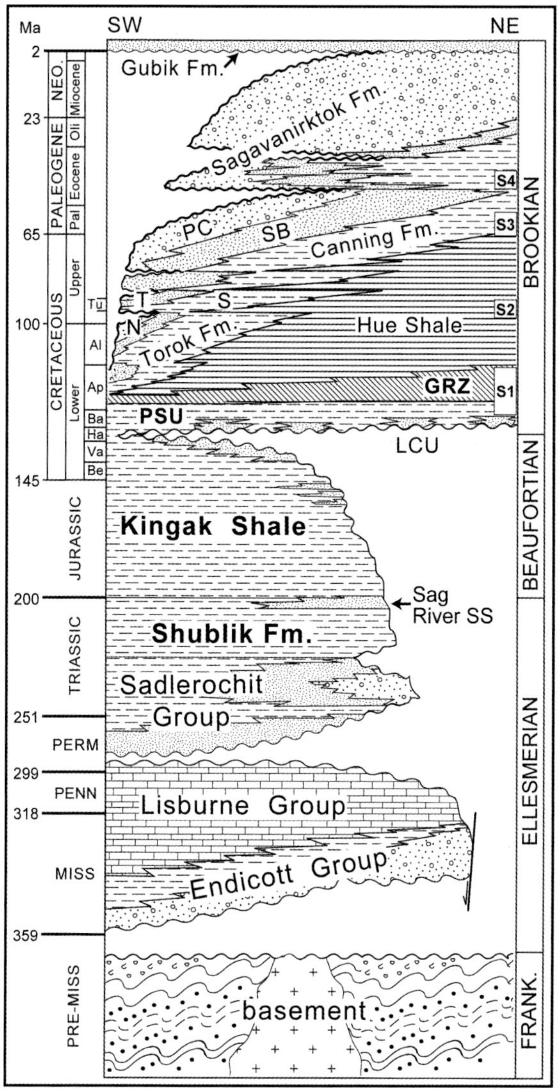

FIG. 2.—Generalized chronostratigraphy of Arctic Alaska. Series names are shown only for Cretaceous, Paleogene, and Neogene systems. Only stage names pertinent to text are shown. Known source rocks shown in bold include Shublik Formation, Kingak Shale, pebble shale unit (PSU), and gamma-ray zone (GRZ) of Hue Shale. Potential source rocks in Canada basin: S1, Lower Cretaceous; S2, Upper Cretaceous; S3, lower Paleogene; S4, upper–lower Paleogene (~*Azolla*). LCU, Lower Cretaceous unconformity; Be, Berriasian; Va, Valanginian; Ha, Hauterivian; Ba, Barremian; Ap, Aptian; Al, Albian; Tu, Turonian. Arctic Alaska stratigraphy modified from Bird (1985, 2001) and Mull et al. (2003); ages from Gradstein et al. (2004).

Slope (ANS) to Tertiary beneath the eastern ANS, and the Beaufort rift shoulder was progressively overstepped and buried from west to east by foreland-basin depositional systems.

Although the Chukchi shelf is stratigraphically similar to the ANS, the presence of a Paleozoic failed rift (Hanna Trough) filled by a thick succession of the Ellesmerian sequence is a distinguishing feature (Sherwood et al. 2002). Ellesmerian strata in the Hanna Trough thin eastward and grade into the passive margin succession of the ANS and thin westward to an onlap pinchout against the Chukchi platform, an ancestral ridge of Franklinian basement (Fig. 1). The Beaufortian sequence beneath the Chukchi shelf is similar to that of the ANS, except that it is punctuated by several unconformities that appear to be more significant than those to the east. The Brookian sequence beneath the Chukchi shelf comprises Lower Cretaceous (mostly Aptian to Albian) and Tertiary successions separated by a significant unconformity. The Lower Cretaceous succession displays a regional geometry indicating influences of both a sag basin developed above the Hanna Trough and a foreland basin related to the Herald arch and western Brooks Range. Lower Cretaceous strata grade eastward into the foreland basin fill of the ANS and—to the northeast—overstep the Beaufort rift shoulder to form a sediment wedge in the southwestern corner of the Canada basin. The Tertiary succession, in contrast, is thickest in a number of syn-sedimentary grabens that open northward into the deep (>30,000-foot) North Chukchi basin (Fig. 1; Lothamer 1994). South of the rift shoulder, no apparent stratigraphic continuity exists between the Tertiary successions beneath the Chukchi shelf and the eastern ANS. Evidence indicates broad uplift of the western ANS during the Paleogene, segmenting the foreland into a Chukchi depocenter influenced by wrench tectonics and an eastern ANS depocenter influenced by contractional tectonics associated with the eastern Brooks Range.

The stratigraphy north of the rift shoulder is poorly known. Jurassic to Lower Cretaceous syn-rift deposits (Beaufortian sequence) have been penetrated by drilling and are evident in seismic data along parts of the rift shoulder. However, any syn-rift deposits to the north are deeply buried beneath an expanding wedge of Brookian strata, which thicken dramatically by growth-faulting north of the rift shoulder (Fig. 3). Data from exploration wells and seismic surveys indicate that significant volumes of sediment initially overstepped the rift shoulder during the Albian along the northeast Chukchi and western ANS margin and during the Paleogene along the central and eastern ANS. Voluminous sediment influx from the Mackenzie delta was the predominant regional influence on the Cenozoic succession in the Canada basin (Grantz et al. 1990b). Although no direct evidence exists for Hauterivian to Aptian deposits in the Canada basin, the presence of generally condensed mudstones of this age on the rift shoulder and to the south indicates that similar deposits are likely present (Fig. 3).

An exception to this general pattern occurs in eastern Alaska, where Brooks Range deformation during the Tertiary migrated northward across the foreland basin, impinged upon the Beaufort rift shoulder, and propagated northward beneath the Beaufort shelf; this deformation remains active, as indicated by earthquakes beneath the coastal plain of the Arctic National Wildlife Refuge (ANWR) and adjacent shelf. Thus, beneath the Beaufort shelf of eastern Alaska, growth-faulted Tertiary strata deposited north of the rift shoulder are deformed by contraction associated with the Brooks Range orogenic front (Grantz and May 1982, Grantz et al. 1990a, Dietrich and Lane 1992, McMillen and O'Sullivan 1992).

The Chukchi borderland (Fig. 1) is a fragment of attenuated continental crust that separated from the Arctic Alaska microplate either during or following opening of the Canada basin (Grantz et al. 1998, 2007). The stratigraphy of the Chukchi borderland is thought to include Ellesmerian strata similar to those of the conjugate margin along the Arctic Islands of Canada (Sverdrup basin), as well as a thin succession of Beaufortian- and Brookian-aged strata (Grantz et al. 1998) deposited as the borderland became increasingly isolated from regional sediment dispersal systems.

Rift-Shoulder Geometry and Evolution

The geometry of the Beaufort rift shoulder varies along strike, as indicated by the Franklinian subcrop beneath the LCU (Fig. 1). Two

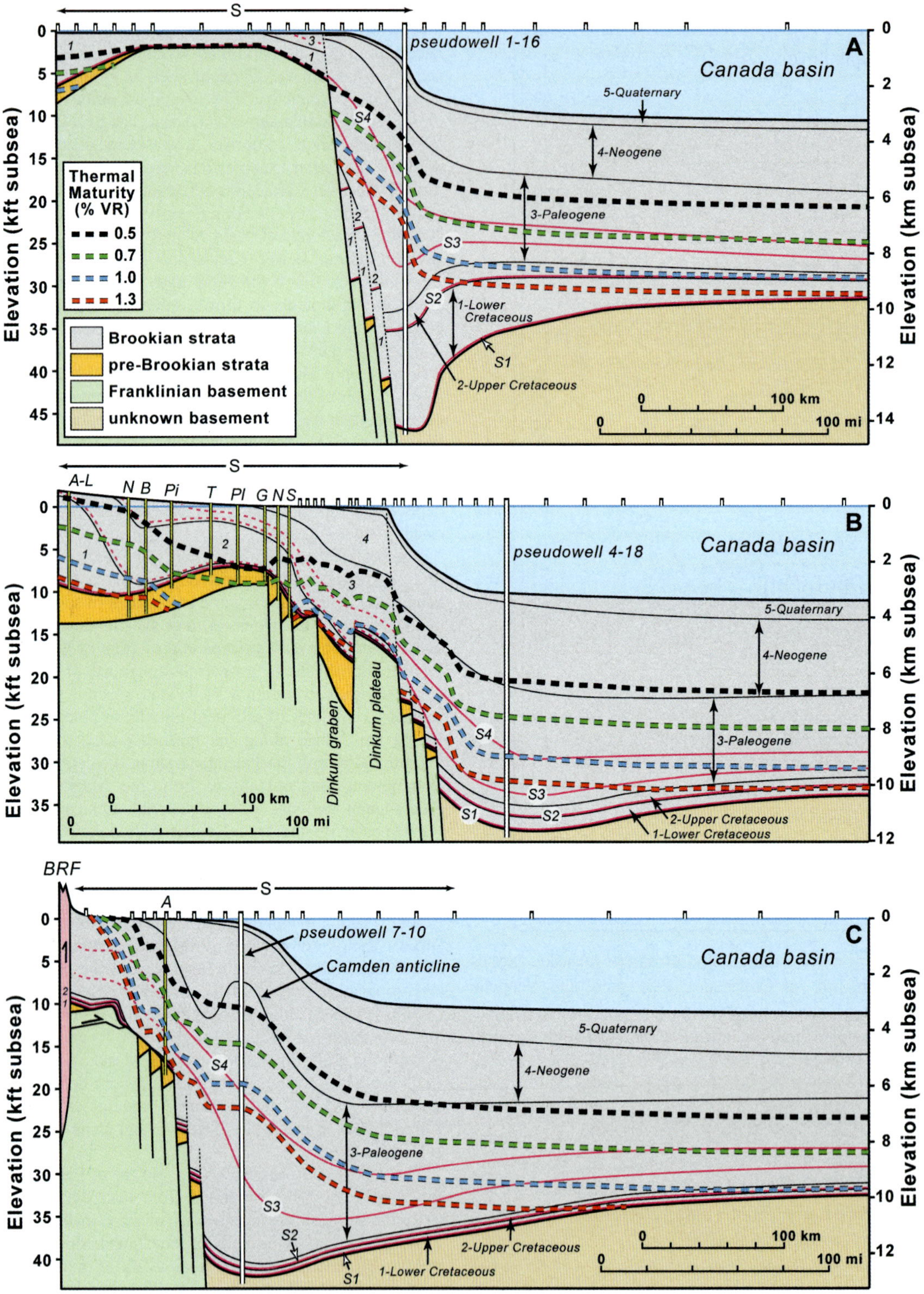

FIG. 3.—Generalized representative cross sections across Beaufort rift shoulder and southern Canada basin showing simplified stratigraphy, potential source rocks, and vitrinite reflectance contours calculated using kinetics of Burnham and Sweeney (1989). Locations of pseudowells shown by black-and-white ticks; labeled pseudowells shown in Figure 10. The following exploration wells indicated by black-and-yellow lines; A-L, Aufeis and Lupine; N, Nora; B, Bush; Pi, Pipeline; T, Toolik 2; Pl, Placid; G, Gwydyr Bay State 1; N, Northstar 1; S, Sandpiper 1; A, Aurora. Key to Brookian strata: 1, Lower Cretaceous; 2, Upper Cretaceous; 3, Paleogene; 4, Neogene; 5, Quaternary. Source rock intervals are labeled S1, S2, S3, and S4, as in Figures 2 and 10. Double-headed arrows labeled "S" show extent of seismic control along each cross section. BRF, Brooks Range front. Locations of cross sections shown in Figure 1.

areas along the rift shoulder, west of ~157° W (northeast Chukchi shelf) and east of ~147° W (ANWR coastal plain), are characterized by large and continuous basement subcrops, indicating that these were areas of broad uplift and erosion during rifting. Seismic data in these areas display relatively little faulting south of a distinct "hingeline," a narrow zone of northward-downstepping normal faults across which basement drops to depths below seismic resolution (Figs. 1, 3A, 3B). In contrast, the central part of the rift shoulder (~157° W to ~147° W) is characterized by a broad zone of normal faults with numerous graben basins, and basement subcrops the LCU only on relatively small and isolated horsts (Figs. 1, 3B). The Dinkum graben (DG in Fig. 1) is best documented (Hubbard et al. 1987), although seismic data and exploration drilling reveal the presence of many other grabens and horsts in this central area. The stratigraphy of the grabens indicates that normal faulting and sedimentation were coeval, with two distinct phases of extension in the Early Jurassic and the Late Jurassic to Earliest Cretaceous (Hubbard et al. 1987).

Along the entire rift shoulder, the LCU is capped by a transgressive systems tract of Hauterivian to Barremian age (localized shallow-marine sandstone overlain by marine mudstone of the pebble shale unit; Fig. 2); the overlying stratigraphy is indicative of relatively deep-marine conditions (condensed marine mudstone of the Hue Shale overlain by turbidites at the toe of clinoforms of the Torok and Canning formations) during the remainder of the Early Cretaceous (Molenaar 1983, Bird and Molenaar 1992, Houseknecht and Schenk 2001, Mull et al. 2003), likely indicating subsidence of the rift shoulder due to thermal contraction. The eastern part of the rift shoulder further subsided during the Paleogene, apparently as the result of tectonic loading by the northward advancing Brooks Range, and perhaps by loading of sediment shed from the growing uplifts (Coakley and Watts 1991, Moore et al. 1994, Potter et al. 2004). The resultant eastward tilting of the rift shoulder was sufficient to spill much of the trapped oil from the Prudhoe Bay accumulation (Jones and Speers 1976, Masterson 2001). There also is evidence of Paleogene uplift along the western part of the rift shoulder (west of ~156° W), although the cause of this uplift is unknown.

The evolution of the Beaufort rift shoulder exerted a significant influence on sediment dispersal and accommodation during deposition of the Brookian sequence. In the west, Lower Cretaceous strata were deposited on the rift shoulder and in the Canada basin to the north. Renewed uplift of the rift shoulder during the Paleogene resulted in truncation of much of the Lower Cretaceous section above the rift shoulder and deposition of thick Paleogene strata on the north flank of the rift shoulder (Fig. 3A). The central part of the study area lies mostly east of the Early Cretaceous depocenter (Bird and Molenaar 1992), and Brookian strata above the rift shoulder include Upper Cretaceous, Paleogene, and Neogene deposits (Fig. 3B). Seismic evidence indicates that most sediment deposited north of the rift shoulder is Cenozoic in age (Fig. 3B). In the east, Cretaceous deposits are thin, and most of the Brookian sequence is Cenozoic (Fig. 3C). Thick Paleogene deposits were accommodated above the rift shoulder, and even thicker, coeval deposits accumulated in the Canada basin to the north (Fig. 3C).

Potential Source Rocks

Oil exploration in Arctic Alaska has focused on petroleum systems related to three main source rocks: the Triassic Shublik Formation, the Jurassic lower Kingak Shale, and the Lower Cretaceous gamma-ray zone (GRZ) of the Hue Shale (Bird 1985, 2001). Although all of these continue to be important for exploration south of the Beaufort rift shoulder, the Shublik Formation and Kingak Shale likely are absent in most areas north of the rift shoulder, as the Canada basin had not yet opened when they were deposited, although both may occur in isolated rift basins immediately north of the rift shoulder.

Therefore, analysis of petroleum systems north of the rift shoulder must focus on potential source rocks in Brookian strata and must be based on indirect evidence from areas adjacent to the Canada basin. Three stratigraphic intervals appear to have source-rock potential, as follows.

(1) The Lower Cretaceous pebble shale unit (PSU) and GRZ (also known as highly radioactive zone) of the Hue Shale together represent a condensed section that occurs at the base of the Brookian sequence throughout Arctic Alaska. These units generally are leaner (lower total organic carbon, TOC) and more gas-prone (lower hydrogen index, HI) in the west (Chukchi shelf and western ANS) and south (near Brooks Range) and richer (higher TOC) and more oil-prone (higher HI) toward the east (eastern ANS) and north (rift shoulder) (Peters et al. 2006). Across Arctic Alaska, the PSU and GRZ range from ~1 to >5% TOC, contain a mixture of type II and III kerogen, and are considered oil-prone over wide areas (Masterson 2001, Peters et al. 2006). Petroleum accumulations in Arctic Alaska that have been documented to contain oil sourced either exclusively or partly from these basal Brookian source rocks include Endicott, Prudhoe Bay, and Tarn (Wicks et al. 1991, Masterson 2001, Peters et al. 2008).

The age of the PSU and GRZ (Hauterivian to Albian) includes the time of Canada basin seafloor spreading, which is inferred to have ended no later than Barremian (Grantz et al. 2007). Thus, even though the Canada basin was the most distal, and perhaps deepest marine, part of the depositional system in which these strata were deposited, source-rock properties may have been influenced by seafloor spreading in the middle of the Canada basin and by thermal contraction of the rift shoulder along the southern margin of the basin. Erosion of sediment from horsts or uplifted margins of half-grabens in the Canada basin and overstepping of the subsiding Beaufort rift shoulder by depositional systems associated with the Colville foreland basin may have diluted the organic content of strata deposited in the older part of this succession. Therefore, the younger (post-Barremian) strata may have the best source-rock potential. Based on this line of reasoning, we include in our modeling the potential for a Lower Cretaceous source rock containing type II kerogen at the base of the Brookian section in the Canada basin.

(2) The lower part of the Upper Cretaceous section (mostly Turonian-aged strata) includes a condensed section that displays source-rock quality in Arctic Alaska (Seabee Formation) and Arctic Canada (Boundary Creek, Smoking Hills, and Kanguk formations). These formations commonly range from 2 to >10% TOC, contain a mixture of type II and III kerogen, and are oil-prone across wide areas (Snowdon 1980, Dixon et al. 1992, Dixon 1996, Threlkeld et al. 2000). The Seabee and Kanguk formations are immature beneath the ANS and the Canadian Arctic Islands, respectively, and have not been inferred as the source of any known oil accumulations. The Boundary Creek–Smoking Hills Formation is inferred to have sourced the Imnak, Mayogiak, Kugpik, and Atkinson oil accumulations in the Mackenzie delta (Snowdon 1980, Dixon 1996).

The widespread occurrence of Turonian source rocks in Alaska and Canada, south of the Beaufort rift shoulder, indicates that coeval source rocks also may be present north of the rift shoulder. We therefore include in our modeling the potential for a source rock containing type II kerogen in the lower part of the Upper Cretaceous section north of the rift shoulder.

(3) The lower Paleogene is a condensed section on Lomonosov Ridge, where it was cored in 2004 during the International Ocean Drilling Program Arctic Coring Expedition (ACEX cores, Fig. 1). Analyses of ACEX cores indicate the Arctic Ocean was a warm, ice-free, brackish, and bioproductive basin during the early Paleogene (Sluijs et al. 2006, Stein et al. 2006), a time of globally high temperature and generally high sea level. The ACEX cores include a Paleocene to Early–Middle Eocene succession that is organic-rich (TOC mainly measuring 1–5%, up to 14%) and locally contains oil-prone (type II) kerogen (Stein 2007). Although most of the Paleocene

kerogen in the ACEX cores is gas-prone, it was deposited prior to and during rifting of the Lomonosov Ridge away from Asia, and it is possible that the kerogen is mostly terrigenous organic matter derived from the continent. We surmise that coeval strata in the more distal reaches of the Canada basin, which had fully opened by the Barremian (Grantz et al. 1990b, 2007), may be richer in organic matter and may include oil-prone kerogen. An Early–Middle Eocene *Azolla* fern facies occurs at the top of the organic-rich succession in the ACEX cores, indicating cooler temperatures and probably fresh surface water (seasonal) in the Arctic Ocean (Brinkhuis et al. 2006). This *Azolla* facies is both organic-rich (mainly 2–5% TOC) and oil-prone (Stein 2007). Thus, evidence from the ACEX cores indicates the potential for organic-rich—and at least partly oil-prone (type II kerogen)—in lower Paleogene strata in the Canada basin.

Numerous oil occurrences along the Beaufort shelf of eastern Alaska and in the Mackenzie delta of Canada have been typed to Tertiary-aged kerogen using biomarker geochemistry (Magoon et al. 1999, Snowdon et al. 2004), and yet viable source rocks of this age have not been penetrated by drilling. It is possible that the widely occurring Tertiary oils were generated from early Paleogene source rocks—likely coeval with those cored on Lomonosov Ridge—deeply buried beneath Brookian strata north of the Beaufort rift shoulder. Although direct evidence to support this inference is lacking, we assume in our modeling the presence of two source rocks, one near the base of the Paleogene and one in the upper part of the lower Paleogene (~*Azolla*).

METHODS

The regional geologic framework for this study is based on data from more than 400 exploration wells and 50,000 miles of two-dimensional seismic data, as well as published reports. Formation tops from 112 exploration wells (Fig. 1) were used for burial history modeling; formation tops of strata below the total depth of each well were estimated from seismic data. Geologic ages were constrained using published ages of stratigraphic units (Fig. 2) and biostratigraphic and radiometric data. The timescale of Gradstein et al. (2004) was used. Laterally gradational ages were assigned to reflect the time-transgressive nature of certain stratigraphic units (for example, progradational clinoform sequences; Houseknecht and Schenk 2001, Houseknecht et al. 2009). Formation lithology was estimated by quantitative analysis of wire-line logs and digital mud logs (Hayba et al. 2004, Hayba and Burns 2006). Estimates of the timing and magnitude of uplift and erosion events for each well were based on regional isopach and structural maps (Saltus and Bird 2003), sonic-log analysis (Burns et al. 2007), and apatite fission track analysis (O'Sullivan 1993, 1996, 1999; O'Sullivan et al. 1993, 1997).

Approximately 300 pseudowells were generated to provide burial history constraints for parts of the study area where data from exploration wells are sparse or unavailable. Most pseudowells are located offshore, where few geologic constraints exist. A series of highly generalized cross sections was constructed extending from Arctic Alaska into the Canada basin (Figs. 1, 3) using public domain digital topographic and bathymetric data, regional maps of key surfaces (Saltus and Bird 2003), total sediment isopach maps of the Canada basin (May and Grantz 1990, Jackson 1990, Jackson and Oakey 1990), and reflection seismic data—most of which are restricted to southern parts of the cross sections (Figs. 1, 3). Thickness and geometry of several stratigraphic units were inferred based on projection of well and seismic data and on regional considerations of sediment dispersal patterns and isopach trends. In addition, the stratal positions of four potential source-rock horizons were estimated, based primarily on known and inferred ages. Although the stratigraphic successions and potential source-rock horizons in the Canada basin were interpreted based on all available data and consistent regional relations, they must be considered highly speculative.

Vitrinite reflectance (VR) data from 95 wells and hundreds of outcrops were used to constrain thermal maturity patterns. Most of the VR data from exploration wells were collected by a small number of commercial, petroleum-industry laboratories and analyzed and published elsewhere (Claypool and Magoon 1985; Magoon et al. 1987; Bayliss and Magoon 1988; Johnsson et al. 1993, 1999; Johnsson and Howell 1996). Additional VR data from exploration wells, also collected by commercial laboratories, were compiled from public-domain well files available from the Alaska Department of Natural Resources and the US Minerals Management Service. Although these data no doubt include variation inherent in most VR data sets, the small number of commercial laboratories involved in collecting the data and the spatial and vertical consistency in VR patterns indicate that they are reliable for our regional thermal maturity modeling. Claypool and Magoon (1985) found that interlaboratory variability in VR measurements in samples from wells in the National Petroleum Reserve in Alaska (NPRA) was ±15%.

Bottom-hole temperature (BHT) data from 56 wells were used to estimate current heat flow. Some of the BHT data previously were corrected using a Horner-plot procedure (Blanchard and Tailleur 1982; Deming et al. 1992, 1996; Deming 1994). Bottom-hole temperature data from additional wells were corrected by us using a Horner-plot technique.

Current surface temperature for each model location was estimated from published information, including the data of Lachenbruch et al. (1982, 1988) for onshore locations and the data of Shimada et al. (2005) for offshore locations, where the Arctic halocline results in little variation in temperature with water depth. Past surface temperatures (Fig. 4) were estimated from paleoclimate and paleolatitude data, including those of Askin and Spicer (1995), Carter and Hillhouse (1992), Spicer and Corfield (1992), Spicer and Parrish (1990), Basinger et al. (1994), Golonka et al. (1994), Nelson and Carter (1992), and Sluijs et al. (2006). Sea-level data used in the burial history models include data from Miller et al. (2005), extracted at 5-my intervals for 65 to 0 Ma and at 10-my intervals for 170 to 65 Ma, and from the long-term curve of Haq and Al-Qahtani (2005) extracted at 10-my intervals for 400 to 170 Ma (Fig. 4). Site-specific estimates of water depth through time were based on our published and ongoing studies of Arctic Alaska seismic stratigraphy and facies (e.g., Houseknecht and Bird 2004, Houseknecht et al. 2009). Particular emphasis was placed on variation in temperature, sea level, and water depth during the past 120 my, the time interval of peak maturation for strata of interest in this study.

All data discussed above were integrated and analyzed using a commercial burial history modeling application (BasinMod 1-D*). Depth-VR trends were calculated using "Easy %R_o" kinetics (Burnham and Sweeney 1989, Sweeney and Burnham 1990) and compared to plots of measured depth-VR. Generation of oil was modeled using either kinetic values for specific source rocks (Lillis et al. 1999, Masterson 2001) or default kinetic values for inferred kerogen types (Table 1).

Following development of one-dimensional (1-D) models for individual wells and pseudowells, regional integration of model results was conducted using software (BasinView*) to batch-process all 1-D models and construct contour maps of calculated results for various stratigraphic horizons at specified time steps.

BURIAL AND THERMAL HISTORY MODELING

Heat Flow Analysis

The initial step in modeling the thermal history of this region was analysis of current and past heat flow, which involved several steps, as follows:

*[Note: Use of trade names in this publication is for descriptive purposes only and does not constitute endorsement by the US Government.]

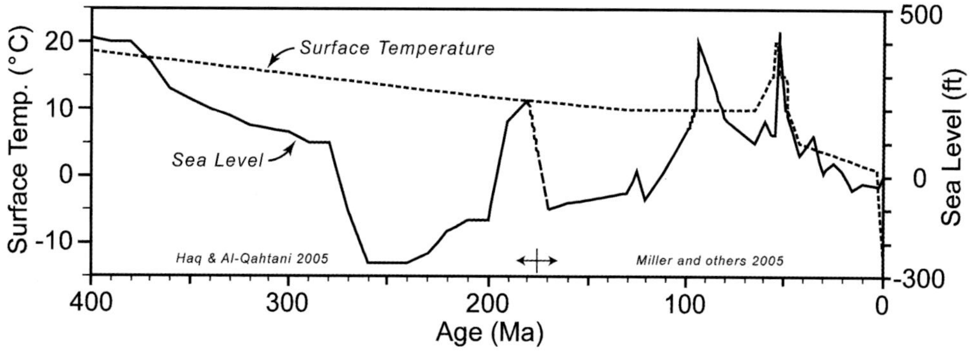

FIG. 4.—Plots of age–surface temperature and age–sea level used in burial history models. Estimates of ancient surface temperature based on Askin and Spicer (1995), Carter and Hillhouse (1992), Spicer and Corfield (1992), Spicer and Parrish (1990), Basinger et al. (1994), Golonka et al. (1994), Nelson and Carter (1992), and Sluijs et al. (2006). Sea-level data from Miller et al. (2005), extracted at 5-my intervals for 65 to 0 Ma and at 10-my intervals for 170 to 65 Ma, and from the long-term curve of Haq and Al-Qahtani (2005), extracted at 10-my intervals for 400 to 170 Ma.

(1) We used BHT data from 56 exploration wells across the North Slope and the Chukchi and Beaufort shelves to compare heat-flow estimates calculated using two different algorithms for addressing the effects of compaction on thermal conductivity—and therefore on estimates of heat flow based on BHT data. The result is a positive, linear correlation (Fig. 5A) between estimates of heat flow at the base of the sedimentary section determined using a "fluid-flow compaction" algorithm (porosity calculated as a function of both pressure and depth, taking into account compaction disequilibrium due to overpressure) and a "mechanical compaction" algorithm (porosity calculated as a function of depth using the empirical relation of Sclater and Christie [1980]). The "mechanical compaction" algorithm yields values that are larger—by an average of 7%—than those calculated using the "fluid-flow compaction" algorithm. In essence, the "mechanical compaction" algorithm tends to estimate higher compaction, lower porosity, higher thermal conductivity, and higher basal heat flow (although these parameters are dependent on the rate of sediment accumulation, which varies both temporally and spatially across study area, and on lithology) than does the "fluid-flow compaction" algorithm. Although this comparison indicates that these two methods yield comparable estimates of basal heat flow, we opted to use the "fluid-flow compaction" method because of the widespread occurrence of overpressure encountered in exploration wells throughout the region (Hayba et al. 1999, Sherwood et al. 2002) and because preliminary modeling indicated the presence of overpressure in the Colville foredeep during rapid subsidence and sediment accumulation in the Early Cretaceous.

(2) We compared our heat-flow estimates to published estimates of heat flow in NPRA and nearby areas. Among the 56 wells used in our analysis of heat flow, Deming et al. (1992, 1996) used 13 of the same wells to estimate both deep and shallow heat flow, five to estimate deep heat flow only, and seven to estimate shallow heat flow only. We omitted from analysis two wells included by Deming et al. (1992, 1996): the East Simpson #2 was omitted because of its proximity and geologic similarity to the East Simpson #1, and the Lisburne #1 was omitted because our regional modeling does not consider the area of significant thrust–fault burial near the Brooks Range mountain front. Our estimates of heat flow calculated with the "fluid-flow compaction" method display a statistically significant, positive correlation to the deep heat-flow estimates of Deming et al. (1992), although the latter are higher, by an average of 3% (Fig. 5B). In contrast, there is no statistical correlation (Fig. 5C) between our heat-flow estimates and the shallow heat-flow estimates of Deming et al. (1992, 1996). This comparison indicates that our heat-flow estimates are not inconsistent with the deep heat-flow estimates of Deming et al. (1992, 1996), even though different techniques were used. Many geological parameters and differences in methodology may contribute to variability in heat-flow estimates (Deming 1994), and determining the origin of contrasts between our heat-flow estimates and those of Deming et al. (1992, 1996) is beyond the scope of this work.

(3) Considering the complex tectonic history of the study area, we estimated the temporal values of heat flow in wells for which both BHT

TABLE 1.—Kinetics and calculation options used in modeling. Kinetic parameters used to calculate transformation ratio based on 1-D burial history models calibrated with vitrinite reflectance data, as discussed in text. Kinetic values for the Shublik, Kingak, and GRZ/Hue were determined directly from samples of those source rocks. Type II "default" kinetic values were used for potential source rocks in the Canada basin.*

		Kerogen primary cracking		
Source rock	Fraction	Activation energy (kcal/mol)	Arrhenius constant (1/s)	Source of information
GRZ/Hue	1.0	52.880	3.31E+13	Lillis et al. 1999
Kingak	1.0	52.456	2.00E+13	Masterson 2001
Shublik	1.0	50.098	7.17E+12	Masterson 2001
Type II	0.05	49.000	3.01E+13	BasinMod 2006†‡
	0.20	50.000	3.01E+13	
	0.50	51.000	3.01E+13	
	0.20	52.000	3.01E+13	
	0.05	53.000	3.01E+13	

* Modeling calculation options: Compaction: BasinMod 1-D† "fluid-flow" method (compaction disequilibrium); Permeability: modified Kozeny-Carman (Ungerer et al. 1990); Geothermal: transient heat flow.

† Use of trade names in this publication is for descriptive purposes only and does not constitute endorsement by the US Government.

‡ BasinMod 2006 default type II kerogen kinetics "developed with technical support from Lawrence Livermore National Laboratory" (BasinMod 1-D Manual, August 2006 release).

and VR data are available and then modeled the thermal maturation history of those wells using various scenarios of heat-flow history. For example, the East Simpson #1 (Fig. 1) is located ~40 miles south of the Beaufort hinge in an area of extension (indicated by local horsts and grabens) during rift-opening of the Canada basin; BHT and VR data are available; and both deep and shallow heat-flow estimates were published by Deming et al. (1992, 1996). The burial, uplift, and erosion history of the East Simpson #1 reveals a complex tectonic evolution of the Beaufort rift shoulder (Fig. 6A), and we inferred three tectonic events that may have increased the heat flow along the rift shoulder at this location: (A) Early Carboniferous extension (Fig. 6A, E1) across much of Arctic Alaska resulted in the development of several large grabens, including the Hanna Trough (Sherwood et al. 2002) and the Ikpikpuk–Umiat basin (Hubbard et al. 1987, Bird 2001). (B) Initial extensional opening of the Canada basin ("failed rift" of Hubbard et al. [1987]) during the Early to Middle Jurassic (Fig. 6A, E2) resulted in the development of the Dinkum and several other grabens along the trend of the future rift shoulder (Grantz and May 1982; Hubbard et al. 1987; Grantz et al. 1990a, 1998, 2007). (C) Late Jurassic to Early Cretaceous rifting of the Canada basin resulted in maximum uplift of the Beaufort rift shoulder (Fig. 6A, R) and complete opening of the Canada basin (Grantz and May 1982; Hubbard et al. 1987; Grantz et al. 1990a, 1998, 2007; Houseknecht and Bird 2004). For the purpose of evaluating the influence of heat-flow history on thermal maturation, we assumed that heat flow may have increased significantly during these events and decreased exponentially following these events (Fig. 6B). The magnitude and duration of increased heat flow is based on typical values for extensional and marginal rift settings (Allen and Allen 1990) and the finite rifting model of Jarvis and McKenzie (1980).

(4) The East Simpson #1 burial history model was run using three different heat-flow assumptions: 79 mW/m^2 (deep heat flow of Deming et al. [1992]), 65.5 mW/m^2 (our estimate based on BHT data), and the temporally variable heat flow based on inferred tectonic influence (Fig. 6B). In each case, the calculated depth-VR trend is a poor match to empirical data (Fig. 6E), plotting higher than measured VR data, particularly in the deeper part of the well. The calculated depth-VR trend for a heat flow of 65.5 mW/m^2 is the closest match to the empirical data, that for the temporally variable heat flow is a poorer match, and that for a heat flow of 79 mW/m^2 is the poorest match. We next adjusted the heat-flow value downward in increments of 2.5 mW/m^2 until the calculated depth-VR trend best matched the empirical data. The best match was obtained with a heat-flow value of 57.5 mW/m^2 (Fig. 6B, E).

(5) The influence of heat flow on thermal maturation was further analyzed by plotting the age–temperature and age–VR outputs of the burial history model run with the four heat-flow values described above. Both temperature and calculated VR increased abruptly and by the largest magnitude during a relatively short interval of time (~102 to ~96 Ma; Fig. 6C, D) that corresponds to the most rapid subsidence and burial of the reference horizon (base of GRZ). Temperature and calculated VR continue to rise modestly until the onset of Paleogene uplift at ~65 Ma (Fig. 6A, C, D). The abrupt increase in temperature

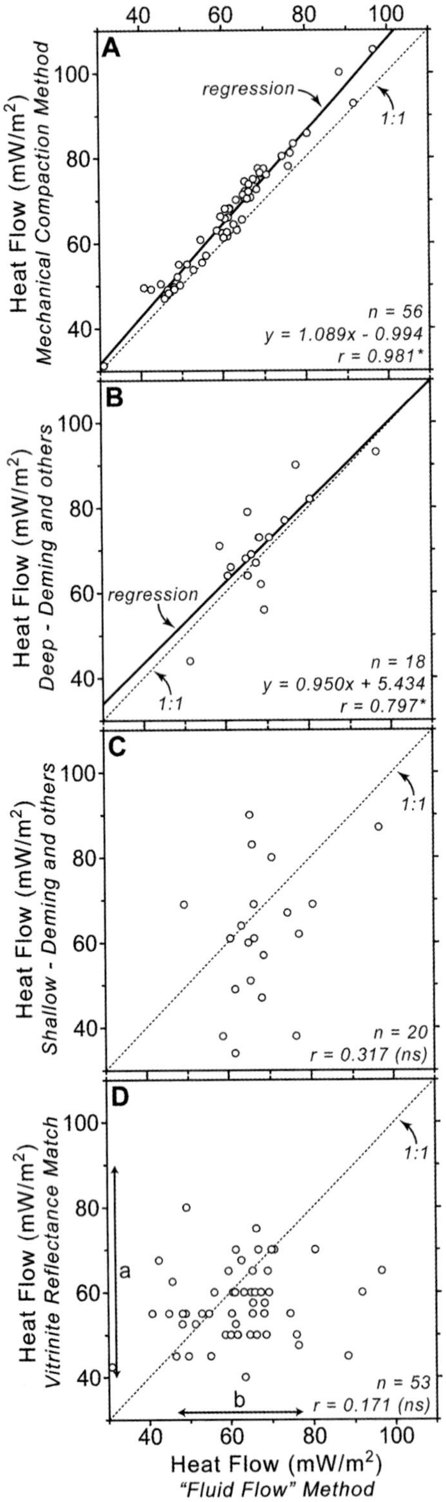

FIG. 5.—Plots of heat flow estimated from corrected BHT data using the BasinMod "fluid flow compaction method" vs. other heat-flow estimates, including (A) heat flow estimated from corrected BHT data using the BasinMod "mechanical compaction" method, (B) deep heat flow estimated by Deming et al. (1992), (C) shallow heat flow estimated by Deming et al. (1992), and (D) heat flow estimated by matching calculated depth-VR to empirical VR data. In plot (D), double-headed arrows represent (a) range of estimated heat flow for wells for which VR data are available and BHT data are not and (b) range of estimated heat flow for wells for which BHT data are available and VR data are not. Each plot includes number of wells (n); correlation coefficient (r), with asterisk if statistically significant and (ns) if not statistically significant; and—for those plots that display a statistically significant r—a regression equation and plot of that equation. A 1:1 line also is included in each plot.

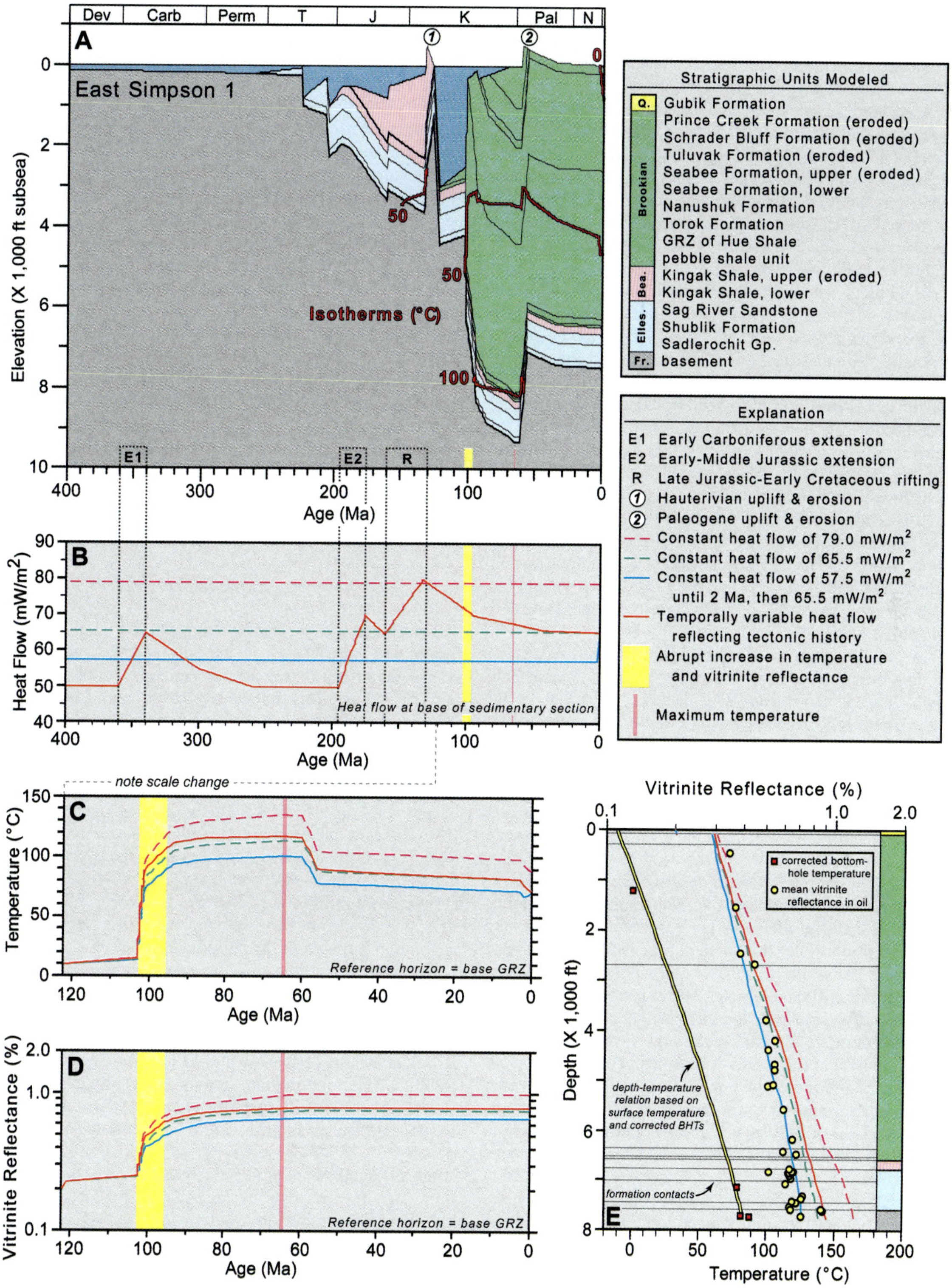

FIG. 6.—Plots for the East Simpson #1 well illustrating how burial history and heat flow were analyzed. (A) Burial history, including estimates of paleo-water depth and topographic relief during significant uplifts. For ease of presentation, estimates of paleo–sea level (Fig. 4) are not shown, and sea level with zero elevation is used as datum. Red lines are calculated isotherms using constant 57.5 mW/m^2 heat-flow assumption. (B) Age vs. heat flow for the four model assumptions evaluated as described in text. (C) Age vs. calculated temperature at base of GRZ for the four model assumptions. Note change of age scale relative to B. (D) Age vs. calculated VR at base of GRZ for the four model assumptions. Yellow bars in A through D represent time interval of greatest increase in calculated temperature and VR, and pink bars represent time interval of maximum calculated temperature. (E) Depth vs. corrected BHT (red squares), temperature gradient (yellow line), measured mean VR (yellow circles), and calculated depth-VR trends for the four model assumptions.

and thermal maturity occurs at ~30 my, and maximum temperature occurs at ~65 my, following maximum uplift and—presumably—peak heat flow associated with rifting. This result indicates that temporally variable heat flow associated with extension and rifting exerted little influence on maximum temperature and thermal maturation of the GRZ (consistent with the conclusions of Deming et al. [1992]) and that the Brookian rapid subsidence and sedimentation event in a foreland-basin tectonic setting (typically associated with lower heat flow; Allen and Allen 1990) was the predominant influence.

(6) Analyses similar to those described in steps 3 through 5 above were conducted for representative wells in contrasting tectonic settings within the study area, including a well in the Colville foredeep, far from the Beaufort rift shoulder; a well near the Hanna Trough in the Chukchi Sea; and a well in the tectonically active area offshore from ANWR. In each case, we found that neither heat flow calculated by us from BHT data nor temporally variable heat flow inferred from tectonic history yields model results that closely match empirical VR data. Therefore, we used our estimates of heat flow calculated from BHT data using the "fluid-flow compaction" method as a starting point for each modeled well for which VR data are available and then adjusted the heat flow by increments of 2.5 mW/m^2 until the calculated depth-VR trend best matched the empirical data. VR data from some wells display dog-legged depth-VR trends, which we did not attempt to model. Rather, we focused on the lower part of the Brookian section in matching model results to empirical data. We then constructed a contour map of the resultant heat-flow values (Fig. 1) and used that map to determine heat-flow input for modeled wells and pseudowells for which no VR data are available.

(7) Heat-flow values generated by matching model results to empirical VR data, as described in steps 4 and 6 above, do not correlate with heat flow calculated by us from BHT data (Fig. 5D). This result indicates that vitrinite reflectance is not in equilibrium with current heat flow. Although the cause of this incongruence is uncertain, evidence indicates that heat flow across Arctic Alaska may have changed during the Cenozoic. Deming et al. (1992, 1996) attributed the regional pattern of low heat flow in the Brooks Range foothills grading to higher heat flow in northern NPRA and the inverse relation between depth-to-basement and heat flow to the gravity-driven flow of groundwater caused by the topographic uplift of the Brooks Range. They inferred that patterns of heat flow, modern temperature gradients, and vertical thermal-maturity gradients are consistent, leading to the suggestion that the groundwater flow system may have been active at the time of thermal maturation, probably during the Cenozoic but perhaps as long ago as 100 Ma. However, recent interpretations of the tectonic evolution of the Brooks Range indicate that an early phase of deformation (Late Jurassic–Early Cretaceous) involved relatively modest structural relief and migration of high-temperature (250–300° C) fluids and a later phase of deformation (Paleogene) involved greater structural relief and migration of lower temperature (~150° C) fluids (Moore et al. 2004). Thus, it is likely that any groundwater flow system driven by topography of the Brooks Range would have been active during the Cenozoic.

Stratigraphic and seismic evidence in northwestern NPRA indicates Paleogene uplift, which has been interpreted to represent a peripheral bulge associated with tectonic loading of the eastern North Slope (Coakley and Watts 1991), and apatite fission track analysis of the Walakpa #1 well (Fig. 1) indicates that ~4000-feet of section was eroded at ~60 Ma (O'Sullivan 1999). Additional apatite fission track analysis of the South Meade, Topagoruk, and Ikpikpuk wells (Fig. 1) confirms similar amounts of erosion at ~60 Ma, indicates additional erosion during the Miocene at the South Meade and Topagoruk wells, and indicates a significant increase in temperature at the South Meade well during the past 5 my (PB O'Sullivan, unpublished data, 2005). The region also has undergone significant variation in surface temperature during the Cenozoic, from the Paleocene–Eocene thermal maximum (surface temperature perhaps as high as 20° C; Sluijs et al. 2006) to the onset of permafrost conditions ~2 Ma (Carter and Hillhouse 1992). Current surface temperatures on the Alaska North Slope vary from –4 to –12° C (Lachenbruch et al. 1988).

Collectively, these observations indicate the potential for significant modification of the subsurface thermal regime during the Cenozoic, perhaps including temporally and spatially variable heat flow, advective heat transport, significant variation in surface temperature, and perturbation of thermal conductivity related to the onset of permafrost conditions. It appears that all these thermal effects postdate peak temperature and maturation of vitrinite (Fig. 6).

Local radiogenic heat sources also are possible across the study area. The East Teshekpuk well (Fig. 1) bottomed in granite of probable Devonian age (Dumoulin 2001), and high heat flow is indicated by BHT data from this and nearby wells and from VR data from nearby wells (no VR data are available from the East Teshekpuk well). The nearby Ikpikpuk well (Fig. 1) bottomed in a Mississippian or older, quartz-veined breccia that may indicate either contact metamorphism or fault-related hydrothermal activity (Dumoulin 2001). No other granitic rocks are known from Arctic Alaska, but the basement complex is sparsely sampled, and others may occur. Although these observations indicate the potential for local granite and related rocks in the basement, we did not incorporate radiogenic sources of heat flow in our modeling.

Exploration Wells

To achieve our goal of modeling the burial and thermal history of this region, we focused on the base of the Brookian sequence as a reference horizon that is present across the entire area. Although this approach generalizes certain aspects of burial history (for example, Ellesmerian rift basins such as the Hanna Trough and Beaufortian rift basins such as the Dinkum graben), it accounts for most hydrocarbon generation from known source rocks. The Shublik Formation, widely considered the most prolific oil source rock south of the rift shoulder, occurs just 0 to ~4500 feet below the base of the Brookian sequence (Houseknecht and Bird 2004), and source rocks in the lower Kingak Shale are even closer to the base of the Brookian. Hydrocarbon generation in the Shublik and lower Kingak likely did not occur prior to Brookian deposition (see below); therefore, using the base of the Brookian as a modeling horizon addresses the timing of most oil generation throughout the region.

Geohistory models for 112 exploration wells were used to characterize the burial and thermal history south of the rift shoulder (Fig. 1). Our modeling extends southward to the northern foothills of the Brooks Range, where moderate structural thickening of the Brookian sequence is evident, but does not extend into the southern foothills, where Brookian strata were tectonically buried by thrust sheets involving Ellesmerian strata (Moore et al. 2002, 2004). Representative examples (Fig. 7) illustrate common aspects of geohistory, including relatively slow accumulation of Ellesmerian strata during late Paleozoic to Triassic time, deposition of variably thick Beaufortian strata during Jurassic to Earliest Cretaceous time, and rift shoulder uplift followed by rapid subsidence and deposition of Brookian strata in the Early Cretaceous. Contrasts in the Early Cretaceous burial histories of these example wells reflect greater uplift closer to the rift shoulder, greater accommodation and sediment accumulation in the Colville foredeep, and earlier accumulation of Brookian sediment in the west and south. Contrasts in the Late Cretaceous and Cenozoic burial histories reflect pulses of uplift in different parts of Arctic Alaska (O'Sullivan 1999; O'Sullivan et al. 1993, 1997; Burns et al. 2007) and the west-to-east filling of the foreland basin by Brookian sediments (Bird and Molenaar 1992).

Our geohistory models yield calculated depth-VR trends that closely match measured data in many wells (Fig. 8A). However, data collected

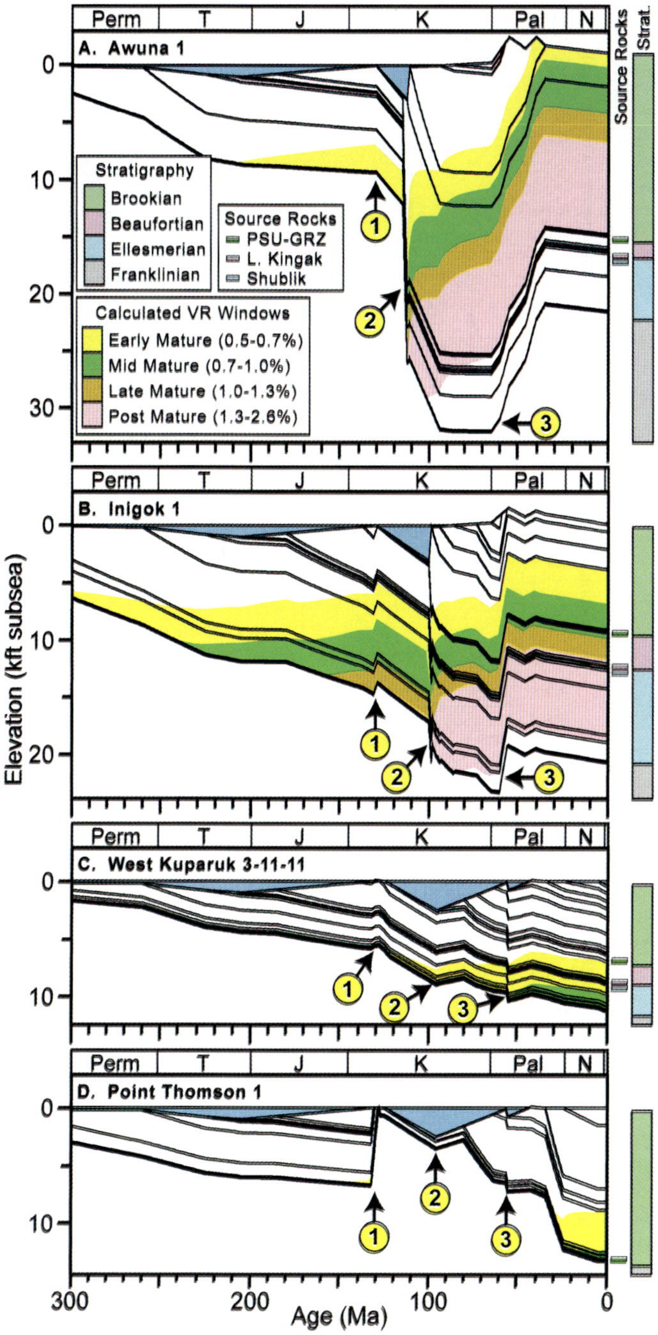

FIG. 7.—Geohistory plots of representative exploration wells, Alaska North Slope, showing burial, uplift, and thermal maturation history. (A) Awuna #1; (B) Inigok #1; (C) West Kuparuk #3-11-11; (D) Point Thomson #1. Timing of main tectonic events is shown by numbered arrows; (1) acme of Beaufort rifting; (2) abrupt subsidence of Colville foreland basin; (3) initial uplift in fold-thrust belt of Brooks Range foothills. Note different influences each event exerts on geohistory of these wells. Generalized stratigraphic columns at right depict stratigraphy, and small colored rectangles show stratigraphic positions of the Shublik, lower Kingak, and PSU-GRZ source rocks. Current sea level used as datum for simplification of plot. Inferred water depth shown in blue. Well locations shown in Figure 1.

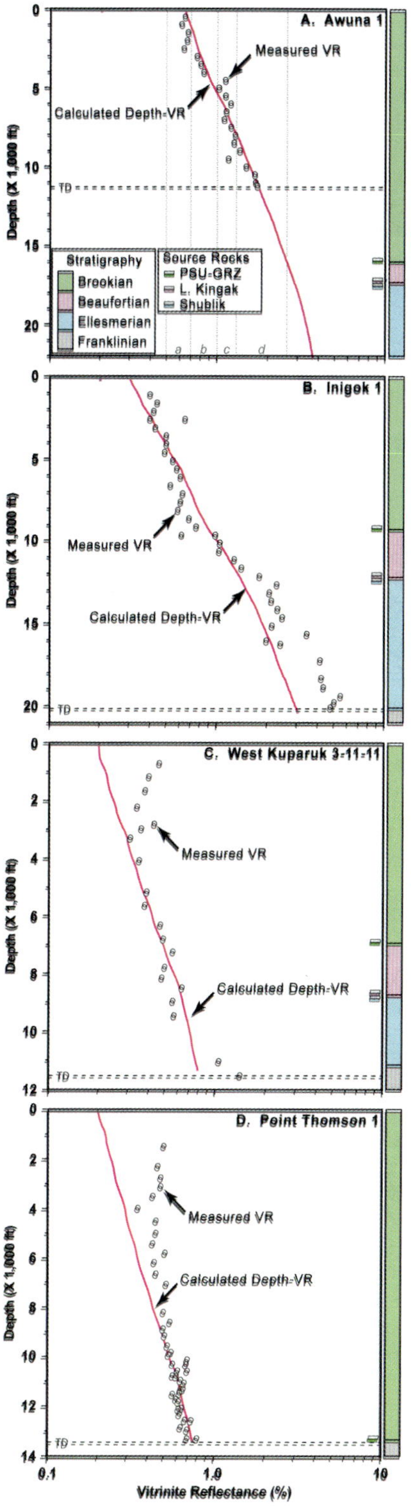

FIG. 8.—Plots of depth-measured VR and depth-calculated VR for wells whose burial histories are shown in Figure 7. Well locations shown in Figure 1.

from some wells display either dog-legged depth-VR trends or so much scatter in measured VR that only approximate matches are possible. Three types of dog-legged depth-VR trends are observed: (1) an increase in log VR per-unit increase in depth (increased thermal maturity/depth gradient) below the LCU (Fig. 8B) or another significant unconformity that may reflect temporal changes in heat flow; (2) an increase in log VR per-unit increase in depth within the Brookian section and near the top of overpressure that may reflect the influence of high pore-fluid pressure on heat flow and/or thermal conductivity (see examples in Bird et al. [1999]); and (3) a near-vertical depth-VR trend observed in data at shallow depths (Fig. 8C, D) that may represent recycled vitrinite (Houseknecht and Hayba 1999). Interpretation of data that display abundant scatter is more difficult, as the public domain data evidently include interlaboratory variability that we cannot explain. In cases in which a close match between calculated and measured depth-VR was not possible, we focused on the lower part of the Brookian section in refining modeling parameters.

Evaluation of timing of oil generation indicates insufficient burial for all major source rocks (Shublik, lower Kingak, and GRZ) prior to Brookian deposition (Fig. 9). Our modeling indicates that the Shublik, lower Kingak, and GRZ all passed through the oil-generation window rapidly during the late Early Cretaceous as the result of burial by the Brookian sequence in the main part of the Colville foredeep (Fig. 9A). On the northern flank of the foredeep, the Shublik and lower Kingak passed rapidly through the oil-generation window, whereas the GRZ peaked at a transformation ratio near 0.5 prior to uplift and cooling (Fig. 9B). On the Beaufort rift shoulder, the Shublik and lower Kingak are absent owing to truncation beneath the LCU in some areas (Fig. 9D), and even where present, none of the source rocks has entered the main phase of oil generation (Fig. 9C, D).

Pseudowells

A simple stratigraphy—including the tops of basement, pre-Brookian, Lower Cretaceous, Upper Cretaceous, Paleogene, Neogene, and Quaternary strata—was defined for each of the ~300 pseudowells used for modeling (Figs. 1, 3). Four potential source-rock horizons were defined, including Lower Cretaceous near the base of the Brookian sequence, basal Upper Cretaceous, lower Paleogene, and upper–lower Paleogene (~*Azolla*). All are assumed to contain type II kerogen, based on the characteristics of coeval source rocks in Alaska, Canada, and on the Lomonosov Ridge. Pseudowells constrained by seismic data were added between the regional cross sections in areas of abrupt changes in geology, including along the Beaufort hingeline and across the Dinkum graben. To establish geologic boundaries for modeling, additional pseudowells were included along the eastern margin of the Chukchi borderland (Northwind Ridge), on the northern Chukchi shelf, and in the North Chukchi basin (Fig. 1). Geologic constraints for pseudowells in these areas include seismic data and published reports (Thurston and Theiss 1987, Lothamer 1994, Grantz et al. 1998, Sherwood et al. 2002).

Heat-flow inputs used for modeling pseudowells in the Canada basin are poorly constrained. Values from areas with abundant well data indicate that heat flow decreases northward from >60 mW/m^2 along the highest standing part of the rift shoulder to ~50 mW/m^2 near the hingeline, where the rift shoulder is buried by 5000 to 20000 feet of Brookian strata (Figs. 1, 3). The few exploration wells drilled in the Tertiary deformed area (eastern Beaufort Sea) north of the rift shoulder yield heat-flow estimates of 50 to 80 mW/m^2 from BHT data and of 45 to 60 mW/m^2 from depth-VR trends. We also modeled wells in the Canadian Mackenzie delta for which refined vitrinite reflectance measurements are available (Stasiuk et al. 2005) and estimated heat flow of ~40 mW/m^2 based on depth-VR trends for wells in the thick, overpressured part of the sediment prism. Seafloor measurements from the deep Canada basin are generally ~55 mW/m^2 (Langseth et al.

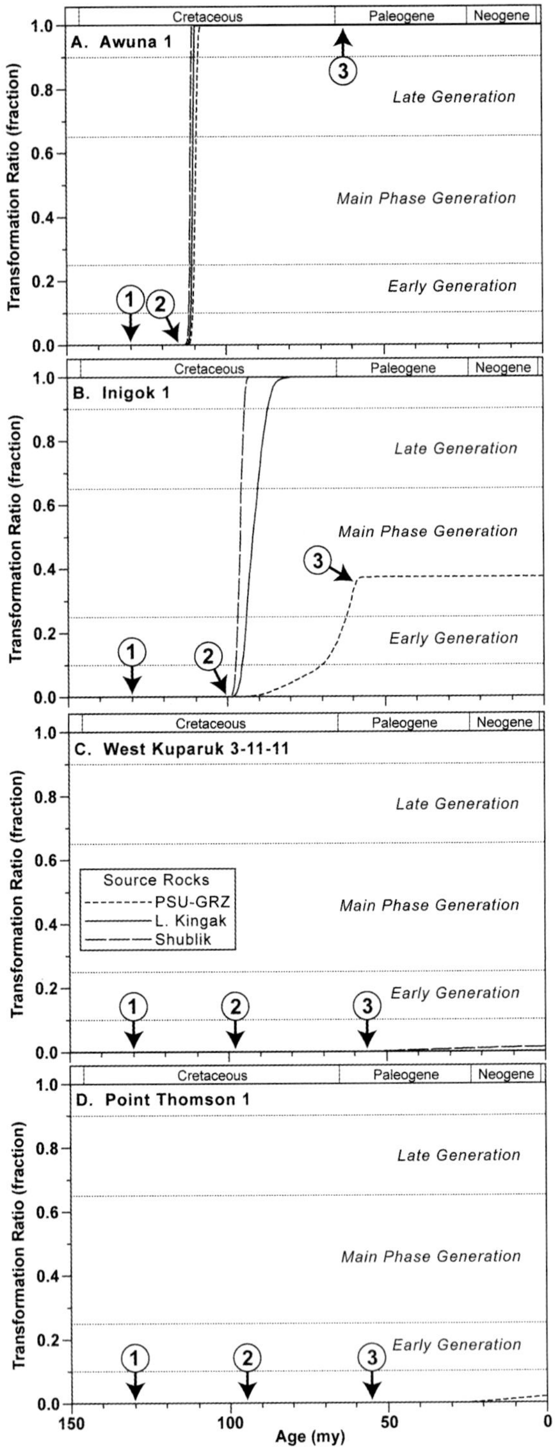

FIG. 9.—Plots of time vs. transformation ratio for source rocks penetrated by wells shown in Figures 7 and 8. Numbered arrows depict timing of the same tectonic events shown in Figure 7. Transformation ratios calculated using kinetics listed in Table 1. Well locations shown in Figure 1.

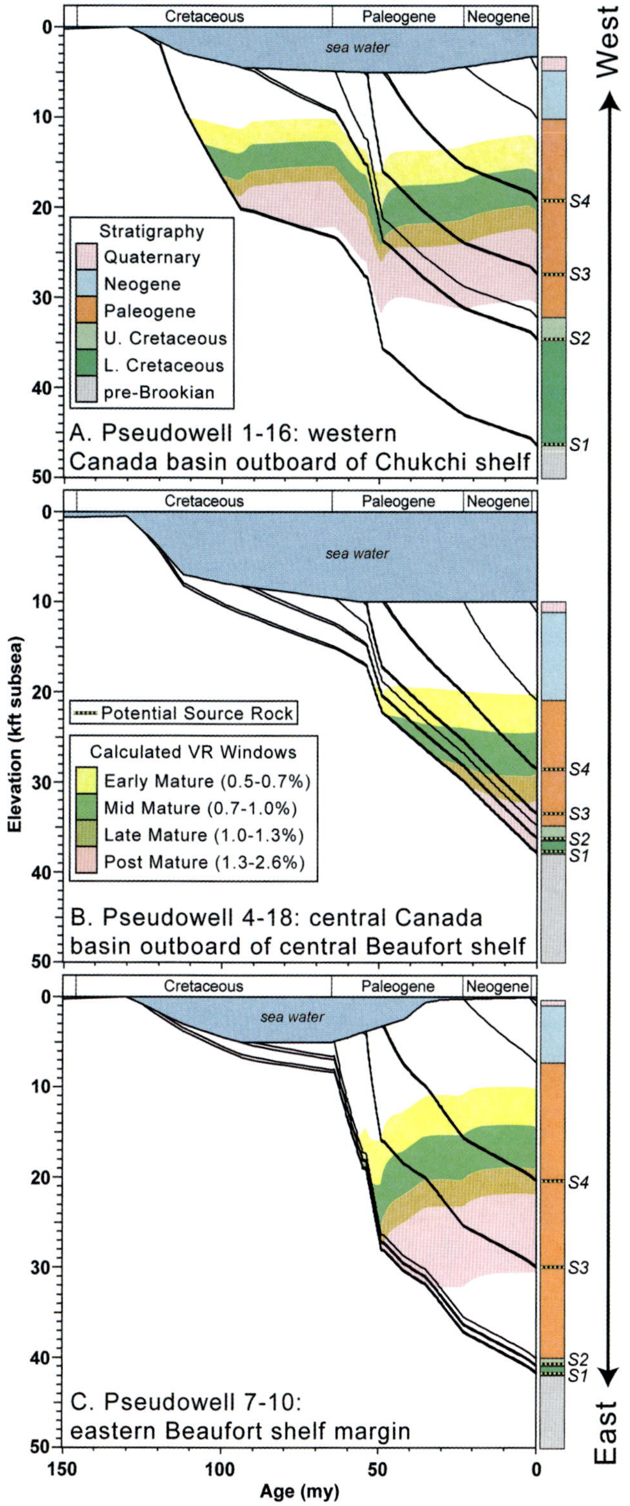

FIG. 10.—Burial history plots of representative pseudowells in depocenters north of the Beaufort rift shoulder showing burial and thermal maturation history, assuming a heat flow of 45 mW/m^2. Generalized stratigraphic columns at right depict stratigraphy, and black-and-yellow bars show inferred stratigraphic positions of potential source rocks in the Lower Cretaceous (S1), Upper Cretaceous (S2), lower Paleogene (S3), and upper–lower Paleogene (S4; ~*Azolla*). Current sea level used as datum for simplification of plot. Inferred water depth shown in blue. Location of pseudowells shown in Figures 1 and 3.

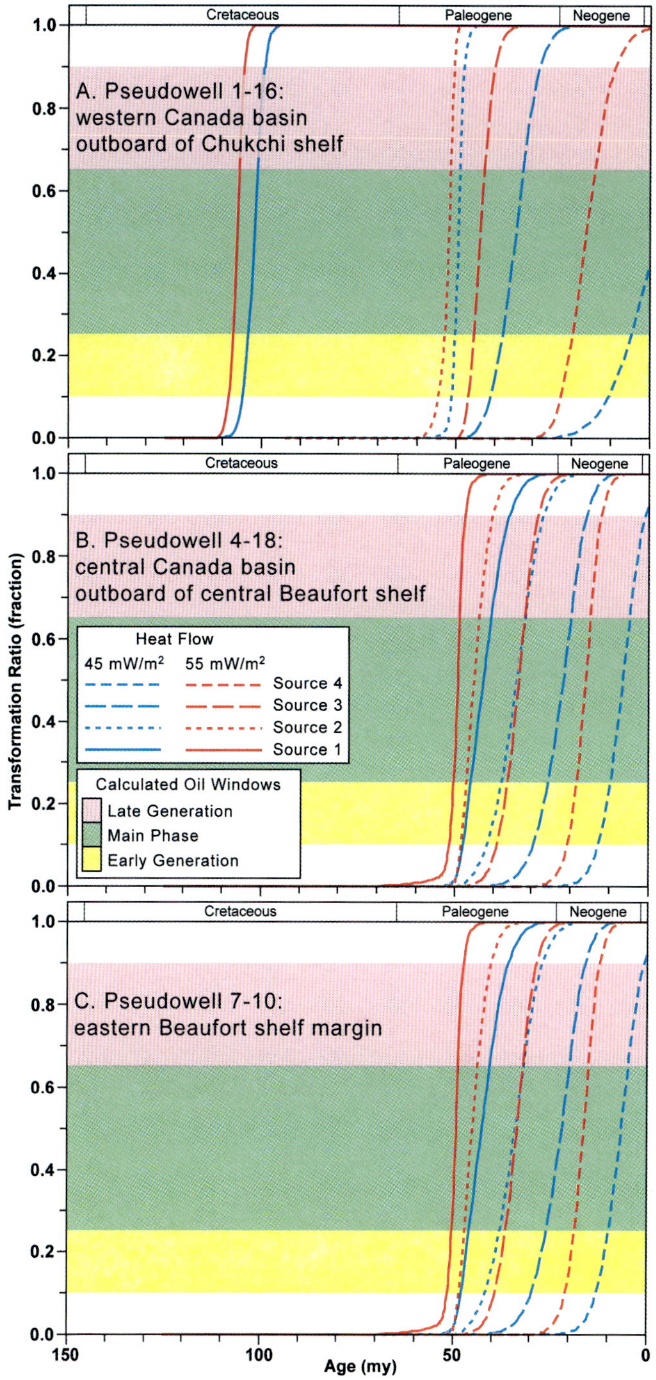

FIG. 11.—Plots of time-transformation ratio for potential source rocks shown in Figure 10. Transformation ratios calculated using default type II kinetics (Table 1). Results are shown for each source rock at heat flows of 45 and 55 mW/m^2.

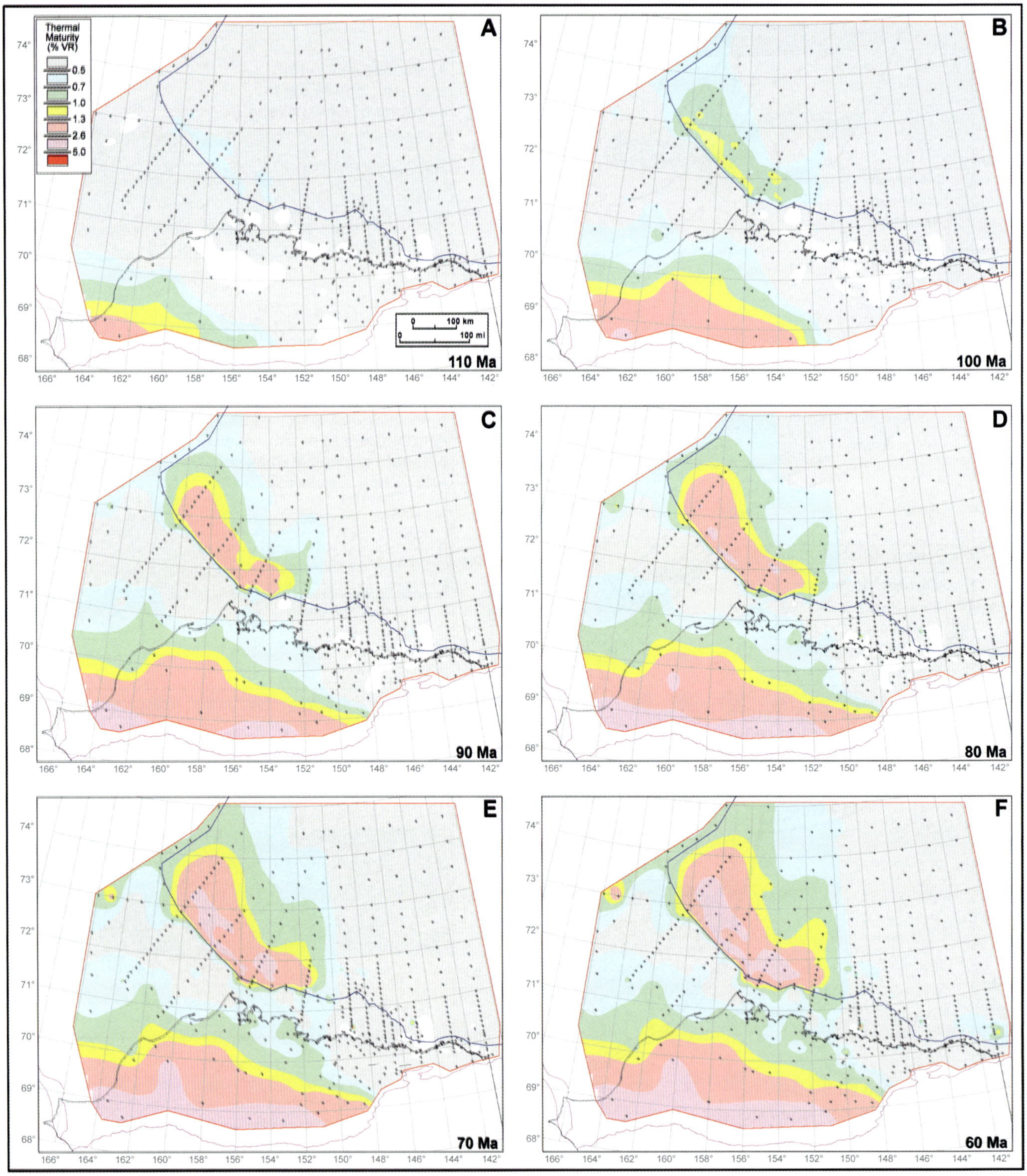

FIG. 12.—Map of the entire study area showing calculated vitrinite reflectance at the base of the Brookian sequence at 10-my time steps from 110 Ma to the present. Model results are clipped (red perimeter line) to eliminate contouring edge effects. Black crosses, locations of exploration wells and pseudowells used in modeling. Calculated values of VR > 2.6% should be considered only as general indicators of increased temperature–time exposure because "Easy %R_o" kinetics (Burnham and Sweeney 1989, Sweeney and Burnham 1990) become asymptotic at elevated levels of thermal maturity. White areas inside red perimeter have not reached temperature-time kinetic threshold for VR calculation.

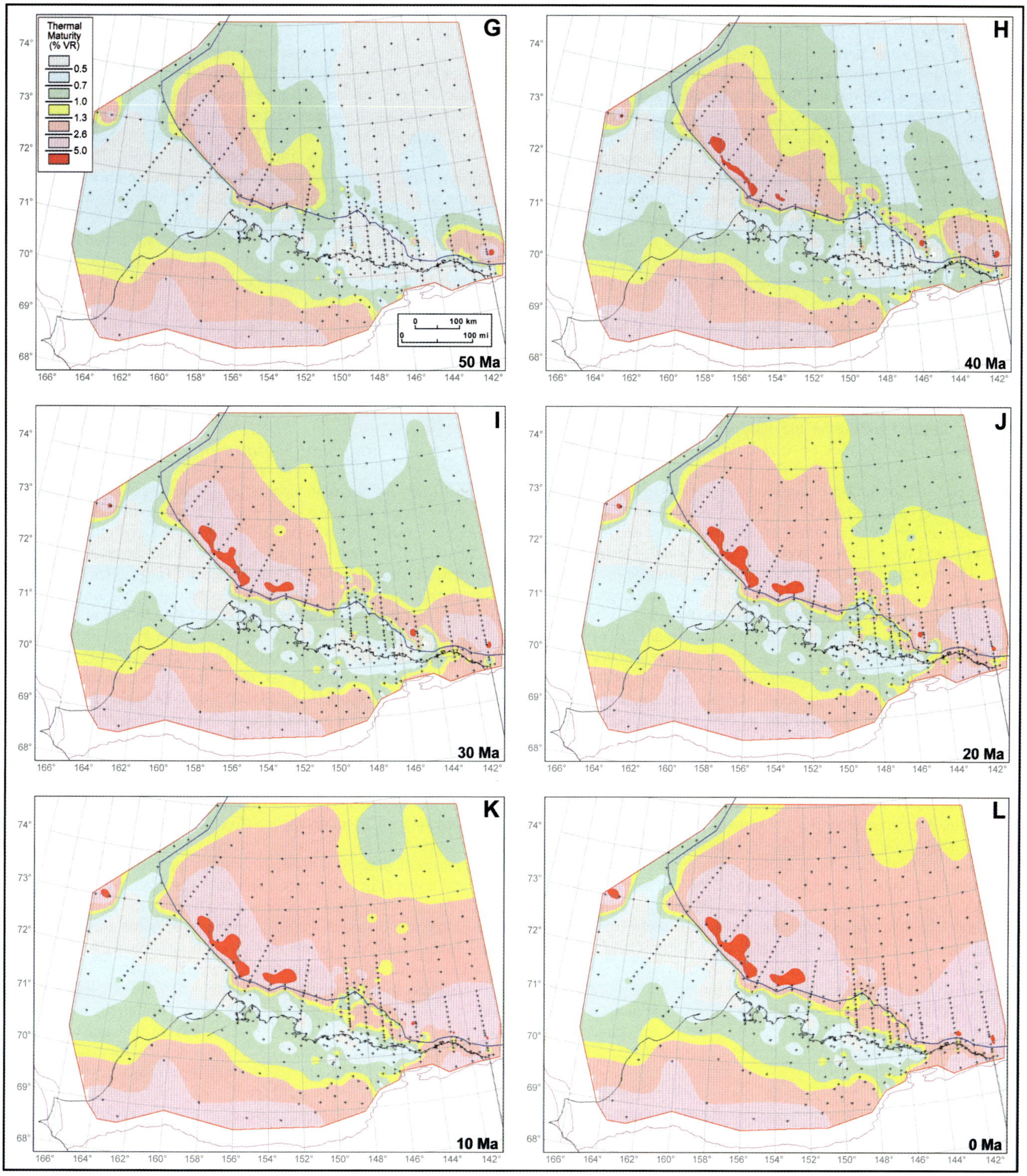

FIG. 12.—*Continued.*

1990, Louden et al. 1990). Based on these observations, we decrease heat-flow values from 55 to 45 mW/m^2 from south to north over a distance of ~40 miles north of the rift shoulder.

Three pseudowells from just north of the rift shoulder illustrate the contrasts in timing of burial and thermal maturation along the length of the Arctic Alaska margin (Figs. 3, 10). In the west, where the Brookian sequence is >40,000 feet thick and two main pulses of sedimentation (Early Cretaceous and Cenozoic) are inferred in the depocenter north of the rift shoulder, the Lower Cretaceous source rock matured rapidly during the initial pulse of sedimentation (Fig. 10A), and oil generation was completed by ~100 Ma (Fig. 11A). Upper Cretaceous and lower Paleogene source rocks matured during Paleogene burial (Fig. 10A), and oil generation was completed by ~50 and ~30 Ma, respectively (Fig. 11A). The upper–lower Paleogene source rock (~*Azolla*) is currently near peak generation (Figs. 10A, 11A).

In the central area, the Brookian sequence is <28,000 feet thick, and most of the section is Cenozoic in age (Fig. 10B). All four source rocks matured during accelerated burial that began in the early Paleogene (Fig. 10B). Oil generation was completed in the Lower and Upper Cretaceous source rocks during the late Paleogene, in the lower Paleogene source rock during the early Neogene, and in the upper–lower Paleogene source rock (~*Azolla*) during the late Neogene (Fig. 11B).

In the east, where the Brookian sequence is >40,000 feet thick and the section is predominately Cenozoic, the Lower and Upper Cretaceous source rocks matured rapidly during the early Paleogene (Fig. 10C), and oil generation was completed by mid-Paleogene (Fig. 11C). The lower Paleogene source rock matured during middle Paleogene to early Neogene, and the upper–lower Paleogene source rock (~*Azolla*) matured during the Neogene (Figs. 10C, 11C).

To evaluate the stratigraphic and spatial variability of thermal maturation among the pseudowells, depths of four modeled VR values commonly used to define oil-generation windows are plotted on the cross sections (Fig. 3): 0.5% (onset of early mature, oil), 0.7% (onset of mid-mature, oil), 1.0% (onset of late mature, oil), and 1.3% (onset of gas generation). In the west (Fig. 3A), most of the Brookian section south of the hingeline is immature, and the base of the Brookian section is just above 0.5% VR. Stepping northward, thermal maturity increases abruptly as Brookian strata expand across the hingeline; much of the Paleogene section is in the oil or gas window, and all older strata are well above 1.3% VR. Thermal maturity decreases gradually northward in the Canada basin as the stratigraphic section becomes thinner and generally younger, but much of the Paleogene section and all older strata are in the oil window to the northern limit of our modeling (Fig. 3A). Among the potential source rocks inferred to be present, the older three (Lower Cretaceous, Upper Cretaceous, and lower Paleogene) are in the oil or gas window everywhere north of the hingeline (Fig. 3A). In contrast, the upper part of the lower Paleogene (~*Azolla*) source rock is mid-mature in a narrow zone near the thickest part of the depocenter, approximately beneath the modern shelf margin, and is early mature everywhere else north of the hingeline (Fig. 3A).

In the central area (Fig. 3B), the northernmost exploration wells provide evidence that Upper Cretaceous and older strata have VR > 0.5% and that the base of the Brookian section is mid-mature (0.7–1.0% VR). Northward to the hingeline, VR contours generally increase in depth and subtly mimic the horst-and-graben structure of the basement (Fig. 3B). Across much of the Dinkum graben and the entire Dinkum plateau, most of the Paleogene section and all older Brookian strata are in the oil window, and the base of the Brookian section in the Dinkum graben is at, or slightly above, 1.3% VR (Fig. 3B). Thermal maturity increases northward across the hingeline, but not as abruptly as in the west. Along this central cross section, the 0.5% VR contour plots near the top of the Paleogene section, and the base of the Brookian is >1.3% VR everywhere north of the hingeline (Fig. 3B). All four potential source rocks are in the oil or gas window north of the hingeline.

In the east (Fig. 3C), a single exploration well near the hingeline provides evidence that much of the Brookian section is mature and that the base of the Brookian section is slightly above 1.3% VR (Paul et al. 1994). To the south, VR of outcrop and shothole samples from the ANWR coastal plain reveals an abrupt increase in thermal maturity toward the Brooks Range front (Bird et al. 1999). Together, these empirical data indicate that the influence of tectonic burial on thermal maturation of the Brookian section is limited to a narrow zone (~20 miles wide) near the mountain front. Northward from the exploration well, thermal maturity increases as Brookian strata expand across the hingeline; the 0.5% VR contour plots near the top of the Paleogene section, and the base of the Brookian is well above 1.3% VR (Fig. 3C). Thermal maturity decreases northward in the Canada basin, but much of the Paleogene section and all older strata are in the oil window to the northern limit of our modeling (Fig. 3C). In the depocenter north of the hingeline, the two Paleogene source rocks are largely in the late–mature oil window, whereas the two Cretaceous source rocks are well into the gas window (Fig. 3C).

Regional Maturation History

The history of thermal maturation for the base of the Brookian sequence, based on more than 400 1-D burial history models for exploration wells and pseudowells, is depicted as contour maps of calculated VR at 10-my intervals from 110 Ma to the present (Fig. 12). The map depicting present thermal maturity (Fig. 12L) is consistent with empirical vitrinite reflectance data from exploration wells, outcrops, and shothole samples (Bird et al. 1999; Johnsson et al. 1999).

Thermal maturation of basal Brookian strata began in the western Colville foredeep at ~110 Ma (Fig. 12A) and spread quickly eastward and more slowly northward as the foredeep accommodated thick Lower Cretaceous strata (Fig. 12B, C). Basal Brookian strata matured by a modest additional amount during the Late Cretaceous (Fig. 12D, E), when deposition occurred in the central to eastern part of the foreland basin. Additional maturation occurred during the Cenozoic as the result of stratigraphic burial during the Paleogene in the eastern part of the foreland basin (Fig. 12F–H) and structural burial during the Paleogene (Fig. 12I) and Neogene (Fig. 12J, K) on the ANWR coastal plain.

The overall pattern of thermal maturity presently observed in the foreland basin (Fig. 12L) was essentially complete by ~90 Ma (Fig. 12C). At that time, basal Brookian strata were in the gas window (VR > 1.3%) in the foredeep, in the oil window along the northern flank of the foreland basin, and immature on the rift shoulder (Fig. 12C). These results indicate that oil generation in basal Brookian (PSU-GRZ) and older source rocks (including the Shublik Formation, lower Kingak Shale, and their distal equivalent, the Otuk Formation) began in the western Colville foredeep and moved eastward and northward from ~110 to ~90 Ma (Fig. 12A–C). Modest additional oil generation may have been associated with the gradual northward advancement of thermal maturation up the northern flank of the foreland basin during the Cretaceous beneath the western ANS and Chukchi Sea (Fig. 12D, E). A second pulse of oil generation from these same source rocks likely occurred in a smaller, eastern part of the foreland basin from ~60 to ~40 Ma (Fig. 12F–H).

Maturation also occurred during the Early Cretaceous in the southwestern corner of the Canada basin, where a prism of sediment accumulated as depositional systems overstepped the rift shoulder along the northeast Chukchi–western ANS margin (Fig. 12). The volume of this sediment prism is poorly constrained, but seismic data on the northeast Chukchi shelf and regional paleogeographic reconstructions provide evidence for significant sediment bypass of the shelf during the Aptian, Albian, and perhaps Cenomanian (~125–

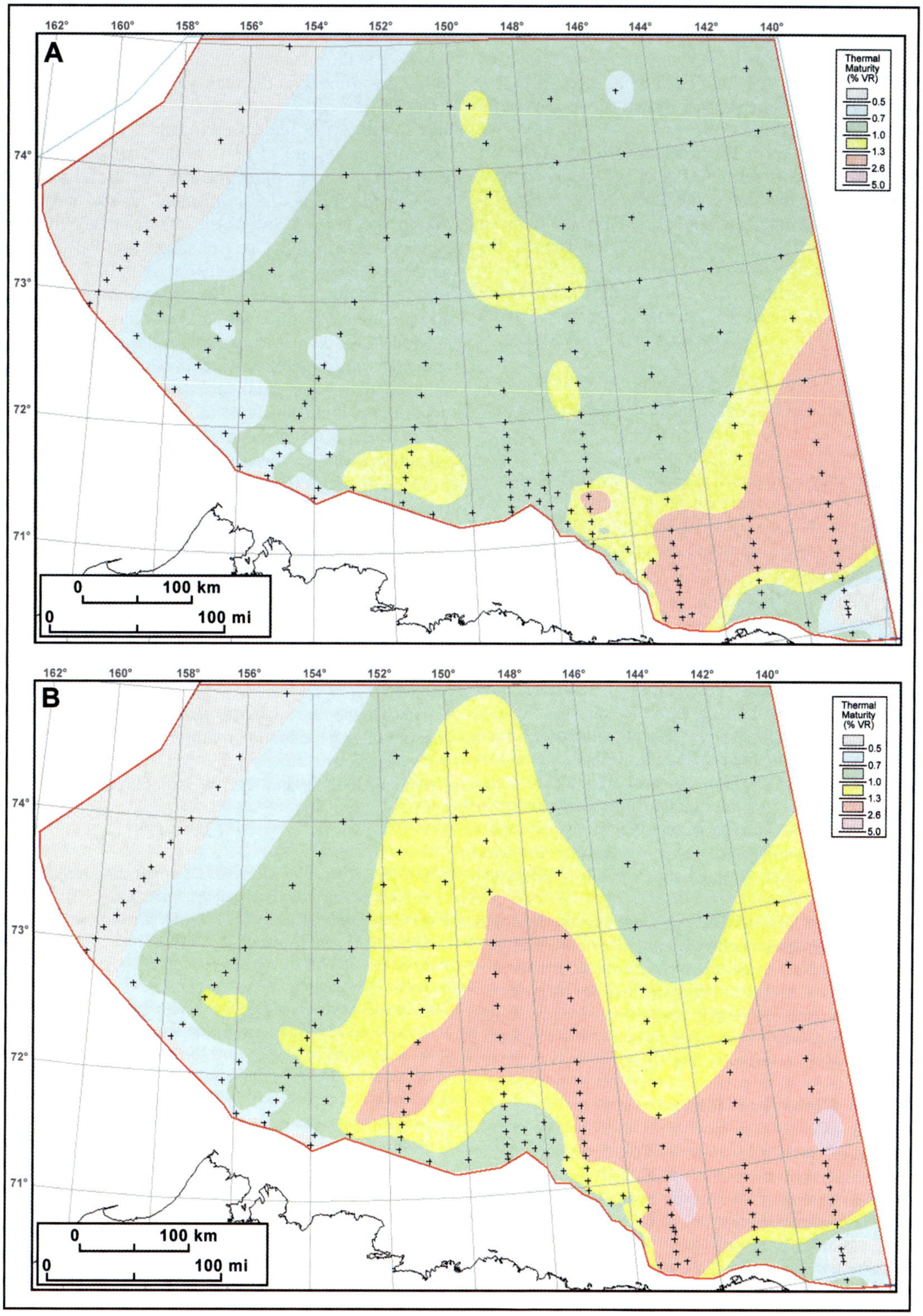

FIG. 13.—Map of the southern Canada basin showing calculated vitrinite reflectance of the upper–lower Paleogene (~*Azolla*) source rock using different heat flows. (A) 45 mW/m^2; (B) 55 mW/m^2. Model results are clipped (red perimeter line) at hingeline along southern and western margins to eliminate edge effects, as discussed in text. Black crosses, locations of exploration wells and pseudowells used in modeling. Calculated values of VR > 2.6% should be considered only as general indicators of increased temperature–time exposure because "Easy $\%R_o$" kinetics (Burnham and Sweeney 1989, Sweeney and Burnham 1990) become asymptotic at elevated levels of thermal maturity.

94 Ma). Our modeling indicates that basal Brookian strata in part of this sediment prism passed completely through the oil window (VR > 1.3%) by ~90 Ma (Fig. 12C). An incremental increase in thermal maturity likely occurred during the Late Cretaceous (Fig. 12D, E), when a modest volume of sediment accumulated along the entire southern rim of the Canada basin.

Most thermal maturation of basal Brookian strata north of the rift shoulder occurred during the Cenozoic (Fig. 12), when sediment accumulation was focused in two depocenters that merge north of a salient formed by the Beaufort hingeline (Figs. 1, 12). Perhaps as much as 30,000 feet of Paleogene and Neogene strata accumulated in the western depocenter, which is spatially coincident with the Early Cretaceous depocenter in the southwestern corner of the Canada basin. This burial significantly increased the thermal maturity of basal Brookian strata and more than doubled the area in which VR exceeds 1.3% (Fig. 12F–K).

The eastern depocenter received sediment both directly from the Brooks Range to the south and from the Mackenzie delta to the east. The accumulation of as much as 40,000 feet of Paleogene and Neogene strata pushed basal Brookian strata rapidly through the oil window across an area that expanded progressively northward and westward (Fig. 12F–K). Areas of similar thermal maturity in the two depocenters eventually merged near the hingeline salient, and this remains an area where thermal maturity is lower and where maturation occurred later than in the adjacent depocenters.

Our modeling indicates that the thermal maturity of the youngest potential source rock (~*Azolla*) north of the Beaufort hingeline generally grades from <0.5% VR in the western Canada basin (east of the Chukchi borderland) to >1.3% VR north of the eastern ANS (Fig. 13A), reflecting the eastward increase in thickness of Cenozoic strata toward the Mackenzie delta. This horizon is in the early to mid- oil-generation window (0.5–1.0% VR) for type II kerogen across a broad area of the southern Canada basin and is in the gas-generation window (1.3–2.6% VR) in the east (Fig. 13A). An area of lower thermal maturity is modeled immediately north of the hingeline adjacent to the ANWR coastal plain, where the horizon is buried at shallower depths owing to mid-Eocene and younger tectonic uplift (Grantz and May 1982, Grantz et al. 1990a, Dietrich and Lane 1992, McMillen and O'Sullivan 1992).

Use of a higher heat flow (55 mW/m^2) in modeling the southern Canada basin produces an increase in thermal maturity of the youngest potential source rock (~*Azolla*) across the central and eastern parts of the study area (Fig. 13B) and yields earlier generation for all potential source rocks (Fig. 11). The differences in timing of modeled generation among the four potential source rocks range from ~2 my (source 1, Fig. 11A) to more than 12 my (source 3, Fig. 11B).

CONCLUSIONS

Thermal maturity of oil- and gas-prospective strata in Arctic Alaska and the adjacent Canada basin reflects the complex geologic elements that formed during rift-opening of the Canada basin (Jurassic to Early Cretaceous) and collision-induced evolution of the Brooks Range orogenic belt and Colville foreland basin (Jurassic to Tertiary). Analysis of bottom-hole-temperature and vitrinite reflectance data indicates that vitrinite reflectance is not in equilibrium with current heat flow across much of the study area. Therefore, empirical vitrinite reflectance data were used to estimate ancient heat flow for modeling purposes. Modeling indicates that oil generation in all regionally important source rocks occurred as the result of burial by the Brookian sequence during Cretaceous to Tertiary time; thus, the base of the Brookian section is a key reference surface for regional interpretations.

Basal Brookian strata on the Beaufort rift shoulder grade from immature in the west (an area of renewed uplift during the Tertiary) to early mature in the central area and to overmature in the east (an area of stratigraphic and structural burial during the Tertiary). Thermal maturity increases southward from the rift shoulder; basal Brookian strata are in the oil window on the north flank of the foreland basin and in the gas window in the foredeep. The area of immature and mature strata on the rift shoulder and north flank of the foreland basin is significantly wider in the west (Chukchi shelf) because of the eastward spatial convergence of the rift shoulder and Brooks Range. Basal Brookian and older source rocks in the foredeep and northern flank of the Colville foreland basin matured rapidly and likely generated oil during the Early Cretaceous, with relatively modest additional oil generation during the Late Cretaceous and Paleogene. Maturation and generation began in the southwestern part of the foredeep and moved rapidly eastward and more slowly northward in response to accumulation of a thick Lower Cretaceous succession.

Models of thermal maturation in the Canada basin north of the rift shoulder, which are more uncertain as a result of sparse geologic data, indicate that basal Brookian strata grade from overmature in the depocenters immediately north of the Beaufort hingeline to mature at the northern limit of modeling, ~200 miles north of hingeline. The youngest potential source rock (upper part of the lower Paleogene, ~*Azolla* horizon) grades from immature in the west to overmature in the east and is in the mid- to late-mature oil window across much of the southern Canada basin. Maturation in the Canada basin mostly occurred during Paleogene and Neogene burial, although Early Cretaceous maturation of basal Brookian strata occurred in the southwestern corner of the Canada basin, adjacent to the northeastern Chukchi shelf margin.

ACKNOWLEDGMENTS

Dan Hayba and Elisabeth Rowan contributed to preliminary work that provided a foundation for this study. The manuscript was improved by constructive reviews by Mark Johnsson, Janet Pitman, Jim Coleman, Joe Curiale, Kate Weissenburger, and Nick Harris.

REFERENCES

Allen PA, Allen JR. 1990. *Basin Analysis—Principles and Applications*: Blackwell Scientific Publications, Oxford, UK. 451 p.

Askin RA, Spicer RA. 1995. *The Late Cretaceous and Cenozoic History of Vegetation and Climate at Northern and Southern High Latitudes; A Comparison, Effects of Past Global Change on Life*: National Academy Press, Washington, DC. p. 156–173.

Basinger JF, Greenwood DR, Sweda T. 1994. Early Tertiary vegetation of Arctic Canada and its relevance to paleoclimatic interpretation. *In* Boulter MC, Fisher HC (Editors). *Cenozoic Plants and Climates of the Arctic*, NATO ASI Series, Vol. 27: Berlin, Springer-Verlag. p. 175–198.

Bayliss GS, Magoon LB. 1988. Organic facies and thermal maturity of sedimentary rocks in the National Petroleum Reserve in Alaska. *In* Gryc G (Editor). *Geology and Exploration of the National Petroleum Reserve in Alaska, 1974 to 1982*: US Geological Survey Professional Paper 1399, Washington, DC. p. 489–518.

Bird KJ. 1985. The framework geology of the North Slope of Alaska as related to oil-source rock correlations. *In* Magoon LB, Claypool GE (Editors). *Alaska North Slope Oil/Rock Correlation Study*, Studies in Geology 20: American Association of Petroleum Geologists, Tulsa, Oklahoma. p. 3–29.

Bird KJ. 2001. Alaska: A twenty-first-century petroleum province. *In* Downey MW, Threet JC, Morgan WA (Editors). *Petroleum Provinces of the Twenty-First Century*, Memoir 74: American Association of Petroleum Geologists, Tulsa, Oklahoma. p. 137–165.

Bird KJ, Burruss RC, Pawlewicz MJ. 1999. Thermal maturity. *In The Oil and Gas Resource Potential of the Arctic National Wildlife Refuge 1002 Area, Alaska*: US Geological Survey Open-File Report 98-34, Washington, DC. 64 p. http://energy.cr.usgs.gov/OF98-34/VR.pdf. Accessed January 19, 2008.

Bird KJ, Molenaar CM. 1992. The North Slope foreland basin, Alaska. *In* Macqueen RW, Leckie DA (Editors). *Foreland Basins and Foldbelts*, Memoir

55: American Association of Petroleum Geologists, Tulsa, Oklahoma. p. 363–393.

Blanchard DC, Tailleur IL. 1982. Temperatures and interval geothermal gradients determinations from wells in National Petroleum Reserve in Alaska: US Geological Survey Open-File Report 82-391, Washington, DC. 79 p.

Brinkhuis H, Schouten S, Collinson M, Sluijs A, Sinninghe Damste JS, Dickens GR, Huber M, Cronin TM, Onodera J, Takahashi K, Bujak JP, Stein R, van der Burgh J, Eldrett JS, Harding IC, Lotter AF, Sangiorgi F, van Konijnenburg-van Cittert H, de Leeuw JW, Matthiessen J, Backman J, Moran K, and the Expedition 302 Scientists. 2006. Episodic fresh surface waters in the Eocene Arctic Ocean. *Nature* 441:606–609.

Burnham AK, Sweeney JJ. 1989. A chemical kinetic model of vitrinite maturation and reflectance. *Geochimica et Cosmochimica Acta* 53:2649–2657.

Burns WM, Hayba DO, Rowan EL, Houseknecht DW. 2007. Estimating the amount of eroded section in a partially exhumed basin using geophysical logs: An example from the Alaska North Slope: US Geological Survey Professional Paper 1732-D, Washington, DC. 18 p. http://pubs.usgs.gov/pp/pp1732/pp1732d/. Accessed September 18, 2008.

Carter LD, Hillhouse JW. 1992. Age of the Late Cenozoic Bigbendian marine transgression of the Alaskan Arctic coastal plain: Significance for permafrost history and paleoclimate. *In* Bradley DC, Ford AB (Editors). *US Geological Survey Bulletin 1999*. Washington, DC. p. 44–51.

Claypool GE, Magoon LB. 1985. Comparison of oil-source rock correlation data for Alaskan North Slope: Techniques, results, and conclusions. *In* Magoon LB, Claypool GE (Editors). *Alaska North Slope Oil/Rock Correlation Study*, Studies in Geology 20: American Association of Petroleum Geologists, Tulsa, Oklahoma. p. 49–81.

Coakley BJ, Watts AB. 1991. Tectonic controls on the development of unconformities: The North Slope, Alaska. *Tectonics* 10:101–130.

Deming D. 1994. Overburden rock, temperature, and heat flow. *In* Magoon LB, Dow WG (Editors). *The Petroleum System—From Source to Trap*, Memoir 60: American Association of Petroleum Geologists, Tulsa, Oklahoma. p. 165–186.

Deming D, Sass JH, Lachenbruch AH. 1996. Heat flow and subsurface temperature, North Slope of Alaska. *In* Johnsson MJ, Howell DG (Editors). *Thermal Evolution of Sedimentary Basins in Alaska*: US Geological Survey Bulletin 2142, Washington, DC. p. 21–44.

Deming D, Sass JH, Lachenbruch AH, DeRito RF. 1992. Heat flow and subsurface temperature as evidence for basin-scale ground-water flow, North Slope of Alaska: *Geological Society of America Bulletin* 104:528–542.

Dietrich JR, Lane LS. 1992. Geology and structural evolution of the Demarcation subbasin and Herschel high, Beaufort-Mackenzie basin, Arctic Canada. *Bulletin of Canadian Petroleum Geology* 40:188–197.

Dixon J (Editor). 1996. *Geologic Atlas of the Beaufort–Mackenzie Area*, Miscellaneous Report 59: Geological Survey of Canada, Calgary, Alberta. 173 p.

Dixon J, Dietrich JR, McNeil DH. 1992. *Upper Cretaceous to Pleistocene Sequence Stratigraphy of the Beaufort–Mackenzie and Banks Island Areas, Northwest Canada*, Bulletin 407: Geological Survey of Canada, Calgary, Alberta. 90 p.

Dumoulin JA. 2001. Lithologies of the basement complex (Devonian and older) in the National Petroleum Reserve—Alaska. *In* Houseknecht DW (Editor). *NPRA Core Workshop—Petroleum Plays and Systems in the National Petroleum Reserve—Alaska*, Core Workshop 21: Society for Sedimentary Geology, Tulsa, Oklahoma. p. 201–214.

Golonka J, Ross MI, Scotese CR. 1994. Phanerozoic paleogeographic and paleoclimatic modeling maps. *In* Embry AF, Beauchamp B, Glass DJ (Editors). *Pangea: Global Environments and Resources*, Memoir 17: Canadian Society of Petroleum Geologists, Calgary, Alberta. p. 1–47.

Gradstein FM, Ogg JG, Smith AG, Bleeker W, Lourens LJ. 2004. A new geologic time scale, with special reference to Precambrian and Neogene. *Episodes* 27:83–100.

Grantz A, Clark DL, Phillips RL, Srivastava SP. 1998. Phanerozoic stratigraphy of Northwind Ridge, magnetic anomalies in the Canada basin, and the geometry and timing of rifting in the Amerasia basin, Arctic Ocean. *Geological Society of America Bulletin* 110:801–820.

Grantz A, Hart PE, Childers VA. 2007. Seismic and geologic evidence for opening of the Amerasia basin, Arctic Ocean by four Early Jurassic to Paleocene extensional events (Abstract): Abstracts and Proceedings of the Geological Society of Norway, p. 68–69.

Grantz A, May SD. 1982. Rifting history and structural development of the continental margin north of Alaska. *In* Watkins JS, Drake CL (Editors). *Studies in Continental Margin Geology*, Memoir 34: American Association of Petroleum Geologists, Tulsa, Oklahoma. p. 77–100.

Grantz A, May SD, Hart PE. 1990a. Geology of the Arctic continental margin of Alaska. *In* Grantz A, Johnson L, Sweeney JF (Editors). *The Arctic Ocean Region*, The Geology of North America L: Geological Society of America, Boulder, Colorado. p. 257–288.

Grantz A, May SD, Taylor PT, Lawver LA. 1990b. Canada basin. *In* Grantz A, Johnson L, Sweeney JF (Editors). *The Arctic Ocean Region*, The Geology of North America L: Geological Society of America, Boulder, Colorado. p. 379–402.

Haq BU, Al-Qahtani AM. 2005. Phanerozoic cycles of sea-level change on the Arabian platform. *GeoArabia* 10:127–160.

Hayba DO, Burns WM. 2006. Estimating lithologic compositions from digital mud logs across the North Slope of Alaska (Abstract): *Geological Society of America, Abstracts with Programs* 38:87.

Hayba DO, Burns WM, Rowan EL. 2004. An objective, spreadsheet-based approach to parameter estimation for basin-scale numerical modeling (Abstract): *American Association of Petroleum Geologists, Annual Meeting Expanded Abstracts* 13:60.

Hayba DO, Houseknecht DW, Rowan E. 1999. Stratigraphic, hydrogeologic, and thermal evolution of the Canning River region, North Slope, Alaska. *In The Oil and Gas Resource Potential of the Arctic National Wildlife Refuge 1002 Area, Alaska*: US Geological Survey Open-File Report 98-34, Washington, DC. 45 p. http://energy.cr.usgs.gov/OF98-34/FF.pdf. Accessed September 18, 2008.

Houseknecht DW, Bird KJ. 2004. Sequence stratigraphy of the Kingak Shale (Jurassic–Lower Cretaceous), National Petroleum Reserve in Alaska. *American Association of Petroleum Geologists Bulletin* 88:279–302.

Houseknecht DW, Bird KJ, Schenk CJ. 2009. Seismic analysis of clinoform depositional sequences and shelf-margin trajectories in Lower Cretaceous (Albian) strata, Alaska North Slope. *Basin Research* 21:644–654.

Houseknecht DW, Hayba DO. 1999. Modeling oil generation in the undeformed part of the Arctic National Wildlife Refuge 1002 Area. *In The Oil and Gas Resource Potential of the Arctic National Wildlife Refuge 1002 Area, Alaska*: US Geological Survey Open-File Report 98-34, Washington, DC. 45 p. http://energy.cr.usgs.gov/OF98-34/HG.pdf. Accessed September 18, 2008.

Houseknecht DW, Schenk CJ. 2001. Depositional sequences and facies in the Torok Formation, National Petroleum Reserve—Alaska (NPRA). *In* Houseknecht DW (Editor). *NPRA Core Workshop—Petroleum Plays and Systems in the National Petroleum Reserve—Alaska*, Core Workshop 21: Society for Sedimentary Geology, Tulsa, Oklahoma. p. 179–199.

Hubbard RJ, Edrich SP, Rattey RP. 1987. Geologic evolution and hydrocarbon habitat of the "Arctic Alaska microplate." *Marine and Petroleum Geology* 4:2–34.

Jackson HR. 1990. Evolution and regional stratigraphy of the northeastern Canadian polar margin. *Marine Geology* 93:179–192.

Jackson HR, Oakey GN. 1990. Sedimentary thickness map of the Arctic Ocean. *In* Grantz A, Johnson L, Sweeney JF (Editors). *The Arctic Ocean Region*, The Geology of North America L: Geological Society of America, Boulder, Colorado. plate 5.

Jarvis GT, McKenzie DP. 1980. Sedimentary basin formation with finite extension rates. *Earth and Planetary Science Letters* 48:42–52.

Johnsson MJ, Evans KR, Marshall HA. 1999. *Thermal Maturity of Sedimentary Rocks in Alaska: Digital Resources*: US Geological Survey, Washington, DC. Digital Data Series DDS-54. http://pubs.usgs.gov/dds/dds-54/. Accessed September 18, 2008.

Johnsson MJ, Howell DG. 1996. Thermal maturity of sedimentary basins in Alaska—an overview. *In* Johnsson MJ, Howell DG (Editors). *Thermal Evolution of Sedimentary Basins in Alaska*: US Geological Survey Bulletin 2142, Washington, DC. p. 1–9.

Johnsson MJ, Howell DG, Bird KJ. 1993. Thermal maturity patterns in Alaska:

Implications for tectonic evolution and hydrocarbon potential. *American Association of Petroleum Geologists Bulletin* 77:1874–1903.

Jones HP, Speers RG. 1976. Permo-Triassic reservoirs of Prudhoe Bay field, North Slope, Alaska. *In* Braunstein J (Editor). *North American Oil and Gas Fields*, Memoir 24: American Association of Petroleum Geologists, Tulsa, Oklahoma. p. 23–50.

Lachenbruch AH, Sass JH, Lawver LA, Brewer MC, Marshall BV, Munroe RJ, Kennelly JP Jr, Galanis SP Jr, Moses TH Jr. 1988. Temperature and depth of permafrost on the Arctic slope of Alaska. *In* Gryc G (Editor). *Geology and Exploration of the National Petroleum Reserve in Alaska, 1974 to 1982*: US Geological Survey Professional Paper 1399, Washington, DC. p. 645–656.

Lachenbruch AH, Sass JH, Marshall BV, Moses TH Jr. 1982. Permafrost, heat flow, and the geothermal regime at Prudhoe Bay, Alaska. *Journal of Geophysical Research* 87:9301–9316.

Langseth MG, Lachenbruch AH, Marshall BV. 1990. Geothermal observations in the Arctic region. *In* Grantz A, Johnson L, Sweeney JF (Editors). *The Arctic Ocean Region*, The Geology of North America L: Geological Society of America, Boulder, Colorado. p. 133–151.

Lillis PG, Lewan MD, Warden A, Monk SM, King JD. 1999. Identification and characterization of oil types and their source rocks. *In The Oil and Gas Resource Potential of the Arctic National Wildlife Refuge 1002 Area, Alaska*: US Geological Survey Open-File Report 98-34, Washington, DC. 101 p. http://energy.cr.usgs.gov/OF98-34/OA.pdf. Accessed September 18, 2008.

Lothamer RT. 1994. Early Tertiary wrench faulting in the North Chukchi basin, Chukchi Sea, Alaska. *In* Thurston DK, Fujita K (Editors). *1992 Proceedings International Conference on Arctic Margins*: US Minerals Management Service OCS Study MMS 94-0040, Anchorage, Alaska. p. 251–256.

Louden KE, Leger G, Hamilton N. 1990. Marine heat flow observations on the Canadian Arctic continental shelf and slope. *Marine Geology* 93:267–288.

Magoon LB, Bird KJ, Burruss RC, Hayba D, Houseknecht DW, Keller MA, Lillis PG, Rowan EL. 1999. Evaluation of hydrocarbon charge and timing using the petroleum system. *In The Oil and Gas Resource Potential of the Arctic National Wildlife Refuge 1002 Area, Alaska*: US Geological Survey Open-File Report 98-34, 66 p. http://energy.cr.usgs.gov/OF98-34/PS.pdf. Accessed September 18, 2008.

Magoon LB, Woodward PV, Banet AC Jr, Griscom SB, Daws TA. 1987. Thermal maturity, richness, and type of organic matter of source-rock units. *In* Bird KJ, Magoon LB (Editors). *Petroleum Geology of the Northern Part of the Arctic National Wildlife Refuge, Northeastern Alaska*: US Geological Survey Bulletin 1778, Washington, DC. p. 127–179.

Masterson WD IV. 2001. Petroleum filling history of central Alaskan North Slope fields, [unpublished PhD dissertation]: University of Texas at Dallas, 222 p.

May SD, Grantz A. 1990. Sediment thickness in the southern Canada basin. *Marine Geology* 93:331–347.

McMillen KJ, O'Sullivan P. 1992. Tectonic and eustatic controls on Paleogene sequence stratigraphy, Beaufort Sea, Alaska and Canada. *In* Watkins JS, Zhiqiang F, McMillen KJ (Editors). *Geology and Geophysics of Continental Margins*, Memoir 53: American Association of Petroleum Geologists, Tulsa, Oklahoma. p. 285–301.

Miller KG, Kominz MA, Browning JV, Wright JD, Mountain GS, Katz ME, Sugarman PJ, Cramer BS, Christie-Blick N, Pekar SF. 2005. The Phanerozoic record of global sea-level change. *Science* 310:1293–1298.

Molenaar CM. 1983. Depositional relations of Cretaceous and Lower Tertiary rocks, northeastern Alaska. *American Association of Petroleum Geologists Bulletin* 67:1066–1080.

Moore TE, Dumitru TA, Adams KE, Witebsky SN, Harris AG. 2002. Origin of the Lisburne Hills–Herald Arch structural belt: Stratigraphic, structural, and fission-track evidence from the Cape Lisburne area, northwestern Alaska. *In* Miller EL, Grantz A, Klemperer SL (Editors). *Tectonic Evolution of the Bering Shelf–Chukchi Sea–Arctic Margin and Adjacent Landmasses*, Special Paper 360: Geological Society of America, Boulder, Colorado. p. 77–109.

Moore TE, Potter CJ, O'Sullivan PB, Shelton KL, Underwood MB. 2004. Two stages of deformation and fluid migration in the west-central Brooks Range fold and thrust belt, northern Alaska. *In* Swennen R, Roure F, Granath JW (Editors). *Deformation, Fluid Flow, and Reservoir Appraisal in Foreland Fold and Thrust Belts*, Hedberg Series 1: American Association of Petroleum Geologists, Tulsa, Oklahoma. p. 157–186.

Moore TE, Wallace WK, Bird KJ, Karl SM, Mull CG, Dillon JT. 1994. Geology of northern Alaska. *In* Plafker G, Berg HC (Editors). *The Geology of Alaska*, The Geology of North America G-1: Geological Society of America, Boulder, Colorado, p. 49–140.

Mull CG, Houseknecht DW, Bird KJ. 2003. *Revised Cretaceous and Tertiary Stratigraphic Nomenclature in the East-Central Colville Basin, Northern Alaska*: US Geological Survey Professional Paper 1673, Washington, DC. 51 p. http://pubs.usgs.gov/prof/p1673/. Accessed September 18, 2008.

Nelson RE, Carter LD. 1992. Preliminary interpretation of vegetation and paleoclimate in northern Alaska during the Late Pliocene Colvillian marine transgression. *In* Bradley DC, Ford AB (Editors): US Geological Survey Bulletin 1999. Washington, DC. p. 219–222.

O'Sullivan PB. 1993. Late Mesozoic to Cenozoic thermal and uplift history of the North Slope foreland basin, northern Alaska and northwestern Canada [unpublished PhD dissertation]: La Trobe University, Bundoora, Victoria, 419 p.

O'Sullivan PB. 1996. Late Mesozoic and Cenozoic thermal-tectonic evolution of the North Slope foreland basin, Alaska. *In* Johnsson MJ, Howell DG (Editors). *Thermal Evolution of Sedimentary Basins in Alaska*: US Geological Survey Bulletin 2142, Washington, DC. p. 45–79.

O'Sullivan PB. 1999. Thermochronology, denudation and variations in paleosurface temperature: A case study from the North Slope foreland basin, Alaska. *Basin Research* 11:191–204.

O'Sullivan PB, Green PF, Bergman SC, Decker J, Duddy IR, Gleadow AJW, Turner DL. 1993. Multiple phases of Tertiary uplift and erosion in the Arctic National Wildlife Refuge, Alaska, revealed by apatite fission track analysis. *American Association of Petroleum Geologists Bulletin* 77:359–385.

O'Sullivan PB, Murphy JM, Blythe AE. 1997. Late Mesozoic and Cenozoic thermotectonic evolution of the central Brooks Range and adjacent North Slope foreland basin, Alaska: Including fission track results from the Trans-Alaska Crustal Transect (TACT). *Journal of Geophysical Research* 102:20821–20845.

Paul LE, Choromanski DR, Turner RF, Flett TO. 1994. *Geological, Geochemical, and Operational Summary, Aurora Well, OCS Y-0943-1, Beaufort Sea, Alaska*: Minerals Management Service, OCS Report MMS 94-0001, Anchorage, Alaska. 71 p., 1 plate.

Peters KE, Magoon LB, Bird KJ, Valin ZC, Keller MA. 2006. North Slope Alaska: Source-rock distribution, richness, thermal maturity and petroleum charge. *American Association of Petroleum Geologists Bulletin* 90(2):261–292.

Peters KE, Ramos LS, Zumberge JE, Valin ZC, Bird KJ. 2008. De-convoluting mixed crude oil in Prudhoe Bay field, North Slope, Alaska. *Organic Geochemistry* 39:623–645.

Potter CJ, Grow JA, Perry WJ, Moore TE, O'Sullivan PB, Phillips JD, Saltus RW. 2004. Tertiary thrust systems and fluid flow beneath the Beaufort coastal plain (1002 area), Arctic National Wildlife Refuge, Alaska, USA. *In* Swennen R, Roure F, Granath JW (Editors). *Deformation, Fluid Flow, and Reservoir Appraisal in Foreland Fold and Thrust Belts*, Hedberg Series 1: American Association of Petroleum Geology, Tulsa, Oklahoma. p. 187–214.

Saltus RW, Bird KJ. 2003. *Digital Depth Horizon Compilations of the Alaskan North Slope and Adjacent Arctic Regions*: US Geological Survey Open-File Report 03-230, Washington, DC. 21 p. http://pubs.usgs.gov/of/2003/ofr-03-230/. Accessed September 18, 2008.

Sherwood KW, Johnson PP, Craig JD, Zerwick SA, Lothamer RT, Thurston DK, Hurlbert SB. 2002. Structure and stratigraphy of the Hanna Trough, US Chukchi shelf, Alaska. *In* Miller EL, Grantz A, Klemperer SL (Editors). *Tectonic Evolution of the Bering Shelf—Chukchi Sea—Arctic Margin and Adjacent Landmasses*, Special Paper 360: Geological Society of America, Boulder, Colorado, p. 39–66.

Shimada K, Itoh M, Nishino S, McLaughlin F, Carmack E, Proshutinsky A. 2005. Halocline structure in the Canada basin of the Arctic Ocean. *Geophysical Research Letters* 32:L03605.

Sluijs A, Schouten S, Pagani M, Woltering M, Brinkhuis H, Sinninghe Damste JS, Dickens GR, Huber M, Reichart G-J, Stein R, Matthiessen J, Lourens LJ, Pedentchouk N, Backman J, Moran K, and the Expedition 302 Scientists. 2006. Subtropical Arctic Ocean temperatures during the Palaeocene/Eocene thermal maximum. *Nature* 441:610–613.

Snowdon LR. 1980. Petroleum source potential of the Boundary Creek

Formation, Beaufort–Mackenzie basin. *Bulletin of Canadian Petroleum Geology* 28:46–58.

Snowdon LR, Stasiuk LD, Robinson R, Dixon J, Dietrich J, McNeil DH. 2004. Organic geochemistry and organic petrology of a potential source rock of early Eocene age in the Beaufort-Mackenzie basin. *Organic Geochemistry* 35:1039–1052.

Spicer RA, Corfield RM. 1992. A review of terrestrial and marine climates in the Cretaceous with implications for modeling the "greenhouse earth." *Geological Magazine* 129:169–180.

Spicer RA, Parrish JT. 1990. Late Cretaceous–Early Tertiary palaeoclimates of northern high latitudes; A quantitative view. *Journal of the Geological Society, London* 147:329–341.

Stasiuk LD, Issler DR, Tomica M, Potter J. 2005. *Re-evaluation of Thermal Maturation—Vitrinite Reflectance Profiles for Cretaceous and Tertiary Strata, Beaufort–Mackenzie Basin, Northwest Territories (Adlartok P-09, Amerk O-09, Edlok N-56, Ikhil K-35, Sarpik B-35, Hansen G-07, and Havik B-41)*: Geological Survey of Canada, Open File 4665. Calgary, AB. 14 p.

Stein R. 2007. Upper Cretaceous/lower Tertiary black shales near the North Pole; Organic-carbon origin and source-rock potential. *Marine and Petroleum Geology* 24:67–73.

Stein R, Boucsein B, Meyer H. 2006. Anoxia and high primary production in the Paleogene central Arctic Ocean; First detailed records from Lomonosov Ridge. *Geophysical Research Letters* 33:L18606.

Sweeney JJ, Burnham AK. 1990. Evaluation of a simple model of vitrinite reflectance based on chemical kinetics. *American Association of Petroleum Geologists Bulletin* 64:1559–1570.

Threlkeld CN, Obuch RC, Gunther GL. 2000. *Organic Geochemistry Data of Alaska*: US Geological Survey, Washington, DC, Digital Data Series DDS-59. http://pubs.usgs.gov/dds/dds-059/. Accessed September 18, 2008.

Thurston DK, Theiss LA. 1987. *Geologic Report for the Chukchi Sea Planning Area, Alaska; Regional Geology, Petroleum Geology, and Environmental Geology*: US Minerals Management Service OCS Report MMS 87-0046, Anchorage, Alaska. 193 p.

Ungerer P, Burrus J, Doligez B, Chenet PY, Bessis F. 1990. Basin evaluation by integrated two-dimensional modeling of heat transfer, fluid flow, hydrocarbon generation, and migration. *American Association of Petroleum Geologists Bulletin* 74:309–335.

Wicks JL, Buckingham ML, Dupree JH. 1991. Endicott field-USA, North Slope basin, Alaska. *In* Foster NH, Beaumont EA (Editors). *Structural Traps V, Treatise of Petroleum Geology, Atlas of Oil and Gas Fields*: American Association of Petroleum Geologists, Tulsa, Oklahoma. p. 1–25.